# Bioluminescence and Chemiluminescence

## Basic Chemistry and Analytical Applications

# Bioluminescence and Chemiluminescence

## Basic Chemistry and Analytical Applications

EDITED BY

MARLENE A. DELUCA

WILLIAM D. McELROY

*Department of Chemistry*
*University of California, San Diego*
*La Jolla, California*

ACADEMIC PRESS 1981

A Subsidiary of Harcourt Brace Jovanovich, Publishers

New York London Toronto Sydney San Francisco

*Academic Press Rapid Manuscript Reproduction*

PROCEEDINGS OF THE SYMPOSIUM ON
BIOLUMINESCENCE AND CHEMILUMINESCENCE: BASIC CHEMISTRY AND ANALYTICAL APPLICATIONS
HELD AT THE UNIVERSITY OF CALIFORNIA, LA JOLLA, CALIFORNIA ON AUGUST 26-28, 1980.



ACADEMIC PRESS, INC.
111 Fifth Avenue, New York, New York 10003

*United Kingdom Edition published by*
ACADEMIC PRESS, INC. (LONDON) LTD.
24/28 Oval Road, London NW1 7DX

**Library of Congress Cataloging in Publication Data**

International Symposium on Analytical Applications of Bioluminescence and Chemiluminescence, 2d, University of California, San Diego, 1980.
Bioluminescence and chemiluminescence.

Includes index.
1. Bioluminescence—Congresses. 2. Chemiluminescence—Congresses. I. DeLuca, Marlene A. II. McElroy, William David, 1917- [DNLM: 1. Luciferase—Congresses. 2. Luminescence—Congresses. QH 641 B6151 1980]
QH641.I57 1980 574.19′125 80-28788
ISBN 0-12-208820-4

PRINTED IN THE UNITED STATES OF AMERICA

81 82 83 84 9 8 7 6 5 4 3 2 1

# Contents

*Frontispiece* *ii*
*Contributors* *xv*
*Preface* *xxvii*

## I. CHEMILUMINESCENCE AND EXCITED STATES

Spectroscopy of the Solvent Cage. Generation and Characteristics of the Excited States 3
Michael Kasha, Barry Dellinger, and Craig Brown

Chemiluminescence 17
Karl-Dietrich Gundermann

Chemiluminescence of Oxygenase Reactions 21
H. H. Seliger, J. P. Hamman, and W. H. Biggley

Activated Chemiluminescence of Dimethyldioxetanone 23
Gary B. Schuster and Steven B. Schmidt

Discussion 31

## II. LUMINOL-DEPENDENT CHEMILUMINESCENCE

Preliminary Events Leading to the Production of Luminol-Dependent Chemiluminescence by Human Granulocytes 45
Knox Van Dyke, Cynthia Van Dyke, David Peden, Maria Matamoros, Vincent Castranova, and George Jones

Chemiluminescence Immunoassay for Anti-Human IgG 55
H. R. Schroeder, C. M. Hines, and P. O. Vogelhut

Lucigenin Chemiluminescence: A New Approach to the Study of Polymorphonuclear Leukocyte Redox Activity 63
Robert C. Allen

Measurement of Opsonic Activity of Serum by Granulocyte Chemiluminescence 75
Paul Stevens and William D. Welch

On the Question of Singlet Oxygen Production in Polymorphonuclear Leucocytes 81
Christopher S. Foote, Rosangela B. Abakerli, Roger L. Clough, and Robert I. Lehrer

Discussion 89

## III. BACTERIAL BIOLUMINESCENCE

Fluorescence Properties of Luciferase Peroxyflavins Prepared with Iso-FMN and 2-Thio FMN 97
J. W. Hastings, Sandro Ghisla, Manfred Kurfürst, and Peter Hemmerich

Properties of a Lumazine Protein from the Bioluminescent Bacterium *Photobacterium phosphoreum* 103
John Lee, L. A. Carreira, Robert Gast, R. M. Irwin, Prasad Koka, E. Duane Small, and A. J. W. G. Visser

Purification, Identification, and Synthesis of *Photobacterium fischeri* Autoinducer 113
A. Eberhard, C. Eberhard, A. L. Burlingame, G. L. Kenyon, N. J. Oppenheimer, and K. H. Nealson

Active Center Studies on Bacterial Luciferase: Evidence That the Reactive Cysteinyl Residue Is within the Protease-Labile Region of the $\alpha$ Subunit 121
Thomas O. Baldwin, John J. Dougherty, Jr., Steven K. Rausch, and Margaret V. Merritt

Bacterial Bioluminescence: Accessory Enzymes 129
Edward Meighen, Denis Riendeau, and Andrew Bognar

Bioluminescence Test for Mutagenic Agents 139
Shimon Ulitzur, Irith Weiser, and Shmuel Yannai

A Thermodynamic Explanation for the Kinetic Differences Observed Using Different Chain Length Aldehydes in the *in Vitro* Bacterial Bioluminescent Reaction 147
James E. Becvar and Li-Huey Wu

Active Center Studies on Bacterial Luciferase: Modification with Methyl Methanethiolsulfonate 155
Miriam M. Ziegler and Thomas O. Baldwin

Isolation and Reaction Product Characterization of an Oxygenated Bacterial Luciferase Intermediate Formed with a Flavin Analog 161
Shiao-Chun Tu

Discussion 169

## IV. FIREFLY BIOLUMINESCENCE

The Chemistry and Applications of Firefly Luminescence 179
W. D. McElroy and Marlene DeLuca

Applications of Firefly Luminescence 187
Arne Lundin

Dynamic Measurement of ATP with Firefly Luciferase Bioluminescence 197
John J. Lemasters

Preliminary Report on the Pink-Colored *Cypridina* Luciferase, A Natural Model of the Luciferin–Luciferase Complex 203
Toshio Goto and Hideshi Nakamura

Standard Assay for Total Creatine Kinase and the MB-Isoenzyme in Human Serum with Firefly Luciferase 209
Karl Wulff, Fritz Stähler, and Wolfgang Gruber

Discussion 215

## V. *RENILLA, AEQUOREA,* EARTHWORM, AND LAND SNAIL BIOLUMINESCENCE

*Renilla* and *Aequorea* Bioluminescence 225
Milton J. Cormier

Properties of the Coelenterate Green-Fluorescent Proteins 235
William W. Ward

Applications of Aequorin 243
John R. Blinks

Earthworm Bioluminescence 249
John E. Wampler

Luminescence Activity of the Land Snail *Quantula striata* 257
Yata Haneda

Discussion 267

## VI. APPLICATIONS OF BIOLUMINESCENCE

Overview of Applications of Bioluminescence 275
Philip E. Stanley

Immunoassays Monitored by Chemilumigenic Labels 283
R. C. Boguslaski and H. R. Schroeder

Aspects on the Potentialities of Bioluminescence Assay in Cell Biology and Microphysiology 287
Sven E. Brolin and Gunar Wettermark

Discussion 295

## VII. SUMMARY AND FUTURE APPLICATIONS OF LUMINESCENCE

Overview 303
T. P. Whitehead

## VIII. ABSTRACTS

### A. CHEMILUMINESCENCE AND RELATED ASSAYS

A Preliminary Study of the Measurement of Urinary Pregnanediol-3$\alpha$- Glucuronide by a Solid-Phase Chemiluminescence Immunoassay 311
G. Barnard, W. P. Collins, F. Kohen, and H. R. Lindner

Chemiluminescent Reactions of Alcohols and Aldehydes 319
Ralph J. Bushnell

Defects Associated with the Transmembrane Potential of Granulocytes in Chronic Granulomatous Disease 327
Vincent Castranova, George S. Jones, Ruth M. Phillips, David Peden, and Knox Van Dyke

Room Temperature Phosphorescence of Some Pharmaceutical Important Imidazoles 335
F. Abdel Fattah, W. Baeyens, and P. DeMoerloose

An Oxygen Probe Based on Tetrakis Alkyl Aminoethylene Chemiluminescence 347
Thompson M. Freeman and W. Ruldolf Seitz

An Immunoassay for Urinary Estriol-16α Glucuronide Based on Antibody-Enhanced Chemiluminescence 351
F. Kohen, J. B. Kim, H. R. Lindner, and G. Barnard

Assay of Gonadal Steroids Based on Antibody-Enhanced Chemiluminescence 357
F. Kohen, J. B. Kim, and H. R. Lindner

Dissociative and Non-Dissociative Electron Transfer Mechanisms in Bioluminescence and Chemiluminescence 365
E. M. Kosower

Site-Specific Chemiluminescent Probes Used in the Analysis of pH, Peroxidation, and Free Radicals Internally in Metabolically Active Organelles 373
Richard D. Lippman

Differential Effects of Bacterial and Viral Infections of Granulocyte Chemiluminescence (CL) 383
J. P. McCarthy, P. Z. Sobocinski, P. B. Jahrling, and D. W. Reichard

Archadonate-Based Chemiluminescence in Human Granulocytes and Platelets Using the Monolight 301 (Drug Studies) 385
David Peden, Chris Van Dyke, Cynthia Van Dyke, Knox Van Dyke, Vincent Castranova, George Jones, and Michael Ringrose

Chemiluminescence Derived from 4a-(Alkylperoxy) Flavins 391
Peter T. Shepherd and Thomas C. Bruice

## B. BACTERIAL BIOLUMINESCENCE AND APPLICATIONS

The Use of Bentonite Clays to Increase the Speed and Yield of Harvest of Luminous Bacteria by Continuous Flow Centrifugation 397
James E. Becvar, Amina Mahomedy, Oscar Dominguez, John Hostak, Bean Burr, A. Bryce Campbell, Frank Loudermilk, Daniel Marquez, and Kevin Ayer

The Red Absorbing Flavin Species in the Reaction of Bacterial Luciferase with $FMNH_2$ and $O_2$ 403
J. W. Hastings, Robert Presswood, Sandro Ghisla, Manfred Kurfürst, and Peter Hemmerich

Bacterial Bioluminescence: Applications to Entomology 409
Edward A. Meighen, Keith N. Slessor, and Gary G. Grant

The Use of the Bacterial Luminescent System for the Quantitative Determination of Dehydrogenases 417
H. Watanabe and J. W. Hastings

Activity Coupling between Bacterial Luciferase and Flavin Adenine Dinucleotide-Dependent Salicylate Hydroxylase 425
Shiao-Chun Tu

## C. FIREFLY LUCIFERASE AND APPLICATIONS

Analytical Aspects of the Firefly Luciferase Reaction Kinetics 435
Mushtaq Ahmad and Eric Schram

Bioluminescent Screening for Bacteriuria 443
J. A. Lazaroni and D. Linkley Henry

Purification of Firefly Luciferase by Ammonium Sulphate Precipitation and Isoelectric Focusing 453
Arne Lundin, Arne Myhrman, and Gunilla Linfors

A Convenient Affinity Chromatography-Based Purification of Firefly Luciferase 467
Thomas M. Marschner, Bruce R. Branchini, and Angelina M. Montemurro

Bioluminescence Microassay of Creatine-Kinase: A Comparison with a Spectrophotometric Method 477
Gianni Messeri, Anna L. Caldini, Paola Tozzi, and Mario Pazzagli

Inhibitors of Firefly Luciferase in Clinical Urine Specimens 485
Wright W. Nichols, G. D. W. Curtis, and H. H. Johnston

Use of a Mathematical Representation of the Time-Course of the Firefly Luciferase Light Reaction 491
Eric Schram, Mushtaq Ahmad, and Eric Moreels

Studies on the Sensitivity of ATP Determination Using Commercial Firefly Luciferase 497
JoAnn J. Webster, Phyllis B. Taylor, and Franklin R. Leach

D. GENERAL BIOLUMINESCENCE AND CHEMILUMINESCENCE, INCLUDING DINOFLAGELLATES, WORMS AND FISH

Chemiluminescence of a 6,7-Dihydroflavin and Some Related Pteridines 507
R. Addink

*In Vivo* Spectroscopy of a Bioluminescent Cell: *Pyrocistis lunula* 515
B. Arrio, A. Dupaix, C. Fresneau, B. Lécuyer, and P. Volfin

The Counterlighting Hypothesis: *In Situ* Observations on *Argyropelecus hemigymnus* 517
Fernand Baguet and J. Piccard

Biochemical Studies of the Bioluminescent Polynoid Worms 525
A. Dupaix, C. Fresneau, P. Volfin, B. Arrio, and B. Lécuyer

Red Fluorescence of Fish and Cephalopod Photophores 527
Peter J. Herring

Oxygen-Dependent *Streptococcus faecalis* Chemiluminescence: The Importance of Metabolism and Medium Composition 531
Deborah J. Hunter and Robert C. Allen

Preliminary Ultrastructural Description of Coelomocytes of the Luminescent Oligochaete, *Pontodrilus bermudensis* (Annelida) 543
Barrie G. M. Jamieson, John E. Wampler, and Michael C. Schultz

On the Bacterial-Associated Light Organ in *Chlorophthalmus* 561
Hiroaki Somiya

Circular Polarization Observed in Bioluminescence 569
H. Wynberg, E. W. Meijer, J. C. Hummelen,
H. P. J. M. Dekkers, P. H. Schippers
and A. D. Carlson

E. APPLICATION OF BIOLUMINESCENCE AND CHEMILUMINESCENCE TO CLINICAL AND BIOLOGICAL PROBLEMS

Advantages in Microprocessing Techniques in Bioluminescence Assays 575
Ambjörn Agren, Sven E. Brolin, and Marianne Lindfors

Autolumography: Using Bioluminescence to "Stain" for Enzyme Activities Resolved on Polyacrylamide Gels 583
James E. Becvar

Bioenergetics of Insulin and Glucagon Producing Cells Evaluated in Bioluminescence Assays 591
Sven E. Brolin and Ambjörn Agren

Applications of the Photoprotein Obelin to the Measurement of Free $Ca^{2+}$ in Cells 601
Anthony K. Campbell, Maurice B. Hallett, Richard A. Daw, Malcolm E. T. Ryall, Russell C. Hart, and Peter J. Herring

Chemiluminescence of Cytotoxic Macrophages 609
Martin Ernst, Helmut Lang, Herbert Fischer,
Marie-Luise Lohmann-Matthes, and
Hansjürgen Staudinger

Measurements of Granulocyte and Platelet Chemiluminescence in Small Samples of Whole Blood 617
Herbert Fischer, Taizo Kato, Heinrich Wokalek, Martin Ernst, Hartwig Eggert, and Ernst Th. Rietschel

Flow Injection Analysis with Chemiluminescence Detection; Recent Advances and Clinical Applications 623
Mary Lynn Grayeski, Jerry Mullin, W. Rudolf Seitz, and Elizabeth Zygowicz

Ultrasensitive Assay of NAD/NADH with Firefly Luciferase 627
Lars-Åke Idahl

A Clinical Use of Solid Phase Sepharose-Isoluminol for Sensitive Thiol Assay of Serum Cholinesterase 633
Richard D. Lippmann

Chemiluminescence Experiments on a Model of the Oligodynamic Activity of Silver 639
Jerzy D. Meduski, Jerzy W. Meduski, James De La Rosa, and Alexander Goetz

Chemiluminescent Immunoassay for the Detection of Virus/ Antibody Aggregates 645
Robert J. Miller, Jr., and Douglas W. Reichard

An Immunoassay for Plasma Cortisol Based on Chemiluminescence 651
M. Pazzagli, F. Kohen, J. B. Kim, and H. R. Lindner

Determination of Hemoglobin and Other Hemoproteins in Serum by Luminol Chemiluminescence 659
Thomas Olsson, Kurt Bergström, and Anders Thore

Bioluminescent Immunoassay: A New Enzyme-Linked Analytical Method for the Quantitation of Antigens 667
Douglas W. Reichard and Robert J. Miller, Jr.

Chemiluminescent Labels in Immunoassay 673
J. Steven, A. Simpson, Anthony K. Campbell, J. Stuart Woodhead, Adrian Richardson, Russell Hart, and Frank McCapra

Evaluation of Metabolic Rates Using Bioluminescence Analysis 681
Peter Wersäll, Sven Brolin, Birger Petersson, and Claes-Göran Østenson

1,2-Dioxetanes as Chemiluminescent Probes and Labels 687
H. Wynberg, E. W. Meijer, and J. C. Hummelen

Design of a Chemiluminescent and Bioluminescent Photometer 691
P. Volfin, B. Lécuyer, B. Arrio, A. Dupaix, and C. Fresneau

Bioluminescent Immunoassays: Use of Luciferase-Antigen Conjugates for Determination of Methotrexate and DNP 693
Jon Wannlund and Marlene DeLuca

## F. INSTRUMENTATION AND REAGENTS

Six-Channel Luminescence Analyzer for Phagocytosis Applications 699
Fritz Berthold, Helmut Kubisiak, Martin Ernst, and Herbert Fischer

Utilization of Multi-Injection for Sample Processing and Automatic Internal Standardization (AIS) in Luminometry 705
S. E. Kolehmainen, H. Vanstaen, and J. Vossen

Preliminary Report on the Mechanism of Bioluminescence in the Oceanic Squid *Symplectoteuthis* 709
G. Leisman and F. I. Tsuji

A Novel Rapid-Mixing Apparatus for Studies of Chemiluminescent Reactions 715
Jerzy D. Meduski, Jerzy W. Meduski, and James De La Rosa

Signals of Chemiluminescence Emitted by Spleen Cells and Bone Marrow Macrophages after Stimulations with Mitogens and Particulate Substances 721
Sybille Müller, Sabine Falkenberg, Robert A. Fromtling, Anne M. Fromtling, and Volker Klimetzek

Chemiluminescence of Lucigenin in Micellar Systems 729
Constantine M. Paleos, George Vassilopoulos, and John Nikokavouras

Pico-Zyme F and Pico-EX for Measuring Cellular ATP on a Pico-Lite Luminometer 735
Cindy Sanville

High Quality Bacterial Luminescence Reagents for FMN Measurements 745
Tenlin S. Tsai

A Comparison of Commercial Firefly Luciferases 755
JoAnn J. Webster, Jenq C. Chang, Jeffrey L. Howard, and Franklin R. Leach

Luminescent Oscillations in Microaerobic Batch Cultures of *Photobacterium phosphoreum* 763
Richard A. Wecher and Ralph J. Bushnell

Index of Authors 771
Subject Index 775

# Contributors

Numbers in parentheses indicate the pages on which the authors' contributions begin.

*Rosangela B. Abakerli (81),* Department of Chemistry, University of California, Los Angeles, California 90024

*R. Addink (507),* Biochemical and Biophysical Laboratory, Technological University of Delft, Julianalaan 67, Delft, The Netherlands

*Ambjörn Agren (575, 593),* Department of Histology, University of Uppsala, Box 571, S-751 23 Uppsala, Sweden

*Mushtaq Ahmad (435, 491),* Institute of Molecular Biology, Paardenstraat-65, 1640 St. Genesius Rode, Belgium

*Robert C. Allen (63, 531),* U.S. Army Institute of Surgical Research and Clinical Investigation Service, Brooke Army Medical Center, Fort Sam, Houston, Texas 78234

*B. Arrio (515, 525, 691),* Institut de Biochimie bat 432, Faculte des Sciences, 91 405 Orsay-Cedex, France

*Kevin Ayer (397),* Department of Chemistry, The University of Texas at El Paso, El Paso Texas 79968

*W. Baeyens (335),* Hansbekedorp 11, 9842 Nevele, Belgium

*Fernand Baguet (517),* Laboratoire de Physiologie Animale, Batement Claude Bernard, Place de La Croix du Sud 5, B-1348 Louvain-La-Neuve, Belgium

*Thomas O. Baldwin (121, 155),* Department of Biochemistry, 390A Roger Adams Laboratory, University of Illinois, Urbana, Illinois 61801

*G. Barnard (311, 351),* Department of Obstetrics and Gynecology, Kings College Hospital Medical School, Denmark Hill, London SE 5 8RX, United Kingdom

*James E. Becvar (147, 397, 583),* Department of Chemistry, University of Texas at El Paso, El Paso, Texas 79968

*Kurt Bergström (659),* Department of Clinical Chemistry, Huddings University Hospital, Cl 62, S-141 86 Huddings, Sweden

*Fritz Berthold (699),* Laboratorium Prof. Dr. Berthold, Calmbacher Str. 22, 7547 Wildbad, Germany

*W. H. Biggley (21),* McCollum-Pratt Institute and Department of Biology, The Johns Hopkins University, Baltimore, Maryland 21218

*John R. Blinks (243),* Department of Pharmacology, Mayo Foundation, Rochester, Minnesota 55901

*Andrew Bognar (129),* Department of Biochemistry, Room 906B, McIntyre Building, McGill University, Montreal, Quebec, Canada

*R. C. Boguslaski (283),* Immunochemistry Laboratory, Ames Research and Development Department, Ames Division, Miles Laboratories Inc., Elkhart, Indiana 46515

*Bruce R. Branchini (467),* Department of Chemistry, University of Wisconsin-Parkside, Kenosha, Wisconsin 53141

*Sven E. Brolin (287, 575, 591, 681),* Department of Histology, University of Uppsala, Biomedicum, Box 571, S-751 23 Uppsala, Sweden

*Craig Brown (3),* Department of Chemistry and Institute of Molecular Biophysics, Florida State University, Tallahassee, Florida 32306

*Thomas C. Bruice (391),* Department of Chemistry, University of California, Santa Barbara, California 93106

*A. L. Burlingame (113),* Department of Pharmaceutical Chemistry, University of California, San Francisco, California

*Bean Burr (397),* Department of Chemistry, The University of Texas at El Paso, El Paso, Texas 79968

*Ralph J. Bushnell (319, 763),* Department of Biology, Sonoma State University, Rohnert Park, California 94928

*Anna L. Caldini (477),* Laboratoire Centrale, Areispedale S. M. Nuova, Careggi, 50134-Firenze, Italy

*A. Bryce Campbell (397),* Department of Chemistry, The University of Texas at El Paso, El Paso, Texas 79968

*Anthony K. Campbell (601, 693),* Department of Medical Biochemistry, Welsh National School of Medicine, Heath Park, Cardiff, CF44XN, United Kingdom

*A. D. Carlson (569),* Department of Neurobiology and Behavior, State University of New York, Stony Brook, New York

*L. A. Carreira (103),* Bioluminescence Laboratory, Department of Biochemistry and Department of Chemistry, University of Georgia, Athens, Georgia; and Department of Biochemistry, Agricultural University, Wageningen, The Netherlands

*Vincent Castranova (45, 327, 385),* Department of Physiology (ALDSH), West Virginia University Medical Center, Morgantown, West Virginia 26506

*Roger L. Clough (81),* Department of Chemistry, University of California, Los Angeles, California 90024

*W. P. Collins (311),* Department of Obstetrics & Gynaecology, Kings College Hospital Medical School, Denmark Hill, London SE5 8RX, United Kingdom

*Milton J. Cormier (225),* Department of Biochemistry, Graduate Studies Research Center, 622, University of Georgia, Athens, Georgia 30602

*G. D. W. Curtis (485),* Department of Microbiology, Level 6/7, Regional Public Health Laboratory, John Radcliffe Hospital, Oxford OX3 9DU, Great Britain

*Richard A. Daw (601),* Department of Medical Biochemistry, Welsh National School of Medicine, Heath Park, Cardiff CF44XN, United Kingdom

*H. P. J. M. Dekkers (569),* Department of Theoretical Organic Chemistry, University of Leiden, The Netherlands

*James De La Rosa (639, 715),* 16345 Denley Street, Hacienda Heights, California 91745

*Marlene DeLuca (179, 693),* Department of Chemistry, Basic Sciences Building, University of California, San Diego, La Jolla, California 92093

*P. DeMoerloose (335),* Department of Pharmaceutical Chemistry and Drug Quality Control, State University of Ghent, Belgium

*Barry Dellinger (3),* Department of Chemistry and Institute of Molecular Biophysics, Florida State University, Tallahassee, Florida 32306

*D. O. Dominguez (397),* Department of Chemistry, University of Texas at El Paso, El Paso, Texas 79968

*John J. Dougherty, Jr. (121),* Department of Biochemistry, 390 A Roger Adams Laboratory, University of Illinois, Urbana, Illinois 61801

*A. Dupaix (515, 525, 691),* Institut de Biochemie bat 432, Faculté des Sciences 91 405 Orsay-Cedex, France

*A. Eberhard (113),* Department of Chemistry, Ithaca College, Ithaca, New York

*C. Eberhard (113),* Department of Chemistry, Ithaca College, Ithaca, New York

*Hartwig Eggert (617),* Max-Planck-Institut fur Immunbiologie, Stubeweg 51, D-7800 Freiburg i. Br., West Germany

*Martin Ernst (609, 617, 699),* Max-Planck-Institut fur Immunbiologie, Stubeweg 51, D-7800 Freiburg i. Br., West Germany

*Sabine Falkenberg (721),* Robert Koch-Institut, Bundesgesundheitsamt, Berlin, West Germany

*F. Abdel Fattah (335),* Department of Pharmaceutical Chemistry and Drug Quality Control, State University of Ghent, Belgium

*Herbert Fischer (609, 617, 699),* Max-Planck-Institut fur Immunbiologie, Stubeweg 51, D 7800 Frieburg i. Br., West Germany

*Christopher S. Foote (81),* Department of Chemistry, University of California, Los Angeles, California 90024

*Thompson M. Freeman (347),* Department of Chemistry, University of New Hampshire, Durham, New Hampshire 03824

*C. Fresneau (515, 523, 691),* Institut de Biochimie bat 432, Faculté des Sciences, 91 405 Orsay-Cedex, France

*Anne M. Fromtling (721),* Robert Koch-Institut, Bundesgesundheitsamt, Berlin, West Germany

*Robert A. Fromtling (721),* Robert Koch-Institut, Bundesgesundheitsamt, Berlin, West Germany

*Robert Gast (103),* Bioluminescence Laboratory, Department of Biochemistry and Department of Chemistry, University of Georgia, Athens, Georgia 30602; Department of Biochemistry, Agricultural University, Wageningen, The Netherlands

*Sandro Ghisla (97, 403),* Fachbereich Biologie, Universität Konstanz, Konstanz, West Germany

*A. Goetz (639),* Nutritional Research Laboratory, University of Southern California School of Medicine, Los Angeles, California 90038

*Toshio Goto (203),* Department of Agricultural Chemistry, Nagoya University, Chikusa, Nagoya 464, Japan

*Gary G. Grant (409),* Forest Pest Management Institute, Canadian Forestry Service, Sault Ste. Marie, Canada

*Mary Lynn Grayeski (623),* Department of Chemistry, University of New Hampshire, Durham, New Hampshire 03824

*Wolfgang Gruber (209),* Boehringer Mannheim GmbH, Research-Center, Postfach 120, D-8132 Tutzing, Tutzing, West Germany

*K. D. Gundermann (17),* Organisch-Chemisches Institut Der Technische, Universitat Clausthal, Leibnifitr 6, D-3392 Clausthal-Fellerfeld, Federal Republic of Germany

*Maurice B. Hallett (601),* Department of Medical Biochemistry, Welsh National School of Medicine, Heath Park, Cardiff CF44XN, United Kingdom

*J. P. Hamman (21),* McCollum-Pratt Institute and Department of Biology, The Johns Hopkins University, Baltimore, Maryland 21218

*Yata Haneda (257),* 3-6-43, Numama, Zushi City, Kanagawa-ken 238, Japan

*Russell C. Hart (601, 673),* School of Molecular Sciences, University of Sussex, Brighton, United Kingdom

*J. W. Hastings (97, 403, 417),* The Biological Laboratories, Harvard University, Cambridge, Massachusetts 02138

*Peter Hemmerich (97, 403),* Fachbereich Biologie, Universität Konstanz, Konstanz, West Germany

*D. Linkley Henry (443),* SAI Technology Co., 4060 Sorrento Valley Boulevard, San Diego, California 92121

*Peter J. Herring (527, 601),* Institute of Oceanographic Sciences, Brook Road., Wormley, Surrey, United Kingdom

*C. M. Hines (55),* Immunochemistry Laboratory, Ames Research & Development Department, Ames Division, Miles Laboratories, Elkhart, Indiana 46514

*John Hostak (397),* Department of Chemistry, University of Texas at El Paso, El Paso, Texas 79968

*Jeffrey L. Howard (755),* Department of Biochemistry, Oklahoma State University, Stillwater, Oklahoma

*J. C. Hummelen (569, 687),* Department of Organic Chemistry, University of Groningen, Nijenborgh, g747AG Groningen, The Netherlands

*Deborah J. Hunter (531),* Brooke-Army Medical Center, Box 455, Beach Pavilion, Fort Sam, Houston, Texas 78234

*Lars-Ake Idahl (627),* Department of Histology, University of Umea, S-901 87 Umea, Sweden

*R. M. Irwin (103),* Bioluminescence Laboratory, Department of Biochemistry and Department of Chemistry, University of Georgia, Athens, Georgia 30602; and Department of Biochemistry, Agricultural University, Wageningen, The Netherlands

*P. B. Jahrling (383),* U.S. Army Medical Research Institute of Infectious Diseases, Fort Detrick, Frederick, Maryland 21701

*Barrie G. M. Jamieson (543),* Zoology Department, University of Queensland, St. Lucia, Brisbane, Queensland, Australia 4067

*H. H. Johnston (485),* Department of Microbiology, Level 6/7, Regional Public Health Laboratory, John Radcliffe Hospital, Oxford, OX3 9DU, Great Britain

*George S. Jones (45, 327, 385),* Department of Physiology (ALOSH), West Virginia University Medical Center, Morgantown, West Virginia 26506

*Michael Kasha (3),* Department of Chemistry and Institute of Molecular Biophysics, Florida State University, Tallahassee, Florida 32306

*Taizo Kato (617),*Max-Planck-Institut fur Immunbiologie, Stubeweg 51, D-7800 Freiburg i. Br., West Germany

*G. L. Kenyon (113),* Department of Phamaceutical Chemistry, University of California, San Francisco, California

*J. B. Kim (351, 357, 651),* Department of Hormone Research, Weizmann Institute of Science, Rehovot, Israel.

*Volken Klimetzek (721),* Department of Microbiology, Freie Universität Berlin, Berlin, West Germany

*F. Kohen (311, 351, 357, 651),* Department of Hormone Research, Weizmann Institute of Science, Rehovot, Israel.

*Prasad Koka (103),* Department of Biochemistry, Agricultural University, Wageningen, The Netherlands

*S. E. Kolehmainen (705),* Lumac Systems, Inc., Titusville, Florida 32780

*E. M. Kosower (365),* Department of Chemistry, Tel-Aviv University, Ramat-Aviv, Tel-Aviv, Israel

*Helmut Kubisiak (699),* Laboratorium Prof. Dr, Berthold, Calmbacher Str. 22, 7547 Wildbad, Germany

*Manfred Kürfurst (97, 403),* Fachbereich Biologie, Universität Konstanz, Konstanz, West Germany

*Helmut Lang (609),* Max-Planck-Institut fur Immunbiologie, Stubeweg 51, D-7800 Freiburg i. Br., West Germany

*J. A. Lazaroni (443),* Lazaroni Laboratories, Daly City, California 94015

*Franklin R. Leach (497, 755),* Department of Biochemistry, Oklahoma State University, Stillwater, Oklahoma 74074

*B. Lécuyer (515, 525, 691),* Institut de Biochimie, Faculté des Sciences, 91 405 Orsay-Cedex, France

*John Lee (103),* Department of Biochemistry, University of Georgia, Athens, Georgia 30602

*Robert I. Lehrer (81),* Department of Medicine, University of California, Los Angeles, California 90024

*G. Leisman (709),* A-002, Marine Biology, Scripps Institution of Oceanography, University of California, San Diego, La Jolla, California 92093

*John J. Lemasters (197),* Laboratories for Cell Biology, Department of Anatomy, School of Medicine, University of North Carolina at Chapel Hill, Chapel Hill, North Carolina 27514

*Gunilla Lindfors (453),* National Defense Research Institute, S-901 82 Umea, Sweden

*Marianne Linfors (575),* Department of Histology, University of Uppsala, Box 571, S-751 23 Uppsala, Sweden

*H. R. Lindner (311, 351, 357, 651),* Department of Hormone Research, Weizmann Institute of Science, Rehovot, Israel; Fogarty International Center, Building 16, Room 309, National Institutes of Health, Bethesda, Maryland 20014

*Richard D. Lippmann (373, 633),* Department of Physical Chemistry, The Royal Institute of Technology, Stockholm, Sweden; and Department of Medical Cell Biology, Biomedicum, University of Uppsala, Uppsala, Sweden

*Marie-Luise Lohmann-Matthes (609),* Max-Planck-Institut fur Immunbiologie, Stubeweg 51, D-7800 Freiburg i. Br., West Germany

*Frank Loudermilk (397),* Department of Chemistry, University of Texas at El Paso, El Paso, Texas 79968

*Arne Lundin (187, 453),* Bioluminescence Centre, LKB-Produkter AB, Box 305, S-16126 Bromma, Sweden

*Amina Mahomedy (397),* Department of Chemistry, University of Texas at El Paso, El Paso, Texas 79968

*Daniel Marquez (397),* Department of Chemistry, University of Texas at El Paso, El Paso, Texas 79968

*Thomas M. Marschner (467),* Department of Pharmaceutical Chemistry, University of California, San Francisco, California 94143

*Maria Matamoros (45),* Department of Pharmacology & Toxicology, West Virginia University Medical Center, Morgantown, West Virginia 26506

*Frank McCapra (673),* School of Molecular Sciences, University of Sussex, Brighton, United Kingdom

*J. P. McCarthy (383),* U.S. Army Medical Research Institute for Infectious Diseases, Fort Detrick, Frederick, Maryland 21701

*W. D. McElroy (179),* Department of Chemistry, Basic Science Building, University of California, San Diego, La Jolla, California 92093

*Jerzy D. Meduski (639, 715),* 1066 S. Genesee Avenue, Los Angeles, California 90019

*Jerzy W. Meduski (693, 715),* Nutritional Research Laboratory, University of Southern California School of Medicine, Los Angeles, California 90038

*Edward A. Meighen (129, 409),* Department of Biochemistry, Room 906B, McIntyre Building, McGill University, Montreal, Quebec, Canada

*E. W. Meijer (569, 687),* Department of Organic Chemistry, University of Groningen, Nijengborg, g747AG Groningen, The Netherlands

*Margaret V. Merritt (121),* Physical and Analytical Chemistry Research, The Upjohn Company, Kalamazoo, Michigan 49001

*Gianni Messeri (477),* Laboratory Centrale, Arcispedale S. M. Nudva-Careggi, Firenze-50-134, Italy

*Robert J. Miller (645, 667),* U.S. Army Medical Research Institute for Infectious Diseases, Fort Detrick, Frederick, Maryland 21701

*Angelina M. Montemurro (467),* Department of Chemistry, University of Wisconsin-Parkside, Kenosha, Wisconsin 53141

*Eric Moreels (491),* Institute of Molecular Biology, Paardenstraat-65, 1640 St. Genesius Road, Belgium

*Sybille Müller (721),* Robert Koch-Institut, Nordufer 20, D-1000 Berlin 65, West Germany

*Jerry Mullin (623),* Department of Chemistry, University of New Hampshire, Durham, New Hampshire 03824

*Arne Myhrman (453),* Bioluminescence Centre, LKB-Produkter AB, Box 305, 5-16126, Bromma, Sweden

*Hideshi Nakamura (203),* Department of Agricultural Chemistry, Nagoya University, Chikusa, Nagoya 464, Japan

*K. H. Nealson (113),* Scripps Institution of Oceanography, La Jolla, California 92037

*Wright W. Nichols (485),* Department of Microbiology, Level 6/7, Regional Public Health Laboratory, John Radcliffe Hospital, Oxford, OX3 9DU, Great Britain

*John Nikokavouras (729),* Department of Chemistry, Nuclear Research Center "Demokritos," Aghia Paraskevi Attikis, Athens, Greece

*Thomas Olsson (659),* Department of Clinical Chemistry, Huddings University Hospital, Cl 62, S-141 86 Huddings, Sweden

*N. J. Oppenheimer (113),* Department of Pharmaceutical Chemistry, University of California, San Francisco, California

*Claes-Göran Östenson (681),* Department of Medical Cell Biology, Biomedicum, Box 571, S-751 23 Uppsala, Sweden

*Constantine M. Paleos (729),* Department of Chemistry, Nuclear Research Center "Demokritos," Aghia Paraskevi Attikis, Athens, Greece

*Mario Pazzagli (477, 651),* Endocrinology Unit, University of Florence, Viale Morgani 85, 50134 Florence, Italy

*David Peden (45, 327, 385),* Department of Pharmacology and Toxicology, West Virginia University Medical Center, Morgantown, West Virginia 26506

*Birger Petersson (681),* Department of Medical Cell Biology, Biomedicum, Box 571, S-751 23 Uppsala, Sweden

*Ruth M. Phillips (327),* Department of Pharmacology and Toxicology, West Virginia University Medical Center, Morgantown, West Virginia 26506

*James Piccard (517),* Foundation pour l'Etude et Protection de la Mer et des Lacs, Cully, Switzerland

*Robert Presswood (403),* Biological Laboratories, Harvard University, Cambridge, Massachusetts 02138

*Steven K. Rausch (121),* Department of Biochemistry, 390A Roger Adams Laboratory, University of Illinois, Urbana, Illinois 61801

*Douglas W. Reichard (383, 645, 667),* Physical Sciences Division, U. S. Army, Fort Detrick, Frederick, Maryland 21701

*Adrian Richardson (673),* School of Molecular Sciences, University of Sussex, Brighton, United Kingdom

*Denis Riendeau (129),* Department of Biochemistry, Room 906B, McIntyre Building, McGill University, Montreal, Quebec, Canada

*Ernst Th. Rietschel (617),* Max-Planck-Institut fur Immunbiologie, Stubeweg 51, D-7800 Freiburg i. Br., West Germany

*Michael Ringrose (385),* Analytical Luminescence Laboratory, Inc., Westlake Village, California

*Malcolm E. T. Ryall (601),* Department of Medical Biochemistry, Welsh National School of Medicine, Heath Park, Cardiff CF44XN United Kingdom

*Cindy Sanville (735),* Packard Instrument Co., Downers Grove, Illinois 60515

*P. H. Schippers (569),* Department of Theoretical Organic Chemistry, University of Leiden, Leiden, The Netherlands

*Steven B. Schmidt (23),* Box 58, Roger Adams Laboratory, University of Illinois, Urbana, Illinois 61801

*Eric Schram (435, 491),* Brussels University, Institut voor Molekulaire Biologie, 65, Paardenstraat, B-1640-Sint-Genesius-Rode, Belgium

*H. R. Schroeder (55, 283, 293),* Immunochemistry Laboratory, Ames Research and Development Department, Ames Division, Miles Laboratories, Elkhart, Indiana 46514

*Michael C. Schultz (543),* Zoology Department, University of Queensland, Australia

*Gary B. Schuster (23),* Box 58, Roger Adams Laboratory, University of Illinois, Urbana, Illinois 61801

*W. W. Rudolf Seitz (347, 623),* Department of Chemistry, University of New Hampshire, Durham, New Hampshire

*H. H. Seliger (21),* McCalum-Pratt Institute, and Department of Biology, The Johns Hopkins University, Baltimore, Maryland 21218

*Peter R. Shepherd (391),* Department of Chemistry, University of California, Santa Barbara, California 93106

*John A. Simpson (673),* Department of Medical Biochemistry, Welsh National School of Medicine, Heath Park, Cardiff, Wales

*Keith N. Slessor (409),* Department of Chemistry, Simon Fraser University, Burnaby, British Columbia

*E. Daune Small (103),* Bioluminescence Laboratory, Department of Biochemistry and Department of Chemistry, University of Georgia, Athens, Georgia 30602; Department of Biochemistry, Agricultural University, Wageningen, The Netherlands

*P. Z. Sobocinski (383),* U.S. Army Medical Research Institute of Infectious Diseases, Fort Detrick, Frederick, Maryland 21701

*Hiroaki Somiya (561),* Fisheries Laboratory, Faculty of Agriculture, Nagoya University, Chikusa, Nagoya 464, Japan

*Fritz Stähler (209),* Boehringer Mannheim GmbH, Research-Center Tutzing, Tutzing, West Germany

*Philip E. Stanley (275),* The Department of Clinical Pharmacology, The Queen Elizabeth Hospital, Woodville, South Australia

*Hansjürgen Staudinger (609),* Max-Planck-Institut fur Immunbiologie, Stubeweg 51, D-7800 Freiburg i. Br., West Germany

*Paul Stevens (75),* Department of Medicine, University of California, Los Angeles, California 90024

*Phyllis B. Taylor (497),* Department of Biochemistry, Oklahoma State University, Stillwater, Oklahoma 74074

*Anders Thore (659),* Department of Clinical Chemistry, Huddings University Hospital, Cl 62, S-141 86 Huddings, Sweden

*Paola Tozzi (477),* Lab Centrale, Arcispedale S. M. Nuova-Careggi, Firenze 50134, Italy

*Tenlin S. Tsai (745),* Clinical Investigations, Packard Instrument Co., Inc., 2200 Warrenville Road, Downer Gove, Illinois 60515

*F. I. Tsuji (709),* Marine Biology Research Division A002, Scripps Institution of Oceanography, University of California, San Diego, La Jolla, California 92093

*Shiao-Chun Tu (161, 425),* Department of Biophysical Sciences, University of Houston, Houston, Texas 77004

*Shimon Ulitzur (139),* Department of Food Engineering and Biotechnology, Technion-Israel Institute of Technology, Haifa, Israel

*Chris Van Dyke (385),* Department of Pharmacology and Toxicology, West Virginia University Medical Center, Morgantown, West Virginia 26506

*Cynthia Van Dyke (45, 327, 385),* Department of Pharmacology and Toxicology, West Virginia University Medical Center, Morgantown, West Virginia 26506

*Knox Van Dyke (45, 327, 385),* Department of Pharmacology and Toxicology, West Virginia University Medical Center, Morgantown, West Virginia 26506

*H. Vanstaen (705),* Lumac B. V., Schaesberg, The Netherlands

*George Vassilopoulos (729),* Department of Chemistry, Nuclear Research Center "Demokritos," Aghia Paraskevi Attikis, Athens, Greece

*A. J. W. G. Visser (103),* Department of Biochemistry, Agricultural University, Wageningen, The Netherlands

*P. O. Vogelhut (55),* Immunochemistry Laboratory, Ames Research and Development Department, Ames Division, Miles' Laboratories Inc., Elkhart, Indiana 46514

*P. Volfin (515, 525, 691),* Institut de Biochemie, Faculte des Sciences, 91 405, Orsay-Cedex, France

*J. Vossen (705),* Lumac B. V., Schaesberg, The Netherlands

*John E. Wampler (249, 543),* Department of Biochemistry, University of Georgia, Athens, Georgia 30602

*Jon Wannlund (693),* Department of Chemistry, Basic Sciences Building, University of California, San Diego, La Jolla, California 92093

*William W. Ward (235),* Department of Biochemistry & Microbiology, Cook College, Lipman Hall, Rutgers University, New Brunswick, New Jersey 08903

*H. Watanabe (417),* The Biological Laboratories, Harvard University, Cambridge, Massachusetts 02138

*JoAnn J. Webster (497, 755),* Department of Biochemistry, Oklahoma State University, Stillwater, Oklahoma 74074

*Richard A. Wecher (763),* Department of Biology, Sonoma State University, Rohnert Park, California 94928

*Irith Weiser (139),* Department of Food Engineering and Biotechnology, Technion-Israel Institute of Technology, Haifa, Israel

*William D. Welch (75),* Department of Anesthesiology, University of California, Irvine Medical Center, Orange, California 92668

*Peter Wersäll (681),* Department of Medical Cell Biology, Biomedicum, Box 571, S-751 23 Uppsala, Sweden

*Gunar Wettermark (287),* Department of Physical Chemistry, The Royal Institute of Technology, Fack, S-100 44, Stockholm 40, Sweden

*T. P. Whitehead (303),* Wolfson Research Laboratories, Department of Clinical Chemistry, Queen Elizabeth Medical Centre, Birmingham, B15 2TH, England

*Heinrich Wokalek (617),* Max-Planck-Institut fur Immunbiologie, Stubeweg 51, D-7800 Freiburg i. Br., West Germany

*J. Stuard Woodhead (673),* Department of Medical Biochemistry, Welsh National School of Medicine, Cardiff CF 4 4XN, United Kingdom

*Li-Huey Wu (147),* Department of Chemistry, University of Texas at El Paso, El Paso, Texas 79968

*Karl Wulff (209),* Boehringer Mannheim GmbH, Biochemica Werk Tutzing, Postfach 120, D-8132 Tutzing, West Germany

*H. Wynberg (569, 687),* Department of Organic Chemistry R.Li, Zernikelaan, Groningen, Holland

*Shmuel Yannai (139),* Department of Food Engineering and Biotechnology, Technion-Israel Institute of Technology, Haifa, Israel

*Miriam M. Ziegler (155),* Department of Biochemistry, 304 B Roger Adams Laboratory, Urbana, Illinois 61801

*Elizabeth Zygowicz (623),* United Technologies Corp., 1 Financial Plaza (23rd), Hartford, Connecticut 06101

# Preface

In 1978 the first International Symposium on Analytical Applications of Bioluminescence and Chemiluminescence, organized by Dr. Eric Schram (Belgium) and Dr. Philip Stanley (Australia), was held in Brussels. The success of that meeting and the increasing interest in the field of luminescence indicated that a second symposium would be both interesting and profitable. This second symposium, organized by the editors, was held at the University of California, San Diego on August 26–28, 1980.

This volume contains the collection of papers, both invited and contributed, which were presented at this meeting. The intent of the formal presentations was a review of the basic science of both bioluminescence and chemiluminescence, with informal discussions following each major presentation. A special section on applications was scheduled after the basic science lectures. Many poster sections covered both the basic and applied aspects of luminescent systems, and the corresponding papers are included in this volume.

It is impossible to express our gratitude to all the individuals who contributed to the success of this meeting. However, a special thanks must go to Jean Rippon who acted as the official secretary for the meeting. Her organizational skills, obtainment of abstracts from contributors, coordination of housing arrangements for our guests, and her efficient handling of many other details assured the success of the symposium. Pat Ringrose, Jeanne Ford, Steve Alter, Jon Wannlund, Linda Strause, Jon Bell, Su Chang, Steve Lee, Penny Durrans, Viki Molinarolo, Nicki Ringrose, and Eric McElroy made important contributions during the meeting. Jon Wannlund was responsible for transcribing the discussions; Mike Ringrose assumed the responsibility for the printing of the program and the arrangement of the commercial exhibits. To all these people we express our appreciation.

In addition, we would like to acknowledge the staff support of the University of California. The services provided by the University Housing Center and Mandeville Center were done with great care and understanding. Needless to say our thanks go to the delegates who came from many countries to make important scientific contributions. The session dealing with *Renilla* and *Aequorea* bioluminescence was in honor of Dr. Yata Haneda of Japan.

Finally, we express our appreciation to the editorial staff of Academic Press for their help in rapidly publishing this volume.

We gratefully acknowledge financial support from the National Science Foundation (PCM79-23854); UCSD Chancellor's Associates; American Instrument Company; Ames Division, Miles Laboratories, Inc.; Analytical Luminescence Laboratory, Inc.; Boehringer Mannheim Corporation; Calbiochem-Behring Corporation; Hoffmann-La Roche, Inc.; La Jolla Scientific Company; LKB-Sweden; Lumac B. V.; Packard Instrument Company, Inc.; SAI-Technology Company; and Technicon Instruments Company.

Top to bottom, left to right: J. LeMasters, P. Stanley, and E. Schram; H. Seliger; E. Schram, A. Lundin, and P. Stanley; M. Kasha; Y. Haneda; and P. Stanley and M. DeLuca.

Top to bottom, left to right: J. Schuster; J. W. Hastings and M. Kasha; W. D. McElroy, Y. Haneda, and M. DeLuca; M. Cormier and F. Tsuji; and T. P. Whitehead.

# I
# CHEMILUMINESCENCE AND EXCITED STATES

FRONT ROW- left to right: K.D. Gunderman, E.Schram, P. Stanley, T.P. Whitehead, and M. Kasha

SECOND ROW- left to right: K. Wulff, M. DeLuca, H. Schroeder, and W.D. McElroy

# SPECTROSCOPY OF THE SOLVENT CAGE. GENERATION AND CHARACTERISTICS OF THE EXCITED STATE*

Michael Kasha
Barry Dellinger[1]
Craig Brown[2]

Department of Chemistry and
Institute of Molecular Biophysics
Florida State University
Tallahassee, Florida

## I. INTRODUCTION

Contemporary molecular electronic spectroscopy is founded on three stages of evolution of the discipline. The Characterization of radiative molecular electronic transitions (1) in single isolated molecules marked the first stage, including orbital characterization (2,3) and multiplicity characterization (4). These aspects, which depend on a long history of development of atomic and molecular spectroscopy (5), are now a standard part of spectroscopic description (6,7).

The second phase of molecular spectroscopic development involves the recognition (1) of the dominant role played by radiationless electronic transitions in polyatomic molecules as contrasted with the case of excitation of atoms and

[1] *Present address: Northrop Services, Inc., Research Triangle Park, North Carolina.*

[2] *Present address: Texas Tech University, Lubbock, Texas.*

* *This work was sponsored by a contract between the Division of Biomedical and Environmental Research, Department of Energy, U.S., and the Florida State University.*




ISBN 0-12-208820-4

diatomic molecules. The pathways of excitation and their Z-dependence are easily defined phenomenologically (8). However, the theoretical background of these complex processes is still under development (9), although progress is being made steadily in defining the conditions of the characteristic radiationless transition process in polyatomic molecules (10, 11,12,13).

Multiple excitation in composite molecules can be considered the third stage in evolution of the understanding of molecular electronic excitation phenomena (14). Instead of considering isolated molecule electronic excitation, the biologist, chemist, and physicist deal frequently with molecules in condensed systems. Not only do inter-molecular interactions (solute-solute and solute-solvent) play a dominant role, but multi-photon events may enter the picture, especially with the photon-flux available with laser excitation. Among the phenomena which fall into this category are charge-transfer excitation, molecular exciton effects, simultaneous transitions in molecular pairs, non-linear biphotonic absorption, triplet-triplet annihilation, superradiance, biprotonic phototautomerism, and others.

## II. EXCITONS IN MOLECULAR PAIRS: DIOXETANE DECOMPOSITION

As examples of the last category of multiple-molecule excitation processes, we shall contrast the molecular exciton and simultaneous transition mechanism in molecular pairs.

In the molecular exciton (15) state excitation, a preformed molecular pair (H-bonded or van der Waals' dimer) is singly excited

$$\psi^{E}_{PAIR} = \frac{1}{\sqrt{2}}\left[ (A_1^*)\text{---}(A_2) \pm (A_1)\text{---}(A_2^*) \right],$$

so that resonance interaction or transfer between units of the molecular pair can occur. Thus, the excited states of the unit molecule A will split by molecular exciton resonance interaction in the dimer, the splitting can be large, in the range 500-2000 cm.$^{-1}$, for strong electronic transitions. The consequence of this splitting, depending on the pair geometry, is generally to alter greatly the excitation mechanism, e.g., a greatly enhanced triplet state excitation (16).

As an example of this mechanism, we apply it to the

dioxetane formation of excited acetone, which appears predominantly as triplet.

$$\begin{matrix} >C-O \\ | \quad\; | \\ >C-O \end{matrix} \rightarrow \left[ \begin{matrix} S_1(>C=O)^* & S_0(>C=O) \\ & \pm \\ S_0(>C=O) & S_1(>C=O)^* \end{matrix} \right] \rightarrow \begin{matrix} S_0(>C=O) \\ T_1(>C=O)^* \end{matrix}$$

The energy diagram for this mechanism is

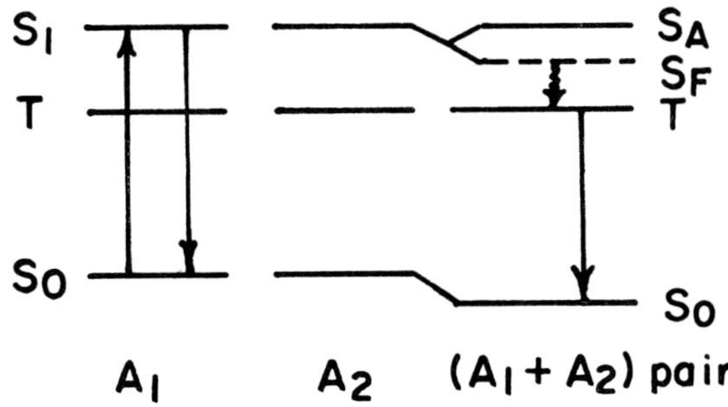


wherein the single acetone molecules, $A_1, A_2$ have predominantly fluorescence emission, the acetone pair formed from the dioxetane would have exciton state splitting with allowed ($S_A$) and forbidden ($S_F$) singlet state components, yielding predominantly triplet state excitation. This is a simpler mechanism than the double or two-stage electron transfer mechanism usually proposed for dioxetane decomposition.

## III. SIMULTANEOUS TRANSITIONS IN MOLECULAR PAIRS: SINGLET MOLECULAR OXYGEN

In simultaneous transitions in molecular oxygen pairs (17,18), a pair of colliding molecules is simultaneously excited

$$\psi_E = (A_1^*)(A_2^*)$$
$$\psi_G = (A_1)(A_2)$$

In this case, there is no possibility of excitation resonance, but instead an <u>excitation superposition</u> occurs. In other words, when two excited singlet oxygen molecules collide, a photon-doubling occurs.

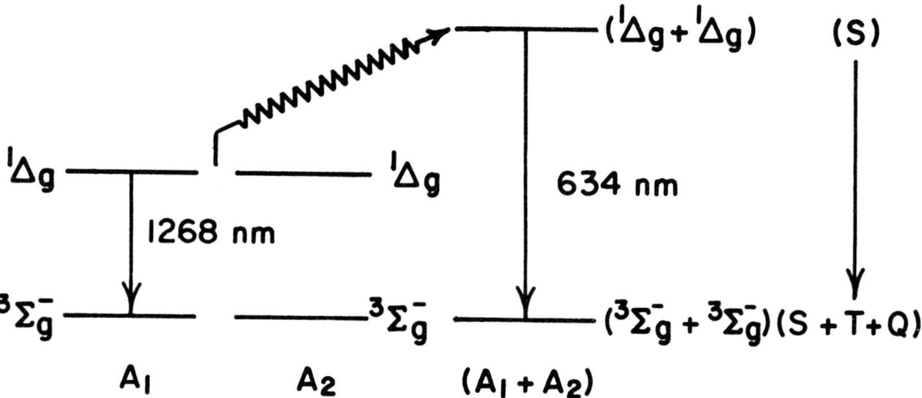


Thus, the near infrared transition at 1268 nm. in the single molecule appears as a 634 nm. single photon (red-orange) two-molecule emission.

The simultaneous transitions dominate the chemiluminescence of molecular oxygen (17), and sensitized chemiluminescences(18) of acceptor molecules present in chemical systems generating singlet molecular oxygen. Because of the importance of detecting singlet molecular oxygen directly, Khan and Kasha (19) have recently described an ultra-sensitive near infrared spectrometer which directly monitors the 1268 nm. singlet delta oxygen emission. This device is sensitive enough to probe delicate biological chemiluminescences.

## IV. SPECTROSCOPY IN SOLVENT CAGES: VISCOUS-FLOW BARRIER

Molecules in condensed media always are surrounded by a solvent cage. The cage may be a liquid solvent, a macromolecular enclosure (enzyme site), a lipid membrane, a multilayer lamellar system, or may be a crystal or surface-adsorption cage. Although photochemical and other kinetic studies have long taken cognizance of the effect of solvent cages on recombination rates, spectroscopic studies of the mechanism of solvent cage action have been neglected in comparison.

The *Mechanical Viscous-Flow Barrier* solvent case has been analyzed on a quantum-mechanical basis by Dellinger and Kasha (20,21). Figure 1 indicates schematically the basis of their model. If a molecule upon excitation involves a rather large distortional motion for equilibrium relaxation, then a large volume of solvent in the cage around the solute must be displaced. For example, in trans-stilbene the torsional relaxation about the double-bond requires that the phenyl groups sweep out a large volume of solvent to reach an equilibrium excited state configuration (21) (upper part, Figure 1). On the other hand, a hydrogen-bonded molecular pair, such as 9,10-diazaphenanthrene complex with t-perfluorobutyl alcohol:

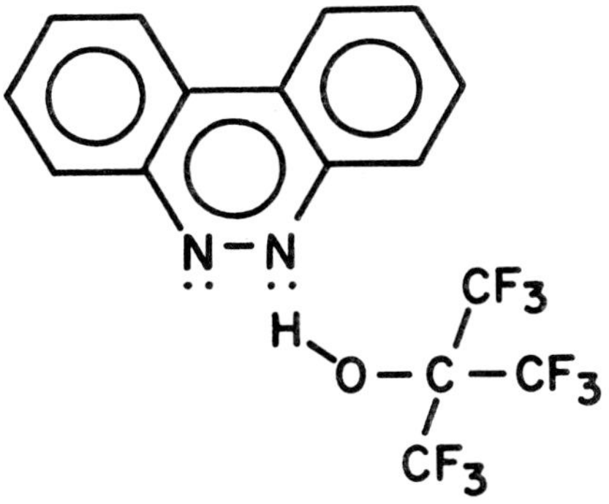

would require displacing a large volume of solvent molecules in the photo-dissociative excitation of this molecular complex (21).

According to the Dellinger-Kasha model, the solvent cage imposes a barrier to the molecular motion, which appears as a gaussian-flow barrier added to the potential function where the latter becomes horizontal, i.e., in the dissociative case, at the dissociative limit. The phenomenological reality of these viscous-flow barriers has been tested by low-temperature spectroscopic studies. Thus, luminescence phenomena reveal that the molecule is trapped in a ground-state configuration in rigid glass solvents.

Recently, Mohammadi and Henry (22) have verified the viscous-flow barrier to molecular motion by an entirely different route. They studied the infrared overtone anharmonicity of C-H vibrations in methyl butanes, and were able to show that the decrease in anharmonicity predicted by the

MECHANICAL VISCOUS-FLOW-BARRIER CAGE

ISOMERIZATIONAL EXCITATION

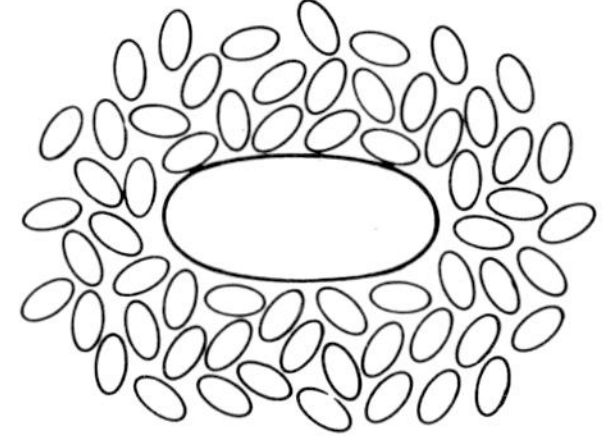

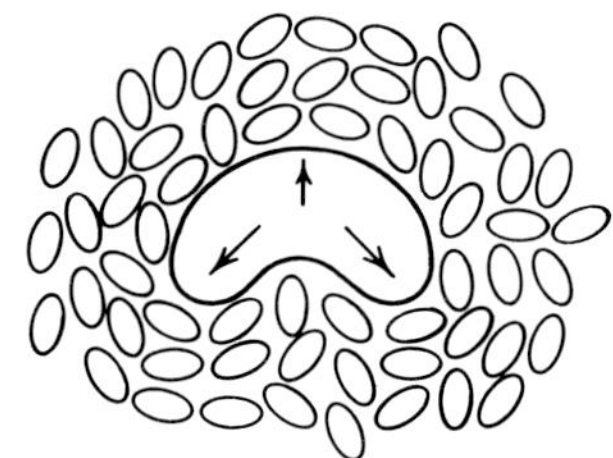

DISSOCIATIVE EXCITATION

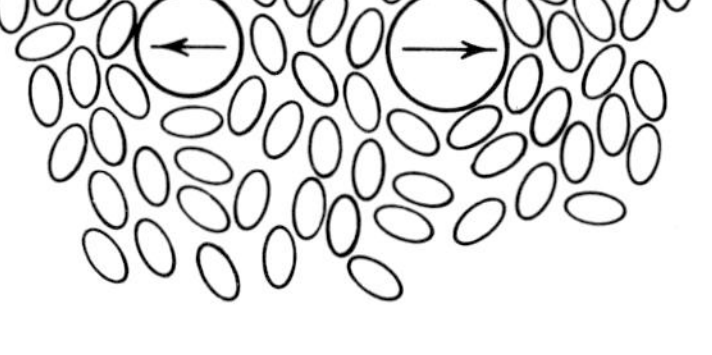

SOLUTE GROUND STATE

SOLUTE EXCITED STATE

*Figure 1*

Dellinger-Kasha model (21) was experimentally verified and could be calculated by the Lennard-Jones potential as a quantitative perturbation of the Morse potential (22).

An interesting case of excited state proton-transfer spectroscopy was discovered recently in flavones by Sengupta and Kasha (23). In 3-hydroxy-flavone

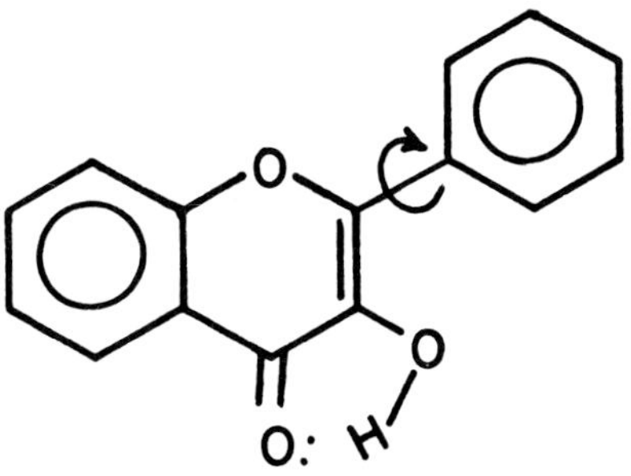

it was discovered that in hydrocarbon solution at room temperature a green fluorescence was observed, unrelated to the ultraviolet absorption of the molecule. It was deduced that an intramolecular proton-transfer had occurred within the internal H-bond to the carbonyl group, and that a pyrilium-like excited state tautomer yielded the green emission. Upon freezing in a rigid glass solvent, the normal violet-ultraviolet fluorescence could be observed. Here, phenyl-torsion could be inhibited by the solvent cage (23), so that the viscous-flow barrier prevented tautomerization. Quercetin, the natural flavone, showed analogous green fluorescence on a TLC-plate. [Interestingly, the 5-hydroxyflavone shows dissipative (non-emissive) proton transfer, and the 7-hydroxyflavone shows dissipative intermolecular proton transfer.]

## V. DIELECTRIC RELAXATION SOLVENT CAGES

The large molecular distortions upon molecular electronic excitation required for the mechanical viscous-flow-barrier solvent cage model are not common. It is true that many fluorescent dyes, especially those used as immunological and cytological probes, exhibit a large viscosity-dependence of fluorescence yield, and probably fit the model well because of the internal torsional potential which is applicable.

However, most molecules would exhibit only a very small skeletal relaxation upon electronic excitation, and not a gross configuration change displacing volumes of cage molecules. Such molecules would be immune to mechanical viscous-flow-barrier cage effects.

On the other hand, many molecules of interest in excitation studies have a moderate to large dipole moment in the ground state, and may have a large dipole moment *reversal* in the excited state. Thus, indole, and 7-azaindole

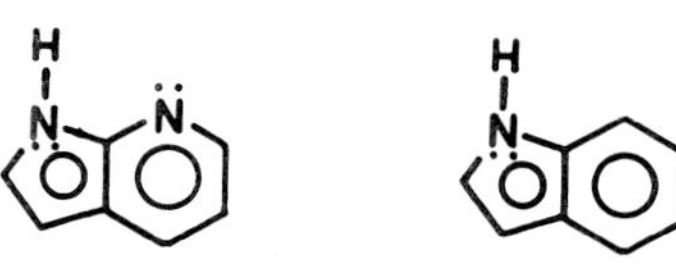

may be expected to be sensitive to dielectric relaxation effects in the solvent cage, but not to exhibit any major mechanical viscous-flow-barrier effects.

## VI. POLARIZATION CAGE EFFECTS

In Figure 2, top, we represent schematically a molecule with a permanent dipole moment in its ground state, surrounded by a non-polar molecule solvent cage. Here we show *induced dipole moments* in the solvent molecules, expected in at least the first shall of cage molecules, and probably beyond. Upon excitation, if the solvent dipole moment strongly reverses, we would expect the induced moments to reverse instantaneously, without perceptible relaxation of the solvent cage molecules. Thus, such a non-polar solvent cage should exhibit little inhibiting effect on the solute excitation.

An example of this case is presented in some current experiments recently completed (24) in which the H-bonded dimer of 7-azaindole was studied for its solvent cage dependence of biprotonic phototautomerism (25).

DIELECTRIC RELAXATION SOLVENT CAGES

POLARIZATION CAGE

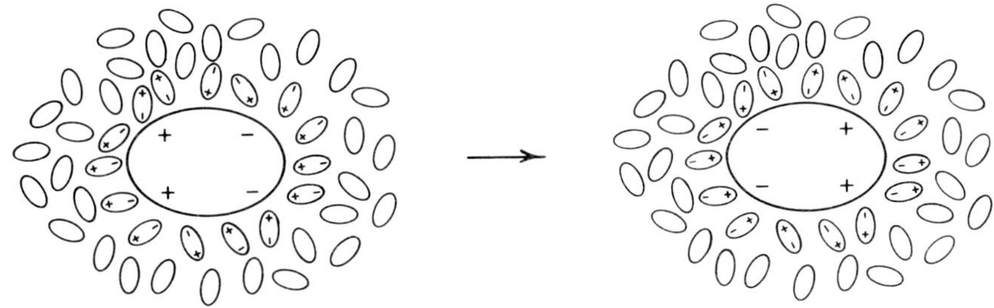

RANDOM-DIPOLE CAGE

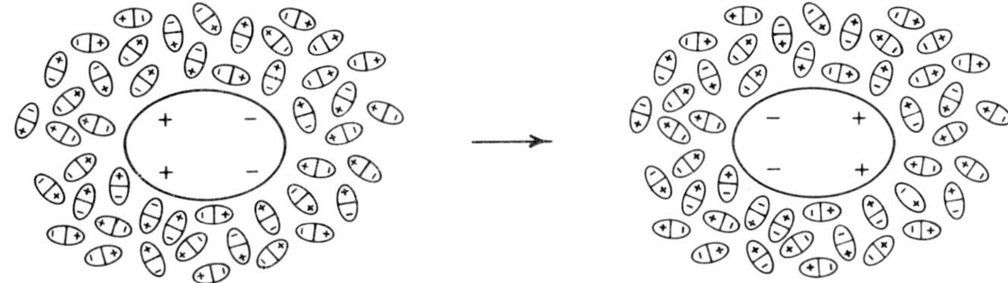

ORDERED-DIPOLE CAGE

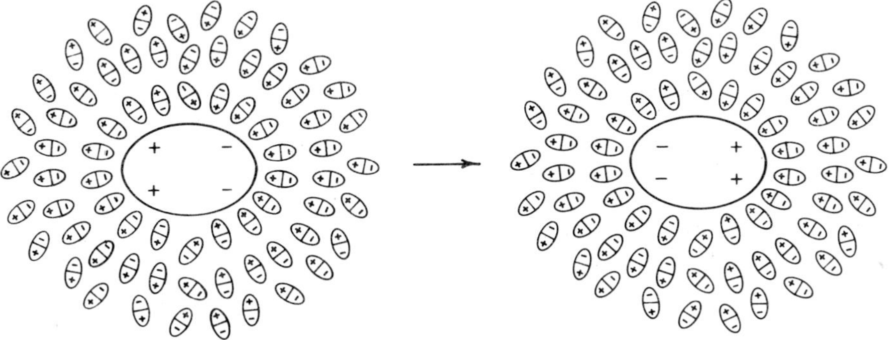

SOLUTE GROUND STATE SOLUTE EXCITED STATE

*Figure 2*

Here, there is no gross molecular configurational change expected, upon excited state proton-transfer. Upon fast-cooling of a $10^{-3}$ M solution of 7-azaindole in 2-methylbutane to make a rigid glass solvent at 77K, the previously observed green fluorescence of the tautomer ($F_2$) and the violet fluorescence ($F_1$) of the normal molecule was observed. Upon slow-cooling with a Cryodyne apparatus to 25K, in which the dimer equilibrium could be maximized, and the solvent cage relaxation (for the ground state solute electronic configuration) optimized, a high ratio of $F_2/F_1$ green to violet fluorescence was observed. Thus, there was no barrier imposed by the solvent cage to solute tautomerization in the dimer, even at 25K.

## VII. RANDOM VS. ORDERED-DIPOLE CAGE EFFECTS

In the middle and lower part of Figure 2, we show solvent cages made up of molecules possessing a permanent dipole moment, surrounding a solute molecule with a permanent dipole moment, which reverses strongly upon molecular excitation.

For the random-dipole cage we choose ethanol or ethanol-methanol solvent mixtures. At room temperature we expect some ordering of the dipoles about the 7-azaindole solute ground state dipole configuration, but considerable randomness because of the temperature. Upon fast freezing we anticipate little chance for solvent-relaxation, because the hydrogen-bonded chains of the solvent reach a high-viscosity quickly as the sample tube is quickly immersed in a liquid nitrogen bath at 77°K. Extensive H-bonding of alcohol chains to the pyridino- and pyrollo-sites will be present, disfavoring tautomerization of the 7-azaindole. We observe after fast-cooling to 77°K, normal violet fluorescence and a small component of green phosphorescence of 7-azaindole in alcohol glasses, as previously observed (25). We assume in the random-dipole solvent cage, that the excited state electronic distribution in the solute is not greatly inhibited by the reaction field of the random-dipole solvent cage, since the frozen-in quasi-random dipole distribution is present.

For the ordered-dipole cage we use the same chemical solution as above, a ~$10^{-3}$M solution of 7- azaindole in ethanol (EPA) or ethanol-methanol (1:1) solvent, but use the Cryodyne low-temperature probe in a slow-cooling regime. Upon cooling to low temperature several hours instead of 1 minute, we anticipate the possibility that the ground state solvent-dipole orientation with respect to the ground state solute-dipole orientation and electronic distribution pattern will be favored.

We model schematically in the lower part of Fig. 2 the idea of a long-range solvent-dipole ordering, roughly schematically (we might preferably orient the solvent dipoles along the lines of force of the solute dipole). If the solute dipole is strong, and naturally reverses upon excitation, the retardation effects of the highly-oriented solvent dipole cage could be very large. We are investigating such effects in indole and other polar molecules of biological interest.

It could be pointed out that the study of excited properties of highly-polar molecules is of considerable current interest (26). The ordered-dipole array can be likened to a solvent electret (27), and such phenomena as we have described could be developed as a molecular-electronic probes for electret regions of molecular films. Secondly, biological membranes, especially neuronal membranes, may be brought to an electret state. The detailed study of the effect of media reaction field perturbations of molecular electronic phenomena could become of direct biological interest.

The prominence of proton-transfer spectroscopy in current research will be evident in the presentation made here. A wide variety of molecules is under current investigation (29-33) including lumichrome (34) and other molecules of biological interest. These proton-transfer tautomerizations are found to be generally in the pico-second range, and are spawning a whole new era in excited state spectroscopy, photochemistry, and chemistry as a basis of interpretation of photobiological phenomena.

## REFERENCES

1. Kasha, M., *in* "Characterization of Electronic Transitions in Complex Molecules," Faraday Soc. Discussion No. *9*, 14 (1950).

2. Kasha, M., *in* "The Nature and Significance of n → π* Transitions" in *Light and Life* (W. D. McElroy and B. Glass, eds.), pp. 31-64. John Hopkins University Press, Baltimore (1961).
3. Kasha, M., and H. R. Rawls, *in* "Correlation of Orbital Classification of Molecular Electronic Transition with Transition Mechanism: The Aromatic Amines," *Photochem. Photobiol. 7,* 561 (1968).
4. Lewis, G. N., and M. Kasha, *in* "Phosphoresence and the Triplet State," *J. Am. Chem. Soc. 66,* 2100 (1944).
5. Ramsay, D. A., and J. Hinze, *Selected Papers of Robert S. Mulliken,* 1127, University of Chicago Press, Chicago (1975).
6. McGlynn, S. P., T. Azumi, and M. Kinoshita, *Molecular Spectroscopy of the Triplet State,* 434, Prentice-Hall, Inc., Englewood Cliffs, N.J. (1969).
7. Becker, R. S., *Theory and Interpretation of Fluorescence and Phosphorescence,* 283, Wiley Interscience, New York (1969).
8. Kasha, M., *in* "Paths of Molecular Excitation" in *Bioenergetics* (L. G. Augenstine, ed.), 243, Radiation Research, Supplement 2, (1960).
9. Henry, B. R., and M. Kasha, *in* "Radiationless Molecular Electronic Transitions," *Ann. Reviews Phys. Chem. 19,* 161 (1968).
10. Lin, S. H., *J. Chem. Phys. 44,* 3759 (1966).
11. Bixon, M., and J. J. Jortner, *J. Chem. Phys. 50,* 3284 (1969).
12. Rhodes, W., B. R. Henry, and M. Kasha, *in* "A Stationary State Approach to Radiationless Transitions. Radiation Bandwidth Effect on Excitation Processes in Polyatomic Molecules," *Proc. Nat. Acad. Sci. 63,* 31 (1969).
13. Rhodes, W., "Molecular Excited State Relaxation Processes," *in J. Chem. Ed. 56,* 562 (1979).
14. Kasha, M., and M. A. El-Bayoumi, "Survey of Multiple Excitation in Composite Moleculre" *in Physical Mechanisms in Radiation Biology* (R. D. Cooper and R. W. Wood, eds.), 326, U.S. Atomic Energy Commission, CONF-721001, Oak Ridge, Tennessee (1974).
15. Kasha, M., "Molecular Excitons in Small Aggregates" *in Spectroscopy of the Excited State* (B. DiBartolo, ed.) 337-363, Plenum Press, New York (1976).
16. McRae, E. G., and M. Kasha, "The Enchantment of Phosphorescence Ability Upon Aggregation of Dye Molecules," *J. Chem. Phys. 28,* 721 (1958).

17. Khan, A. U., and M. Kasha, "Chemiluminescence Arising from Simultaneous Transitions in Pairs of Singlet Oxygen Molecules," *in J. Am. Chem. Soc. 92,* 3293 (1970).
18. Kasha, M., and D. E. Brabham, "Singlet Oxygen Electronic Structure and Photosensitization," *in Singlet Oxygen* (H. H. Wasserman and R. W. Murray, eds.), 1-33, Academic Press, New York (1979).
19. Khan, A. U., and M. Kasha, "Direct Spectroscopic Observation of Singlet Oxygen Emission at 1268 nm Excited by Sensitizing Dyes of Biological Interest in Liquid Solution," *in Proc. Nat. Acad. Sci. 76,* 6047 (1979).
20. Dellinger, B., and M. Kasha, "Intermolecular Perturbation of Molecular Potentials," *in Chem. Phys. Letters 36,* 410 (1975).
21. Dellinger, B., and M. Kasha, "Phenomenology of Solvent Matrix Spectroscopic Effects," *in Chem. Phys. Letters 38,* 9 (1976).
22. Mohammadi, M. A., and B. R. Henry, "Nonbonded Interaction Potentials in Methyl Substituted Butanes Obtained from High Energy Overtone Spectra," *in Proc. Nat. Acad. Sci.* (in press).
23. Sengupta, P. K., and M. Kasha, "Excited State Proton-Transfer Spectroscopy of 3-Hydroxyflavone and Quercetin," *in Chem. Phys. Letters 68,* 382 (1979).
24. Brown, C., and M. Kasha, Unpublished Work, Florida State University, Tallahassee, Florida (1980).
25. Taylor, C. A., M. A. El-Bayoumi, and M. Kasha, "Excited-State Two-Proton Tautomerism in Hydrogen-Bonded N-Heterocyclic Base Pairs," *in Proc. Nat. Acad. Sci. 63,* 253 (1969).
26a. Carsey, T. P., G. L. Findley and S. P. McGlynn, *J. Am. Chem. Soc. 101,* 4502 (1979).
26b. Findley, G. L., T. P. Carsey, and S. P. McGlynn, *J. Am. Chem. Soc. 101,* 4511 (1979).
27a. Mascarenhas, S., "Electrets in Biophysics," *J. Electrostatics 1,* 141 (1975).
27b. Mascarenhas, S., "The Electret Effect in Bone and Biopolymers and the Bound-Water Problem," *Ann. N.Y. Acad. Sci. 23d,* 36 (1974).
28. Collins, S., and M. Kasha, Unpublished Work, Florida State University, Tallahassee, Florida, (1980).
29. Moomaw, W. R., and M. R. Anton, *J. Phys. Chem. 80,* 2243 (1976).
30. El-Bayoumi, M. A., *J. Phys. Chem. 80,* 2259 (1976).
31. Smith, K. K., and K. J., Kaufmann, *J. Phys. Chem. 82,* 2286 (1978).
32. Hetherington, W. M. III, R. H. Micheels, and K. B. Eisenthal, *Chem. Phys. Letters 66,* 230 (1979).

33. Shizuka, H., K. Tsutsumi, H. Takeuchi, and I. Tanaka, *Chem. Phys. Letters 62*, 408 (1979).
34. Choi, J. D., R. D. Fugate and P. S. Song, *J. Am. Chem. Soc. 102*, 5293 (1980).

# CHEMILUMINESCENCE

Karl-Dietrich Gundermann

Organisch-Chemisches Institut
der Technischen Universität
Clausthal, Clausthal-Zellerfeld
Fed. Rep. Germany

The present state of chemi- and bioluminescence may be characterized by three features:

1. The gap concerning the efficiencies between enzyme-catalyzed and non-enzymatically catalyzed chemiluminescence appears to be vanishing: the chemiluminescence of special substituted oxamides reaching quantum yields of 34% thus surpassing bacterial bioluminescence.

2. The development of mechanistic chemiluminescence theories, especially in respect of the formation of electronically excited products focuses on two general mechanisms:

a. one-electron transfer processes resulting in radical ion pairs

b. energy transfer from singlet oxygen.

3. The practical application of chemi- and bioluminescence phenomena appears to be extending more and more in analytical fields as the very high sensitivity of modern light measuring devices allows estimation down to femtomoles e.g. of ATP.

Luminol chemiluminescence mechanisms are taken as example for outlining the main topics in the mechanistic field beginning with concerted multiple bond cleavage (1) (the problem being: are the WOODWARD-HOFFMANN rules involved in chemiluminescence or are there multistep processes to be taken into account.) That cyclic hydrazides of o-dicarboxylic acids of




ISBN 0-12-208820-4

aromatic hydrocarbons (the dianions of which are sufficiently fluorescent) only exhibit strong chemiluminescence is explainable on the basis of the assumption of an o-xylylene peroxide intermediate (2, 3); as a consequence the chemiluminescence of the substituted maleic hydrazides 1 and 2 is only ca 5% that of DPA-2,3-dicarboxylic hydrazide 3 (4) in the aqueous alkali/$H_2O_2$/hemin system:

1: R = -CH=CH-C(=CH-C(=O)-NH-NH-C(=O)-), R′ = H

2: R = -C(=CH-C(=O)-NH-NH-C(=O)-), R′ = H

3: R,R′ = -C(=O)-NH-NH-C(=O)-

Substituent effects in the amino group of luminol are described: the chemiluminescence light yields of the tetrapeptide 5 and of the heptapeptide 6 are only 12.0% and 1.6% that of luminol, respectively (5), although the chemiluminescence of the lysine derivative 4 (ca 40% that of luminol) is essentially not affected by addition of polylysine, polyglutamic acid, ovalbumin, or collagen. (6)

4: R = H ; R′ = OH

5: R = H ; R′ = Val→Val→ILe·OEt

6: R = ←ILe←Val←Val–Z

R′ = Val→Val→ILe·OEt

Z = –C(=O)–OCH$_2$–C$_6$H$_5$

Ten sides such as cetyltrimethylammonium bromide (CTMB) were found to enhance the light yield of short-chained (e.g. n-butyl) N-alkylamino luminol derivatives up to 75% in the hydrogen perioxide/persulfate system (6); hemin catalysis is inhibited by CTMB (7).

REFERENCES


1. Rauhut, M.M., *Accounts chem. Res. 2,* 80 (1969).
2. Michl, J., *Photochem. Photobiol. 25,* 141 (1977).
3. Schuster, G.B., *Accounts chem. Res. 12,* 366 (1979).
4. Brinkmeyer, H., Dissertation, TU Clausthal 1979.
5. Dübbert, U., Dissertation, TU Clausthal 1977.
6. Rauhut, M.M., A.M. Semsel, and B.G. Roberts, *J. Org. Chem. 31,* 2431 (1966).
7. Massau, A., Diplomarbeit, TU Clausthal 1978.

# CHEMILUMINESCENCE OF OXYGENASE REACTIONS

H. H. Seliger
J. P. Hamman
W. H. Biggley

McCollum-Pratt Institute and Dept. of Biology
The Johns Hopkins University
Baltimore, Maryland

Since the important mechanistic studies of McCapra, and Koo and Schuster it has been recognized that the electron exchange decomposition of dioxetanes to intramolecular diradicals is a major evolutionary pathway in biochemiluminescent oxygenation reactions. The radical recombination results in an electronically excited state of the monomer product whose formation is not governed by photoselection rules. In bioluminescence the chemiluminescent intermediate appears to be a dioxetane, except for the bacteria. In the latter, a peroxy hemiacetal intermediate has been proposed. Therefore a specific although non-functional and presumably low quantum yield chemiluminescence should be observable in peroxidative reactions, in oxidative decarboxylation and in dioxygenase reactions that involve dioxetane intermediates and in epoxidative reactions where rearrangement of perepoxides to dioxetanes may be possible. This chemiluminescence is named "specific" since it results from oxygenation of a substrate molecule involved in a specific biochemical pathway as contrasted with "non-specific" chemiluminescence which results from inadvertent release of $O_2^{\dot{-}}$, $H_2O_2$ or $'O_2^*$ precursors and their subsequent chemiluminescent reactions with susceptible molecules in their vicinity. The specific chemiluminescence of the parent carcinogen, benzo[a]pyrene, concomitant with its metabolism by 3-methylcholanthrene-induced rat liver microsomes, is due mainly to the production of a dioxetane on the 9,10 bond of its proximate carcinogenic metabolite, 7,8-diol-B[a]P, during the further enzymatic epoxidation of the diol to the ultimate carcinogenic metabolite, 7,8-diol-9,10-




ISBN 0-12-208820-4

epoxy-B[a]P. This dioxetane undergoes intermolecular electron exchange decomposition resulting in exciplex luminescence. The regiospecificity of bay region diols to produce bay region diol epoxides enzymatically is paralleled by their regiospecificity to produce bay region diol dioxetanes chemically in the presence of singlet oxygen. Chemiluminescence concomitant with phagocytizing polymorphonuclear leucocytes appears to be non-specific in that the PMNL's produce $H_2O_2$, $O_2^{\dot{-}}$ and possibly singlet oxygen precursors that are responsible for the observed chemiluminescence. The chemiluminescence can be enhanced many thousand-fold by the addition of Luminol and Lucigenin. The 1,4-endoperoxide and the 1,2-dioxetane respectively, of these chemiluminescent probes are intermediates that decompose to produce monomer product excited states.

# ACTIVATED CHEMILUMINESCENCE OF DIMETHYLDIOXETANONE

Gary B. Schuster
Steven B. Schmidt

Department of Chemistry
University of Illinois
Urbana, Illinois

The chemiluminescence of dioxetanones is of particular interest due to their postulated intermediacy in several bioluminescent reactions (1), including that of the firefly (2). An early report by Adam (3) noted that the addition of rubrene to solutions of dimethyldioxetanone 1 gave a yield of light twenty times that when 9,10-diphenylanthracene was added. Only recently, however, was a satisfactory explanation advanced by us (4) and later Adam (5).

$$R_1R_2\text{-dioxetanone} \xrightarrow{\Delta} R_1C(=O)R_2 + CO_2 \qquad (1)$$

The addition of any of several easily oxidized, emitting aromatic hydrocarbons or amines to benzene or dichloromethane solutions of 1 results in greatly enhanced chemiluminescence. Moreover, addition of these molecules accelerates the rate of reaction of 1. The reaction is first order in both 1 and aromatic hydrocarbon (or amine) and the hydrocarbon (amine) is not consumed in the reaction, but rather serves as a catalyst for the decomposition of the dioxetanone. These catalyst molecules will subsequently be referred to as catalytic chemiluminescence activators (ACT). The observed kinetic behavior is described by the simple rate law of eq 2, where $k_1$ is the rate constant for unimolecular reaction and $k_2$ is the biomolecular rate constant for the activator catalyzed reaction



ISBN 0-12-208820-4

(Figure 1).

$$k_{OBS} = k_1 = k_2\ [ACT] \tag{2}$$

The data of Fig. 1 indicate a relationship between the one electron oxidation potential of the activator and the magnitude of $k_2$. In general the more easily oxidized activator (lowest $E_{OX}$) has a larger $k_2$ associated with it.

The luminescence observed when activators are added to benzene solutions of **1** corresponds to fluorescence from the excited singlet state of the activator. The spectrum of the chemiluminescence emission in all cases is identical with the photoexcited fluorescence spectrum of the activator. The relative initial instantaneous chemiluminescence intensity is highly dependent upon the nature of the activator employed; after correcting for differences in fluorescence quantum yields photomultiplier tube and monochromator efficiencies, a hundred thousand fold range in intensity is observed. The corrected relative intensity is uniquely predicted by the one electron oxidation potential of the activator. This relationship, shown in Figure 2, holds for nearly all activators investigated. The exceptions to this general relationship are

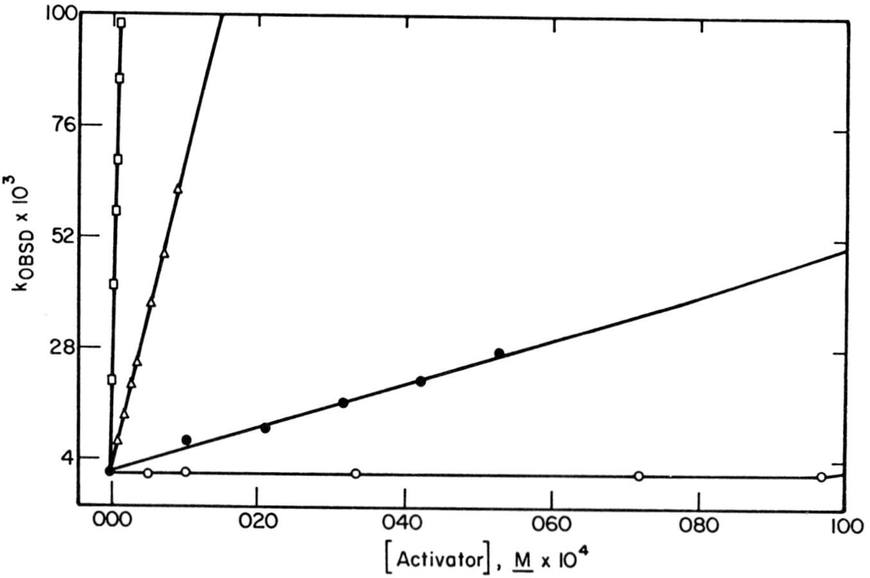


*FIGURE 1. Effect of activator identity and activator concentration on the observed first-order decay of 2a in argon purged benzene at 24.5°C: ( □ ) DMP; ( Δ ) DMBI; ( ● ) dimethyl-6-aminochrysene; ( ○ ) rubrene.*

certain metalloporphyrins. For these unusual activators the observed intensity is several orders of magnitude greater than that predicted by the oxidation potential.

The findings described herein on the activator catalyzed chemiluminescence of dimethyldioxetanone are fully consistent with the generalized mechanism for chemical light formation which we have recently identified as chemically initiated electron-exchange luminescence (CIEEL) (6). The proposed sequence as applied to dimethyldioxetanone is shown in Scheme I. In short, the light-generating sequence is initiated by electron transfer from the activator (ACT) to the dioxetanone in an encounter complex. Subsequent decarboxylation gives acetone radical anion. Annihilation of acetone radical anion and activator radical cation generates the excited state of the activator.

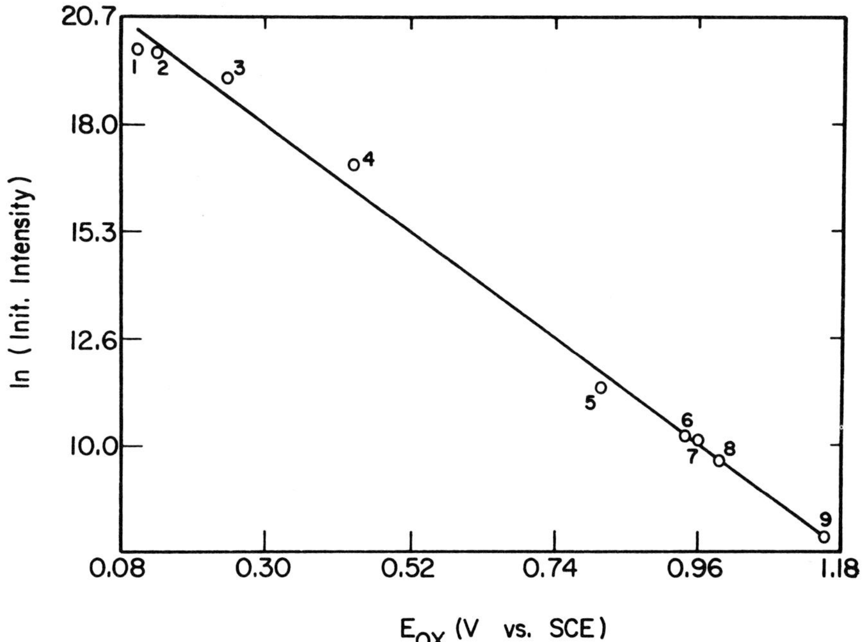


*FIGURE 2. Correlation of the initial chemiluminescence intensity in benzene at 24.5°C, corrected for fluorescence efficiency and photomultiplier tube and monochromator response, with the oxidation potential of activator: (1) DMP; (2) DMAC; (3) DMPP; (4) DMBI; (5) rubrene; (6) tetracene; (7) BPET; (8) perylene; (9) BPEA; (10) DPBF.*

*SCHEME I.*

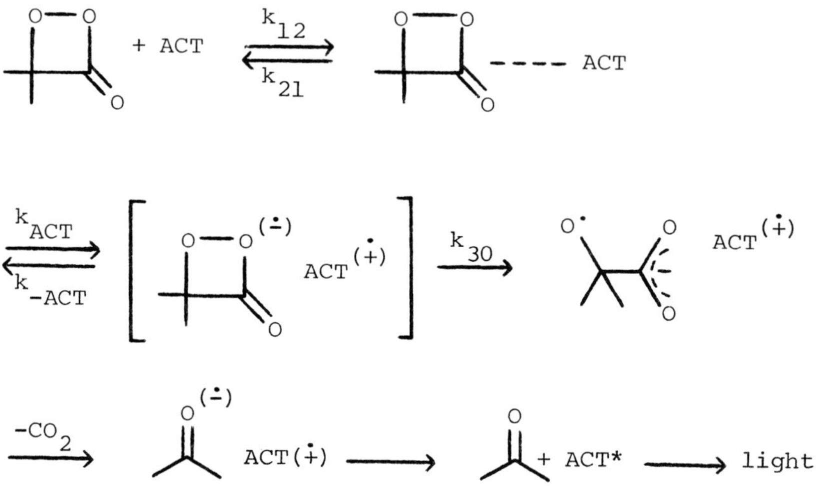

(dioxetanone) + ACT $\underset{k_{21}}{\overset{k_{12}}{\rightleftharpoons}}$ (dioxetanone) ---- ACT (3)

$\underset{k_{-ACT}}{\overset{k_{ACT}}{\rightleftharpoons}}$ [ (dioxetanone)$^{(\dot{-})}$ $ACT^{(\dot{+})}$ ] $\xrightarrow{k_{30}}$ (radical anion) $ACT^{(\dot{+})}$ (4)

$\xrightarrow{-CO_2}$ (acetone)$^{(\dot{-})}$ $ACT(\dot{+})$ $\longrightarrow$ (acetone) + ACT* $\longrightarrow$ light (5)

Formation of an encounter complex by diffusion together of reactants (rate constant $k_{12}$) is the initial event of any bimolecular reaction. The reverse reaction, $k_{21}$, is diffusion limited as well in the Weller model for electron transfer (7), which we adopt for amine and hydrocarbon activators. Thus the encounter complex of eq 3 is expected to enjoy only a fleeting lifetime for these activators.

A key feature of the proposed pathway is the electron transfer ($k_{ACT}$) and bond cleavage ($k_{30}$) steps. The correlation of the initial chemiluminescence intensity, which is proportional to the magnitude of the catalytic rate constant $k_{CAT}$ (the experimentally observed rate constant for electron transfer), with the oxidation potential of the activator, follows directly from the above considerations (8). We can express $k_{CAT}$ in terms of the oxidation potential of the activator ($E_{ox}$), the reduction potential of the peroxide ($E_{red}$), and the coulombic attraction between the developing oppositely charged radical ions ($E_{coul}$), as in eq 6, where K is a constant. The factor $\alpha$ is similar to the well- known transfer coefficient

$$\ln k_{CAT} = \ln(K) + \frac{\alpha(E_{red} + E_{coul})}{RT} - \frac{\alpha(E_{ox})}{RT} \qquad (6)$$

which generally takes a value between 0.3 and 0.7 for electrode reactions (9). Only a fraction ($\alpha$) of the total free energy change, as measured by the thermodynamic quantities

$E_{ox}$ and $E_{red}$, is realized as an increase in the activation barrier. Since the chemiluminescence intensity is proportional to $k_{CAT}$, the plot of ln(Intensity) vs. $E_{ox}$ in Fig. 2 yields a straight line of slope $-\alpha/RT$, where $\alpha = 0.3$.

The unusual chemiluminescence from the metalloporphyrins can also be explained within this framework. The ratio of initial chemiluminescence intensity ($I_0$) to the fluorescence efficiency of ACT ($\phi_f$) can be related to the magnitude of $k_{CAT}$, and the initial concentrations of **1** and ACT according to:

$$I_0/\phi_f = k_{CAT}\ [1]_0[ACT] \tag{7}$$

Thus under conditions where the concentrations of **1** and ACT are constant during the time required for the experiment, $I_0/\phi_f$ is directly proportional to $k_{CAT}$. In general, for amine and aromatic hydrocarbon ACT their $E_{ox}$ accurately predict the magnitude of $I_0/\phi_f$. The exceptions to this correlation are some metalloporphyrins. It should be noted the the exceptionally large values $I_0/\phi_f$ for the zinc and magnesium porphyrins are consistent with independently measured values for $k_{CAT}$.

In order to determine if the unusual catalytic ability of the zinc and magnesium porphyrins is due to the metal ion or to the porphyrin ring system we investigated the behavior of some free-base porphyrins. In particular, we determined $I_0/\phi_f$ and $k_{CAT}$ for tetraanisylporphyrin (H2TAP). H2TAP does not behave unusually. That is, its $E_{ox}$ is a reliable predictor of both $I_0/\phi_f$ and $k_{CAT}$. Similarly, tetraphenylporphyrin (H2TPP) does not appear to catalyze the decomposition of **1** or its chemiluminescence beyond that which would be predicted by its $E_{ox}$. Evidently, the metal ion plays a crucial role in causing the unusually efficient catalysis by MgTPP and ZnTPP.

To investigate further the mechanism of this metal ion dependent catalysis we examined the effect of changing the metal ion on $I_0/\phi_f$ and $k_{CAT}$ for tetraphenyl- and tetraanisylporphyrins. Cyclic voltammetric measurements have revealed that for zinc, cadmium, and magnesium porphyrins the first oxidation is ligand centered (10). On the other hand, for silver and cobalt porphyrin the first oxidation is metal centered (11). Interestingly, zinc, cadmium, cobalt, and magnesium porphyrin (including chlorophyll a) exhibit the unusually effective catalysis, but AgTPP does not. Evidently, the identity of the metal ion matters greatly for the catalysis. But, the catalysis can occur with metalloporphyrins that undergo either metal-or ligand-centered oxidation.

It is quite well known that certain metalloporphyrins form stable complexes with nitrogen bases such as pyridine (12).

Moveover, it is known that free-base porphyrins do not form these complexes (13). Recently the results of a thermodynamic study of the complex formation of ZnTPP with several neutral donors were reported (14). For metalloporphyrins it has been observed that the maximum of the Soret band shifts to longer wavelengths upon complex formation. We investigated the possibility that the formation of a ground state complex between dioxetanone 1 and certain metalloporphyrins is the cause of the unusual catalysis. We used tetramethyldioxetane (2) as a model for 1. In cyclohexane solution the maximum of the Soret band for ZnTPP is at 416.0 nm. In the presence of 0.35 M 2 this maximum shifts to 417.2 nm. A similar result is obtained with MgTPP. Interestingly, AgTPP, which does not readily expand its coordination sphere (14), and for which no special catalysis is observed, does not give evidence of complex formation with dioxetane 2. This is in line with our observation that even pyridine does not shift the absorption maximum or oxidation potential of this metalloporphyrin. Evidently the ability of the metalloporphyrin to form detectable complexes with dioxetane 2 and other donors predicts reliably the unusual catalytic behavior.

The magnitude of $k_{CAT}$ is related to several of the rate constants specified in Scheme 1. Simple analysis of the kinetics shows that:

$$k_{CAT} = K_{12}\ k_{ACT} \qquad (8)$$

where $K_{12} = (k_{12}/k_{21})$ is the equilibrium constant for complex formation (13). Evidently, ZnTPP, MgTPP, CoTPP, CdTPP, etc. form ground state complexes with peroxide 1 and the magnitude of $k_{CAT}$ is therefore the product of K for the complex and $k_{ACT}$.

In sum, our findings show that the chemically initiated electron-exchange luminescence pathway outlined in Scheme I easily accounts for the activator catalyzed chemiluminescence from dimethyldioxetanone. We are currently working to expand this understanding to other peroxides and other catalysts.

## REFERENCES

1. McCapra, F., *Acc. Chem. Res.* 9, 201 (1976).
2. Shimomura, O., T. Goto, F. H. Johnson, *Proc. Nat'l. Acad. Sci. USA.* 74, 2799 (1977).
3. Adam, W., O. Cueto, F. Yany, *J. Phys. Chem.* 78, 2559 (1974).

4. Schmidt, S. P., G. B. Schuster, *J. Am. Chem. Soc.* 100, (1978); *ibid.* 102, 306 (1980).
5. Adam, W. O. Cueto, F. Yany, *J. Am. Chem. Soc.* 100 (1978); *ibid.* 101, 6511 (1979).
6. Koo, J. -Y., G. B. Schuster, *J. Am. Chem. Soc.* 100, 4496 (1978).
7. Rehm, D. A. Weller, *Isr. J. Chem.* 8, 259 (1970).
8. Schuster, G. B., *J. Am. Chem. Soc.* 101, 5851 (1979).
9. Delahay, P., *in* "Double Layer and Electrode Kinetics," Chap. 7, Wiley, New York (1965).
10. Davis, D. G., *in* "The Porphyrins," (D. Dolphin, ed.) Vol. V, chap. 4. New York (1978).
11. Kadish, D. M., L. A. Bottomley, *Inorg. Chem.* 19, 832 (1980).
12. Hambright, P., *Coord. Chem. Rev.* 6, 247 (1971).
13. Miller, J. R., G. D. Dorough, *J. Am. Chem. Soc.* 74, 3977 (1952).
14. Vogel, G. C., J. R. Stahlbush, *Inorg. Chem.* 16, 950 (1977).

## DISCUSSION

Dr. McElroy

I wanted to ask Dr. Schuster, dealing with the oxidation potentials, you didn't mention the color of the light and whether there were changes in the color of the light as you change their potential.

Dr. Schuster

In the case of the dioxetanone the emission we detect is in all cases identical to the fluorescence of the activator so that the color of the light corresponds to that of the fluorescence. For example, with rubrene we observe emission at about 550nm and with diphenyl anthracene we have emission at 420 nm. Other activators start in the ultraviolet, span the visible and then go into the infrared region. With some of the other peroxides we have looked at, the chemiluminescence we observe is not necessarily the fluorescence of the activator but an exciplex of the activator is formed; and in non-polar solvents the lowest excited state is the exiplex and we see the exiplex emission rather than monomer fluorescence.

Dr. McElroy

Mike, I want to ask three questions. 1) On the slow freezing, where you observed the phosphorescence and the disappearance of phosphorescence, when you warm this up slowly again do you get delayed phosphorescence or fluorescence like you do in a crystal?

Dr. Kasha

The answer is no. Let me elaborate on that "no" slightly. The "no" means there was no emission seen upon warming up, but we had no reason to expect any trapping of excitation. In systems where there is trapping of the 'phosphorescence' phenomenon, usually there has been electron transfer to the solvent. The recombination allows thermal glow which occurs in some organic molecules. But, not with triplet state excitation. One cannot trap the excitation in that case.

Dr. McElroy

Would you also tell me a little bit more about what you mentioned in passing on the proton tunneling barrier?



ISBN 0-12-208820-4

Dr. Kasha

On the tunneling, the barrier in 7-azaindole is supposed to be very low and no deuterium effect is found, and so it is assumed that there is no proton tunneling there. Now, on the other hand, in the case of the 3-hydroxyflavone, a deuterium effect on phototautomerism is very pronounced. The barrier is very substantial in the low temperature glass, as tantomerization can be virtually eliminated.

Dr. McElroy

Concerning the hydroxy flavone and the changes in fluorescence as it is cooled down, you are talking about singlet emission and not triplet emission?

Dr. Kasha

No triplets are observed at all, for the reason that the picosecond rise time (I. Tanska) of the tautomerized molecule absolutely competes with all other excitation paths. I would like to elaborate on the hydroxyflavones. Quercetin and kaempferol have 3-5- and 7-hydroxylated positions. But the 3-position in a natural flavone is glycosilated, so there is no possibility of proton transfer from that position. What we study on TLC plates are the hydrolyzed flavones. I mention this because Dr. Toshio Goto has suggested that the 5-hydroxyflavone has a dissipative proton transfer as a plant protective function. In my laboratory Dr. Pradeep Sendupta has verified spectroscopically that indeed the 5-hydroxy does proton transfer internally in a dissipative cycle; there is no luminescence. The 7-hydroxy has anintermolecular proton transfer (to solvent) which is also dissipative. The last is a more familiar type of proton transfer, which is found in many other molecules.

Dr. Campbell

I have some difficulty in understanding the differences between the oxygen moieties which attach some of the potentially chemiluminescence of polymorphs using luminol and various acridinium compounds synthesized for us by Prof. McCapra and his group. We wonder if there is simply a kinetic difference in the way different moieties attack these compounds which might partly explain the fact that a response with luminol can be detected within a few seconds whereas the acridinium compounds require considerably longer time periods to show any response.

Dr. Gunderman

It is that according to my knowledge luminol is far more sensitive to a special degree of alkalinity, but you know

Dr. Gunderman

that there is a very distinct optimum of alkalinity a PH about 11. So maybe this could explain, but I have no other explanation, I must confess.

Dr. Campbell

I was trying to get from the chemists what is the precise mechanism by which the oxygen moiety attacks for example luminol, lucigenin or acridinium compounds? Is there a difference in rates, whether it is superoxide anion or hydrogen peroxide itself or even oxygen, or is there an absolute requirements for only one of those moieties in producing the dioxetane?

Dr. Schuster

Well, I am not exactly sure I understand the question. In the studies that we have carried out the peroxides have been preformed, and as I understand your experiment you are forming peroxide and trying to gauge the amount of peroxide formed or oxidizing equivalents formed in a crystal, is that right?

Dr. Campbell

No, in solution. Dr. Schuster, I don't think I can offer an explanation for your observations right now. Maybe we should talk about it later.

Dr. Foote

I would like to make a comment on that if I could. There are several oxidizing species present in leukocytes. Superoxide is one; hypochlorous acid is another, hydrogen peroxide is a third. Each of these has different reactivities, different pH-rate profiles and different substrate specificities, which could easily explain the differences you observed.

Dr. Seliger

I would like to point out a slight additional complexity to this rather pretty mechanism that Gary Schuster has developed for firefly bioluminescence. This relates to the question that Dr. McElroy initially asked. First of all the keto product is not the native bioluminescent emitter. The keto excited state emits red light. In the firefly there is an excited state enolization that is catalyzed by the very same luciferase molecule that results in a di-anion which emits the yellow green light normally seen from the firefly *Photimus pyralis*. The question that Dr. McElroy related to is the fact that *in vitro* extracts of luciferases

Dr. Seliger

from different species of fireflies emit different colors of bioluminescence using the identical (synthesized) luciferin molecule. This means, therefore, that the intra-molecular mechanism which Dr. Schuster proposed, while valid, is not complete, and requires the "cage interaction" that Dr. Kasha referred to, between the enzyme molecule and the excited state product. It is assumed, although not completely verified, that the different colors of light are due to the different interactions of the excited state enol product bound with the enzyme. Therefore different species luciferase molecules will create different types of hydrophobic environments and therefore different energy levels for the same product excited state.

Dr. Schuster

In our mechanism we deal with the generation of the first formed excited state. Obviously there are other reactions that occur after the excited state is formed. One of these may be deprotonization by the enzyme giving the di-anion which is responsible for the yellow-green emission of the adult firefly. Although, I believe that the larval stage of the firefly emits red. Is that correct?

Dr. Seliger

The larval stage of the Railroad Worm *Phrixothrix* emits red.

Dr. Kasha

I mentioned to Howard Seliger that if one has cage effects, one possible test or probing of those would be pressure-dependence of the spectroscopic emission. By simply varying the external pressure, one changes the cage effects. I think the word cage in application to biology has a very general connotation. It can mean enzyme sites. It could mean a protein environment; it could mean a surface macro-molecular or cellular surface. Anything which allows specific interaction between some spectroscopically studied molecule and its environment can give a cage effect. We have brilliant examples of adsorbed dyes on macro-molecular surfaces. A cage could be a lamellars system, etc. Therefore, the word cage doesn't necessarily mean literally a solvent inclusion or crystalline inclusion.

Dr. McElroy

Let me follow up on Dr. Seliger's comment. I think in your proposed mechanism where decarboxylation occurs, it's

Dr. McElroy

clearly the adenylate derivative of the firefly luciferin. The adenylate must be released before decarboxylation occurs. We know that the bound adenylate affects the color of the light; which is in effect the same as what you describe as the cage effect. If you have the epsilon AMP bound to the luciferase, you obtain a red emission. Ordinary adenylate bound to the enzyme gives yellow-green emission. The adenylate also has a profound effect on the rate of the reaction. Concerning your earlier question, Dr. Kasha, the effect of pressure on the firefly light emission was studied some years ago and there actually is no change in the color under high pressures of at least up to 5,000 pounds per square inch which is really quite surprising with what we now know about the hydrophobic nature of that enzyme.

Dr. Kosower

We proposed a dissociative electron transfer to explain the mechanism the emitter in the bacterial luciferase becomes an excited state. Based on a communication from Dr. Hastings who found that the emitter is more properly identified as the 4 α hydroxy adduct, I would like to mention that the dissociative electron transfer mechanism produces an excited proteinated flavin, an OH ion, and carboxylate ion; and in that local environment in the cage in which the species are produced the OH ion should add to the proteinated flavin and produce the 4 α excited flavin. This is an example of a reaction which succeeds the production of an excited state.

Dr. Eneroth

I would like to address Dr. Seliger on the subject on microsomal oxidation. Could you please specify what kind of media you used and did you see any background activity? I am referring to the fact that endogenous cholesterol will probably undergo an oxidation in the 5 - 6 position.

Dr. Seliger

The technique for the extraction of microsomes from methylcholanthrene or benzpyrene-induced liver systems follows the techniques that have been developed by the NIH group for their examination of the metabolism of benzpyrene. So, all of the chemical techniques are identical. There is always a low level chemiluminescence even in non-induced microsomal preparations where the particular P 448 cytochrome is not present. Other cytochrome P 450 will produce epoxidations. In the microsomal system you have the disrupted endoplasmic reticular system and an artificial micelle. You have a rather inefficient process, and so there is always some

Dr. Seliger

significant release of intermediate states of oxygen, $O_2$ or hydrogen peroxide. These will interact with lipids and with other saturated fatty hydrocarbons to produce a background chemiluminescence. The distinction that we make is that these latter are inhibitable by superoxide dismutatase and by catalase. That makes the distinction between the specific chemiluminescence that we were talking about and non-specific chemiluminescence. We usually are one to three orders of magnitude higher in intensity than this non-specific background.

Dr. Eneroth

Could I also address the discussion group here terms of the statements of Dr. Smith and Dr. Teng and their collaborators that when you find 5-chydroperoky-$\Delta^6$-cholester-3β-ol this is an indicator of the generation of singlet oxygen.

Dr. Seliger

You want to know whether I believe that?

Dr. Eneroth

Yes.

Dr. Selinger

No.

Dr. Foote

I'd like an amplification on that last comment, please.

Dr. Seliger

I don't believe it, Chris.

Dr. Foote

If you challenge this technique, you would have to provide some other mechanism by which the 5α-hydroperoxycholesterol could be formed. No such other mechanisms have been developed.

Dr. Seliger

I just believe that chemical trapping techniques have proved to provide very poor circumstantial evidence for implicating singlet oxygen.

Dr. Foote

Well, it's the best there is right now because most of the other systems which have been used to trap singlet oxygen suffer from lack of specificity. This one system, messy

Dr. Foote
as it is, seems to be the best indicator that we have at the present.

Dr. Seliger
Well, I believe that the spectroscopic system that Khan has developed in Mike Kasha's laboratory is the only unequivocal method for verifying the presence of singlet oxygen. I think that the chemical techniques are sufficiently diffuse so that they can't be unequivocal.

Dr. Foote
Well, I partially agree with you on that point. The luminescence method is superb when you can focus on a specific wave length as Krasnovsky, Kasha, Khan, Pitts, and Krinsky have shown. However, that requires a very substantial amount of light to be formed, and you just simply don't have it in many cases.

Dr. Seliger
Mike, do you have any figure on the ultimate sensitivity and your signal to noise ratio for your germanium crystal?

Dr. Kasha
No, I am afraid I don't have specific characteristics of it.

Dr. Seliger
If we were to assume that the quantum efficiency for the crystal that Khan is using in Mike Kasha's laboratory is equivalent to what one might see with a phototube, taking into account the rather small geometry of the crystal, it should be possible to detect the presence of the order of femtomoles of singlet oxygen.

Dr. Foote
This may well be true, although the sensitivity at 1.27 is very low with the best systems. However, you also have to take into account the fact that singlet delta oxygen has a radiative lifetime on the order of 45 minutes, but a lifetime of about a microsecond in solution, which makes the maximum luminescence quantum yield for luminescence on the order $10^{-9}$.

Dr. Seliger
Well, that's a valid point no matter which reaction you're working with. Isn't that true, Chris?

Dr. Foote

No. Chemical trapping can be efficient, can trap essentially all of the singlet $O_2$ under the right conditions.

Dr. Schuster

Do you think that it's necessarily true that the radiative lifetime in solution is the same as the radiative lifetime in the gas phase for singlet oxygen?

Dr. Foote

A most interesting point on which I would like to see some evidence.

Dr. Kasha

There is a point, Chris, about that which I don't have any quantitative information at hand, but the electronic mechanism for exciting singlet delta oxygen changes. The radiative lifetime that we are quoting, the 45 minute lifetime, is for the magnetic-dipole mechanism. This does not apply at all to the singlet oxygen that we observe. The radiation mechanism quickly changes to an electric-dipole mechanism in oxygen studied in perturbed systems. That lifetime is very much shorter than the magnetic-dipole lifetime, so that when you do a calculation of the quantum yield based on that change in lifetime you are using the wrong reference point. You are switching mechanisms. Do I make that clear?

Dr. Foote

This is an important point, but I am not sure that is of interest to the whole audience. Is that still a radiative lifetime?

Dr. Kasha

Oh, yes. In fact, there is work in the literature which shows the change of mechanism, e.g. as the pressure of gaseous oxygen changes. What one should do is take the integrated absorption data at conditions where the transition is electric-dipole, and then compare that with solution measured lifetime; then you would have the proper quantum yield ratio. I think that magnetic-dipole data has nothing to do with oxygen as we ordinarily see it under chemical systems. A factor of 1000 could easily exist between the two mechanisms.

Dr. Addink

I would like to ask Dr. Schuster about the CIELL mechanism. You explained it by means of the dioxetanones, but actually you developed this CIELL mechanism from reactions

Dr. Addink

of other peroxides. Could you please give the general requirements that you think are necessary for the CIEEL mechanism, and I would like to ask if the CIEEL mechanism is also applicable to the decomposition of tetra-alkyldioxetanes?

Dr. Schuster

A key requirement for operation of CIEEL mechanism in the chemiluminescence of peroxides is related to the reduction potential of the peroxide. It is a requirement that the peroxide be relatively reduced, which means structurally that the peroxide needs to be substituted with electron withdrawing groups. The electron withdrawing groups most available are carbonyl groups. Thus, most of our studies center diacyl peroxides which have two electron withdrawing carbonyl groups, or secondary peroxyesters which have one electron withdrawing carbonyl group. A second requirement for operation of the CIEEL mechanism is that on reduction of the peroxide an irreversible chemical reaction must occur. The irreversible reaction, in the case of the dioxetanone is cleavage of the oxygen-oxygen bond and decarboxylation. Part of the requirement for the irreversible reduction comes from the fact that most of these electron transfers are energetically uphill; and, if the electron transfer was reversible, there would be no net chemistry because the electron would just go back to the activator. The third requirement for the CIEEL path is that there be a chemical route available for the release of energy. The chemical path in the case of the dioxetanone is the loss of carbon dioxide to generate a carbonyl compound. This reaction essentially is pumping the electron up in energy so when it goes back to the activator radical cation there is sufficient energy released to make an electronically excited state. The third requirement, then, would be sufficient energy released in the ion annihilation to form an electronically excited state. Now to get to the second part of your question. With tetramethyldioxetane we have not seen direct evidence for electron exchange chemiluminescence. We believe the reason is that it does not meet the first requirement. The reduction potential of the tetraalkyl dioxetanes is high and the transfer of the electron does not compete with the normal thermal decomposition. Now, Paul Schaap has done some important experiments in which he has shown that in protonating media like acidic alcohols or silica gel he does see evidence for electron transfer with some dioxetanes. In those cases the reduction potential of the dioxetane is changed.

Dr. Schaap

As Dr. Schuster noted, we have looked at a number of substituted dioxetanes and I think we are able to turn on the electron transfer mechanism: either by lowering the reduction potential of the dioxetane by changing the medium or lowering the oxidation potential of the substituent.

Dr. Wampler

Many of the comments on the dioxetane mechanism are, of course, pertinent to bioluminescence. But a few of the bioluminescent systems are less clearly associated with that kind of mechanism. I was wondering if any of you gentlemen would make comments on linear peroxide or hydroperoxides as precursers to the excited state.

Dr. Schuster

This subject is dear to my heart because we recently developed a chemiluminescent system based on a linear peroxide. The system involves thermodysis of secondary peroxyesters, and the precursers to these are hydroperoxides. The linear peroxides are a series of secondary peroxybenzoates substituted. The substituent ranges from dimethylamine to nitro. The mechanism of chemiluminescence varies with the substituent. Direct chemiluminescence occurs with the dimethylamino substituted compound.

Dr. Foote

It occured to me that these proton transfers allow for a mechanism by which the firefly can modulate the light emission in a way which has always seemed wonderous to me. The firefly produces (if I remember correctly) a train of light pulses, and with the dioxetanone intermediate, with a short but probably longer than millisecond lifetime, I didn't understand how it could be modulated. If you simply modulate the proton transfer, that could account for the modulation.

Dr. Seliger

The shortest interval between flashes, the firefly signaling, is approximately 60-70 milliseconds.

Dr. Hastings

I would like to add a comment. I think, Dr. Foote, that in this your assumption is that the decay of a flash of a firefly relates to the lifetime of one of the intermediates rather than to some other kinetic aspect, such as enzyme turnover. Am I correct in that assumption?

Dr. Foote

Yes.

Dr. Hastings

Then some experiments we recently carried out together with Drs. Wulff and Presswood may be relevant. We measured the lifetime of intermediates in the firefly luciferase system by quenching the reaction either with EDTA to remove $Mg^{++}$, Apyrase to remove ATP, or glucose/glucose oxidase to remove oxygen. Half-decay times were in the 100 msec range, values similar to the decay of the luminescence in the *in vivo* firefly flash. The results may thus be consistent with your assumption, that is, the kinetics of a flash may reflect the lifetime of a reaction intermediate such as, for example, peroxy-luciferyl AMP.

Dr. Branchini

I would like to get back to an earlier point on the color of bioluminescence and state some things about the observations we have made, and I would like to have your comments on these in terms of the CIEEL mechanism. With the firefly bioluminescence system, we have examined three amino analogs of luciferin--aminoluciferin, the monomethylaminoluciferin, and the N,N-dimethylaminoluciferin. What we found, just looking preliminarily at the emssion spectra, was that the amino compound had lambda max of about 600 nanometers, the monomethyl amino of about 625 nanometers, and the N,N-dimethylamino was further red shifted to about 650 nanometers. Also, it looks like at the same time we are getting a corresponding decrease in the amount of light emitted from these.

Dr. Schuster

Do the bioluminescent emission spectra correspond to the fluorescent spectra of the carbonyl compound product?

Dr. Branchini

We haven't yet measured these.

Dr. Schuster

That most likely cause of the shift is that you're shifting the fluorescence spectra of the product. Also, the change in the light intensity could be caused by the change of fluorescence efficiency of the substituted amino group.

# II
# LUMINOL-DEPENDENT CHEMILUMINESCENCE

# PRELIMINARY EVENTS LEADING TO THE PRODUCTION OF LUMINOL-DEPENDENT CHEMILUMINESCENCE BY HUMAN GRANULOCYTES

Knox Van Dyke
Cynthia Van Dyke
David Peden
Maria Matamoros

Department of Pharmacology & Toxicology
West Virginia University Medical Center
Morgantown, West Virginia

Vincent Castranova
George Jones

Department of Physiology (ALOSH)
West Virginia University Medical Center
Morgantown, West Virginia

## I. INTRODUCTION

Since the initial discovery by Allen *et al.* (1) that human neutrophils (granulocytic white cells) produced chemiluminescence (CL) when engulfing opsonized bacteria, there have been a variety of observations (2-5) of soluble stimulants (membrane pertubants) producing CL with granulocytes. The chemiluminescence so produced can be amplified using luminol thereby allowing the freedom of using fewer granulocytes. It has been noted by our group as well as others that CL does not begin immediately with either particulate or soluble stimulants. Therefore, there must be a triggering mechanism which eventually turns on the metabolism and produces light. Although the details are not totally understood, this review will encompass ideas backed by experimental data.




ISBN 0-12-208820-4

## II. BINDING OF STIMULANTS

The first event that should occur is the binding of the stimulating substance to its binding point or receptor on the external surface of the granulocyte. A variety of different chemicals is known to produce a CL response. The question to pose is: are all these responses receptor mediated? Some of them definitely are mediated by a specific site (receptor). The chemotactic peptide, formyl methioninyl-leucinyl phenylalanine (FMLP) is known to bind to a specific receptor on the surface of the granulocyte. These receptors have been quantified (6). Indirect evidence for receptor mediation could come from an early perturbation of the membrane, i.e., depolarization, by activators. If the change in membrane potential difference (transmembrane potential) occurs before the generation of the metabolic events (and CL) as measured by determination of superoxide $[.O_2]^-$ concentration, these changes in electrical potential could be a link between the binding of the stimulant ligand (or perturbant) and the metabolic events and CL.

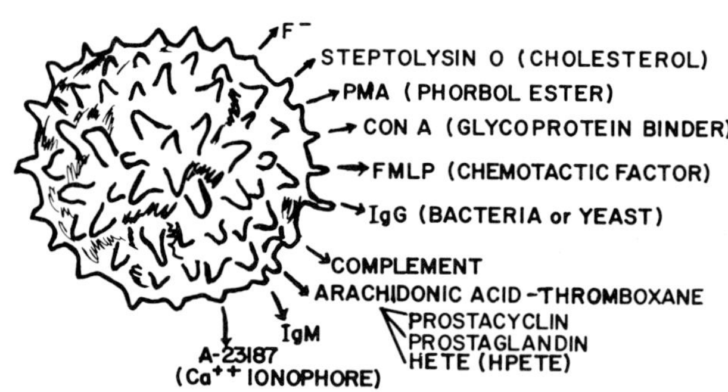


*FIGURE I. STEP I - Binding of Perturbant to Granulocyte Membrane*

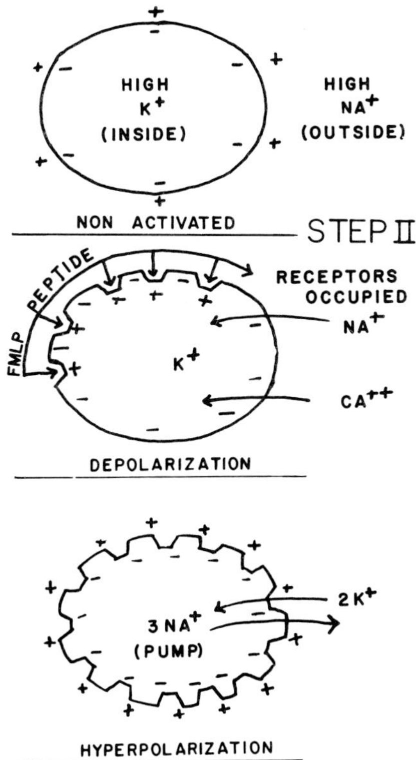


*FIGURE 2. Three states of membrane electrical activity*

When 1mm $Ca^{++}$ is present in the outside solution, the $Ca^{++}$ ionophore A23187 causes an initial depolarization, but hyperpolarization is not observed. Calcium may assist sodium in the depolarization of the cell (8).

## III. LINKAGE BETWEEN IONIC AND METABOLIC EVENTS

The linkage between ionic (electrical) events and metabolic events (in membrane stimulated granulocytes) can be traced to the inward movement of $Ca^{++}$ and $Na^{+}$ (9,10) and the breakdown of phospholipids (11) by phospholipase $A_2$ is a $Ca^{++}$ dependent event. (See Figure 3).

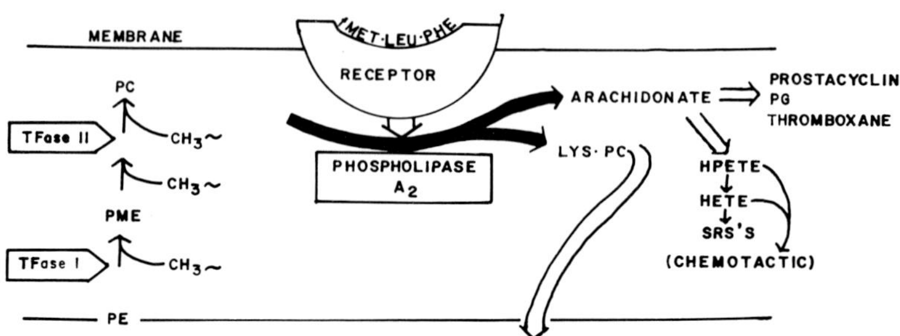


*FIGURE 3. Hypothetical binding of chemotactic peptide (FMLP) to granulocyte membrane triggering metabolism of sodium arachidonate. TFase I and II, phospholipid methyltransferase I and II; PE, phosphatidylethanolamine, PME, phosphatidyl-N-monomethylethanolamine; PC, phosphatidyl choline, LysPC, lysophosphatidylcholine; PG, prostaglandin.*

Phospholipid breakdown produces arachidonate, which can be metabolized with addition of oxygen via lipoxygenase or cyclooxygenase producing a variety of products (See Peden, this symposium). The intracellular binding of calcium occurs at least in part via calmodulin (12) (a calcium-ion-binding protein). CL can be almost completely inhibited by a $2 \times 10^{-5}$ Molar trifluoroperazine (TFP), a drug which blocks the oxidative burst. This effect can be reversed by increasing the external concentration of calcium (12,13).

## IV. MEASUREMENT OF RESTING TRANSMEMBRANE POTENTIAL

The resting transmembrane potential ($E_m$) of human granulocytes was measured by Jones, Van Dyke and Castranova (7) (See Table I). These determinations were accomplished using the lipophilic dye (Di-S-$C_3$ (5)) and measuring a fluorescence end point. These measurements produced the highest known transmembrane potential of any human cell familiar to the authors.

When 10mm $K^+$ is added to granulocytes incubated in $K^+$ free medium with the dye, a hyperpolarization occurs which can be inhibited by the addition of ouabain ($6.7 \times 10^{-5}$M), which poisons the $Na^+$ - $K^+$ electrogenic pump. This is a clear cut indication that such a pump exists, is electrogenic and contributes to the high membrane potential.

*TABLE I. The Transmembrane Potential ($E_m$) of Human Granulocytes*

| *K null point (mM)* | *$E_m$(mV)* |
|---|---|
| *2.7± 0.6$^n$* | *-101.7 ± 4.8$^n$* |

*n = 5 determinations - where intracellular $K^+$ concentration was 120 millimolar.*

## V. EFFECTS OF STIMULANTS ON MEMBRANE POTENTIAL

Addition of soluble stimulants (f-met-leu-phe), concanavillin A or phorbol myristate acetate to granulocytes causes an initial depolarization which is mostly dependent on sodium and to some extent calcium ions. This is followed by a hyperpolarization with the time course varying somewhat with the stimulant type and concentration.

The intracellular calmodulin concentration of neutrophils is 7 micromolar which represents 0.04% of the total soluble protein (12). NADPH-oxidase-rich-membrane preparations display 50% inhibition at 20 micromolar TFP. Whitin et al. (14) demonstrated that phorbol myristate acetate (PMA) depolarized granulocytes in 10 seconds while superoxide $[.O_2]^-$ production was detected in 45 to 60 seconds. Korchak and Weissman (15) showed that changes in membrane potential using concanavalin A and FMLP peptide precede the metabolic response.

## VI. METABOLIC EVENTS AND CHEMILUMINESCENCE

Allen has done extensive work in discovery of reactions leading to chemiluminescence (16-18). Basically, they involve the production of $[.O_2]^-$ from NADPH oxidase and subsequently $H_2O_2$.

Consequently, there occurs the degradation of $H_2O_2$ in the presence of chloride ion producing the bacteria killing system (hypochlorite ion) and its reaction with reactive groups, possibly producing excited state carbonyls and therefore CL.

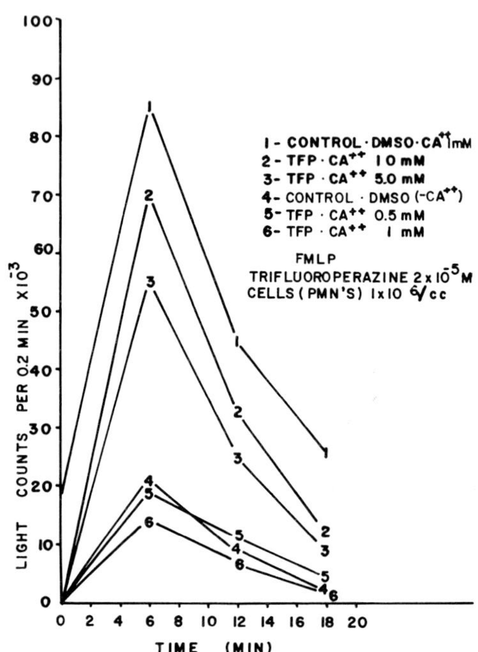


*FIGURE 4. Calcium-reversal-trifluoroperazine inhibition of chemiluminescence in response to chemotactic peptide (FMLP).*

## VII. SECONDARY CHEMILUMINESCENCE AND ARACHIDONATE METABOLISM

Granulocytes have the potential of producing arachidonate based chemiluminescence either from the usual soluble or particulate stimulant or from added arachidonate (19,20). We have reported that inflammatory drugs inhibit the CL from these systems and we are presenting a poster session (this symposium) on arachidonate based CL in granulocytes and platelets. The CL which is luminol enhanced probably occurs from peroxidases, lipoxygenases and cyclooxygenase (21).

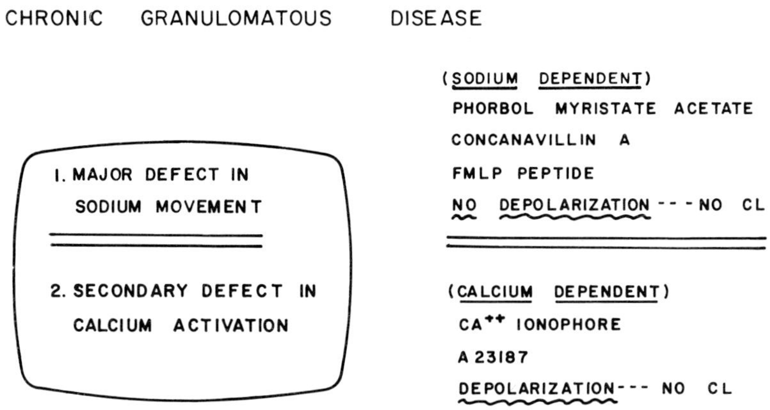

CHRONIC GRANULOMATOUS DISEASE

1. MAJOR DEFECT IN SODIUM MOVEMENT

2. SECONDARY DEFECT IN CALCIUM ACTIVATION

(SODIUM DEPENDENT)
PHORBOL MYRISTATE ACETATE
CONCANAVILLIN A
FMLP PEPTIDE
NO DEPOLARIZATION --- NO CL

(CALCIUM DEPENDENT)
$CA^{++}$ IONOPHORE
A 23187
DEPOLARIZATION --- NO CL

*FIGURE 5. Depolarization defects in granulocytes of chronic-granulomatous disease patient.*

## VIII. DEPOLARIZATION DEFECTS IN DISEASE

Chronic Granulomatous Disease granulocytes (CGD) have been shown to be defective in depolarization of the granulocyte membrane (14). In a poster session, Castranova et al. (22) detail the probable defects in sodium movement with a secondary defect in $Ca^{++}$ activation of metabolism and chemiluminescence. See summary.

## IX. SUMMARY OF SOME EARLY EVENTS AND CHEMILUMINESCENCE OF GRANULOCYTES

The following diagram depicts a partial summary of these cellular events: depolarization (initiated by a perturbant binding to the granulocytes), metabolic activation and, finally, chemiluminescence.

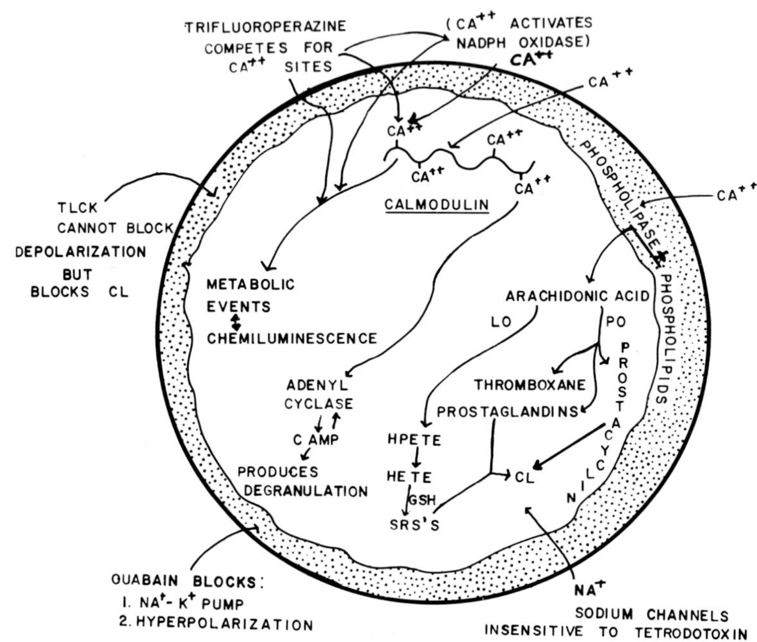


*Summary of some of the early events which may produce chemiluminescence.*

REFERENCES


1. Allen, R. C., Stjernholm, and R. H. Steele, *Biophys. Res. Res. Comm. 47,* 679 (1972).
2. Wilson, M. E., M. A. Thruse, K. Van Dyke, and W. Neal, *FEBS Letters 94,* 287 (1978).
3. Anderson, B. R., and J. L. Duncan, *J. Infect. Dis. 141,* 680 (1980).
4. Hatch, G. E., D. E. Gardner, and D. B. Menzel, *J. Infect. Med. 147,* 182 (1978).
5. DeChatelet, L. R., P. S. Shirley, and R. B. Johnston, *Blood 47,* 545 (1976).
6. Williams, L. T., R. Snyderman, M. C. Pike, and R. J. Lefkowitz, *Proc. Natl. Acad. Sci. 74,* 1204 (1977).
7. Jones, G., K. Van Dyke, and V. J. Castranova, *Cell Physiology,* in press 1980.
8. Jones, G., K. Van Dyke, and V. J. Castranova, *Cell Physiology,* in press 1980.
9. Simchowitz, L. and I. Spilberg, *J. Immunol. 123,* 2428 (1979).
10. Simchowitz, L. and I. Spilberg, *J. Lab Clin. Med. 93,* 583 (1979).
11. Hirata, F., B. A. Corcoran, K. Venkatasubramanian, E. Schiffman, and J. Axelrod, *J. Proc. Nat'l. Acad. Sci. 76,* 2640 (1979).
12. Jones, H. P., W. F. Petrone, J. M. McCord, *Fed. Proc. 39,* 1626 (1980).
13. Matamoros, M. and K. Van Dyke (unpublished observations)
14. Whitin, J. C., C. E. Chapman, E. R. Simons, M. E. Chovaniec, and H. J. Cohen, In preparation.
15. Korchak, H. M. and G. Weissman, *Proc. Nat'l. Acad. Sci. 75,* 3818 (1978).
16. Allen, R. C., *Photochem. and Photobiol. 30,* 157 (1979).
17. Allen, R. C., *Biochem. Biophy. Res. Comm. 63,* 675 (1975).
18. Allen, R. C., *Biochem. Biophys. Res. Comm. 63,* 684 (1975).

# CHEMILUMINESCENCE IMMUNOASSAY FOR ANTI-HUMAN IgG

H. R. Schroeder
C. M. Hines
P. O. Vogelhut

Immunochemistry Laboratory
Ames Research and Development Department
Ames Division
Miles Laboratories, Inc.
Elkhart, Indiana

## I. INTRODUCTION

Many clinically significant substances are measured at very low levels in protein binding radioassays. We have substituted chemilumigenic compounds as labels in these procedures because they are stable and eliminate the inconvenience associated with radiolabels. The aminophthalhydrazide labels we used were derivatives of these compounds. They participate in simple oxidation reactions to produce light with high quantum yield which allows their detection within seconds at pM levels (1). Homogeneous assays for biotin (2) and progesterone (3) as well as heterogeneous assays for thyroxine (4) and testosterone (5) have been developed. In addition, recent publications have described heterogeneous immunoassays for proteins using luminol (6,7) and a highly efficient isoluminol derivative (8) as label. This paper describes development of a model double antibody, solid phase (sandwich) immunoassay with automated read-out of chemiluminescence.




ISBN 0-12-208820-4

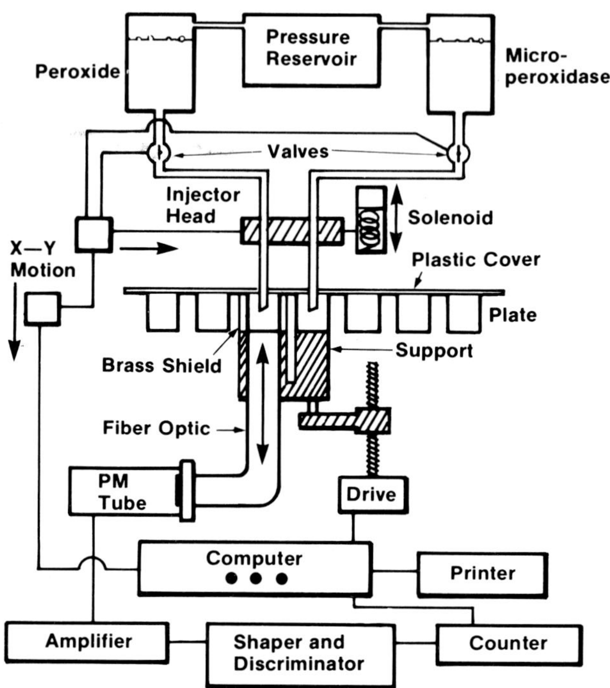


*FIGURE 1. Schematic of the chemiluminescence reader.*

## II. MATERIALS AND METHODS

### *A. Instrumentation*

An experimental, computer-directed instrument was developed in-house to provide an automated read-out for the immunoassays carried out in a multiwell plastic plate. The plastic plate fits a carrier that is moved by motors and lead screws in both X and Y directions to position individual wells over the photoreceptive area (Fig. 1). Microperoxidase and peroxide are added to wells in sequence from pressurized vessels to produce light which is measured by photon counting (EMI-9789A, low noise, housing amplifier 3262/$AD_4$ PMT with discriminator S/N 323 from Pacific Precision Instruments, Concord, CA), displayed by a printer.

### *B. Reagents*

Synthesis of 6-[Carboxymethoxyacetyl-N-(6-aminohexyl)-N-ethylamino]-2,3-dihydrophthalazine-1,4-dione (Fig. 2, II). Four mmoles of 6-[N-(6-aminohexyl)-N-ethylamino]-2,3-dihydro-phthalazine-1,4-dione (Fig. 2, I) (1), 4.4 mmole diglycolic anhydride and 4 mmole triethylamine were dissolved in 20 ml of dry dimethylformamide (DMF) under nitrogen. After 17 hours the solvent was removed in vacuo at 35°, 0.1 mm, and the oily residue was dissolved in 2.5 ml of hot methanol. The yield of first crop crystals formed after 3 weeks at 4° was 1.067 g (63%) with m.p. 181-182°C. Analysis: Calc. for $C_{20}H_{28}N_4O_{16}$: C, 57.13; H, 6.71; N 13.33 Found: C, 56.62; H, 6.95; N, 13.22. The mass spectrum confirmed the structure with a P-18 peak (loss of water) at m/e of 402 which was 8.2% of the base peak (m/e 218).

Preparation of the Chemilumigenic Human IgG Conjugate (Fig. 2). The chemilumigenic label (Fig. 2, II) was coupled to human IgG (Pentex, Fraction II, Miles Laboratories, Inc.) by a two step procedure. First, 20 μmole of II and 20 μmole

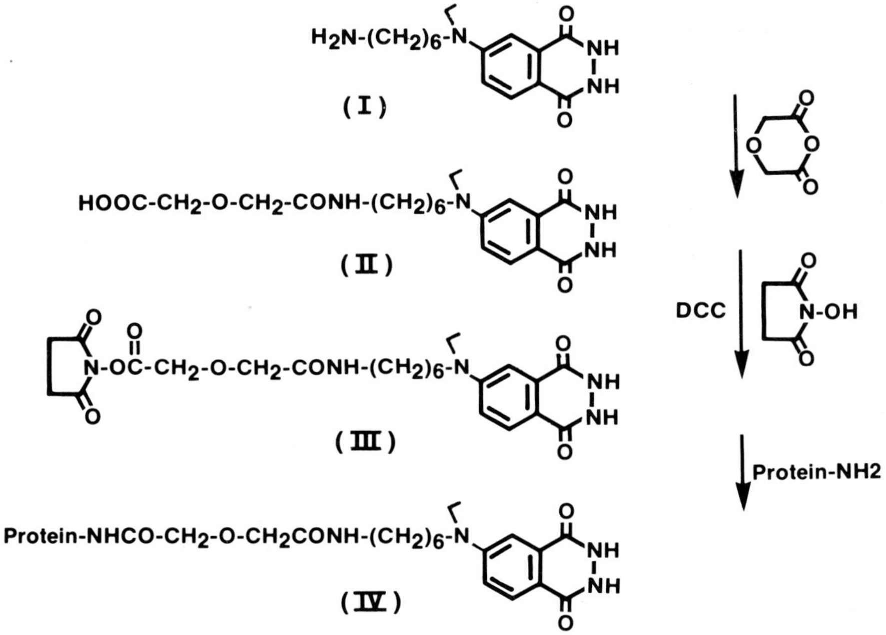

*FIGURE 2. Synthetic scheme for the preparation of a 6-[N-(6-aminohexyl)-N-ethylamino]-2,3-dihydrophthalazine-1,4-dione-protein conjugate.*

of N-hydroxysuccinimide were dissolved in 204 µl of dry DMF under nitrogen. The solution was cooled to -10° and 46 µl (23 µmole) of N,N-dicyclohexylcarbodiimide in DMF were added. Aliquots of the active ester formed at 0° by 15 hours (Table I) were added to human IgG in 2 ml of 0.1 M sodium phosphate - 0.15 M NaCl at pH 8. After 2.5 hr at 4° and 0.5 hr at 20° Tween 20 was added to 0.25% concentration to solubilize the protein. The mixture was placed on a 1.5 x 48 cm column packed with Sephadex G-25M and eluted with 25 mM Tris-HCl-250 mM NaCl, pH 8.6.

### C. *Sandwich Immunoassay for Anti-human IgG*

All immunoassays were carried out in 96 well polyvinylchloride plates (#1-220-29, Dynatec Laboratories). Individual wells (200 µl) were coated overnight with 100 µl of 2 µg/ml IgG in 0.1 M $Na_2CO_3$ at pH 9.5, 4°. Excess protein binding sites on the plastic were then saturated with 1% BSA. Prior to use wells were washed three times with 180 µl each of 0.01 M phosphate - 0.15 M NaCl, pH 7 (PBS) and aspirated using a Miniwash apparatus (Dynatec).

The sandwich assay was initiated by adding contrived samples consisting of various amounts of rabbit antihuman IgG (Cappel Laboratories) in 100 µl of 1% BSA-PBS to the IgG precoated wells. After 1.5 hr at 40° plates were washed three times with PBS. Then the sandwich was completed by addition of 100 µl of conjugate B (Table I) at 0.7 µg/ml in 30% normal rabbit serum/PBS and incubation for 1.5 hr at 40°. The wells were washed five times with PBS and three times with water. Then the plates were placed in the chemiluminescence reader which added 100 µl of 2 µM microperoxidase (Sigma Chem. Corp.) in 50 mM borate, pH 8.6, followed by 20 µl of 5 mM $H_2O_2$ in 10 mM Tris-HCl, pH 7.4. Photons generated by the reaction were counted for 2 sec.

## III. RESULTS AND DISCUSSION

A solid phase sandwich immunoassay for rabbit anti-human IgG was developed to determine the sensitivity obtainable with this format when monitored by chemiluminescence. We chose active ester chemistry for labeling IgG because this approach has been successful for coupling small molecules to proteins (9,10) and avoids the unwanted side reactions of bifunctional coupling agents. The label (I) was derivatized to provide a carboxyl group necessary for this method as shown

TABLE I. *Coupling of Label to Human IgG Via Active Ester*

| Reaction | IgG (mg) | Active Ester[a] (μl) | pH | Protein Recovery (%) | Incorporation[b] | |
|---|---|---|---|---|---|---|
| | | | | | Spectral | Chemiluminescent |
| A | 2.47 | 5 | 8.0 | 88 | 5.87 | 0.34 |
| B | 2.47 | 10 | 8.0 | 83 | 8.17 | 0.63 |
| C | 2.47 | 20 | 8.0 | 60 | 12.02 | 0.59 |
| D | 2.47 | 40 | 8.0 | 36 | 17.04 | 1.50 |
| E | 2.47 | 80 | 8.0 | 13 | 31.30 | 7.50 |
| F[c] | 2.47 | 40 | 8.0 | 37 | 17.20 | 0.66 |
| G | 2.47 | 20 | 7.7 | 64 | 9.84 | 0.59 |
| H | 2.47 | 20 | 8.3 | 59 | 14.40 | 0.57 |
| I | 0.62 | 20 | 8.0 | 47 | 17.09 | 1.82 |
| J | 1.24 | 20 | 8.0 | 49 | 15.74 | 1.05 |
| K[d] | 4.32 | 20 | 8.0 | 64 | 9.84 | 0.28 |
| L[d] | 8.64 | 20 | 8.0 | 66 | 7.32 | 0.22 |
| M[e] | 2.47 | 20 | 8.0 | 53 | 14.38 | 0.78 |

*Coupling reactions were conducted as described earlier. The amount of (I) incorporated into the conjugate was calculated from the $A_{325}$ and an extinction coefficient of 12.5 at 325 nm since this measurement was unaffected by low levels of protein. Also, the incorporation of (I) was determined from the light yield of the conjugate in the $H_2O_2$-microperoxidase system at pH 13 (1) and a label (I) reference at several concentrations. Protein was precipitated at 10% trichloroacetic acid and determined according to Lowry procedure.*

[a]*Eighty μmole/ml of II (Fig. 2) was activated.*
[b]*Molar ratio of (I)/IgG.*
[c]*Coupling was for 1 hr at 20° instead.*
[d]*Columns (1.5 x 120 cm) filled with Sephadex G-25M were used for isolation of conjugate.*
[e]*Reaction volume was 1 ml.*

in Fig. 2. The carboxy derivative (II) was first activated under anhydrous conditions and then combined with the aqueous protein. The resulting conjugate was isolated by chromatography.

A number of conjugates were prepared under a variety of coupling conditions and analyzed for amount of label incorporated into the conjugate (Table I). Incorporation of label was greatly influenced by the relative levels of active ester and protein, as well as pH and temperature. An overnight activation of II at 0° and a 3 hr coupling at about 4° and pH 8 were most favorable for incorporation of label and improved agreement of chemiluminescence and spectral estimates. The discrepancy between these two evaluations may arise from quenching of the label by protein of self coupling of II with the phthalhydrazide ring nitrogens which abolishes the light yielding capacity. Although the highly labeled conjugates gave best agreement, their recovery was poorest because of increased hydrophobicity. Furthermore, use of such conjugates in sandwich assays gave lower sensitivity because of greater nonspecific binding and possibly some loss in immunoreactivity.

The chemilumigenic human IgG conjugates were used in a sandwich assay for anti-human IgG. This format was adopted since it has inherently greater sensitivity than the competitive protein binding assay because reagents can be used in excess (11). Also, this procedure includes wash steps at various points that eliminate substances (such as proteins) that interfere in the chemiluminescent monitoring reaction (4).

The assay was carried out in multiwell plastic plates which have demonstrated utility and convenience in similar RIA (12,13) and EIA (14-16) procedures for protein. The partially optimized assay monitored with the best chemilumigenic conjugate (B, Table I) gave a lower detection limit of 0.5 ng of anti-human IgG per well (Fig. 3) and produced a broad dose response over more than two orders of magnitude (data not shown). The specificity of the procedure was demonstrated by using wells coated with BSA only or substituting the label (I) for the conjugate which resulted in only low nonspecific binding and the absence of a dose response.

The model chemiluminescence immunoassay developed in this study shows considerable merit. Although samples must be added individually, wash steps can be performed simply and rapidly on an instrument. Labeled conjugate is added with ease using a 12 tip multichannel pipettor (Dynatec). The automatic read-out at only 5 sec intervals is much more rapid than the 1 min counting time of radioassays and avoids the usual 0.5 hr incubation with substrate in enzyme immunoassays. Thus, this type of assay monitored by chemiluminescence may find clinical application in large scale screening of numerous samples and where a highly sensitive, non-radioisotopic method is desired.

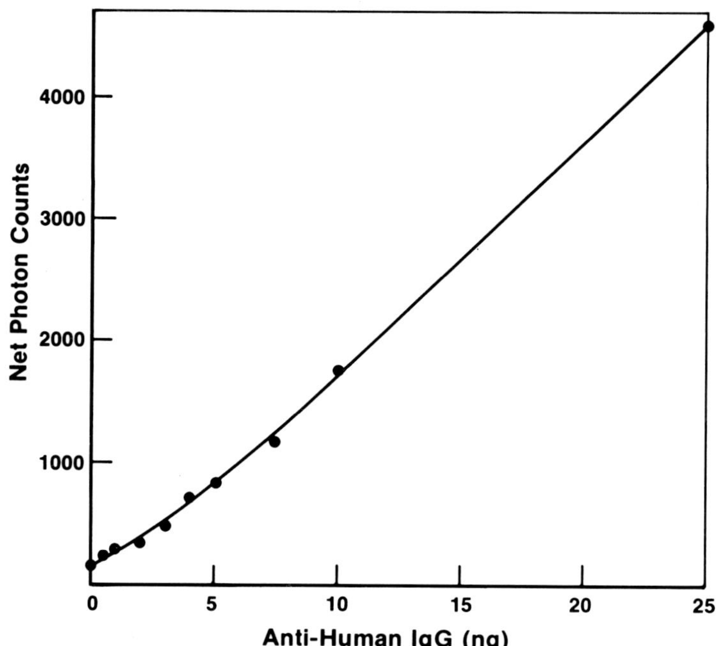


*FIGURE 3. Immunoassay for anti-human IgG monitored by chemiluminescence.*

## ACKNOWLEDGEMENTS

We are grateful for the invaluable help of R. Rogers, R. Hurtle, F. Wogoman and S. Hout in developing and building the chemiluminescence reader.

## REFERENCES

1. Schroeder, H. R. and F. M. Yeager, *Anal. Chem.* 50, 1114 (1978).
2. Schroeder, H. R., P. O. Vogelhut, R. J. Carrico, R. C. Boguslaski, and R. T. Buckler, *Anal. Chem.* 48, 1933 (1976).
3. Kohen, F., M. Pazzagli, J. B. Kim, H. R. Lindner, and R. C. Boguslaski, *FEBS Letters* 104, 201 (1979).
4. Schroeder, H. R., F. M. Yeager, R. C. Boguslaski, and P. O. Vogelhut, *J. Immun. Meth.* 25, 275 (1979).

5. Pratt, J. J., M. G. Woldring, and L. Villerius, *J. Immun. Meth.* 21, 179 (1978).
6. Hersh, L. S., W. P. Vann, and S. A. Wilhelm, *Anal. Biochem.* 93, 267 (1979).
7. Simpson, J. S. A., A. K. Campbell, M. E. T. Ryall, and J. S. Woodhead, *Nature* 279, 646 (1979).
8. Schroeder, H. R., D. D. Osborn, and R. C. Boguslaski, Future Perspectives in Clinical Laboratory Immunoassays, Scripps Institute, San Diego, CA. (1980).
9. Cuatrecasas, P. and M. D. Hollenberg, *Adv. in Prot. Chem.* 30, 227 (1976).
10. Hosoda, H., Y. Sakai, H. Yoshida, and T. Nambara, *Chem. Pharm. Bull.* 27, 2150 (1979).
11. Schuurman, H. J., and C. L. de Ligny, *Anal. Chem.* 51, 2 (1979).
12. Purcell, R. H., D. C. Wong, H. J. Alter, and P. V. Holland, *Appl. Microb.* 26, 478 (1973).
13. Trent, D. W., C. L. Harvey, A. Qureshi, and D. LeStourgeon, *Inf. and Immun.* 13, 1325 (1976).
14. Wolters, G. and L. P. C. Kuijpers, *J. Infect. Diseases* 136, 8311 (1977).
15. Voller, A., D. E. Bidwell, A. Bartlett, D. G. Fleck, M. Perkins, and B. Oladehin, *J. Clin. Path.* 29, 150 (1976).
16. Bartlett, A., K. M. Dormandy, C. M. Hawkey, P. Stableforth, and A. Voller, *Br. Med. J.* 1, 994 (1976).

# LUCIGENIN CHEMILUMINESCENCE: A NEW APPROACH TO THE STUDY OF POLYMORPHONUCLEAR LEUKOCYTE REDOX ACTIVITY

*Robert C. Allen*[1]

U.S. Army Institute of Surgical Research and
Clinical Investigation Service
Brooke Army Medical Center
Fort Sam Houston, Texas

## I. NATIVE AND SUBSTRATE-AMPLIFIED PHAGOCYTE CHEMILUMINESCENCE

Chemiluminescence is a product of polymorphonuclear leukocytes and other phagocytes activated by either phagocytosis or chemical stimuli (1-3). Postphagocytic stimulation of PMNL[2] metabolism is characterized by increased glucose utilization via the hexose monophosphate shunt, and increased nonmitochondiral $O_2$ consumption (4,5). Both activities are required to generate the oxidants necessary for microbicidal action, and both activities correlate with the yield of CL[3] (6). The native photon emitters responsible for CL are postulated to be electronically excited carbonyl functions of oxidation products (7).

The efficiency of oxidation reactions yielding excited products is relatively low, and consequently, native CL from stimulated PMNL is relatively weak. However, sensitivity in

[1]*The opinions or assertions contained herein are the private views of the author and are not to be construed as reflecting the views of the Department of the Army or the Department of Defense.*

[2]*PMNL: Polymorphonuclear leukocyte*

[3]*CL: Chemiluminescence*




ISBN 0-12-208820-4

measurement of PMNL redox activity by CL can be increased by several orders of magnitude through introduction of a chemilumigenic substrate whose oxidation gives a high yield of electronically excited product. This concept serves as the basis for using luminol, 5-amino-2,3-dihydro-1,4-phthalazinedione, to measure phagocyte oxidative activity by CL (8,9).

Under appropriate conditions, lucigenin, 10,10'-dimethyl-9,9'-biacridinium dinitrate, can serve as substrate in $O_2$-redox reactions yielding electronically excited N-methylacridone as product (10.11). Such reactions are characterized by intense CL. In the present study lucigenin, $DBA^{++}$, [4] is successfully employed as a chemilumigenic probe for the study of stimulated PMNL redox activity.

## II. EXPERIMENTAL METHODS

Human whole blood was obtained from volunteers by venipuncture. The leukocyte-rich plasma was isolated following dextran-sedimentation of erythrocytes (12). After hypotonic lysis of the remaining erythrocytes (0.2% saline for 15 sec), and two additional washes with Dulbecco's phosphate buffered saline without $Ca^{++}$ or $Mg^{++}$, pH 7.2, total and differential counts were done on the leukocyte suspension, and the volume was adjusted to yeild a concentration of 1,000 PMNL/µl. The desired quantity of the PMNL suspension was added to each sterile, siliconized glass vial (24 ml capacity) containing 1.75 ml of barbital (veronal) buffered saline with $Ca^{++}$, $Mg^{++}$, (13) and also albumin (0.1% w/v) and glucose (0.1% w/v), pH 7.2. Each vial also received 20 µl of fresh autologous serum as a source of alternative pathway complement for zymosan opsonification. Serum was not required for PMNL stimulation by PMA[5]. A 5 mM solution of lucigenin (Aldrich Chem. Co.) was prepared in distilled water, and the desired quantity was added to each vial just prior to measurement of background CL.

Chemiluminescence was quantified at room temperature (22°C) using the single phonton counting capacity of a Beckman LS-150 scintillation counter equipped with EMI 9829A photomultiplier tubes, and operated in the out-of-coincidence mode

[4] *$DBA^{++}$: 10,10'-dimethyl-9,9'-biacridinium*
[5] *PMA: phorbol-12-myristate-13-acetate*

using the tritium channel settings. The CL intensity from the unstimulated PMNL suspension was measured for 3 cycles. Each individual vial was counted for 0.2 min once each cycle. Therefore, the time per cycle defines the time period between measurements of CL intensity for any particular vial. Either the particulate stimulant, zymosan, or the chemical stimulant, PMA, was added to the vial 20 to 30 sec period to the fourth cycle.

Zymosan was prepared by suspending 250 mg of zymosan A (Sigma Chem. Co.) in 100 ml of normal saline at 90°C for 30 min. PMA (Sigma Chem. Co.) was prepared as a 5 mM stock solution in dimethyl sulfoxide and further diluted with distilled water.

## III. LUCIGENIN-DEPENDENT CHEMILUMINESCENCE FROM POLYMORPHONUCLEAR LEUKOCYTES

No CL above background was detected from unstimulated PMNL incubated with micromolar concentrations of lucigenin at a pH of 7.4. However, background CL from the PMNL suspensions was observed to increase with increase in buffer pH. Therefore, pH was maintained between 7.1 and 7.4 for all experiments described.

A large CL response was detected following phagocytosis or chemical stimulation of the PMNL suspension in the presence of lucigenin. As few as one thousand PMNL could be measured by this method. However, heat-killed suspensions of PMNL (i.e. 56°C for 10 min) yielded no CL when stimuli were added in the presence of lucigenin.

### A. *Concentration-dependence of Lucigenin-CL from Chemically Stimulated PMNL*

In order to quantify the effect of lucigenin concentration on the CL response, the number of PMNL, the volume and type of buffer, and the quantity and type of stimulus were held constant, and the concentration of lucigenin was varied. The upper portion of figure 1 depicts the results plotted as net CL intensity, in relative counts per minute, against time. The CL intensity measurements for the first three cycles reflect the CL activities from unstimulated PMNL incubated with

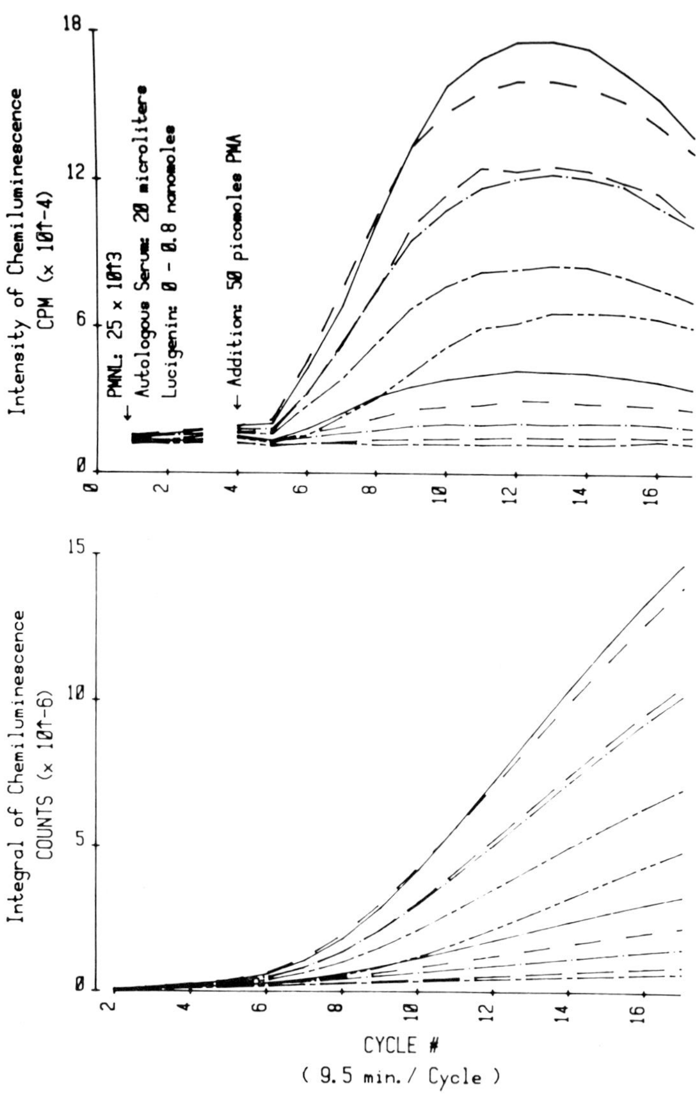


*FIGURE 1. Lucigenin-dependent CL from PMA-stimulated PMNL. The CL intensity and the integral CL responses are plotted against time in the upper and lower graphic, respectively. Each vial contained 25,000 PMNL and 20 μl of autologous serum in a final volume of 2.0 ml barbital buffer, pH 7.2. PMA, 50 picomoles in 20 μl, was added to each vial 20 to 30 sec prior to the fourth cycle count. In the graphics, a vertical arrow preceding a number indicates that the number is an exponent, i.e. 10↑3 =* $10^3$.

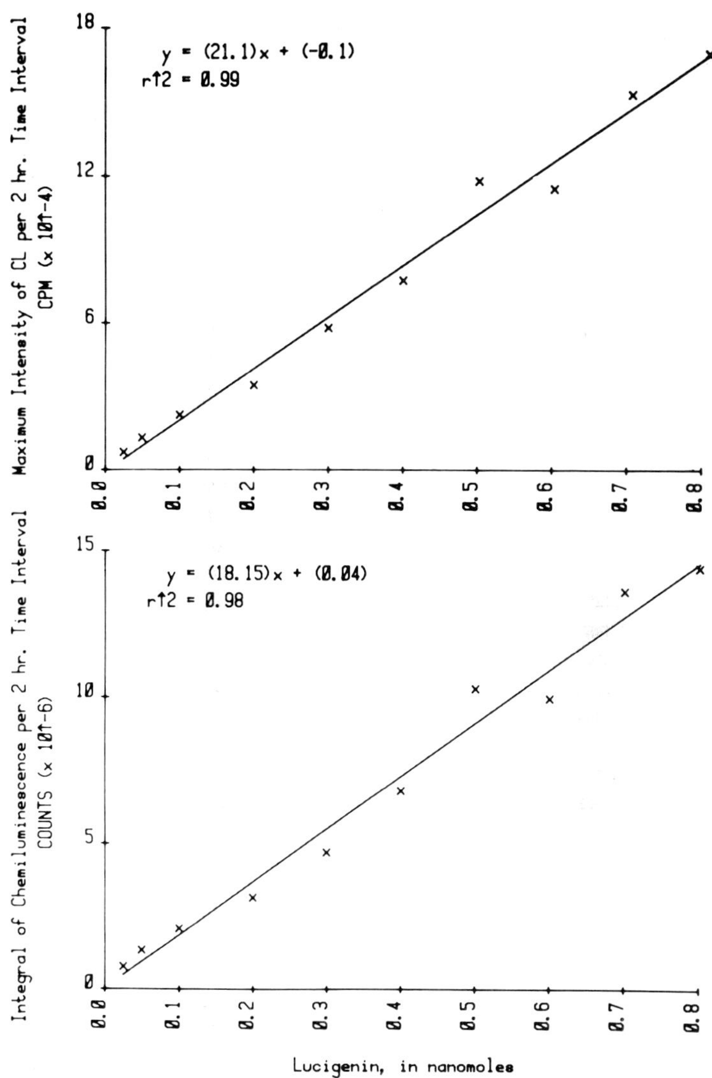


*FIGURE 2. Plot of maximum CL intensity, upper graphic, and two hour integral CL response, lower graphic, against the quantity of lucigenin added. The experimental conditions are those described in figure 1, and the values are those calculated from the data of figure 1.*

various concentrations of lucigenin. PMA was added to each vial just prior to the fourth cycle count, and the change in CL intensity was measured for an additional 13 cycles, approximately two hrs. A cycle is defined as the time period from one CL measurement to the next.

The lower graphic of figure 1 is the plot of cumulative or integral CL against time. The integral values were calculated from the intensity data by trapezoidal approximation. Many biologists and biochemists are not familiar with intensity measurements. Consequently, in some of the published research relating CL to other measurable PMNL activities, correlations have been drawn between maximum CL intensity and integral measurements. As depicted in figure 2, the maximum CL intensity and the integral CL are both linear functions of the concentration of lucigenin employed, an integral value. However, in situations of greater kinetic complexity attempts to find correlation between integral and intensity values can result in erroneous conclusions. It is therefore suggested

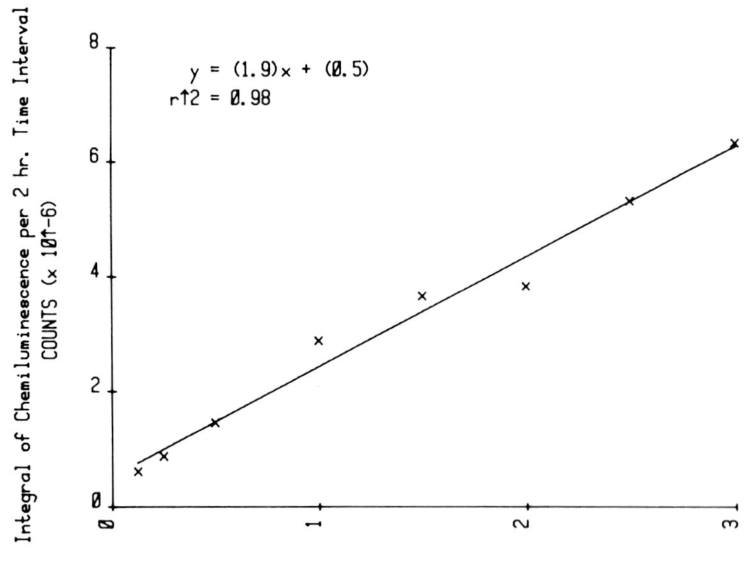


*FIGURE 3. Lucigenin-dependent CL from opsonified zymosan-stimulated PMNL. Plot of the two hour integral CL response against the quantity of lucigenin added. With the exception that stimulation was by addition of 50 micrograms of zymosan suspended in 20 μl of saline, the conditions were the same as described in figure 1.*

that the integral CL for the same time interval of measurement be used when comparing CL with integral data.

*B. Lucigenin-dependent Chemiluminescence from PMNL Following Phagocytosis of Opsonified Zymosan*

In additional experiments of similar design, opsonified zymosan was employed as a phagocytosable particulate stimulus. The plot of integral CL responses against the concentration of lucigenin used is presented in figure 3. Once again the integral response from the PMNL was related to the

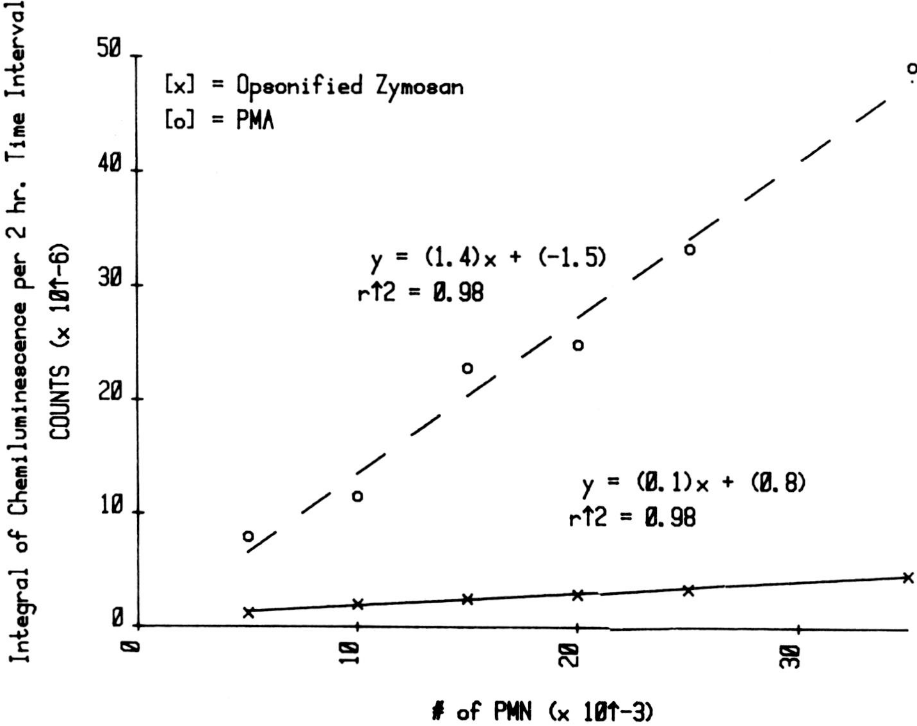


*FIGURE 4. Plot of CL against the number of PMNL stimulated. When zymosan was employed as the stimulus, 5.0 nanomoles of lucigenin were added; when PMA was the stimulus, 1.0 nanomole of lucigenin was added per vial. the PMNL tested were suspended in a final volume of 2.0 ml of barbital buffer containing 20 microliters of autologous serum, pH 7.2.*

concentration of lucigenin employed by a linear function. However, the total CL response using complement-opsonified zymosan was approximately an order of magnitude below that obtained by PMA stimulation.

### C. *Lucigenin-dependent Chemiluminescence and Quantification of the Number of PMNL*

To test the corollary, the concentration of lucigenin, the volume and type of buffer, the quantity and type of stimulus were held constant, and the number of PMNL was varied. Two different stimuli, PMA and opsonified zymosan, were tested. The results are depicted in figure 4, a plot of integral CL response against the number of PMNL present. CL was a linear function of the number of PMNL present regardless of the stimulus employed. However, the degree of stimulation of lucigenin-dependent CL was much greater for PMA than for opsonified zymosan.

## IV. MECHANISTIC PROPOSALS AND GENERAL COMMENTS

Lucigenin yields CL when reduced in base in the presence of oxygen (14), and alkaline solutions of lucigenin have been reported to react with $H_2O_2$ to yield CL (10,11). McCapra proposed that reduced lucigenin, 10,10'-dimethyl-9,9'-biacridilidene, could react with singlet molecular oxygen to yield a dioxetane product, and that decomposition of the dioxetane yielded excited carbonyl groups responsible for photon emission (15). Based upon these previous observations and the present data, the following mechanism is presented in explanation of the lucigenin-dependent amplification of PMNL-CL.

The increased redox activity of stimulated PMNL is reflected by an increased reduction of the added lucigenin. The univalent reduction of lucigenin can be effected by flavin or metal catalysis:

$$DBA^{++} + e^- \longrightarrow \cdot DBA^+$$

Superoxide anion may directly serve as the univalent reducing agent.

$$DBA^{++} + \cdot O_2^- \longrightarrow \cdot DBA^+ + O_2$$

Radical-radical annihilation reaction between univalently reduced lucigenin, $\cdot DBA^+$, and an additional superoxide anion, $\cdot O_2^-$, can yield a dioxetane product.

$$\cdot DBA^+ + \cdot O_2^- \longrightarrow DBA\text{-}O_2$$

Cleavage of the dioxetane yields two molecules of N-methylacridone. One of the NMA[6] will be electronically excited, NMA*.

$$DBA\text{-}O_2 \longrightarrow NMA + NMA^*$$

The relaxation of NMA* to ground state by phonton emission results in CL.

One alternative possibility is that $H_2O_2$ may react with lucigenin by a non-alkaline catalyzed mechanism. An additional but more remote possibility is that lucigenin is divalently reduced prior to reacting with $O_2$. The susceptibility of reduced lucigenin to electrophilic attack by $^1O_2$[8] has been considered (15). However, the role of free $^1O_2$ as a reactant in solution is questionable, and possibly a more stable singlet-like oxidant, such as a molecular complex of $O_2$, may serve as the oxidant responsible for dioxetane production (Seliger and Allen, unplublished discussions).

All of the mechanisms proposed to explain lucigenin-dependent CL from PMNL require the participation of PMNL $O_2$-redox activity. Thus, regardless of mechanism, lucigenin CL can be employed as a sensitive probe for assessment of function using a small number of PMNL. The data presented indicate that CL response is a linear function of the quantity of lucigenin present when lucigenin is the variable; CL response is also a linear function of PMNL activation when the number of PMNL is the only variable.

Although both luminol and lucigenin can serve as bystander chemilumigenic substrates for assessment of PMNL redox activity, the two probes differ in physical and chemical properties and reactivities, and each may measure a different activity occurring at a compartmentally different locus. For example, the large differences in CL responses to stimulation by PMA and opsonified zymosan are not observed when

[6] *NMA: N-methylacridone*
[7] *NMA*: Electronically excited methylacridone*
[8] *$^1O_2$ : single multiplicity molecular oxygen*

luminol is employed as the chemilumigenic probe. Analysis and comparison of the difference and similarities between luminol and lucigenin relative to PMNL-CL is presently under investigation

## SUMMARY

Chemiluminescence is a product of the postphagocytic metabolic activity of polymorphonuclear leukocytes and other phagocytes. The yield of luminescence is dependent upon the number and degree of activation of phagocytes present, and the presence of substrates susceptible to oxidations yielding electronically excited products. Redox reactions involving lucigenin give high yields of excited products and are characterized by intense luminescence. Use of lucigenin as a chemilumingenic probe provides a highly sensitive, nondestructive, and quantitative approach to the study of phagocyte $O_2$- redox activity. The reactive role of phagocyte generated superoxide anion in lucigenin-dependent luminescence is considered along with alternative mechanistic possibilities.

## ACKNOWLEDGMENTS

The author wishes to thank G. L. Strong, D. J. Hunter and J. L. Kelly for their technical assistance, and Mrs. D. S. Bratten for her assistance in preparing this manuscript and Dr. Basil A. Pruitt, Jr., for advice and discussion.

This research was supported by Clinical Investigation Service sponsored project #C-5-79.

## REFERENCES

1. Allen, R. C., Stjernholm, R. L. and Steele, R. H. (1972) *Biochem. Biophys. Res. Commun. 47*, 679.
2. Allen, R. C., Yevich, S. J., Orth, R. W. and Steele, R. H. (1974) *Biochem. Biophys. Res. Commun. 60*, 909.
3. DeChatelet, L. R., Shirley, P. S. and Johnston, R. B., Jr. (1976) *Blood 47*, 545.
4. Sbarra, J. J. and Karnovsky, M. L (1959) *J. Biol. Chem. 234*, 1355.

5. Rossi, F. and Zatti, M. (1964) *Brit. J. Exptl. Pathol. 45,* 548.
6. Allen, R. C. (1979) *Photochem. Photobiol 30,* 157.
7. Allen, R. C. (1979) *In* Lysosomes in Applied Biology and Therapeutics. Vol. 6. (eds. J. T. Dingle, P. J. Jacques and I. H. Shaw), p. 197, North Holland Publ. Co., Amsterdam.
8. Allen, R. C. and Loose, L. D. (1976) *Biochem. Biophys. Res. Comm. 69,* 245.
9. Wilson, M. E., Trush, M. A., Van Dyke, K., Kyle, J. M., Mullett, M. D. and Neal, W. A. (1978) *J. Immunol. Meth. 23,* 315.
10. Totter, J. R. (1974) *Photochem. Photobiol. 3,* 231.
11. Totter, J. R. (1975) *Photochem. Photobiol. 22,* 203.
12. Stjernholm, R. L., Allen, R. C., Steele, R. H., Waring, W. W. and Harris, J. A. (1974) *Infect. Immun. 7,* 313.
13. Marucci, A. A. and Fuller, T. C. (1971) *Appl. Microbiol. 29,* 260.
14. Gleu, K. and Petsch, W. (1935) *W. Angew. Chem. 48,* 57.
15. McCapra, F. and Hann, R. A. (1969) *Chem. Commun.* 442.

# MEASUREMENT OF THE OPSONIC ACTIVITY OF SERUM BY GRANULOCYTE CHEMILUMINESCENCE[1]

Paul Stevens

Department of Medicine
University of California, Los Angeles
Los Angeles, California

William D. Welch

Department of Medicine
University of California, Irvine
Irvine, California

## I. INTRODUCTION

The production of light or chemiluminescence (CL) from polymorphonuclear leukocytes (PMN) during phagocytosis which was initially described by Allen et al. (1) has been shown to be related to increased glucose oxidation via the hexose monophosphate shunt (1) and most likely occurs from the formation and subsequent relaxation of excited carbonyl groups of the particle being phagocytized and oxidized (1,2). Since CL reflects direct metabolic activity of the PMN, it has been used to determine intracellular abnormalities of PMN function such as chronic granulomatous disease (CGD) (3) and MPO deficiency (4) and is currently the only method to accurately determine the carrier state of CGD (5). In addition, CL is a useful technique to determine the opsonic activity of normal human serum (4,6,7) and the capacity of specific sera to opsonize certain bacteria (8,9). We will briefly review this latter use of CL.

[1]*This work was supported by NIH Grant #AI-13518 and AI-11271.*




ISBN 0-12-208820-4

## II. METHODOLOGY

The methodology of the CL system to assess the opsonic activity of serum has varied but basically consists of a leucocyte preparation free of red blood cells, the microorganism of interest, and the serum source to be tested. Opsonization can be carried out either prior to the mixing of all the reactants or the serum can be simultaneously added to the leukocyte-microorganism suspension. The only major methodological difference when comparing the initial work in this area with more recent reports has been the introduction of luminol as an additional reactant (4,10). Since luminol will emit light upon oxidation it was introduced as a modification to the basic CL system to serve as an amplifier to detect low levels of the various oxidative species formed during phagocytosis. The use of luminol has greatly increased the sensitivity of CL so that very few PMN ($2x10^4$ or less) are necessary per assay. Because of this increased sensitivity one may also use diluted whole blood instead of purified leucocytes thereby eliminating the time consuming step of cell separation. Data for the CL response has been presented in two major ways: (1) the maximum initial slope of the CL response calculated by the linear regression least squares method and expressed as counts per minute squared or $CP(M)^2$ and (2) the area under the curve or the integral of the temporal CL response between selected time periods. The units for integral are in counts and can be calculated by various methods.

## III. RELATIONSHIP OF SERUM OPSONIC ACTIVITY TO LUMINOL-DEPENDENT CHEMILUMINESCENCE

Table I demonstrates the relationship of the opsonic activity of normal human serum to two different clinically isolated *Escherichia coli* as determined by luminol-dependent CL while keeping the number of PMN constant. Two different *E. coli* isolates at $10^9$ colony-forming units/ml were opsonized with either varying amounts of serum or serum heated at 56°C for thirty minutes. With increases in the amount of serum per organism there are significant increases in CL as measured by both slope and integral demonstrating a direct correlation between the degree of CL and the amount of serum used for opsonization. There was no CL when using heated serum thereby demonstrating the necessity of heat

*TABLE I. Effect of Variations in the Amounts of Serum on the CL of Normal Human PMN Undergoing Phagocytosis of E. Coli.*

| Bacterium | Amount of serum per $10^9$ bacteria (ml) | Mean slope ± SD[a] ($CPM^2 \times 10^3$) (n = 2) | Mean CL integral ± SD[a] (counts x $10^6$) (n = 2) |
|---|---|---|---|
| *E. coli* | Serum heated | 0 | 0 |
| | 0.5 | 0.45 ± 0.02 | 5.90 ± 0.25 |
| | 5 | 2.63 ± 0.05 | 8.79 ± 0.31 |
| | 10 | 6.18 ± 0.11 | 16.10 ± 0.20 |
| *E. coli* | Serum heated | 0 | 0 |
| | 0.5 | 0.39 ± 0.01 | 7.07 ± 0.28 |
| | 5 | 2.20 ± 0.04 | 10.41 ± 0.40 |
| | 10 | 6.11 ± 0.09 | 15.52 ± 2.08 |

[a]*All slopes and CL integrals were significantly different from each other at $P < 0.001$ (paired t test).*

labile serum components, most likely complement in the opsonization process.

## IV. OPSONIC ACTIVITY OF VARIOUS SPECIFIC SOMATIC AND CAPSULAR K ANTISERA FOR ESCHERICHIA COLI 073K92

As shown in Table II luminol-dependent CL represented by initial slope had an excellent correlation with the degree of phagocytosis as measured by light microscopy and PMN bactericidal killing both of which represent more traditional methods to assess the extent of opsonization. Table II also illustrates that 0 somatic antibody as represented by 073K(-) has no opsonic activity for *E. coli* 073K92 since there was no significant CL or bacterial killing. It is clear that the active opsonic factor is antibody specifically directed against the K92 component of the cell envelope since there was increased CL when using either 072K92 or the heterologous 016K92 antisera. In addition, since group C meningococcal polysaccharide has been shown to crossreact with K92 (11), it is apparent that antisera raised to group C meningococcus

*TABLE II. Opsonic Activity of Homologous 073:K92, Somatic 073:K(-), Capsular 016:K92 and Anti-Group C Meningococcal Antisera Against Escherichia coli 073:K92 As Measured by % Bacterial Killing, Luminol-Dependent Chemiluminescence and Microscopic Phagocytosis*

| *Antisera* | *% of bacteria killed in 1 hr.* | *Mean CL slope ± SD ($cpm^2$)* | *Phagocytic microscopic observations mean number of bacteria per PMN* |
|---|---|---|---|
| *Pooled normal rabbit serum* | *None* | *290 ± 50* | *13.6 ± 2.8* |
| *Anti-073:K(-)* | *None* | *240 ± 20* | *14.2 ± 2.9* |
| *Anti-073:K92* | *97.2 ± 2.2* | *1600 ± 100* | *57.6 ± 7.1* |
| *Anti-016:K92* | *95.2 ± 1.3* | *1740 ± 110* | *58.0 ± 4.2* |
| *Anti-Group C Meningococcus* | *97.6 ± 2.0* | *1340 ± 130* | *58.8 ± 7.0* |

has similar activity to the K92 antiserum. Fig. 1 more graphically illustrates some of the CL data presented in Table II. Only in the presence of K92 antiserum and not 073 antiserum is CL greater than in the absence of specific opsonins. The lack of opsonic activity observed with the somatic 0 antibody antisera may be due to the inability of the antisera to penetrate the K92 capsular polysaccharide or if bound to these bacteria its ability to interact with the PMN because of steric hindrance from the surrounding cell wall components.

## V. SUMMARY AND CONCLUSIONS

The measurement of CL from phagocytosing granulocytes either in the absence (6, 7) or presence of luminol (4) has been shown to be a useful method to detect the opsonic activity of serum to various microorganisms. However, the use of luminol has been an important modification since luminol-dependent CL is independent of the particle as a direct source of light and therefore eliminates possible variations in CL due to various substrate compositions. In addition the CL is more sustained than in the absence of luminol (4) and more

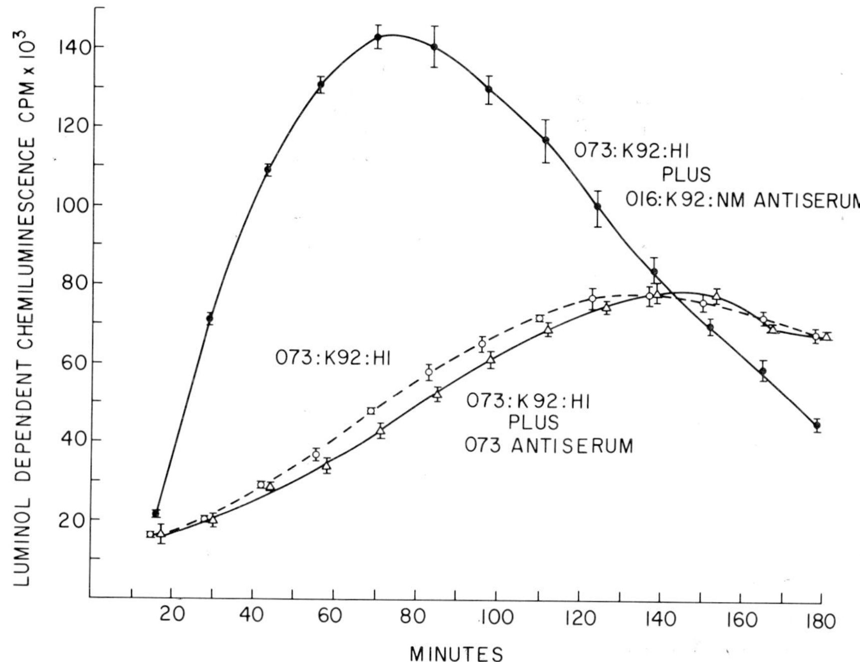


*FIGURE 1. Temporal luminol-dependent CL response of the opsonophagocytosis of E. coli 072:K92:H1 in the absence ---o---o---o, and presence of specific 073 —Δ—Δ—Δ, or 016:K92 antiserum —o—o—o.*

likely represents measurement of oxidative species throughout the duration of phagocytosis and killing. Since CL was shown to be dependent on the presence of specific antibody (Table II) and correlated well with phagocytosis and bacterial killing, we conclude there are no significant non-specific bacterial-PMN interactions that result in CL in the absence of phagocytosis. Since many of the other methods to determine rates of specific opsonic activity can be very tedious and expensive (12, 13), luminol-dependent CL in the proper controlled setting offers the distinct advantages of low cost, simplicity and rapidity.

REFERENCES


1. Allen, R. C., R. L. Stjernholm, and R. H. Steele, *Biochem. Biophys. Res. Commun.* 47, 679 (1972).
2. Cheson, B. D., R. L. Christensen, R. Sperling, B. E. Kohler, and B. M. Babior, *J. Clin. Invest.* 58, 789 (1976).
3. Allen, R. C., R. L. Stjernholm, M. A. Reed, T. B. Harper, S. Gupter, R. H. Steele, and W. W. Waring, *J. Infect. Dis.* 136, 510 (1977).
4. Stevens, P., D. J. Winston, and K. Van Dyke, *Infect. Immun.* 22, 41 (1978).
5. Mills, E. L., K. S. Rholl, and P. G. Quie, *J. Clin. Microbiol.* 12, 52 (1980).
6. Stevens, P. S., and L. S. Young, *Infect, Immun.* 16, 796 (1977).
7. Allen, R. C., *Infect. Immun.* 15, 828 (1977).
8. Anderson, D. C., M. S. Edwards, and C. J. Baker, *J. Infect. Dis.* 141, 370 (1980).
9. Welch, W. D., and P. Stevens, *Curr. Microbiol.* 2, 245 (1979).
10. Van Dyke, K., M. Trush, M. Wilson, P. Stealey and P. Miles, *Microchem. J.* 22, 463 (1977).
11. Glode, M. P.,J. B. Robbins, T. -Y. Liu, E. C. Gotschlich, I. Ørskov, and F. Ørskov, *J. Infect. Dis.* 135, 94 (1977).
12. Klebanoff, S. J., R. A. Clark, *J. Lab. Clin. Med.* 89, 675 (1977).
13. Verhoef, J., P. K. Peterson, and P. G. Quie, *J. Immunol. Methods* 14, 303 (1977).

# ON THE QUESTION OF SINGLET OXYGEN PRODUCTION IN POLYMORPHONUCLEAR LEUCOCYTES[1]

Christopher S. Foote
Rosangela B. Abakerli
Roger L. Clough

Department of Chemistry
University of California
Los Angeles, California

Robert I. Lehrer

Department of Medicine
University of California
Los Angeles, California

The microbicidal activity of polymorphonuclear leucocytes (PMN's) is a subject of much current interest (1). On phagocytosis, molecular oxygen consumption followed by $O_2^{\cdot-}$ release and $H_2O_2$production are well documented (2). Either $O_2^{\cdot-}$ or $H_2O_2$ could be bactericidal by itself. However, the lack of production of $O_2^{\cdot-}$ and $H_2O_2$ in chronic granulomatous disease (CGD) PMN's (2, 3) which are deficient in bactericidal activity suggests an important role for the neutrophil myeloperoxidase, since CGD cells can show bactericidal activity towards microorganisms that generate $H_2O_2$ as a metabolic product (1b).

There have been several suggestions for the involvement of $^1O_2$ in the PMN bactericidal activity, based on the observation of chemiluminescence of cells after phagocytosis (4), or on the possible formation of $^1O_2$ from dismutation of

[1]*This work was supported by Public Health Service grant No. GM 20080. R. B. Abakerli thanks the Fundacao de Amparo a Pesquisa do Estado de Sao Paulo for a fellowship.*




ISBN 0-12-208820-4

superoxide (5) or its reactions with other substrates (6, 7). The reported production (8) or $^1O_2$ products in a cell free system of myeloperoxidase, $H_2O_2$ and chloride strengthened previous suggestions of the role of $^1O_2$ as the microbicidal agent. However, the $^1O_2$ traps used (furans) lack specificity, as pointed out by many workers (9). No specific test for $^1O_2$ production has yet been applied to phagocytic neutrophils. Cholesterol gives a characteristic product with singlet oxygen, the 5α-hydroperoxide (1a), which is distinct from the products of radical autoxidation, which include the 7α- and 7β-hydroperoxides (2a) (10). Since cholesterol is virtually insoluble in water, a system was designed using cholesterol (4-$^{14}$C) supported on polystyrene latex microbeads (Ⓟ-chol) (11). The use of this system as a singlet oxygen trapping system has been described (12).

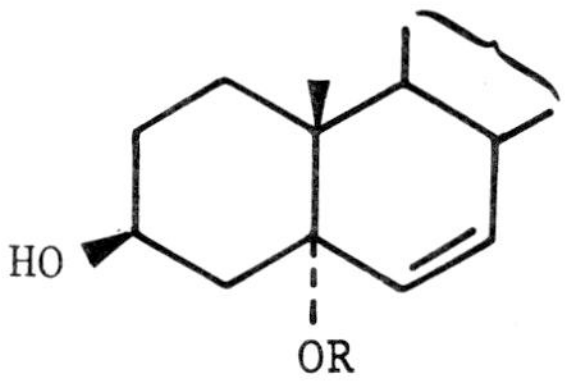

HO OR 1 ($^1O_2$ Product)

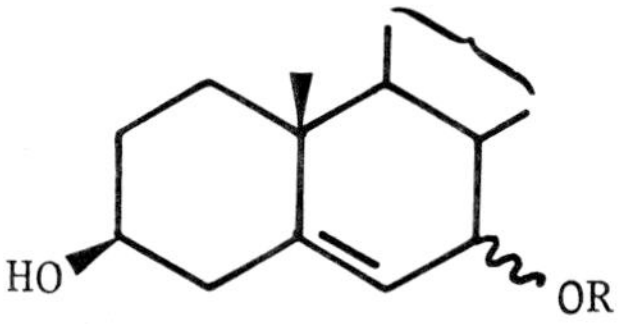

a, R = OH
b, R = H

HO OR 2 + complex product mixture (Radical Products)

We have used this system and a related one to test for the production of $^1O_2$ in phagocytizing leucocytes. Ⓟ-$^{14}$C-cholesterol beads (11) were incubated with a suspension of freshly isolated human PMN's (13) in 2 ml phosphate buffer (pH 7.4) for 65 min. The oxygen uptake (14) rate at 37°C was 48.6 nmoles/$10^7$ cells/min. After separation of ingested from uningested particles (15) the organic products were extracted and analyzed by a procedure similar to that described in ref. 12; unlabelled standards (cholesterol, 3β,5α-diol, 3β,7α and 7β-diols, 5α,6α-epoxide and 7-ketone) were synthesized by published methods (16). Autoradiograms of the TLC plates (17) were also carried out. The products formed in one of many runs are shown in Table I.

In the fraction isolated from the cells and assumed to be inside the phagocytic vacuoles, 1.0 x $10^{-8}$ moles of labelled material were recovered, of which no more than 3.1 x $10^{-11}$ moles were 3β,5α-diol. It is very probably that not

*TABLE I. Radioactivity Distribution in TLC (17a) from PMN's.*

| Product | % Total Recovered in | |
|---|---|---|
| | Extracellular Fraction | Intracellular Fraction |
| 3β,7α-diol | 0.03 | 0.07 |
| 3β,7β-diol | 0.13 | 0.52 |
| 3β,5α-diol | 0.27 | 0.31 |
| 7-ketone | 0.58 | 0.59 |
| 5,6-epoxide[+] | 2.26 | 2.10 |
| cholesterol | 95.7 | 95.0 |

[+]α and β epoxides not resolved

all of the radioactive material collected in this spot was the 5α product since there is a substantial background of radioactivity where the known products do not occur. Although we attempted to carry out a quantitative study of the upper limits for singlet oxygen production in this system, the trapping efficiency of the beadlet system depends strongly on concentration. Since the beads are concentrated locally after ingestion, but in an unknown fashion, we decided that it was meaningless to calculate the upper limit for singlet oxygen production. However, the following statement can be made: *although new products are formed under some conditions, the singlet oxygen products are at most a minor constituent.* In Table I, the upper limit for the 5α product is 0.3%; many runs had up to an order of magnitude less. Thus, it seems very unlikely that singlet oxygen is a major contributor to the oxidation of unsaturated materials in the membranes of phagocytized organisms (18).

A very similar series of experiments was carried out using [4-$^{14}$C]-cholesterol dissolved in a mineral oil dispersion (19). The technique was quantitated by the method used for the beadlets (12). If the efficiency is assumed to be independent of the local concentration of the dispersion, an upper limit of 2 ± 2% of the oxygen taken up by the PMN's during ingestion of the mineral oil droplets was found to be singlet. Although we did not carry out experiments to test the concentration dependence with the mineral oil dispersion, it is likely that localization effects would also affect the efficiency of this system.

The major product (over 2% of material formed both inside and outside the cells) appears at an $R_f$ value identical to

*TABLE II. Readioactivity Distribution in TLC (17b) from HOCl Reaction*[a]

| Product | % of Total Radioactivity | |
|---|---|---|
| | Recovered | Control |
| Unknown | 1.53 | 0.14 |
| 3β,7α-diol | 0.29 | 0.19 |
| 3β,7β-diol | 0.19 | 0.16 |
| 5α-Cl-6β-OH-Chlorhydrin | 3,09 | 0.13 |
| 3β,5α-diol | 0.43 | 0.14 |
| 5β,6β-epoxide | 20.04 | 2.03 |
| Unknown | 8.8 | 0.25 |
| Cholesterol | 50.64 | 95.09 |
| 6α-Cl,5α-OH-Chlorhydrin[b] | 6.71 | 0.29 |

[a]In 0.05 M acetate buffer, pH 4.9; [HOCl] = $1.0 \times 10^{-5}$M.
[b]Regiochemistry not completely certain.

that of the 5β,6β-epoxide and is not separated from it on any this layer system investigated. Although its identity has not been established completely, we believe that this material is the epoxide. Smaller amounts of other unknown materials are also formed. Indeed, the TLC pattern of the total product mixture is similar to that formed on reaction of cholesterol with hypochlorous acid; the product distribution with this reagent is shown in Table II; epoxide, chlorhydrins and unknown compounds are the major products.

In order to study the effect of the PMN peroxidase system alone, we carried out a reaction in which isolated PMN myeloperoxidase (21) and chloride ion were used in the beadlet system, using various amounts of hydrogen peroxide. As will be seen from Table III, a variety of products were formed. At the lowest concentration of $H_2O_2$, the epoxide is the major product; at higher concentrations substantial amounts of chlorhydrins and unknown products were also formed. Interestingly, the yield of these products was much lower at pH 7 than at pH 4.9.

The product spectrum using the myeloperoxidase and $H_2O_2$ in the presence of chloride closely resembles that formed on reaction with hypochlorous acid; indeed, the TLC patterns are virtually superposable. This strongly suggests that the major species formed in the myeloperoxidase-$H_2O_2$-$Cl^-$ system which is capable of attacking unsaturated molecules is, in fact, hypochlorous acid, as has been suggested by some authors (9a, b, 23).

*TABLE III. Radioactivity Distribution in TLC (17b) from Myeloperoxidase, $Cl^-$ and $H_2O_2$* [a]

| *Product* | *% of Total Radioactivity Recovered at $H_2O_2$ =* | | | *Control* |
|---|---|---|---|---|
| | *1.18x10⁻⁴ M* | *1.18x10⁻⁵ M* | *1.18x10⁻⁶ M* | |
| *Unknown* | *1.66* | *0.95* | *0.17* | *0.08* |
| *3β,7α-β-diols* | *0.86* | *0.32* | *0.28* | *0.11* |
| *5α-Cl-6β-OH-Chlorhydrin* | *17.67* | *2.73* | *0.11* | *0.06* |
| *Unknown* | *4.86* | *0.79* | *0.10* | *0.04* |
| *5,6-epoxide* [b] | *14.12* | *10.14* | *1.54* | *0.32* |
| *Cholesterol* | *16.45* | *75.05* | *95.0* | *93.00* |
| *6α-Cl,5α-OH-Chlorhydrin* [c] | *23.38* | *4.32* | *0.21* | *0.68* |

[a] *MPO = 21.4 μg/ml; [$Cl^-$] = 0.1 M in 0.05 M acetate buffer, pH 4.97.*
[b] *Believed to be β-epoxide.*
[c] *Regiochemistry not completely certain.*

The leucocyte results using the beadlet system do not differ greatly from those with myeloperoxidase/$Cl^-$/$H_2O_2$ at the lowest concentration used; the products in the leucocyte system are formed at very low concentration and it is possible that the other hypochlorous acid products were overlooked. These results provide suggestive, but by no means conclusive, evidence that hypochlorous acid is capable of attacking unsaturated molecules in the leucocyte system as well. Further work will obviously have to be done to establish this; new systems are being designed which will have improved sensitivity and specificity for hypochlorous acid and other possible reactive intermediates.

REFERENCES


1. For a review see (a) Karnovsky, M. L., *Fed. Proc.* 32, 1527 (1973); (b) Klebanoff, S. J., *Seminars in Hemat,* 12, 117 (1975); (c) Sbarra, A. J., R. S. Selvaraj, B. B. Paul, P. F. K. Poskitt, J. M. Zgliczynski, G. W. Mitchell, Jr., and F. Louis, *Int. Rev. Exp. Path.* 16, 249 (1976).
2. Babior, B. M., R. S. Kipnes, and J. T. Curnutte, *J. Clin. Invest.* 52, 741 (1973).
3. Curnutte, J. T., D. M. Whitten, and B. M. Babior, *N. Engl. J. Med.* 290, 539 (1974).
4. Allen, R. C., R. L. Stjernholm, and R. H. Steele, *Biochem. Biophys. Res. Commun.* 47, 679 (1972).
5. Stauff, J., V. Sander, and W. Jaeschke, *in* "Chemiluminescence and Bioluminescence" (M. J. Cormier, D. M. Hercules,and J. Lee, eds.), p. 131. Plenum Publishing Corp., New York, (1973).
6. Kellogg II, E. W. and I. Fridovich, *J. Biol. Chem.* 250, 8812 (1975).
7. Khan, A. U., *J. Phys. Chem.* 80 2219 (1976).
8. (a) Klebenoff, S. J., *J. Biol. Chem* 252, 4803 (1977); (b) Rosen, H. and S. J. Klebanoff, *Fed. Proc.* 35, 1391 (1976).
9. (a) Held, A. M. and J. K. Hurst, *Biochem. Biophys. Res. Commun.* 81. 878 (1978); (B) Harrison, J. E., B. D. Watson, and J. Schultz, *FEBS Lett.* 92, 327 (1978); (c) Krinsky, N. I., *in* "Singlet Oxygen" (H. H. Wasserman and R. W. Murray, eds.), Vol. 40, p. 606. Academic Press, New York (1979); (d) Foote, C. S., *in* "Biochemical and Clinical Aspects of Oxygen" (W. S. Caughey, ed.), p. 603. Academic Press, New York (1979).
10. (a) Schenck, G. O., K. Gollnick, and O. -A. Neumüller, *Liebigs Ann.* 603, 46 (1975); Nickon, A. and J. F. Bagli, *J. Am. Chem. Soc.* 81, 6330 (1959); 83, 1498 (1961); (b) Schenck, G. O., *Angew Chem.* 69, 579 (1957); (c) Kulig, M. J. and L. L. Smith, *J. Org. Chem.* 38, 3639 (1973); (d) Teng, J. I. and L. L. Smith, *J. Am. Chem. Soc.* 95, 4060 (1973); (e) Smith, L. L., M. J. Kulig, and J. I. Teng, *Chem.Phys. Lipids* 20, 211 (1977); (f) Smith, L. L.,and M. J. Kulig, *J. Am. Chem. Soc.* 98, 1027 (1975); (g) Smith, L. L. M. J. Kulig, D. Miller, and G. A. S. Ansari, *J. Am. Chem. Soc.* 100, 6206 (1978); (h) Smith, L. L., J. I. Teng, M. J. Kulig, and F. L. Hill, *J. Org. Chem.* 38, 1763 (1973).
11. 0.740 ml of benzene solution of TLC-purified [4-$^{14}$C]-cholesterol (17.6 μCi/ml) were dried under a nitrogen

stream and redissolved in 1 ml methanol. To this solution, 0.5 ml of a 10% polystyrene suspension were added and dried under nitrogen. The residue was resuspended in 1 ml of doubly distilled water and diluted to 2 ml with phosphate buffer saline (PBS). The suspension was filtered through a capillary, homogenized and 10 μl were checked for radioactivity.

12. Foote, C. S., F. C. Shook, and R. A. Abakerli, *J. Am. Chem. Soc.* 102, 2503 (1980).
13. The PMN's were isolated as described by Boyum A., *Scand. J. Clin. Lab. Inv.* 21, 77 (1968), and resuspended in PBS, pH 7.4. A count gave $1.0 \times 10^7$ cells/ml.
14. Followed in a Gilson Oxygraph with a Clark electrode.
15. The separation of the particles [by layering over a density gradient (Ficoll type 400, density 1.06 $g/cm^3$) and centrifuging at 2000 rpm for 10 min] did not pellet the cells completely. This step was followed by several saline solution washing-centrifuging steps. Microscopic examination showed that the supernatants did not contain cells. However, some uningested particles remained in the cell fraction due to agglomeration during centrifuging.
16. The 3β,5α-diol was synthesized by the procedure of L. L. Smith (10c); the 7-ketone by that of Nickon, A. and W. L. Mendelson, *J. Am. Chem. Soc.* 87, 3928 (1965); the 3β,7α, β-diols through the reduction by LAH of the 7-ketone and the 5α,6α-epoxide by reacting cholesterol and m-chlorperbenzoic acid at room temperature. All standards were purified by column chromatography.
17. The TLC's were carried out on 0.25 mm silica gel HF 254 plates (Merck). The solvent system was (a) ether: hexane (9:1) for the two first irrigations and benzene: ethyl acetate (3:1) for the third one or (b) benzene: ethyl acetate (3:1) for all three irrigations. The plates were stained by spraying with 50% methanol-sulfuric acid and heating. SB54 Kodak films and intensifying screens (Cronex Quanta IIB) were used for autoradiography; the system was kept at -70° for one week.
18. Many runs did not yield any detectable products in excess of controls; none gave 5α or 7α or 7β product in excess of controls.
19. The use of mineral oil emulsions to study phagocytosis was developed by Stossel. (20) 50 μCi of purified [4-$^{14}$C]-cholesterol in 0.5 ml of benzene was taken to dryness under $N_2$, redissolved in diethyl ether and added to 0.2 - 1.0 ml of a saturated solution of oil red O in mineral oil. The ether was removed by purging with $N_2$ and the oil was added to saline with 10 mg/ml of lipo-

polysaccharide. This suspension was sonicated. PMN, suspended at 1 x $10^8$ cells/ml in PBS with 10% normal group AB serums, received 100 µl of emulsion/ml of cell suspension and were incubated with rotation (30 rpm) at 37°C for 15-60 min. in various experiments. The phagocytic vesicles were recovered as described by Stossel (20).

20. Stossel, T.,T. Pollard, R. Mason, and M. Vaughn, *J. Clin. Invest.* 50, 1745 (1971).

21. The enzyme was extracted as described by Schultz, J. and H. W. Schmukler, *Biochem.* 3, 1234 (1964), and assayed by the o-dianisidine method (22), 68.4 units/mg.

22. "Worthington Enzyme Manual", p. 43. Worthington Biochem. Corporation, Freehold, N. J. (1972).

23. Strauss, R. R., B. B. Paul, A. A. Jacobs, and A. J. Sbarra, *Infect. Immun.* 3, 592 (1971); Zgliczynski, J. M. T. Stilmaszynska, J. Domanski, and W. Ostrowski, *Biochim. Biophys. Acta* 235, 419 (1971).

## DISCUSSION

Dr. Cormier

I'd like to address this question to Dr. Van Dyke. I agree with you that calmodulin is a logical target for calcium leukocyte activation. In fact, we collaborated with Jim McCord recently and have found as you found that NADPH oxidase is inhibited by trifluoroperpazine and so is the generation of superoxide anions. I wanted to clarify a point relative to the interpretation. I thought you said that trifluoroperpazine competed with calcium for the calcium binding sites on calmodulin. The work of Weiss and Levin have shown that trifluoroperpazine did not compete with calcium for the calcium binding to calmodulin. In fact, calcium binding to calmodulin is a prerequisite to trifluoroperpazine binding.

Dr. Van Dyke

But, it knocks out the site, in any case.

Dr. Cormier

No, not the calcium binding site. In other words, calcium must be bound to calmodulin prior to trifluoroperpazine binding. We have shown this in our own lab by using immobilized Phenothiazine columns, in which it's clear that calcium binding must be a prerequisite prior to Phenothiazine binding. But the fact is that trifluoroperpazine will bind to calmodulin presence of calcium. This binding will then prevent the binding of the calcium calmodulin complex to a variety of enzymes which calmodulin normally activates. That's the mechanism of trifluoroperpazine inhibition of calmodulin function.

Dr. Van Dyke

O.K. The detail is a little incorrect. But, did you measure the chemiluminescence along with it and show that it knocks it out?

Dr. Cormier

No.

Dr. Van Dyke

But, that's clear; and it's also clear that it knocks out the metabolic system, is it not? That was McCord's thought.

Dr. Cormier

That is true.




ISBN 0-12-208820-4

Dr. Van Dyke

The point I am making is that this (inhibition of calmodulin by trifluoroperpazine is a very important step because calcium is obviously crucial to this whole business. If you can knock out that calcium activity presumably, by inhibiting calmodulin, you can separate the chemiluminescent effects from the depolarization events. In other words, it's a one-two effect; 1. depolarization occurs first and then 2. on to metabolic events; that is shown because when you measure super oxide formation in parallel with chemiluminescence, there is a lag time between depolarization when you put the particle into the incubation mix until the time CL occurs. Many researchers have done these reactions; they will tell you the same thing. Now, what is going on in between? I think it's the ionic movements, starting from the membrane with $Na^{+}$ and $Ca^{++}$ going inside and then exerting their effects; sodium and calcium are both involved. This was clearly shown by Simchowitz and Spilberg and others.

Dr. Cormier

You are dealing with a membrane effect and that brings up another point. That is that these phenothiazine drugs, of course, are amphipathic cationic but ionic detergents and they bind profusely to membrane sites.

Dr. Van Dyke

Well, I'm not saying that these things are specific. Any time you use a drug as a probe, you don't know exactly what it does. But, one thing it does do is that it knocks out chemiluminescence, and also knocks out the metabolic events. But, it doesn't effect the depolarization. Therefore, there appears to be a temporal separation of the events.

Dr. Cormier

One other comment, if I may. You mentioned in the diseased cells that you thought that there may be a defect in calmodulin.

Dr. Van Dyke

I wasn't really trying to point to calmodulin per se. I was just saying that that was a possibility based on the fact that although calcium plus the ionophore could depolarize the diseased cell and no CL was produced something was missing. It certainly could be that the missing portion could be calmodulin regulated.

Dr. Cormier

I think it's safer to interpret that data as a defect in one of the calmodulin activatible enzymes. This is simply because calmodulin plays a number of fundamental roles in the cell. You know that the cell would be dead if it contained a defective calmodulin.

Dr. Van Dyke

I agree.

Dr. Brolin

I would like to ask if any efforts are put into making a site specific chemiluminescent probe? This has been achieved for mitochondria as reported from our laboratory by Dr. Lippman and Dr. Agren in the poster session. In one of your diagrams we saw that it might be thought that the chemiluminescent reactions occur inside the cell but it was also suggested that the reactions occurred outside the cell. I feel this problem is interesting from two different aspects. First, if free oxygen radicals are passing from the lysosomes through the cytoplasm, the cytoplasm may be damaged which might explain the comparatively short life time of granulocytes. The other aspect is that electron leakage from the mitochondrial respiratory chain may contribute to the production of free oxygen radicals. I have also a second question. Several cells have now been described which react with chemiluminescence. This concerns various kinds of macrophages. As you know we have various types of lymphocytes. Among the T lymphocytes there are cells which are called killer cells. Do you know if they react by forming free oxygen radicals?

Dr. Allen

I can address the first question which you asked. Yes, there are differences depending on how you select your chemiluminogenic probe molecule. You can get different types of kinetics and likewise you can use different stimulants to selectively stimulate the poly. There are basically two types of activity. One activity involves myeloperoxidase; there is another activity which you can measure with acridinium salts which measure superoxide generation. Proof for that is based on kinetics and on inhibitor studies using reasonable concentration of scavenger enzymes such as catalase and superoxide dismutase. There was a question addressed to that in the morning session. That is with lucigenin or DPA chemiluminescence there is a delay - it is a gross delay and it is also true that you can selectively inhibit luminol light and, in fact, instead of getting

Dr. Allen

a decrease in DPA light, you get an increase in it. The reason for this is that you are not only knocking out myeloperoxidase by azide inhibition, but you are also knocking out other protoporphyrins such as catalase. It's been well illustrated by the inhibitor studies, and also by the kinetic studies.

Dr. Foote

The reaction changes from hypochlorous acid to superoxide.

Dr. Allen

In essence, if you take a cell that has no functional myeloperoxidase you'll still get a light response. With myeloperoxidase-deficient cells, one sees a delay in chemiluminescent response. One can artificially produce a myeloperoxidase-deficient cell with azide. The concentrations of azide that I am talking about are nanomolar quantities. It's not enough so that it would be suspect as a quencher for singlet oxygen. In addition, you can wash the cells essentially free of the remaining azide and you will still get the effect. The second question you addressed was about mitochondria. The mature neutrophil has very few mitochondria. If you look at the TCA cycle activity, it is practically nonexistent. The mitochondrial activity in the neutrophil - not so much in the macrophage and in some of the lymphocytic systems - is not a big contributor. You can also use mitochondrial inhibitors and not significantly affect the metabolism.

Dr. Stevens

We were fortunate to have the opportunity to study a myeloperoxidase deficient patient who had been studied earlier in Seattle by Rosen and Klebanoff. They demonstrated that this patient's granulocytes were capable of generating super-normal amounts of superoxide anion upon stimulation with zymosan. When we studied the same patient's cells using luminol-dependent chemiluminescence and zymosan stimulation we observed virtually no significant chemiluminescent response. In fact, it looked very much like a chronic granulomatous disease in which there is no oxidative burst or chemiluminescence. Without the need to use azide to chemically suppress myeloperoxidase we observed the natural phenomenon of no luminol-dependent chemiluminescence in myeloperoxidase-deficient granulocytes. Our conclusions are that luminol-dependent chemiluminescence is dependent on the presence of peroxidase and that luminol is not oxidized by superoxide anion from phagocytizing granulocytes to produce

Dr. Stevens

light at the conditions of pH of 7.4 and a concentration of $10^{-7}$M luminol.

Dr. Whitehead

My question is very simple. That is, has luminescence, in its various forms, any role to play in phagocytosis -- in the routine investigation of clinical patients?

Dr. Stevens

Absolutely, yes.

Dr. Allen

Yes.

Dr. Van Dyke

Yes, there are still many important clinical correlations to be made.

# III
# BACTERIAL BIOLUMINESCENCE

FRONT ROW- left to right: M. DeLuca, J.W. Hastings, W.D. McElroy, K. Nealson, and M. Zielger

SECOND ROW- left to right: J. Lee, S. Ulitzer, E.A. Meighen, and T. Baldwin

# FLUORESCENCE PROPERTIES OF LUCIFERASE PEROXYFLAVINS PREPARED WITH ISO-FMN AND 2-THIO FMN[1]

J. W. Hastings

The Biological Laboratories
Harvard University
Cambridge, Massachusetts

Sandro Ghisla, Manfred Kurfürst, Peter Hemmerich

Fachbereich Biologie
Universität Konstanz
Konstanz, West Germany

The light emitting reaction catalyzed by luciferase obtained from luminous bacteria is an interesting one, and apparently quite different from several other bioluminescent reactions, such as firefly and *Cypridina* (1, 2). The overall reaction involves the mixed function oxidation of $FMNH_2$ and long chain aldehyde by molecular oxygen, but already in the early studies (3) it was shown that the reaction could be divided into steps, the first involving the reaction of $O_2$ with reduced flavin and the second the oxidation of the aldehyde by the intermediate.

$$FlH_2 + O_2 \longrightarrow HFlOOH$$

$$HFlOOP + RCHO \longrightarrow h\nu + Fl_{ox} + RCOOH + H_2O$$

The intermediate has been isolated and characterized as a luciferase-bound peroxyflavin (4-8). A key question concerns the reaction mechanism of this intermediate with aldehyde, involving the formation of the excited molecular species responsible for light emission. In this abstract, we summarize knowledge and recent experiments relating to the identity

[1]*These studies were supported in part by grants from the U.S. National Science Foundation (PCM 77-19917) and the Deutsche Forschungsgemeinshaft to S.G. J.W.H. was an awardee of the Alexander Van Humboldt Foundation.*




ISBN 0-12-208820-4

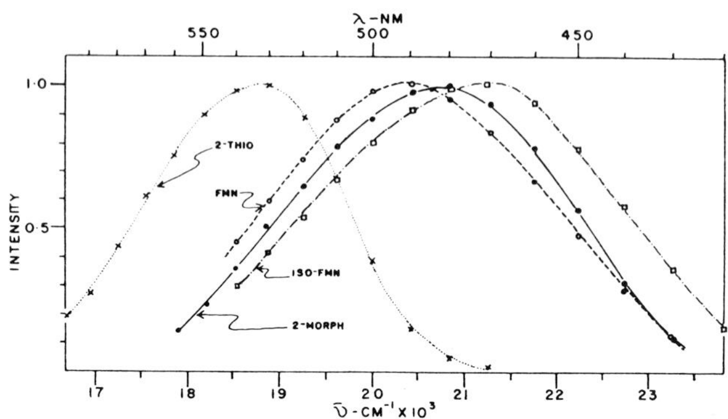


*FIGURE I. Bioluminescence emission spectra for FMN, iso-FMN, 2-morpholino-, and 2-thioFMN. Intensity values are normalized and plotted in the terms of wave numbers (from 9).*

of the emitting species in the reaction.

Since FMN occurs as a product of the reaction, it had seemed at first that its excited singlet might be designated as the emitting species. But the fact that its fluorescence emission is centered at about 525 nm while that of the *in vitro* bacterial bioluminescence is maximal at about 490 nm makes this unlikely (3). In work explicitly concerned with this question, Mitchell and Hastings (9) considered several possibilities. One was that the emission derives from an enzyme-bound excited state of FMN, with the blue shift attributed to the protein environment. A second possibility was that a flavin chemically altered in the course of the reaction was the species responsible for light emission. A third was that the emission results from some luciferase-bound or associated group, even perhaps an aromatic amino acid residue of the luciferase itself (10).

In order to evalute these possibilities we tested a considerable number of flavin isomers and analogues for activity; several were found to exhibit low but authentic activity (9). Since these isomers differ from FMN itself with regard to the wavelength of fluorescence emission, it was of interest to determine if the color of the bioluminescence was similarly changed. Changes were indeed observed (Fig. 1), but they did not correlate with the excited singlet fluorescence of the flavin as such (Table I). The possibility that the flavin cation might be the emitting species was also suggested (11), and although its fluorescence emission correlates with bioluminescence in the case of FMN and possibly 2-thioFMN, it does not in the case of isoFMN (Table I).

*TABLE I. Color of Bioluminescence and Fluorescence Emissions with Different Flavins*

| *Emission* | *Peak wavelength (nm)* | | |
|---|---|---|---|
| | *FMN* | *isoFMN* | *2-thioFMN* |
| *Bioluminescence with luciferase*[a] | *490* | *472* | *534* |
| *Fluorescence of flavin in* $H_2O$[a] | *525* | *543* | *nonfluor.* |
| *Fluorescence of the flavin cation*[b] | *480-493* | *530* | *510-530* |
| *Fluorescence of the luciferase-peroxyflavin intermediate*[c] | *490* | *465-470* | *540-545* |

[a] *from (9)*
[b] *from (11)*
[c] *uncorrected values, Perkin Elmer MPF44*

The fact that the emission spectrum of bioluminescence differed with different flavins implicates a flavin as the emitter in the purified system. Such a result would not be expected if (as in the third alternative above) some other molecule were serving as the emitter and the flavin functioned only as a reductant, since altering the flavin should not alter the emission spectrum.

Nevertheless, the possibility that flavin is not the emitter in the purified *in vitro* system continues to be seriously considered (12), at least in part because of the reported occurrence of secondary emitters in the *in vivo* emission of at least some species and strains of luminous bacteria. The proposal that *in vivo* bacterial "bioluminescence might ... (involve) transfer of reaction energy to a pigment adsorbed on the luciferase ..." was first put forward by van der Burg (quoted in 13), who also showed that emission spectra differ in different species (14). In addition to the "blue shifted" emission of *Photobacterium phosphoreum,* there is the recently discovered "yellow" emitting strain of *P. fischeri* (15). Both of these cases have now been attributed to the *in vivo* occurrence of secondary emitters involving prosthetic fluorescent groups attached to a protein separable *in vitro* from the luciferases (16, 17). In the case of *P. phosphoreum* the emitter has been reported to be a pteridine (18), while in *P. fischeri* it is a flavin (19). These molecules may be designated as secondary emitters; the purified luciferases emit (reportedly with lower quantum yields) in the absence of such secondary emitters, and at wave lengths which are characteristic of the particular flavin used to initiate the reaction.

The luciferase itself may also be determinant with regard to color. This was clearly shown by the fact that in cells possessing luciferase altered by mutation, the color of the light *in vivo* was red-shifted, and that the reaction catalyzed by luciferase purified from these cells was similarly color shifted (20).

Thus the luciferase-flavin intermediate should evidently be designated as the primary emitter in the pure *in vitro* system. What then is its structure? The experiments described below support the postulate that the emitter is an altered flavin generated as a reaction intermediate.

The flavin hydroperoxide formed in the first step of the reaction has an interesting property; its fluorescence closely matches that of the bioluminescence (Fig. 2) (6). But since this intermediate occurs prior to the step responsible for populating the excited state, it cannot be the emitter as such. Its structure might, however, be a clue to that of the emitter. For example the emitter could be a flavin having some other substituent in the 4a position, possibly the 4a-hydroxy species (Fig. 3), which would be expected to have the same fluorescence properties.

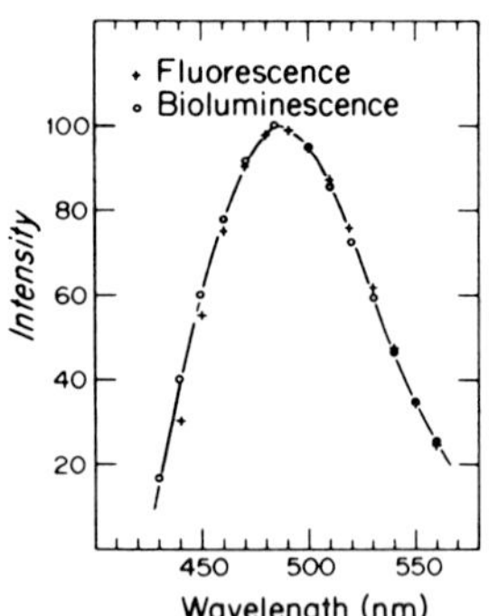


*FIGURE 2. Emission spectrum of bioluminescence (o), measured directly from the cuvet during warming of a luciferase-peroxy FMN preparation in the presence of aldehyde, plotted together with the fluorescence emission spectrum of the peroxy FMN (*). Ordinate, intensity of bioluminescence and fluorescence, normalized at the peak (from 6).*

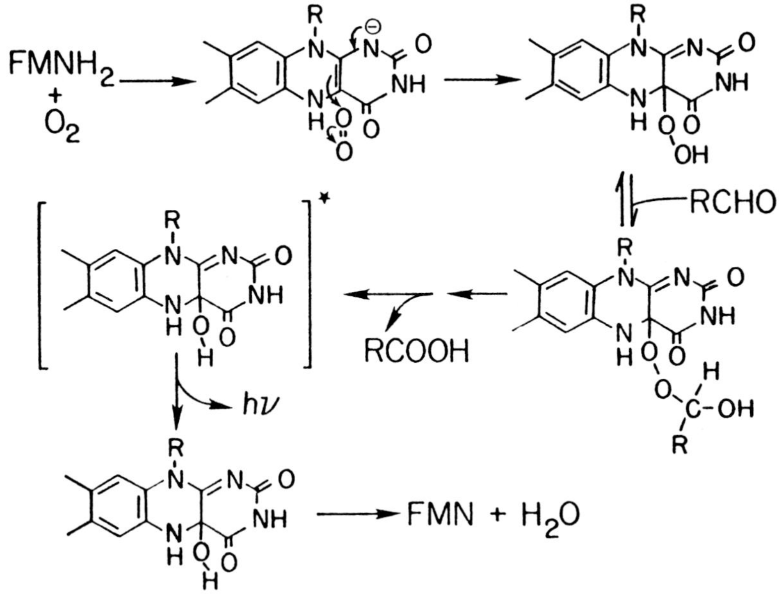

*mediates, including the emitting flavin chromophore in bacterial bioluminescence (from 1).*

To test this idea we examined the fluorescence of the luciferase peroxy adducts using FMN, isoFMN and 2-thioFMN. Luciferase and flavin were mixed at 0°, reduced by dithionite, and chromatographed on Sephadex G-25 at 2° in the presence of 5 μM tetradecanol, which stabilizes the intermediates without altering fluorescence properties (21). The fluorescence emission spectra of these adducts were found to peak at wavelengths similar to the bioluminescence emissions with the same flavins (Table I), supporting the postulate that the primary emitter in luminescence with pure bacterial luciferase is a 4a substituted flavin formed as an intermediate in the reaction.

REFERENCES

1. Hastings, J. W., and K. H. Nealson, *Ann. Rev. Microbiol.* 31, 549 (1977).
2. Hastings, J. W., *Ciba Foundation Symposium* 31, 125 (1975).
3. Hastings, J. W., and Q. H. Gibson, *J. Biol. Chem.* 238, 2537 (1963).
4. Hastings, J. W., C. Balny, C. Le Peuch, and P. Douzou, *Proc. Nat. Acad. Sci.* 70, 3468 (1973).
5. Hastings, J. W., and C. Balny, *J. Biol. Chem.* 250, 7288 (1975).
6. Balny, C., and J. W. Hastings, *Biochemistry* 14, 4719 (1975).
7. Ghisla, S., J. W. Hastings, V. Favaudon, and J. M. Lhoste, *Proc. Nat. Acad. Sci.* 75, 5860 (1978).
8. Becvar, J. E.,S.-C. Tu, and J. W. Hastings, *Biochemistry* 17, 1807 (1978).
9. Mitchell, G., and J. W. Hastings, *J. Biol. Chem.* 244, 2572 (1969).
10. Cormier, M. J., and J. R. Totter, *Annu. Rev. Biochem.* 33, 431 (1964).
11. Eley, E., J. Lee, J. M. Lhoste, C. Y. Lee, M. J. Cormier, and P. Hemmerich, *Biochemistry* 14, 2902 (1970).
12. Hart, R. C., and M. J. Cormier, *Photochem. Photobiol.* 29, 209 (1979).
13. Harvey, E. N., "Bioluminescence," p. 88. Academic Press, N.Y. (1952).
14. Spruit-van der Burg, A., *Biochem. Biophys. Acta* 5, 175 (1950).
15. Ruby, E. G., and K. H. Nealson, *Science* 196, 432 (1977).
16. Gast, R., J. Nearing, and K. H. Lee, *Biochem. Biophys. Res. Commun.* 80, 14 (1978).
17. Leisman, G., and Nealson, unpublished.
18. Koka, P., and J. Lee, *Proc. Nat. Acad. Sci.* 76, 3068 (1979).
19. Leisman, G., K.H. Nealson, and J.W. Hastings, unpublished.
20. Cline, T.W., and J.W. Hastings, *J. Biol. Chem.* 249, 4668 (1974).
21. Tu, S.-C., *Biochemistry* 18, 5940 (1979).

# PROPERTIES OF A LUMAZINE PROTEIN FROM THE BIOLUMINESCENT BACTERIUM *PHOTOBACTERIUM PHOSPHOREUM*[1]

John Lee
L. A. Carreira
Robert Gast
R. M. Irwin
Prasad Koka
E. Daune Small
A. J. W. G. Visser

Bioluminescence Laboratory
Department of Biochemistry, and
Department of Chemistry
University of Georgia
Athens, Georgia, and
Department of Biochemistry
Agricultural University, Wageningen
The Netherlands

The bioluminescent bacteria emit a broad spectral distribution with maxima ranging from 472 to 545 nm depending on the type (1-3). All the *Photobacterium phosphoreum* types have been found at the short wavelength end of the range (1,4). The *in vitro* reactions with purified luciferase, $FMNH_2$, $O_2$ and an aliphatic aldehyde such as dodecanal, have spectral emission maxima again type-dependent but spanning a much narrower range around 495 nm (2). For *P. phosphoreum* for example the *in vivo* to *in vitro* shift is from 476 to 494 nm and no change in extrinsic factors, temperature, pH, aldehyde chain length for example, is able to influence this *in vitro* spectral distribution in any way (2). This would suggest that the molecule that emits the light in the *in vivo* reaction

[1]*This work was supported by grants from the NSF PCM 79-11064, NIH GM-19163 and NATO grant 1912.*




ISBN 0-12-208820-4

is not the same as *in vitro*, at least for those types with *in vivo* spectra at the extremes of the range.

It goes without saying that the identification of the emitting molecule in either case is central to the understanding of the mechanism of the light reaction. Such a species must have several essential qualifications. It must be there in the reaction, either present in the bacterial extracts, or added to or formed as a product of the *in vitro* reaction. It must be efficiently fluorescent, sufficient to account for the maximum bioluminescence quantum yield, around 0.1 (5,6). Finally it must have a fluorescence spectral distribution the same as the bioluminescence.

Terpstra (7) reported in 1962 on the extraction of a compound from *P. phosphoreum* that had a broad fluorescence in the 450 nm region and a stimulatory effect on the *in vitro* reaction with luciferase. Later she reported (8,9) that the addition of $FMNH_2$ to a luciferase preparation followed by UV irradiation, results in the formation of a product with a fluorescence maximum at 470 nm. A similar result was found on addition of $H_2O_2$ and she suggested a relation between these compounds and a function for them in the bioluminescence.

A blue fluorescence protein isolated and purified from *P. phosphoreum* recently (9) is probably the same material observed by Terpstra (7). It has been now completely characterized as a low molecular weight protein, $M_r$ 20,000 (10) containing 6,7-dimethyl-8-ribityllumazine as its fluorescent prosthetic group (11); hence it is named lumazine protein.

The significance of this finding lies firstly in the fact that this metabolite which is the precursor to riboflavin in the biosynthetic pathway, is accumulated in such large amounts by these bacteria, they are classified as lumazine overproducers (11). Lumazine protein is one of the several major proteins made by these cells (10) and is made approximately in proportion to the amount of light they are producing (12). Next, it is found that, when lumazine protein which has no activity for bioluminescence itself, is included in the reaction mixture, the bioluminescence intensity, the rate of decay of intensity and the total light yield, are all increased by several times (9). This is accompanied by a shift back of the *in vitro* bioluminescence spectral maximum from 494 nm to 476 nm, where the spectral distribution is identical to the *in vivo* bioluminescence (9). It is also identical to the fluorescence spectral distribution of lumazine protein itself which emits with high efficiency, $\phi_F$ 0.6 (13).

All these properties would implicate lumazine protein in the bioluminescence in these cells and it is relevant to note that *P. phosphoreum* gives off much more light than the other types (14).

Development of techniques by Koka (11) for rapid and efficient purification of this protein in quantity has now made a detailed study of its properties feasible. It was first only possible to purify small quantities which were initially "associated" with the heavier ($M_r$ 80,000) luciferase fraction. Inclusion of 2-mercaptoethanol in the preparation buffer acts to stabilize labile sulfhydryls (10) and DEAE-Sephacel for ion exchange, Ultrogel ACA-54 for high resolution gel chromatography followed by affinity chromatography on Blue-Sepharose (10-12) provides material of good homogeneity made complete by a final step of ion exchange on DEAE-Sepharose (15).

The purified lumazine protein shows absorption maxima at 265, 275 and 414 nm (15). The absorbance of the two UV peaks is equal and about 2.5 times greater than the 414 nm absorbance which corresponds to a 1:1 ratio of bound lumazine to protein, on a dry weight basis. The visible maximum is shifted from 407 nm of the free chromophore and has about the same extinction coefficient, 10,300 $M^{-1}$ $cm^{-1}$, correcting the earlier guestimate (9).

Affinity chromatography on Blue-Sepharose at low ionic strength provided a means of separating the bound lumazine allowing its characterization by the classical chemical methods (11). The prosthetic group is bound non-covalently to the protein and the dissociation is favored at low ionic strength (13), and evidently the lumazine has some affinity for the Cibacron-Blue dye (11). The characterization procedures included fluorescence and absorption spectra in neutral solution and base, and PMR. Unfortunately the PMR was not completely unambiguous as rapid exchange of the 7-methyl protons during sample preparation removed them from the spectrum. However, riboflavin synthetase which is absolutely specific for the authentic lumazine (16), converts the isolated chromophore to riboflavin and at the same rate as with the regular substrate (11).

The groundwork for further study of lumazine protein has been established and it is the purpose of this present report to describe some properties which suggest a mechanism of its function in bioluminescence and which show it to be a very interesting protein to study from a spectroscopic point of view.

Figure 1 illustrates what happens to the *in vitro* bioluminescence spectrum as lumazine protein is added. All spectra here are fully corrected both for self-absorption and for instrumental distortion. The reaction is the NADPH:FMN coupled bioluminescence of *P. phosphoreum* luciferase, with dodecanal, at 12°. Lumazine protein becomes more effective for producing the spectral shift as the temperature is lowered and it can be seen in Fig. 1 that the shift induced even by 2 μM lumazine protein can easily be discerned. Since the shift is accompanied by an increase in light intensity, the spectra are normalized to constant quantum yield and this reveals isoemissive points at 452 and 506 nm. This implies that the emission occurs from either of two states, one the one produced in the *in vitro* reaction and the other the fluorescent state of the lumazine protein added. The addition of 29 μM lumazine protein is seen to have completely displaced the first emitter and the spectral distribution is now identical to the fluorescence of the bound lumazine (Fig. 2, lower panel).

It has already been pointed out (9) that there is only a small overlap between the absorption spectrum of lumazine protein and the spectral distribution of 494-emission. With lumazine protein purified free of the flavoprotein contamination (10,12) the overlap is reduced even more. Yet the spectral shift is to a higher energy and the only way this can be accounted for is that the bioluminescence reaction generates some species of high energy content which can be decomposed with the release of this energy equivalent to around 400 nm wavelength, sufficient to populate the lumazine fluorescent state, or the unidentified species in its absence.

In Fig. 2, the apoprotein has been separated from the lumazine protein by ultrafiltration at low ionic strength and room temperature. It has only a few percent residual fluorescence and no effect when added to the *in vitro* bioluminescence reaction. However, when 8 μM 6,7-dimethyl-8-ribityllumazine is added to the reaction in the presence of excess apoprotein (Fig. 2, top panel), a blue shift is seen. Note that the isoemissive points 451, 506 nm are almost the same as Fig. 1. The free lumazine itself has no effect. In the lower panel the free lumazine fluorescence is shifted in the presence of excess apoprotein, to become identical to that of the native holoprotein.

These experiments demonstrate reconstitution of the biologically active holoprotein from the authentic chromophore, thereby adding confirmation to the characterization (11,17).

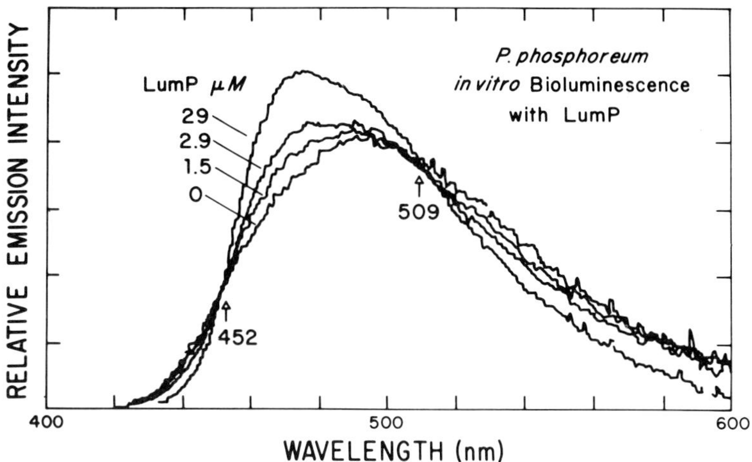


Figure 1. Bioluminescence of *P. phosphoreum* luciferase with NADH:FMN, oxidoreductase and dodecanal (12°). Addition of lumazine protein (LumP) shifts the spectral maximum.

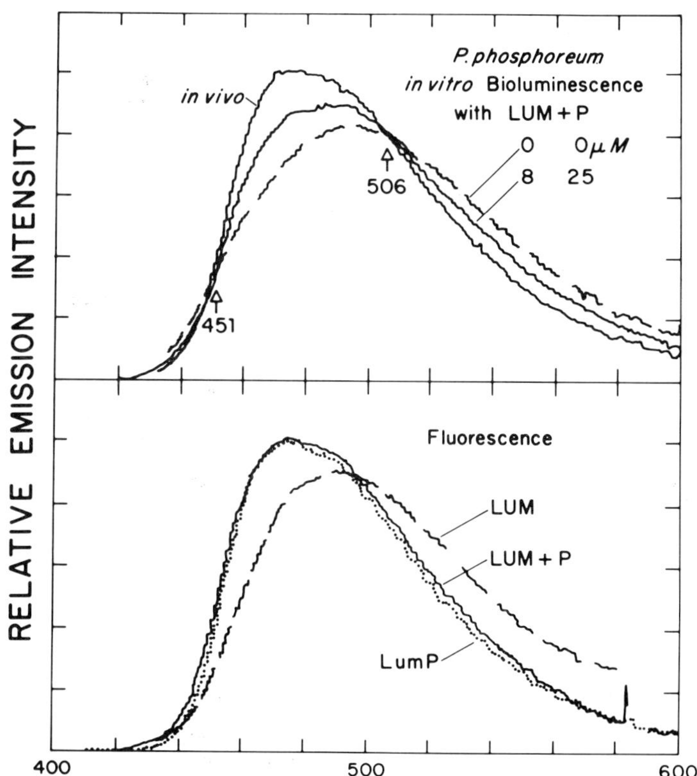


Figure 2. *Top*. Shift of the *in vitro* bioluminescence (dash curve) by 6,7-dimethyl-8-ribityllumazine (LUM) and apo-lumazine protein (P). *Lower*. Fluorescence of LUM, native lumazine protein (LumP) and the reconstituted holoprotein (LUM+P) 12°; Excitation 395 nm.

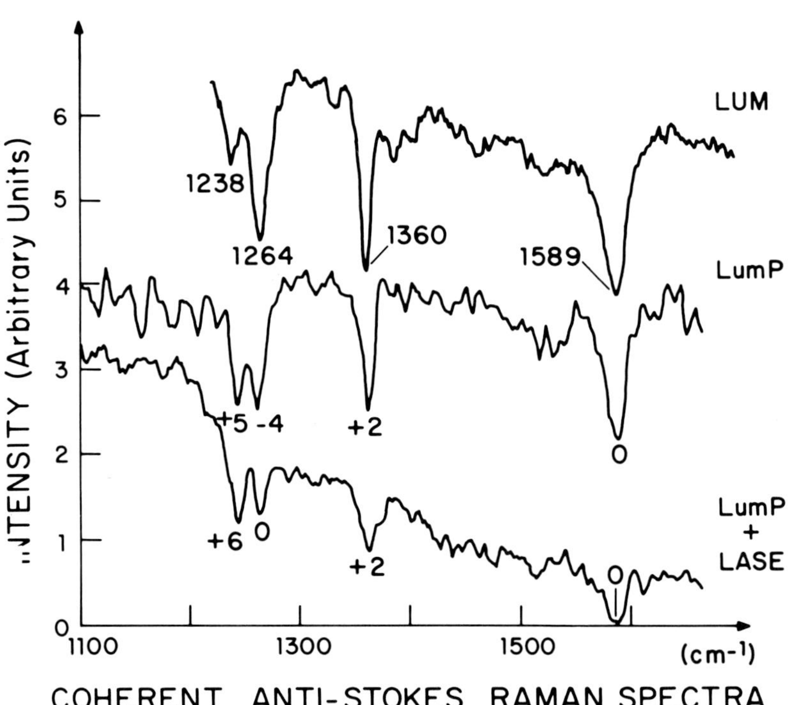


Figure 3. CARS of the free lumazine (LUM), protein-bound LumP, and in the presence of excess luciferase (LUMP + LASE). All at about 5°, 1-2 mM.

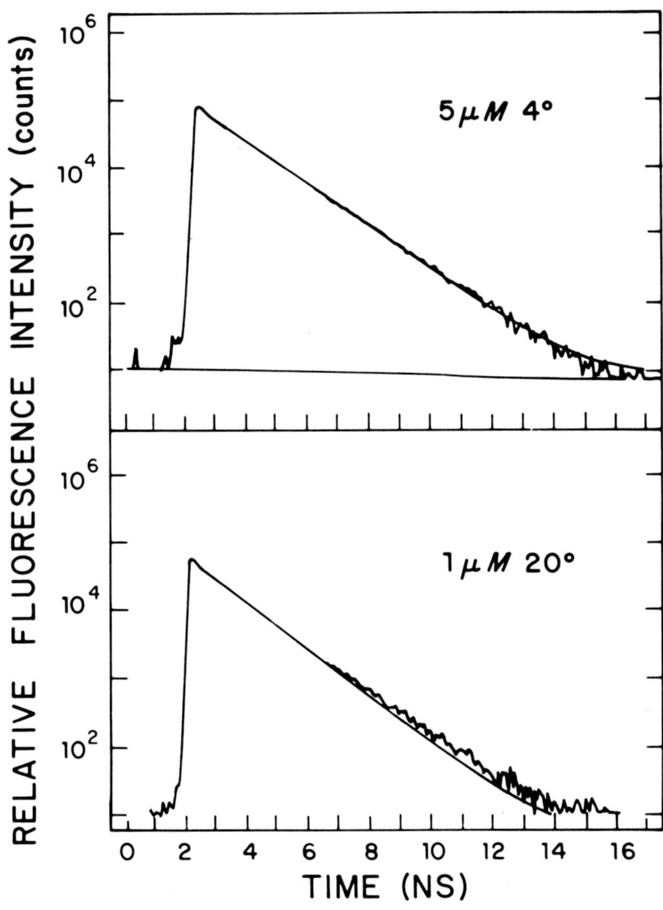


Figure 4. Fluorescence decay of lumazine protein following 100 ps excitation by an Ar laser, 458 nm. Top, $\tau$ = 14.4 ns; Lower, $\tau$ = 12.7 ns, single exponential fit.

One of the methods used for characterization of the isolated fluorophore was a comparison of the Raman spectrum (CARS) with that of the authentic substance. In the fingerprint region, 1200-1700 $cm^{-1}$, the CARS spectrum of lumazine is relatively simple, consisting of only 4 major (negative) peaks (Fig. 3) and there was a good correspondence in the frequency positions between the isolated and the authentic compounds (11). Figure 3 compares the CARS spectra of free 6, 7-dimethyl-8-ribityllumazine (LUM, top spectrum), when bound to protein (LUMP) and in the presence of excess *P. phosphoreum* luciferase (LASE). The frequencies are indicated on the top spectrum and only the changes on the others.

In each of the three states there is seen to be no change in the modes at 1360 and 1589 $cm^{-1}$, assigned respectively to motions of the groups of atoms N(5) - C(4a) - C(8a) - N(1) - C(2) and C(4) - C(4a) - C(8a) (18). The 1238 $cm^{-1}$ is assigned to C(2) - N(3) stretch coupled to N(3)-H in plane motion (18) and the remarkable observation is that the intensity of this mode increases relative to the 1264 $cm^{-1}$ pyrimidine breathing mode, on binding the lumazine to protein. This is suggestive of a proton migration in the excited state in this system (18). In the presence of luciferase this intensity is increased even more and this, and the other slight changes in frequency in the lower spectrum, and indicative of protein-protein interaction possibly taking place. Luciferase alone produces no spectrum.

The spectral properties of lumazine protein make it very suitable for a study of ligand-protein interaction. The fluorophore is highly fluorescent both on and off the protein and a study of the equilibrium

$$LUM + P \rightleftharpoons LUMP$$

by the polarization of fluorescence (19) results in a value $K_d$=5 x $10^{-8}$ M (4°, pH 7), just tight enough that both free and bound lumazine concentrations can be measured accurately.

Another technique that can be used to study this equilibrium is the fluorescence decay. Figure 4 (top) shows that lumazine protein has an unusually homogeneous fluorescence decay profile. It is a single exponential extending over 3 decades. Bound lumazine has a fluorescence lifetime of 14.4 ns (3°) and 10.0 ns (4°) when free (13). A more dilute solution at room temperature has a fluorescence decay profile (Fig. 4, lower) which deviates from a single exponential (the line), because the solution now contains both free and bound species. It has been noted before that the dissociation is responsible for the red-shift of the fluorescence of a lumazine protein solution on dilution (9).

The spectral results together suggest that the lumazine is attached to the protein via its ribityl tail, analogous to the interaction of FMN in flavodoxin (9). This accounts for the only slight frequency shifts on binding seen in the Raman spectra (Fig. 3) and for the strong fluorescence of the bound group.

It is attractive to speculate that the sensitizer mechanism postulated for the operation of lumazine protein (9) might be extended to the other luminous bacteria (11). It is reasonable to suggest that all *P. phosphoreum* overproduce lumazine protein specifically for bioluminescence generation. The major question now to be investigated is the identification of the species emitting the radiation in the *in vitro* reaction, without added lumazine protein and to establish the presence of the postulated high energy intermediate.

REFERENCES

1. Spruit-van der Burg, A. (1950) *Biochim. Biophys. Acta 5*, 175-178.
2. Seliger, H. H. and R.A. Morton, (1968) *in* "Photophysiology" ed. Giese, A.C. (Academic Press, N.Y.), Vol. IV, pp. 253-314.
3. Ruby, E. G., G. T. Reynolds, A. J. Walton, and C. J. Hardy (1976) *Biol. Bull. 151*, 428.
4. Fitzgerald, J. M. and J. Lee (1978) *in* "Microbial Ecology", eds. Loutit, M.W. and Miles, J.A.R. (Springer-Verlag, Berlin) pp. 40-41.
5. Lee, J. (1972) *Biochemistry 11*, 3350-3359.
6. Lee, J. and C. L. Murphy (1975) *Biochemistry 14*, 2259-2268.
7. Terpstra, W. (1962) *Biochim. Biophys. Acta 60*, 580-590.
8. Terpstra, W. (1962) *Biochim. Biophys. Acta 75*, 355-364.
9. Gast, R. and J. Lee (1978) *Proc. Nat'l. Acad. Sci. (USA) 75*, 833-837.
10. Small, E.D., P. Koka and J. Lee, (1980) *J. Biol. Chem.* (in press).
11. Koka, P. and J. Lee (1979) *Proc. Nat'l. Acad. Sci. (USA) 76*, 3068-3072.
12 Lee, J. and P. Koka (1978) *Methods Enzymol. 57*, 226-234.
13. Visser, A.J.W.G. and J. Lee (1980) *Biochemistry*, in press.
14. Lee, J., C.L. Murphy, G.J. Faini and T. L. Baucom (1974) *in* "Liquid Scintillation Counting: Recent Developments", eds. Stanley, P. E. and Scoggins, B. A. (Academic Press, N.Y.) pp. 403-420.
15. Koka, P., E.D. Small and J. Lee (1979) *Am. Soc. Photobiol. Abstr. 7*, 64.

16. Plaut, G. W. E. and R. L. Beach, (1975) *in* "Chemistry and Biology of Pteridines",ed. Pfleiderer, W. (de Gruyter, Berlin) pp. 101-124.
17. Lee, J. and P. Koka (1979) *Am. Soc. Photobiol. Abst.* 7, 64.
18. Irwin, R. M., J. Lee and L. A. Carreira (1980) *Biochemistry,* in press.

# PURIFICATION, IDENTIFICATION AND SYNTHESIS OF *PHOTOBACTERIUM FISCHERI* AUTOINDUCER

A. Eberhard
C. Eberhard

Department of Chemistry
Ithaca College
Ithaca, New York

A. L. Burlingame
G. L. Kenyon
N. J. Oppenheimer

Department of Pharmaceutical Chemistry
University of California
San Francisco, California

K. H. Nealson

Department of Marine Biology
Scripps Institution of Oceanography
University of California
La Jolla, California

## I. INTRODUCTION

The synthesis and activity of the luminescent system in luminous bacteria is subject to a variety of control mechanisms (1), one of which is called autoinduction (2,3,4). Autoinduction involves the production by bacteria of the autoinducer; a low molecular weight molecule that accumulates in the medium and, when it reaches sufficient concentration,



ISBN 0-12-208820-4

causes induction of the luminous system. We describe here the purification, identification and synthesis of the autoinducer for *Photobacterium fischeri*.

## II. MATERIALS AND METHODS

Autoinducer was assayed using strain B-61, an isolate of *P. fischeri* that is dark due to non-production of autoinducer, as a bioassay (4). The method was a slight modification of that described by Nealson (4); samples to be tested were added to 1 ml samples of B-61 cells (about $10^7$/ml in conditioned medium) in scintillation vials. The amount of light produced by the B-61 cells was a linear function of the amount of autoinducer added over a wide range of concentrations. A bioassay of authentic and synthetic autoinducer is shown in Figure 1.

A sea water nutrient medium, SWC, (5) was used for growth of strain MJ-1, a strain of *P. fischeri* that is very bright and produces more autoinducer than any other strain tested (4).

The purification of autoinducer is outlined in Figure 2. The purified material was examined by high resolution mass spectrometry using a direct inlet probe with a modified Kratos-AEI MS9 mass spectrometer on line to a xerox sigma 7-logos II computer system (6). Scanning was at 8 seconds/decade with an ion source temperature of about 250° at 70 E.B. with a mass resolution M/delta M = 10,000. MNR spectra were obtained at 360 Mhz with a bruher hxs-360 MNR spectrometer at the Stanford magnetic resonance laboratory. Chemical shifts were measured from internal standards, using tetramethylsialine when deuterated dimethylsulfoxide as solvent and using 3-trimethylsilylpropioate-2,2,3,3-$D_4$ when deuterium oxide was the solvent. Infra-red spectra were obtained with a Perkin-Elmer model 137 spectrophotometer using NaCl plates.

Synthesis of N-(3-oxohexanoyl)-3-aminodihydro-2(3H)-furanone was accomplished through a series of steps schematically shown in Figure 3. Amino acid analysis was done with a Beckman Model 116 amino acid analyzer after hydrolysis of the purified material.

ASSAY

Assay depends on light production in P. fischeri strain B-61, a natural dim isolate deficient in autoinducer production.

Conditioned medium must be used because fresh complete medium contains an inhibitor. It is removed by growth of a different species of luminous bacteria producing a heat labile, non cross reacting autoinducer.

PROCEDURE:

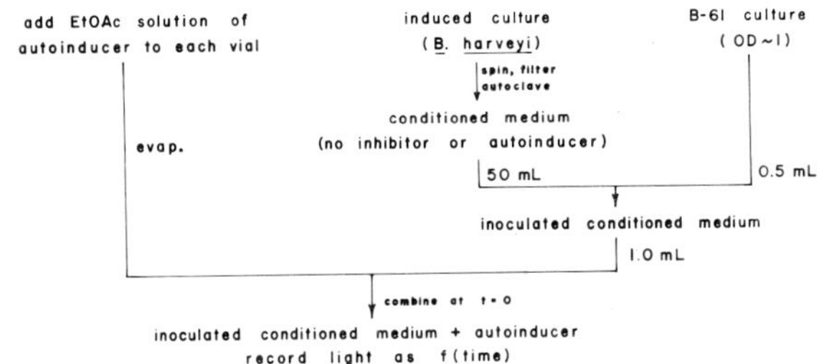


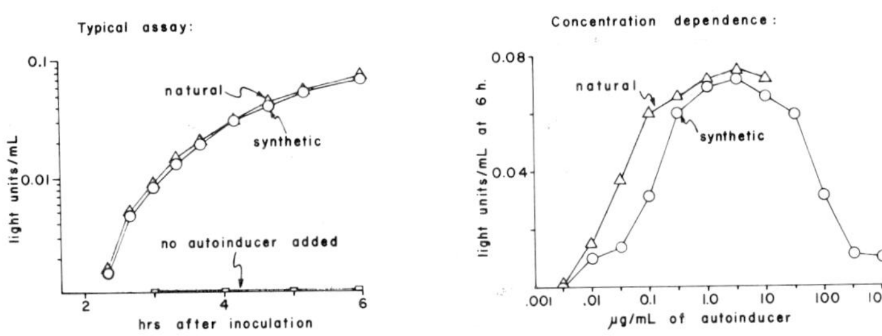


The synthetic material is less active than the natural presumably because only one enantiomer is active.

*FIGURE 1. Assay procedure for P. phosphoreum autoinducer activity.*

## III. RESULTS

The results of the studies presented here, which are discussed in detail in another publication (6), left little doubt of the structure of the autoinducer of *Photobacterium fischeri* strain MJ-1. Based MNR, infra-red, mass spectral, and amino acid analysis, the autoinducer was concluded to be the material shown in Figure 4. Some of its properties are

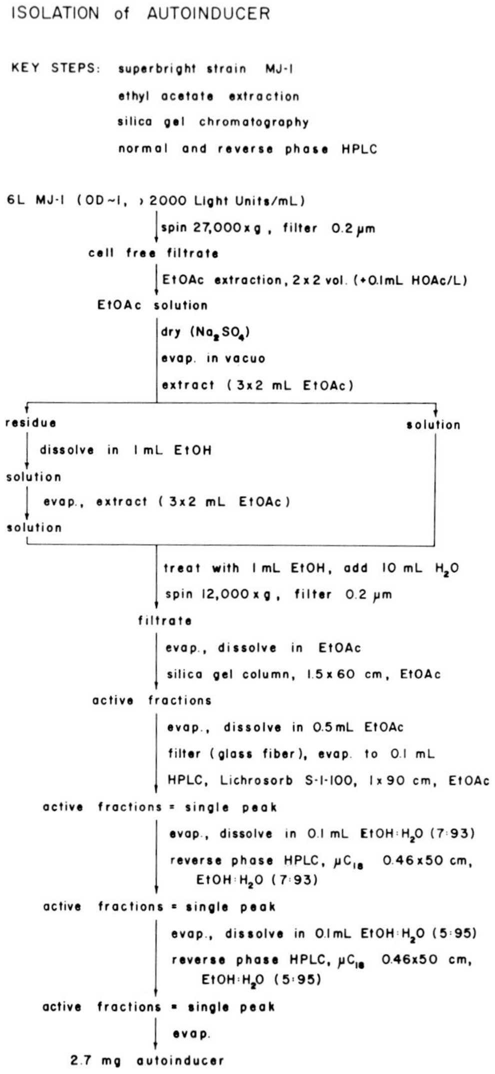

ISOLATION of AUTOINDUCER

KEY STEPS: superbright strain MJ-1
ethyl acetate extraction
silica gel chromatography
normal and reverse phase HPLC

6L MJ-1 (OD~1, > 2000 Light Units/mL)
↓ spin 27,000 x g, filter 0.2 μm
cell free filtrate
↓ EtOAc extraction, 2 x 2 vol. (+0.1 mL HOAc/L)
EtOAc solution
↓ dry ($Na_2SO_4$)
evap. in vacuo
extract (3x2 mL EtOAc)

residue
↓ dissolve in 1 mL EtOH
solution
↓ evap., extract (3x2 mL EtOAc)
solution

solution

↓ treat with 1 mL EtOH, add 10 mL $H_2O$
spin 12,000 x g, filter 0.2 μm
filtrate
↓ evap., dissolve in EtOAc
silica gel column, 1.5 x 60 cm, EtOAc
active fractions
↓ evap., dissolve in 0.5 mL EtOAc
filter (glass fiber), evap. to 0.1 mL
HPLC, Lichrosorb S-1-100, 1 x 90 cm, EtOAc
active fractions = single peak
↓ evap., dissolve in 0.1 mL EtOH:$H_2O$ (7:93)
reverse phase HPLC, $\mu C_{18}$ 0.46 x 50 cm, EtOH:$H_2O$ (7:93)
active fractions = single peak
↓ evap., dissolve in 0.1 mL EtOH:$H_2O$ (5:95)
reverse phase HPLC, $\mu C_{18}$ 0.46x50 cm, EtOH:$H_2O$ (5:95)
active fractions = single peak
↓ evap.
2.7 mg autoinducer

*Figure 2.* *Purification scheme for the autoinducer of P. fischeri.*

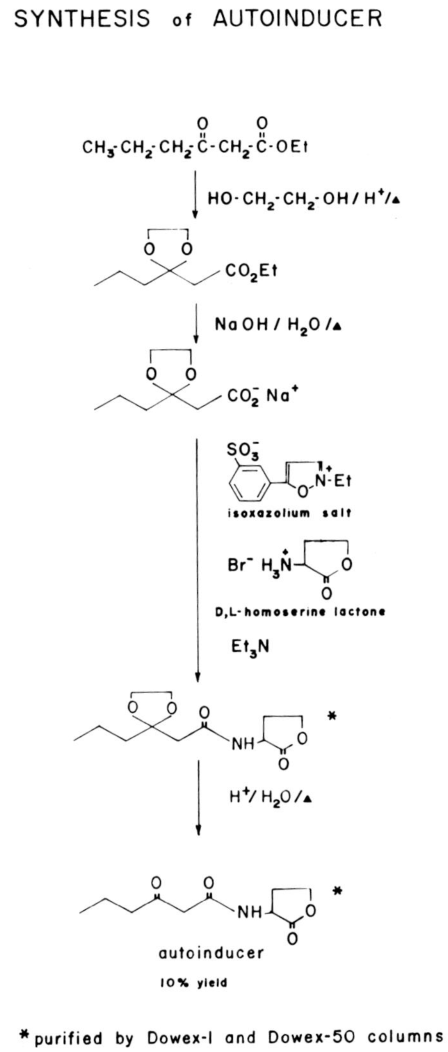

SYNTHESIS of AUTOINDUCER

$CH_3$-$CH_2$-$CH_2$-C(=O)-$CH_2$-C(=O)-OEt

↓ HO-$CH_2$-$CH_2$-OH / $H^+$/▲

$CO_2Et$

↓ NaOH / $H_2O$ /▲

$CO_2^-$ $Na^+$

$SO_3^-$

N-Et

isoxazolium salt

$Br^-$ $H_3\overset{+}{N}$

D,L-homoserine lactone

$Et_3N$

↓

NH *

↓ $H^+$/ $H_2O$ /▲

NH *

autoinducer

10% yield

*purified by Dowex-1 and Dowex-50 columns

*Figure 3. Synthesis of autoinducer of* *P. fischeri*.

shown in Figure 5. Synthesis of this compound gave biologically active material showing that the structure shown was was indeed the autoinducer. Furthermore, when the synthesized material was examined by the criteria used to determine the structure, identical spectra were obtained as had been obtained with the natural autoinducer. When equal concentrations of natural and synthetic autoinducer were tested with strain B-61, the natural material gave a slightly higher response than did the synthetic compound (Fig. 1). The dose response curve for synthetic autoinducer compared to that of natural autoinducer was shifted to a higher concentration by a factor of 2 to 3. At high concentrations, the synthetic inducer caused inhibition.

## IV. DISCUSSION

Although autoinduction has been known in luminous bacteria for over 10 years, the data presented here reveal for the first time the structure of an autoinducer of a species of luminous bacteria. The difficulty in the solution to this problem was due to several factors, including the low levels of autoinducer produced during growth, the chemical complexity of the media needed to grow the bacteria to maximal light production, and the difficulty of performing the bioassays for autoinducer. In these studies we were aided by a very bright strain which produces unusually large amounts of autoinducer and a sensitive assay system using a natural isolate that produces extremely low levels of autoinducer.

$$CH_3\text{-}CH_2\text{-}CH_2\text{-}\overset{\overset{\displaystyle O}{\|}}{C}\text{-}CH_2\text{-}\overset{\overset{\displaystyle O}{\|}}{C}\text{-}NH\text{-}CH\text{(}CH_2\text{-}CH_2\text{-}O\text{-}\underset{\underset{\displaystyle O}{\|}}{C}\text{)}$$

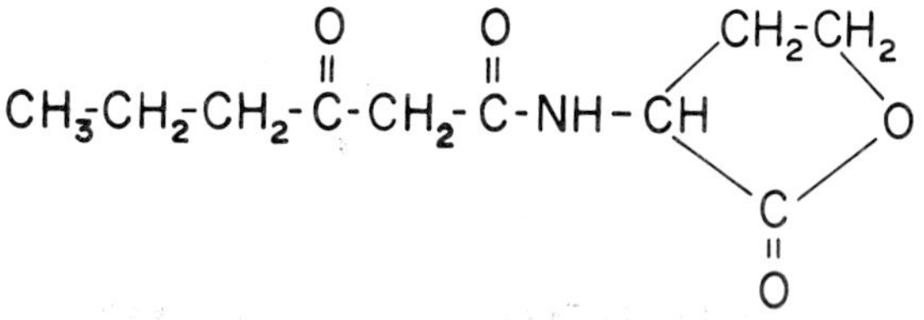

*Figure 4. The structure of autoinducer of P. fischeri N-(3-oxohexanoyl)-3-aminohydro-2(3H)-furanone.*

## CHEMICAL PROPERTIES

Activity, as measured by B-61 assay, was found to be:

- extremely base sensitive
- stable to acid (pH 1.57/100°/10 min) and glacial HOAc (50°/1 h)
- stable to heat (140°/10 min)
- partly stable to 3% $H_2O_2$ or $H_2$/Pt for 1 h/RT
- non volatile
- not bound by anion or cation exchange resins
- soluble in water and polar organic solvents but not in hexane

**Infrared spectrum**

Shows 5-membered ring lactone, amide, ketone

**Mass spectrum**

Mass = 213.1003, consistent with $C_{10}H_{15}O_4N$

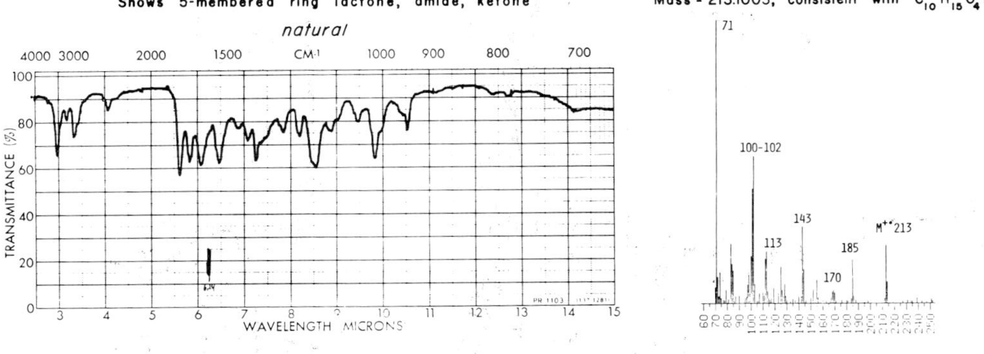


**NMR Spectrum**

*Figure 5. Some properties of the autoinducer.*

Chemical synthesis of the autoinducer confirms that the structure identified in the media extract is indeed the autoinducer. It also suggests a variety of experiments using labelled compounds, in order to examine in detail the exact mechanism by which autoinducer acts at the level of transcription to cause the synthesis of messenger RNA and therefore can be classed a classical inducer. However, the precise mechanism by which this compound acts cannot be examined without purified inducer. Now that the inducer can be synthesized easily, it will be possible to synthesize labelled autoinducer and ask direct questions as to its mechanism of action.

The autoinducer produced by *Photobacterium fischeri* induces luciferase synthesis in all other strains of *Photobacterium fischeri,* but not in any other species of luminous bacteria. Whether or not such species specificity is true for other species of luminous bacteria has not yet been investigated in detail. In fact, whether or not autoinduction is a mechanism common to all species of luminous bacteria has not been established. It seems clear, however, that for *P. fischeri,* autoinduction is a major mechanism responsible for the control of luminescence, synthesis and activity and the availability of pure autoinducer should aid in the elucidation of the mechanism by which this occurs.

## REFERENCES

1. Nealson, K. and J.W. Hastings, *Microbiol. Rev. 43,* 496 (1979).
2. Nealson, K., T. Platt, and J.W. Hastings, *J. Bacteriol. 104,* 300 (1970).
3. Eberhard, A., *J. Bacteriol. 109,* 1101 (1972).
4. Nealson, K.H., *Arch. Microbiol. 112,* 73 (1977).
5. Nealson, K.H., *Methods Enzymol. 57,* 153 (1978).
6. Eberhard, A., A.L. Burlingame, C. Eberhard, G.L. Kenyon, K.H. Nealson, and N.J. Oppenheimer, *Biochemistry* (submitted) (1980).

ACTIVE CENTER STUDIES ON BACTERIAL LUCIFERASE: EVIDENCE THAT THE REACTIVE CYSTEINYL RESIDUE IS WITHIN THE PROTEASE-LABILE REGION OF THE α SUBUNIT[1]

Thomas O. Baldwin, John J. Dougherty, Jr., Steven K. Rausch

Department of Biochemistry
University of Illinois
Urbana, Illinois 61081

Margaret V. Merritt

Physical and Analytical Chemistry Research
The Upjohn Company
Kalamazoo, Michigan 49001

## I. INTRODUCTION

Bacterial luciferase is a heterodimeric (αβ) enzyme with a single active center which appears to reside primarily, if not exclusively, on the α subunit (ca. 42,000 daltons) (1-3); the specific role of the β subunit (ca. 37,000 daltons) is unknown, but it is required for activity (4). Recent studies demonstrate that alterations in the β subunit can affect the binding affinity for $FMNH_2$[2] and suggest the possibility that β subunit residues might contribute directly to the structure of the active center(5-7).

[1]*This work was supported by grants from the National Science Foundation (NSF PCM 79-25335) and the National Institute on Aging (AG-00884).*

[2]*Abbreviations used: FMN and $FMNH_2$, oxidized and reduced riboflavin 5'-phosphate; DTT, dithiothreitol; ESR, electron spin resonance; SDS, sodium dodecylsulfate; NBPS, the N-[p-(2-benzoxazolyl)phenyl]succinimide group.*




ISBN 0-12-208820-4

The present study was undertaken to better evaluate the location of the reactive thiol relative to the active center. To this end, three experiments were performed: 1) the effect of high concentrations (10-100 x $K_d$) of FMN upon reactivity of the thiol was determined, 2) the binding of $FMNH_2$ to luciferase stoichiometrically modified with N-n-octylmaleimide was measured by circular dichroism spectroscopy, and 3) the modification of the enzyme with methyl methanethiolsulfonate, which forms an $-SCH_3$ mixed disulfide with the protein, was analyzed.

## II. MATERIALS AND METHODS

Luciferase was purified from an aldehyde-deficient mutant (M17) of *B. harveyi* and assayed by our published method (8). FMN and iodoacetamide were obtained from Sigma and N-n-octylmaleimide from Pfaltz & Bauer; methyl methane-thiolsulfonate was a generous gift from G. L. Kenyon.

Luciferase solutions were dialyzed to remove reducing agent and inactivation (alkylation) reactions were carried out and monitored as previously described (1, 2). Solutions for circular dichroism measurements were prepared in 0.1M BisTris, pH 7, and reduced with 5 mN $Na_2S_2O_4$ under a layer of mineral oil. Circular dichroism spectra were obtained on a Jasco J-40A spectropolarimeter at 25° in a 2 mm pathlength cuvette (9).

## III. RESULTS AND DISCUSSION

A saturating concentration of FMN completely protects luciferase from inactivation by iodoacetamide (Fig. 1). The partial protection reported earlier (1) was presumably due to the low concentration of FMN relative to the $K_d$ of the complex (10). This finding is particularly interesting when compared with our earlier observation (11) that modification of the reactive thiol with N-n-octylmaleimide had little effect on the binding of FMN.

The binding of $FMNH_2$ to luciferase stoichiometrically modified with N-n-octylmaleimide was monitored by circular dichroism spectroscopy essentially as described by Becvar and Hastings (9) (Fig. 2). The results of these measurements demonstrate two points. First, $FMNH_2$ does bind to the weight 79,000 and $E^{1\%}_{1cm}$ = 9.4 at 280 nm); 6.4% of the initial

## ERRATA

**Bioluminescence and Chemiluminescence**

Basic Chemistry and Analytical Applications

*Marlene DeLuca and William McElroy, Editors*

Chapter by T. O. Baldwin, J. J. Dougherty, Jr., S. K. Rausch, and M. V. Merritt

Substitute the following text for page 122.

Miriam Ziegler (Nicoli) and her co-workers have analyzed in detail a particularly reactive thiol and its environment (8, 9). The thiol resides on the α subunit in the tryptic peptide Phe-Gly-Ile-Cys-Arg (10) and is in a hydrophobic cleft (9, 11) ca. 17 Å in length (11).

The protease susceptibility of the enzyme has received considerable attention (12-16). The published data demonstrate the that α subunits of luciferases from taxonomically distinct species possess a protease-labile region located about 100 residues from one terminus and ca. 20 residues in length (14, 15). Hydrolysis of peptide bonds within this region results in loss of measurable activity, $FMNH_2$ binding and FMN binding (14). Binding of FMN or orthophosphate protects this region from proteases (13, 17, 18). The action of proteases is quite specific; the α subunit is converted to two families of polypeptides, γ (ca. 28,000 daltons) and δ (ca. 14,000 daltons) (14). The molecular weight of the protein measured by sedimentation equilibrium methods does not change upon treatment with trypsin or chymotrypsin, demonstrating that the fragments of the protein that result from the action of these proteases do not dissociate under nondenaturing conditions to an extent measurable by the analytical ultracentrifuge (12, 15). We have proposed that the protease-labile region of the α subunit of luciferase is either close to or an integral part of the active center (14). The experiments described here were designed to determine the relative locations of two structural elements, both thought to be in or near the active center, the reactive thiol and the protease-labile region.

## II. MATERIALS AND METHODS

Bacterial luciferase was purified from *Beneckea harveyi* (strain 392; 19) and the activity assayed by our published procedures (20). Purity of luciferase was estimated by the criterion of polyacrylamide gel electrophoresis by the method of Laemmli (21). Treatment of luciferase with proteases was performed as previously described (12).

Luciferase was modified with N-[1-$^{14}$C]ethylmaleimide (New England Nuclear) as described (8); the reaction was stopped with excess DTT when about 25% of the initial activity remained. Luciferase was spin-labeled using 0.9 moles 3-(maleimidomethyl)-2, 2, 5, 5-tetramethyl-1-pyrrolidinyloxyl (Syva; dissolved in ethanol) per mole luciferase (molecular weight 79,000 and $E^{1\%}_{1\,cm} = 9.4$ at 280 nm); 6.4% of the initial

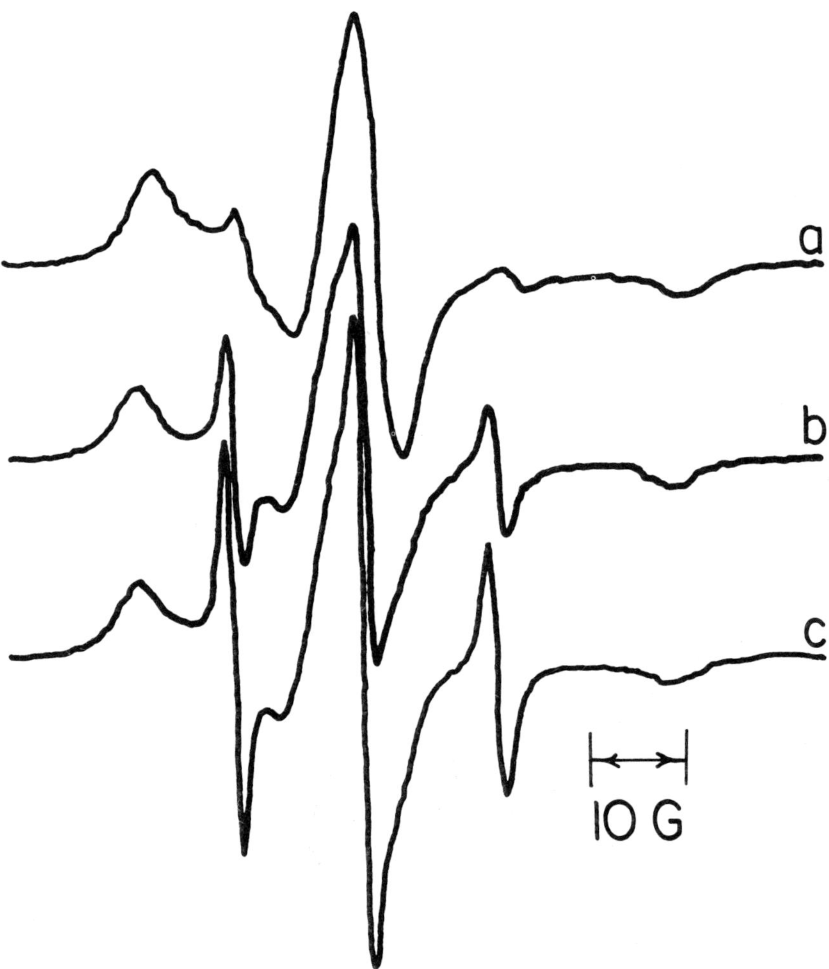


*FIGURE 1. ESR spectra of (a) nitroxide-labeled luciferase (b) nitroxide-labeled luciferase treated with chymotrypsin and (c) nitroxide-labeled luciferase treated with trypsin. The spectra were collected using a modulation amplitude of 2.5G, a power of 50 mW, a scan time of 4 min, and a time constant of 0.3s. The ESR signals are not saturated at this microwave power.*

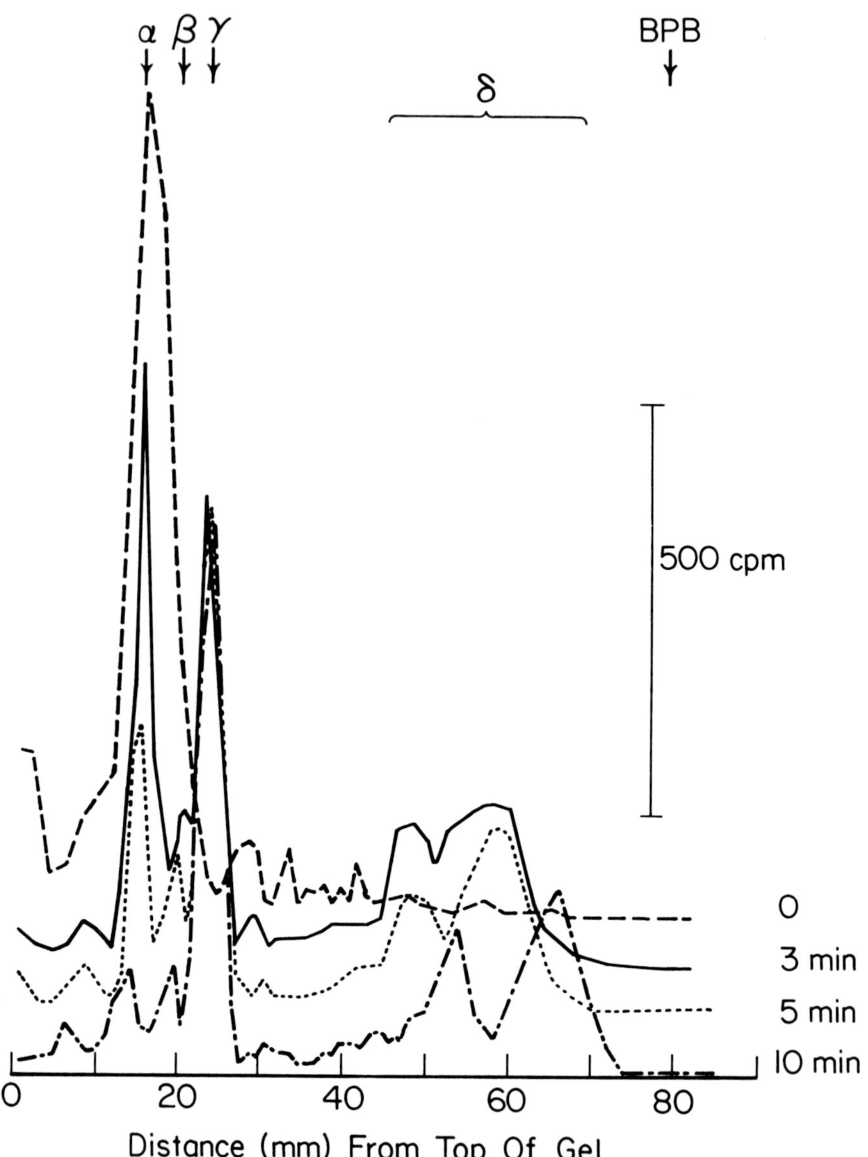


*FIGURE 2. Location of radioactivity by liquid scintillation counting of gel slices. Luciferase was labeled with N[1-$^{14}$C]ethylmaleimide, and treated with trypsin; aliquots were removed for electrophoresis (23) at the indicated times. The gels were not stained; the markings indicate the positions of the protein bands and the bromphenol blue marker.*

activity remained, showing that the reaction was stoichiometric. Luciferase was also labeled with 0.9 moles of the fluorescent probe N-[p-(2-benzoxazolyl)phenyl]maleimide (NBPM) (Eastman; dissolved in methyl cellosolve) per mole luciferase; about 8% of the initial activity remained, indicating that the reaction with NBPM was also stoichiometric.

All ESR measurements were made as previously described (11) using a Varian E-9 ESR spectrometer; samples were maintained at 20 ± 1°. Fluorescence polarization measurements were made on a photon counting apparatus that has been described (22). Measurements were made at 25° using an excitation wavelength of 310 nm. Fluorescence emission due to tryptophanyl residues was minimized through the combined use of a 5 nm excitation bandpass and Corning 0-52 filters at the emission slits. The location of radioactivity in the polyacrylamide gels (23) was determined using Aquasol and a Beckman LS 7000 liquid scintillation counter following dissolution of 1-2 mm thick slices of the gel in 0.4 ml 30% $H_2O_2$.

## IV. RESULTS AND DISCUSSION

The mobility of the reactive thiol following protease treatment was determined using a fluorescent maleimide derivative and a spin-labeled maleimide; spectral properties of both bound to the luciferase have been previously reported (11, 24). Upon treatment with either trypsin or chymotrypsin, the polarization of fluorescence of the NBPS-luciferase drops sharply to a value approximating the polarization of the NBPS-luciferase in 6M urea (Table 1). These observations demonstrate that the action of trypsin or of chymotrypsin upon NBPS-luciferase results in a relatively free rotation of the probe, even though no dissociation of fragments occurs (12, 15).

The effect of chymotrypsin upon the ESR spectrum of the nitroxide-labeled luciferase (Fig. 1) is rather different; there is no apparent effect of the action of chymotrypsin upon the mobility of the probe. On the other hand, following treatment with trypsin, the spectrum of the spin-labeled luciferase shows that a large proportion of the probe has attained rotational freedom.

The results of the experiments with the two spectral probes demonstrate that the action of the proteases causes a change in the structure of the luciferase in the vicinity of the reactive thiol. The fact that little change is observed in the mobility of the nitroxide probe upon treatment with chymotrypsin apparently reflects the fact that trypsin

*TABLE I. The Effects of limited Proteolysis and Urea Denaturation on the Polarization of Fluorescence of NBPS-Luciferase.*

| | % Activity[a] | P |
|---|---|---|
| NBPS-Luciferase | 100.0 | 0.330 |
| Trypsin-treated | 13.2 | 0.244 |
| Chymotrypsin-treated | 3.2 | 0.241 |
| 6 M Urea | 0.0 | 0.234 |

*[a]The 100% activity reported here corresponds to the ca. 8% activity remaining after treatment with N-[p-(2-benzoxazolyl)phenyl]maleimide.*

and chymotrypsin hydrolyze different peptide bonds (there are 5 trypsin-sensitive bonds and 2 chymotrypsin-sensitive bonds; 14), although both attack peptide bonds within the ca. 20-residue protease-labile region. The nitroxide probe is much smaller than the fluorescent probe and would be expected to respond somewhat differently.

While the mobility measurements demonstrate an environmental (i.e., tertiary structural) alteration following proteolysis, these experiments do not yield information regarding the relative positions of the reactive thiol and the protease-labile region in the primary structure. The location of the reactive thiol on the α subunit of *B. harveyi* luciferase relative to the protease-labile region was determined by analysis of radioactivity on SDS gels of the trypsin-generated fragments of luciferase that had been labeled with N-[1-$^{14}$C]ethylmaleimide (Fig. 2). Most of the label incorporated (>90%) was associated with the α subunit, in accord with previous reports (8). In the experiment presented in Fig. 2, the protein was not stained in order to avoid the possibility of selective loss of small polypeptides during the staining and destaining process. Instead, the gels were sliced, solubilized, and counted, and the location of the various protein bands determined by comparison with gels that had been stained (12). While the method of electrophoresis employed (23) shows only a single γ band (ca. 28,000 daltons) and a trace of material in the δ region (ca. 14,000 daltons) rather than resolving γ into five bands as the Laemmli method (21) does, it is clear that the action of trypsin results in formation of fragments with molecular weights of about 28,000 and 14,000; *both* sets of fragments are radioactive. The lower resolving power of the electrophoresis method does not show which bands (visible by the Laemmli method; 21) within the γ and δ families are labeled. The only conclusion consistent with the observation of a unique label within two sets of fragments is that the thiol exists *within* the protease labile region and that hydrolysis of bonds on one side results in label in one set of fragments while hydrolysis of bonds on the other side results in label in the other set of fragments.

## REFERENCES


1. Meighen, E. A., M. Z. Nicoli, and J. W. Hastings, *Biochem.* 10, 4062-68 (1971).
2. Meighen, E. A., M. Z. Nicoli, and J. W. Hastings, *Biochem.* 10, 4069-73 (1971).
3. Cline, T. W., and J. W. Hastings, *Biochem.* 11, 3359-70 (1972).
4. Friedland, J., and J. W. Hastings, *Proc. Natl. Acad. Sci. USA* 58, 2336-42 (1967).
5. Cline, T. W., Ph.D. Thesis, Harvard University, Cambridge MA 02138 (1973).
6. Meighen, E. A., and I. Bartlett, *J. Biol. Chem.*, in press (1980).
7. Welches, W. R., and T. O. Baldwin, *Biochem.*, in press (1980).
8. Nicoli, M. Z., E. A. Neighen, and J. W. Hastings, *J. Biol. Chem.* 249, 2385-92 (1974).
9. Nicoli, M. Z., and J. W. Hastings, *J. Biol. Chem.* 249, 2393-96.
10. Nicoli, M. Ziegler, Ph.D. Thesis, Harvard University, Cambridge, MA 02138 (1972).
11. Merritt, M. V., and T. O. Baldwin, *Arch. Biochem. Biophys.* 202, 499-506 (1980).
12. Baldwin, T. O., J. W. Hastings, and P. L. Riley, *J. Biol. Chem.* 253, 5551-4 (1978).
13. Holzman, T. F., and T. O. Baldwin, *Biochem. Biophys. Res. Commun.* 94, 1199-1206 (1980).
14. Holzman, T. F., P. L. Riley, and T. O. Baldwin, *Arch. Biochem. Biophys.*, in press (1980).
15. Holzman, T. F. and T. O. Baldwin, *Proc. Natl. Acad. Sci. USA*, in press (1980).
16. Ruby, E. G., and J. W. Hastings, *Curr. Microbiol.* 3, 157-59 (1979).
17. Baldwin, T. O. and P. L. Railey, *in* "Flavins and Flavoproteins" (T. Yamano and K. Yagi, eds.), pp. 139-147. Japan Scientific Societies Press, Tokyo (1980.

18. Nicoli, M. Z., T. O. Baldwin, J. E. Becvar, and J. W. Hastings, *in* "Flavins and Flavoproteins" (T. P. Singer, ed.), pp. 94-100. ASP, Amsterdam (1976).
19. Reichelt, J. L., and P. Baumann, *Arch. Mikrobiol.* 94, 283-330 (1973).
20. Hastings, J. W., T. O. Baldwin, and M. Z. Nicoli, *Meth. Enzymol.* 57, 135-152 (1978).
21. Laemmli, U. K., *Nature* 227, 680-685 (1970).
22. Jameson, D. M., G. Weber, R. D. Spencer, and G. Mitchell, *Rev. Sci. Instrum.* 49, 510-514 (1978).
23. Weber, K. and M. Osborn, *J. Biol. Chem* 244, 4406-12 (1969).
24. Tu, S. -C., C. -W. Wu, and J. W. Hastings, *Biochem.* 17, 987-93 (1978).

# BACTERIAL BIOLUMINESCENCE: ACCESSORY ENZYMES[1]

Edward Meighen
Denis Riendeau
Andrew Bognar

Department of Biochemsitry
McGill University
Montreal, Quebec

## I. INTRODUCTION

The elucidation of the role of enzymes other than luciferase in bacterial bioluminescence has centered on the detection and characterization of functional activities involved in the metabolism of the substrates ($FMNH_2$ and aldehyde) in the bioluminescent reaction shown below.

$$FMNH_2 + O_2 + \text{RCHO} \xrightarrow{\text{Luciferase}} \text{RCOOH} + H_2O + FMN + h\nu$$

Enzymes catalyzing the reduction of FMN to FMNH (1-3) and the synthesis (4, 5) as well as the oxidation and reduction (6) of long chain aldehydes have been detected in extracts of bioluminescent bacteria and some of these enzymes have now been purified to homogeneity (7-9).

Although the detection of such functional activities strongly suggests that these enzymes are directly involved in the bacterial bioluminescent system, additional criteria are generally needed to confirm this proposal and to more clearly determine their specific roles. Among the experiments that would support the involvement of a specific enzyme in the bioluminescent system would be establishing that (a) the enzyme is coregulated with luciferase *in vivo* under certain

[1]*Supported by the Medical Research Council of Canada (MT-4314).*




ISBN 0-12-208820-4

physiological conditions, such as during development of bioluminescence, (b) the enzyme's specificity for substrates is restricted primarily to metabolites in pathways leading to or derived from the bioluminescent reaction, (c) the enzyme interacts physically with luciferase or another enzyme in the luminescent system or (d) mutation of the bacteria can specifically and coordinately affect the enzyme function and bioluminescence. Clearly, not all these criteria will be met by any one enzyme, however, such data provides confidence that one is indeed investigating an accessory enzyme in the luminescent system.

Illustrated in Fig. 1 is an experimental approach used to determine whether the synthesis of a protein may be coregulated with luciferase (10). Proteins synthesized before and during induction of luminescence in *Beneckea harveyi* were labelled with $^{3}H$ and $^{14}C$, respectively and after resolution by chromatography and SDS electrophoresis, induced polypeptides were detected from their high isotope ratios ($^{14}C/^{3}H$) compared to those of the constitutive proteins. Three of the seven induced polypeptides detected by this approach are present in the two electrophorograms given in Fig. 1; the two induced polypeptides on the left being the α and β subunits of luciferase. Other experiments have also indicated the presence of

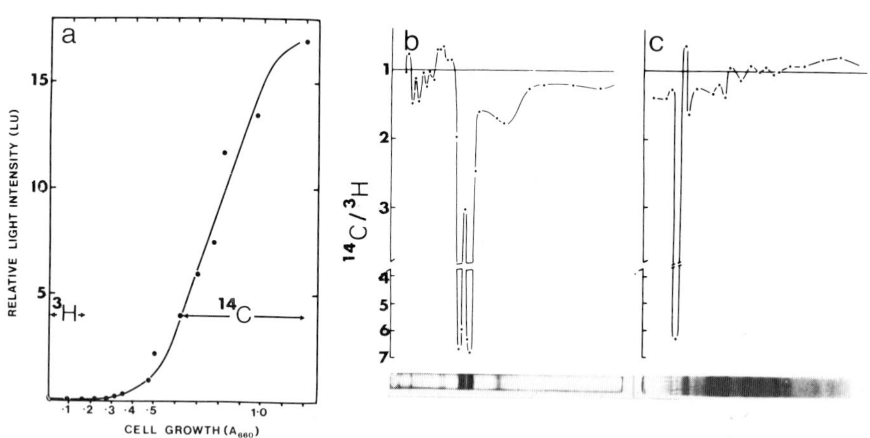


*FIGURE I. a. Radioactive labelling of the proteins in B. harveyi with [$^{3}H$]-leucine and [$^{14}C$]-leucine before and during induction of luminescence. b.c. Detection of induced polypeptides on SDS gel electrophoresis of samples previously resolved by anion exchange chromatography.*

induced proteins in the bacterial membrane (11). These results have demonstrated that a limited number of polypeptides, as well as both subunits of luciferase, are specifically synthesized during induction of bioluminescence in *B. harveyi* (10) and have indicated that the luminescent system may consist, at least in part, of a group of proteins under similar regulation at the gene level. However, not all enzymes involved in bacterial bioluminescence may be coregulated with luciferase in this fashion.

## II. FUNCTIONS OF ACCESSORY ENZYMES

### *A. Reduction of FMN*

Enzymes catalyzing the reduction of FMN to produce free $FMNH_2$ in solution have been demonstrated in a number of luminescent bacteria (1-3). Two of these FMN reductases in *B. harveyi*, one NADH specific and the other NADPH specific, have been purified to homogeneity (7, 8). These enzymes do not appear to be under the same regulation as luciferase (3, 6, 12, 13). The purified NADH:FMN oxidoreductase, isolated from cultures labelled with different isotopes before and during induction of luminescence has the same isotope ratio as the constitutive proteins (Fig. 2). In contract luciferase and a long chain aldehyde dehydrogenase, purified in the same experiment, have high isotope ratios indicating that these enzymes and not NADH:FMN oxidoreductase are induced (13). Recent evidence has suggested that this oxidoreductase may interact with luciferase (14), thus more directly implicating it in the bioluminescent system. However, the mechanism *in vivo* by which the FMN reductases supply $FMNH_2$ to the large molar excess of luciferase molecules still remains unclear.

### *B. Oxidation and Reduction of Aldehydes*

The removal of long chain aldehydes in the bioluminescent bacterium, *B. harveyi*, is catalyzed by an aldehyde dehydrogenase (6) which has now been purified to homogeneity (9). An alternate pathway involving the reduction of aldehyde to alcohol has been detected in *Photobacterium phosphoreum* extracts (see Fig. 3), however, additional evidence will be needed to implicate aldehyde reductases in the regulation of bacterial bioluminescence.

The aldehyde dehydrogenase, however, has been investigated in some detail (9) and the synthesis of this enzyme occurs at the same time as luciferase during induction of luminescence (Fig. 2). Furthermore, its specificity for the chain length of the aldehyde. Table I shows that this specificity is due to a decrease in $K_m$ and not a change in $V_m$. Since attempts to reverse the reaction catalyzed by aldehyde dehydrogenase using ATP and NAD(P)H have been unsuccessful, it appears that the role of this enzyme is to regulate aldehyde levels for the bacterial bioluminescent reaction by removal by oxidation reather than by synthesis.

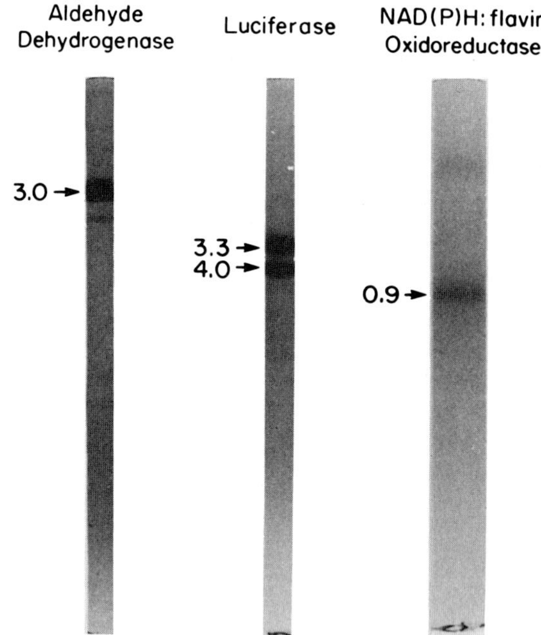


*FIGURE 2. SDS gel electrophoresis of enzymes purified from B. harveyi labelled with [$^3$H]- and [$^{14}$C]-leucine before and during induction of bioluminescence. Isotope ratios are given relative to those for the constitutive proteins.*

*TABLE I. Specificity of Aldehyde Dehydrogenase*[a].

| *Aldehyde* | $K_m$ *(mM)* | *Relative* $V_m$ | *Relative* $V_m/K_m$ |
|---|---|---|---|
| | 0.05 | 12 | 240 |
| | 0.005 | 8 | 1,600 |
| | <0.001 | 10 | >10,000 |

[a]*Analyzed in 0.05 M phosphate, 1.5 mM* $NAD^+$*, pH 7.0.*

*TABLE II. Stimulation of Luminescence by Fatty Acid Reductase*

| *Components added* | *Relative light intensity*[a] |
|---|---|
| --- | 0.2 |
| ATP | 0.2 |
| NADPH | 1.0 |
| ATP, NADPH | 70 |
| ATP, NADPH, Tetradecanoic acid | 100 |
| ATP, NADH, Tetradecanoic acid | 4 |
| GTP, NADPH, Tetradecanoic acid | 4 |

[a]*Relative luminescence on injection of* $FMNH_2$*, after incubation of an extract of* P. phosphoreum *for 10 min. with the indicated components (ATP, GTP, NADH, NADPH, 1 mM; tetradecanoic acid, 10 μM).*

### C. *Biosynthesis of Aldehydes*

Recently, we have shown that extracts of *P. phosphoreum* catalyze the synthesis of long chain aldehydes using an assay system coupled to luciferase (Table II). Both ATP and NADPH are essential for activity and cannot be replaced by GTP or NADH (4). Tetradecanoic acid is required for optimal activity in this coupled assay with the dependence on fatty acids being much more marked after partial purification of the enzyme (5). A ten-fold stimulation of activity with fatty acid can be obtained using conditions designed to convert the majority ( >60%) of the fatty acid into aldehyde (unpublished data). Based on these results, it was concluded that this enzyme is a fatty acid reductase catalyzing the reaction given below.

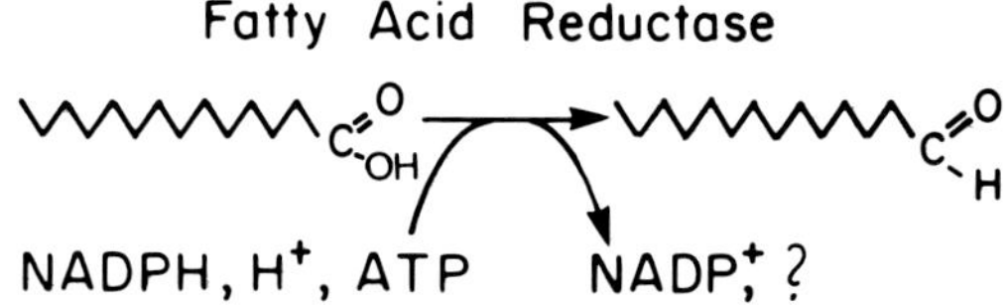


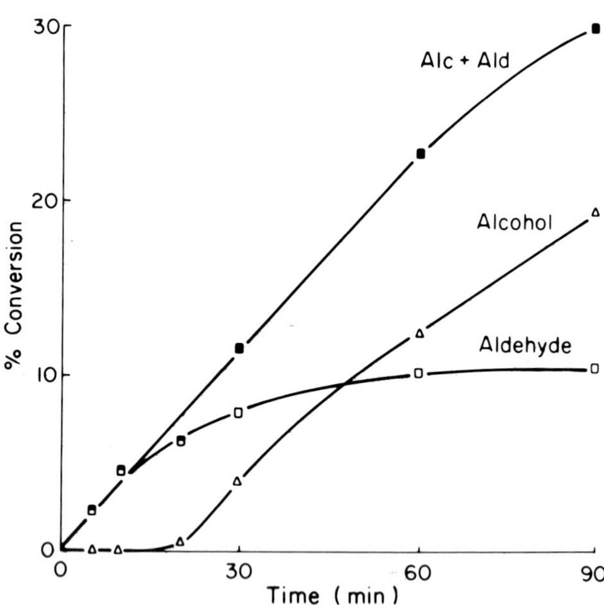


*FIGURE 3. The dependence of the synthesis of long chain aldehyde and alcohol from [$^3H$]-tetradecanoic acid (0.58 μM) incubated with 1 mM ATP and NADPH and* P. phosphoreum *extract.*

It may be noted that the specificity for tetradecanoic acid in the luciferase coupled assay may primarily reflect the specificity of luciferase rather than fatty acid reductase.

An alternate assay has now been developed using [$^3$H]-tetradecanoic acid and measuring the appearance of [$^3$H]-aldehyde after resolution of the labelled compounds by TLC (Riendeau and Meighen, manuscript in preparation). Fig. 2 shows the production of fatty aldehyde from tetradecanoic acid and its subsequent reduction to alcohol providing evidence for the existence of aldehyde reductase(s) in the extract. Furthermore, this luciferase independent assay allows the direct analysis of the level of fatty acid reductase activity in different extracts and was used to demonstrate that fatty acid reductase is co-induced with luciferase (Fig. 4). Although this activity has not yet been demonstrated in extracts of *B. harveyi* (4, 15), *in vivo* studies suggest this enzyme is present and also induced in this bacterium (15).

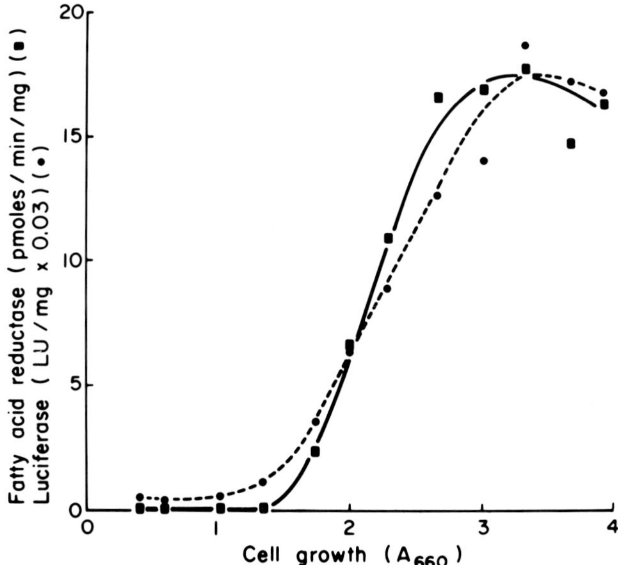


*FIGURE 4. Specific activities of luciferase and fatty acid reductase as a function of growth of P. phosphoreum.*

## III. SUMMARY AND CONCLUSIONS

A summary of the reactions catalyzed by the enzymes implicated in the bacterial bioluminescent system is given in Fig. 5. The determination of the specific roles of these and other enzymes will be necessary to understand the function and regulation of bacterial bioluminescence.

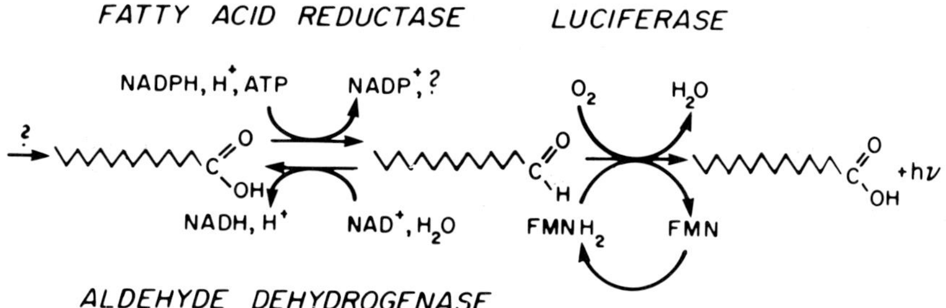


FIGURE 5. *Accessory enzymes in bacterial bioluminescence.*

## REFERENCES

1. Duane, W. and J. W. Hastings, *Mol. Cell. Biochem.* 6, 53 (1975).
2. Gerlo, E. and J. Charlier, *Eur. J. Biochem.* 57, 461 (1975).
3. Puget, K. and A. Michelson, *Biochimie.* 54, 1197 (1972).
4. Meighen, E., *Biochem. Biophys. Res. Commun.* 87, 1089 (1979).
5. Riendeau, D. and E. A. Meighen, *J. Biol. Chem.* 254, 7488 (1979).
6. Meighen, E. A., I. Bogacki, A. Bognar, and G. A. Michaliszyn, *Biochem. Biophys. Res. Commun.* 69, 423 (1976).
7. Jablonski, E. and M. Deluca, *Biochem.* 16, 2932 (1977).
8. Michaliszyn, G. A., S. S. Wing, and E. A. Meighen, *J. Biol. Chem.* 252, 7495 (1977).
9. Bognar, A. L. and E. A. Meighen, *J. Biol. Chem.* 253, 446 (1978).
10. Michaliszyn, G. A. and E. A. Meighen, *J. Biol. Chem.* 251, 2541 (1976).

11. Ne'eman, Z., S. Ulitzur, D. Branton, and J. W. Hastings, *J. Biol. Chem.* 14, 5150 (1977).
12. Jablonski, E. and M. Deluca, *Biochemistry* 17, 672 (1978).
13. Bognar, A., G. Michaliszyn, and E. A. Meighen, *Can. J. Biochem.* 56, 605 (1978).
14. Tu, S. C. and J. W. Hastings, *Proc. Nat'l. Acad. Sci. USA* 77, 249 (1980).
15. Ulitzur, S. and J. W. Hastings, *J. Bacteriol.* 137, 854 (1979).

# BIOLUMINESCENCE TEST FOR MUTAGENIC AGENTS

Shimon Ulitzur
Irith Weiser
Shmuel Yannai

Department of Food Engineering
and Biotechnology
Technion-Israel Institute of Technology
Haifa, Israel

There is increasing evidence that the initiation of human cancer involves mutational events (1). This has led to the concept that mutagenic chemicals are also likely to be carcinogenic. Therefore tests for chemical mutagenicity might reveal their carcinogenic potential. Currently, the most widely used test for mutagenicity is the Ames Test (1), in which an increase in the rate of reversion of certain strains of *Salmonella typhimurium* from auxotrophy for histidine to prototrophy is indicative of the mutagenic activity of the compound in question.

Based on the same established concept, we have recently developed a new, fast and sensitive bioluminescence assay for testing mutagenic potential (2). This new assay determines the capability of a test compound to cause an increase in the reversion rate of a dark variant of a luminous bacterium to the luminescent state. For this test we selected a dark variant of the luminous bacterium *Photobacterium leiognathi*, which exhibits a low frequency of spontaneous reversion to the luminescent state. The dark variant (designated 8SD18) exhibits *in vivo* luminescence corresponding to only $1:10^6$ of that of the wild-type cells and *in vitro* luciferase activity corresponding to about $1:10^4$ of that of the wild type. Different base-substitution and frame-shift agents increase considerably the *in vivo* luminescence and the reversion rate of this dark variant to a genetic-hereditary luminous form. This test is capable of detecting nanogram quantities of different mutagens i.e., it is about 100 times more sensitive




ISBN 0-12-208820-4

than the Ames Test. Moreover, different procarcinogenic chemicals that are known to require liver microsomal activation in the Ames Test, are highly active in the bioluminescence test even without activation. This self-activating system could be attributed to the presence of P450-cytochrome system in this type of luminous bacteria (3).

In Table 1 all the mutagenic agents that have been tested so far are listed. It can be seen that all the compounds are active even without microsomal activation.

The mode of action of the tested chemicals fall into four main categories: 1) base-substitution agents; 2) frame-shift agents; 3) DNA intercalating agents; 4) DNA synthesis inhibitors, known also as error-prone or SOS-inducing agents.

A detailed kinetic study, made with chemicals of each group revealed that these agents also differ in their action on the luminescence of 8SD18 cells (see Fig. 1).

The intercalating agents (e.g. acriflavin) act promptly on the luminescence of 8SD18 cells, both *in vivo* luminescence and *in vitro luciferase* activity begin to increase 10-20

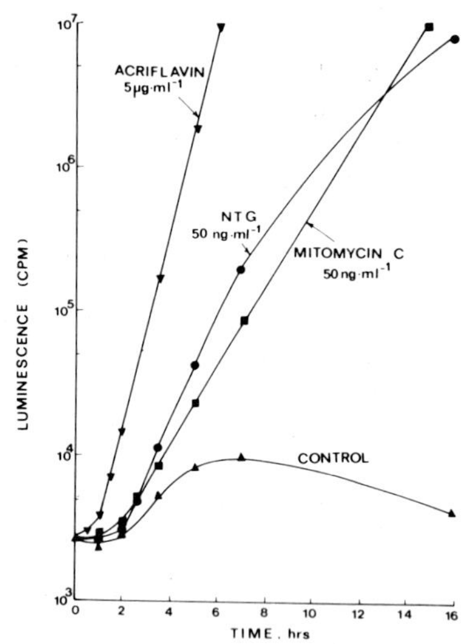


*FIGURE 1. The effect of acriflavin, NTG and mitomycin C on the reversion rate of 8SD18 cells. 1ml of 8SD18 suspension ($10^5$ cells. $ml^{-1}$) were incubated with different chemical agents at the shown concentrations in ASWRP medium (2). The luminescence was determined with time in a scintillation counter as already described (2).*

*TABLE I. The minimal concentrations of different mutagens shown to be detectable by the Bioluminescence Test (2) without microsomal activation.*

| | $\mu g.ml^{-1}$ |
|---|---|
| *(1) Base-substitution agents* | |
| *N-methyl-N-nitro-N-nitrosoguanidine (NTG)* | *0.002* |
| *Ethyl methane sulfonate (EMS)* | *0.005* |
| *Hydroxylamine (HA)* | *0.1* |
| *Hydrazine* | *0.007* |
| *(2) Frame-shift agents* | |
| *20-methyl cholanthrene* | *0.200* |
| *2-anthramine* | *0.600* |
| *Quinacrine HCl* | *0.600* |
| *Emodine* | *0.700* |
| *Nitrofluorene* | *0.7* |
| *2-amino biphenyl* | *25* |
| *9,10-dimethyl-1,2-benzanthracene* | *0.1* |
| *(3) Intercalating Agents* | |
| *Ethidium bromide* | *2.5* |
| *Acriflavin* | *0.1* |
| *9-amino acridine* | *0.2* |
| *Proflavine-$SO_4$* | *0.2* |
| *Caffeine* | *20* |
| *Theophylline* | *20* |
| *(4) DNA-synthesis inhibitors* | |
| *Mitomycin C* | *0.5* |
| *Novobiocin* | *0.2* |
| *Nalidixic acid* | *5* |
| *Coumermycin* | *2* |

minutes after the addition of these agents. However, in spite of the high level of the *in vivo* luminescence no genetic hereditary luminous revertants could be isolated from the treated culture.

The direct mutagens belonging to either the base-substitution group or to the frameshift agents resulted in the appearance of genetic hereditary luminous revertants. The effect of these agents (e.g., NTG) on the *in vivo* luminescence of 8SD18 cells was first observed about 5 hours after the introduction of the mutagens. In this case it is possible to isolate genetic hereditary luminous cells from the treated culture.

The chemicals belonging to the third group, that includes the DNA inhibitors such as mitomycin C, increase the *in vivo* luminescence of the treated cells about 4 hours after the introduction of the tested agent to 8SD18 culture. The appearance of genetic hereditary revertants however has not been demonstrated.

The primary genetic failure of 8SD18 cells has not yet been investigated. Generally, the dark variant of 8SD18 resembles some spontaneous dark mutants of luminous bacteria (K-mutants) described by Hastings and Nealson (4). The *in vivo* luminescence of different spontaneous dark variants of luminous bacteria belonging to the *P. phosphoreum* and *Beneckea harveyi* species have also been shown to be similarly affected by different mutagenic agents such as acridine dyes or NTG.

The phenotypic reversion to the luminescence state by different agents, including acridine dyes and other curing agents, gave us reason to believe that this phenomenon was caused by the presence of a plasmid carrying a repressor (2). This assumption should now be excluded after it has been found that there is no correlation between the appearance of plasmids in luminous bacteria and the bright or dark features of these cells. The hypothetic repressor has not been isolated either, but its involvement is still suggested to explain the high sensitivity and the wide generality of the test.

The effect of the different chemical agents on the reversion to luminescence may be due to their specific effect on the synthesis, stability or activity of this repressor. According to this assumption, the direct mutagens (the base-substitution of the frameshift agents) directly affect the synthesis of the repressor, resulting in appearance of genetic hereditary revertants. The increase in the *in vivo* luminescence observed with these agents appears only when the remainder of the repressor is diluted out among the offspring, a process that takes 4 to 5 hours (i.e., 8 to 10 generations).

The other mutagenic agents may directly affect the stability or activity of the hypothetic repressor. The intercalating agents (including acridine dyes) are known to affect the physical state of DNA, resulting in its relaxation state (5). This may enable transcription of the luciferase genes, either directly or through altering the ability of the DNA to bind the repressor. The transcription of the luciferase cistron under these circumstances will depend on the continuous presence of the intercalating agent. This may explain both the phenotypic reversion of the *in vivo* luminescence and the absence of genetic hereditary luminous revertants.

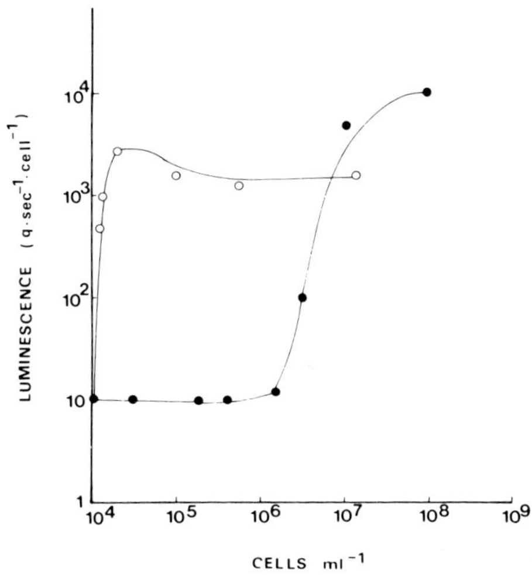


*FIGURE 2. Induction of the luminescence system in wild-type cells of P. leiognathi (BE-8) by proflavine.*

*P. leiognathi cells were inoculated into ASWRP liquid medium (2), to final cell density of 5-10 cells per ml. After a growth period of a few hours with shaking at 30°C the culture was divided into two and proflavine-$SO_4$ (5 $\mu g.ml^{-1}$), was added to one flask. The cultures were incubated with shaking and the in vivo luminescence, as well as cell counts were determined in samples withdrawn at different times. The figure shows the specific luminescence ($q.sec^{-1}.cell^{-1}$) in the control cells (●——●) and in the acridine-treated cells (○——○). It was later found that the in vivo luminescence in the acridine-treated cells was accompanied by a simultaneous increase in the in vitro luciferase activity (data not shown).*

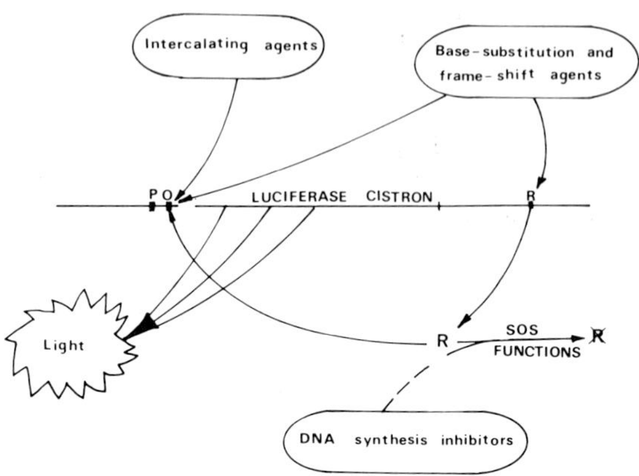


*SCHEME I. Proposed mode of action of mutagens in restoring of luminescence in dark variants of P. leiognathi 8SD18 cells.*

The DNA synthesis inhibitors (SOS-inducing agents) are known to be involved in the formation of a specific protease of cellular repressors (6,7) (e.g., the repressor of lysogenic bacteriophages). Destruction of the repressor is expected to allow the synthesis of the luciferase cistron. Here again, however, upon the removal of the active agents, the cells regain the ability to repress the luciferase cistron, thus preventing the appearance of genetic hereditary luminous revertants (see Scheme I). According to this model, the luminescence system of both wild-type and the dark variant cells is repressed during the preinduction period. Although both types produce an equally active autoinducer, only the wild-type cells undergo induction, while the luminescence system of the dark variant remains repressed all along the growth cycle. This difference between the two types may stem from the presence of an excess of the repressor molecules in the dark variant cells, or due to alteration in the transcription control mechanism. This assumption implies that the preinduced wild-type cells' behavior will be similar to that of the dark variant. Indeed we have shown that acridine dyes act as an inducer and increase promptly the *de-novo* synthesis of the luciferase and the *in vivo* luminescence in preinduced wild-type cells of *P. leiognathi* (Fig. 2).

Confirmation of this working hypothesis requires further analysis and experimental work. Great difficulties are inevitably encountered in such a study, due to the little available knowledge on the genetic nature of luminous bacteria in general and of the luciferase system in particular.

The new bioluminescence test is sensitive, general, fast, simple, and cheap. It is also capable of detecting many procarcinogenic agents, without microsomal activation, and we consider it therefore, an ideal short-term first screening test for testing suspected carcinogens. The suspected chemicals showing positive response should then be tested by a battery of *in vivo* and *in vitro* tests, in order to confirm their carcinogenicity potential. This approach makes it possible to test more chemicals, saves money and may afford better use of the limited resources for a comprehensive study of the suspected carcinogens.

## REFERENCES


1. Ames, B. N., J. McCann and E. Yamasaki, *Mutation Res.* 31, 347-364 (1975).
2. Ulitzur, S., I. Weiser and S. Yannai, *Mutation Res.* 74, 113-124 (1980).
3. Ismailov, A. D., H. C. Egorov, and B. C. Danilov, *SSSR* 249, 482-485 (1979).
4. Hastings, J. W. and K. H. Nealson, *Ann. Rev. Microbiol.* 31, 549-595 (1977).
5. Wang, J. C., DNA: Bihelical Structure, Cold Spring Harbour Symposium 43, 29-52 (1979).
6. Roberts, J. W., C. W. Roberts, and D. W. Mount, *Proc. Nat. Acad. Sci.*, 74, 2283-2286 (1977).
7. Oishi M., C. L. Smith and B. Friefeild, Cold Spring Harbour Symposium 43, 897-907 (1979).

# A THERMODYNAMIC EXPLANATION FOR THE KINETIC DIFFERENCES OBSERVED USING DIFFERENT CHAIN LENGTH ALDEHYDES IN THE *IN VITRO* BACTERIAL BIOLUMINESCENT REACTION[1]

James E. Becvar
Li-Huey Wu[2]

Department of Chemistry
The University of Texas at El Paso
El Paso, Texas

## I. INTRODUCTION

Bacterial luciferase, under the single turnover condition of the standard *in vitro* assay, catalyzes light emission which has different kinetic parameters when aldehydes of different alkyl chain length are used (1-3). For example, when *B. harveyi* luciferase is assayed using decanal, both a higher initial miximum intensity, $I_o$, and a larger first-order rate constant, k, for emission decay are found than when the same amount of enzyme is assayed with octanal or dodecanal. Moreover, the pattern of $I_o$ and k value differences observed with different aldehydes are different for luciferases from different species of luminous bacteria (1-4). As an interesting application, Nealson (5) has suggested the use of differences in luciferase decay kinetics with dodecanal as a rapid and easy means of taxonomic distinction between species of *Beneckea* and of *Photobacterium*.

Although a molecular explanation is presently lacking for the differences in $I_o$ and k values observed for different

[1]*This research was supported by grant AH-777 from the Robert A. Welch Foundation, Houston, Texas.*

[2]*Present address: University of Houston, Houston, Texas.*




ISBN 0-12-208820-4

aldehydes in the *in vitro* reaction catalyzed by any of the bacterial luciferases, a thermodynamic rationale can now be offered as a possible explanation for these differences at least with *Beneckea harveyi* luciferase. Arrhenius analyses of the emission decay kinetics for the aldehydes from 8 to 12 carbons yield energies of activation for the rate limiting step in the reaction which are consistent with the $I_o$ and k differences observed. Aldehydes which produce large $I_o$ and k values (nonanal and decanal) correspond to those with the lowest $E_a$ values. Aldehydes which produce smaller initial intensities and slower emission decay kinetics are the aldehydes showing large $E_a$ values.

## II. EXPERIMENTAL

Standard *in vitro* luciferase assays were performed using a photometer (6) calibrated with the light standard of Hastings and Weber (7). Emission reactions were initiated by injection of one ml of 5 x $10^{-5}$M $FMNH_2$ in 1 mM phosphate buffer, pH7.0, into one ml of assay buffer containing 0.2% bovine serum albumin, 0.02M phosphate, pH7.0, 10 µl luciferase sample (approximately 1 µg), and the appropriate volume (10 to 25 µl) of stock 0.1% aldehyde solution or suspension. Both the reduced flavin stock solution and the vial containing the assay buffer were maintained at the required temperature by a refrigerated, constant temperature circulator. Temperature control and measurement to better than a degree celsius was achieved during all manipulations of the assay and during the emission measurements. The *B. harveyi* luciferase used in all assays was purified by methods previously described (8,9) and had a specific activity of $1.4x10^{14}$ q $sec^{-1}$ $ml^{-1}A_{280}{}^{-1}cm^{-1}$ at 25° C with decanal. FMN (Sigma) was catalytically reduced by hydrogen gas in the presence of platinized asbestos. The aldehydes (Aldrich) used were purified by gas chromatography prior to the preparation of 0.1% (v/v) solutions in ethanol or sonicated suspensions in water.

## III. RESULTS AND DISCUSSION

The $I_o$, k, and Q values seen for *B. harveyi* luciferase catalyzed emission with a given aldehyde at a single temperature are dependent on the concentration of the aldehyde used in the assay (3). Figure 1 illustrates the dependence of

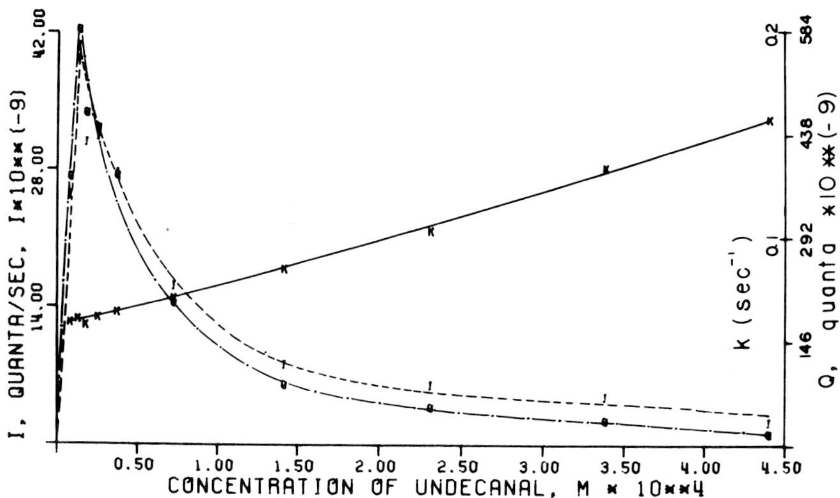


*Figure 1. Dependence of the bioluminescence emission parameters $I_o$, k, and Q on undecanal concentration.*

these parameters on undecanal concentration when this aldehyde is present in the assays. Results (not shown) for assays using other aldehydes from 8 to 12 carbons in length are similar but show a range in variation in the value of k over the same concentration range which is smaller than the nearly two fold range seen in Figure 1. The results of similar experiments examining the effect of tetradecanal concentration on these parameters is radically different (10). Over a similar concentration range, the k values for tetradecanal vary by more than an order of magnitude while Q values vary to a much smaller extent than seen in Figure 1. The reason for discussing the concentration dependence of k values here is to bring to the attention of the reader that in order to examine the temperature dependence of the k values for the different aldehydes, a reasonable choice of aldehyde concentration must be made at the outset. Examination of all results like those in Figure 1 shows that in the region of maximum Q value for a given aldehyde, a relatively small variation in k value is observed at a given temperature (This, of course, is not true for tetradecanal). For the present studies a reasonable, but arbitrary, decision was made to use in all assays the aldehyde concentration which produced the maximum Q value for that aldehyde at 20° C. This represented approximately 10 to 25 µl of stock 0.1% ethanolic solution or aqueous sonicated suspension per assay.

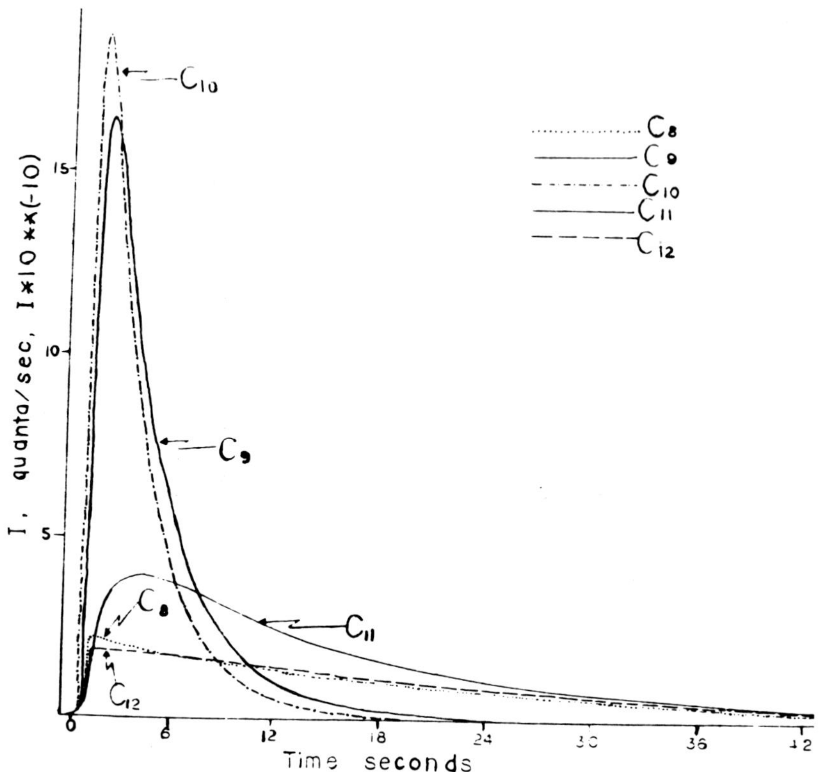


*Figure 2. Single Turnover emission kinetics at 23° for B. harveyi luciferase using different chain length aldehydes.*

Figure 2 shows the remarkably different emissions catalyzed by identical amounts of *B. harveyi* luciferase in standard assays at 23° C using aldehydes of different chain length. Each assay represents a single turnover of the enzyme molecules present and in all cases the decay of emission intensity decreases exponentially over considerably more than a single decade. The emissions for nonanal and decanal can be grossly characterized as more flash-like while those for octanal, undecanal, and dodecanal are more glow-like. The relative areas under these curves are not identical but are nearly so, indicating nearly the same quantum yield in each case. Although luciferase has three substrates, $FMNH_2$, oxygen, and aldehyde, the decay of emission is a first-order process characterized by a rate constant $k = .693/t_{1/2}$, where $t_{1/2}$ is the half-time of intensity decrease observed.

The temperature dependence of the value of k for the different aldehydes was examined. Several assays were conducted at each of five or more temperatures over the range from 10 to 30° C for each aldehyde and the Arrhenius analyses performed on the k values obtained. The results of these analyses are summarized in Table I. The thermodynamic activation energies shown in the table are in fair agreement with the values which Tu (11) obtained on similar studies with the even numbered aldehydes from 8 to 14 carbons. For example, his $E_a$ values for the 8, 10, 12, and 14 carbon aldehydes were 16.1, 13.8, 18.6 and 26.5 kcal/mol respectively.

*TABLE I. Thermodynamic Parameters of the In Vitro Standard Assay of B. harveyi Luciferase Using Different Aldehydes.*

| *Aldehyde* | $E_a$ (kcal/mol) | $\Delta H^{\ddagger}$ (kcal/mol) | $\Delta S^{\ddagger}$ (e.u.) | $\Delta G^{\ddagger}$ (kcal/mol) |
|---|---|---|---|---|
| *Octanal*[a] | 17 | 16 | -11 | 19 |
| *Nonanal*[a] | 10 | 10 | -28 | 18 |
| *Decanal*[a] | 13 | 13 | -17 | 18 |
| *Undecanal*[a] | 17 | 16 | -9 | 19 |
| *Dodecanal*[a] | 18 | 17 | -8 | 19 |
| *Tetradecanal*[b] | 24 | 23 | +14 | 19 |
| *Octanal*[c] | 19 | 18 | -4 | 19 |
| *Nonanal*[c] | 8 | 8 | -35 | 18 |
| *Decanal*[c] | 11 | 11 | -25 | 18 |
| *Undecanal*[c] | 21 | 20 | +5 | 19 |
| *Dodecanal*[c] | 18 | 18 | -6 | 19 |

[a] *Added as 0.1% (v/v) solution in ethanol*

[b] *Added as 0.1% (v/v) solution in ethyl acetate*

[c] *Added as 0.1% (v/v) sonicated aqueous suspension*

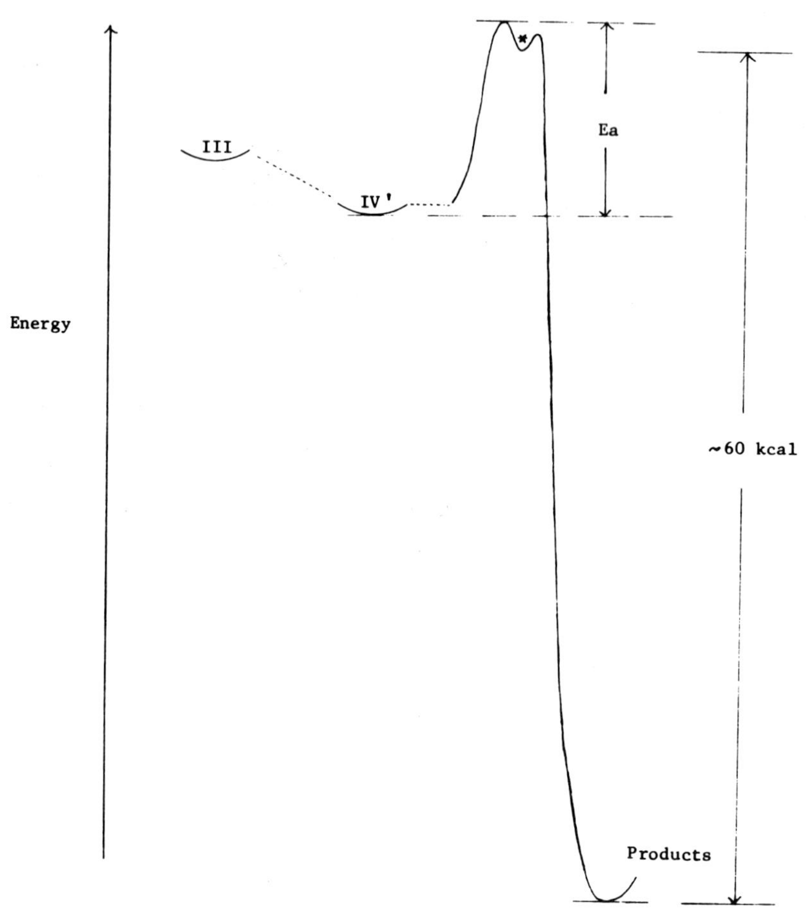


*Figure 3. Energy diagram for the B.harveyi emission reaction near the rate limiting step.*

With the exception of the data for tetradecanal, which can produce under certain conditions an $I_o$ value greater than that for decanal, a rather good correlation can be seen between the aldehyde reactions characterized by the lowest $E_a$ or $\Delta G^{\ddagger}$ values and those with highest initial intensities at 23°C. The simple thermodynamic diagram shown in Figure 3 may be able to explain the $I_o$ and k differences observed for reactions of aldehydes of different chain length. The first-order decay of emission indicates the existence of a rate limiting step or steps in the reaction after all substrates have bound to the enzyme but prior to emission. Since any enzyme molecule in the population during an assay can turn over only once, the instantaneous value of intensity in a reaction is proportional to the amount of enzyme in the population which has not traversed the rate limiting step shown here as IV' → (*). Until it gains sufficient thermal energy to pass over the activation energy barrier, an enzyme is temporarily "trapped" at the level of IV'. Aldehyde reactions showing low $E_a$ (or $\Delta G^{\ddagger}$) values have, at all times after initiation of the reaction, a greater proportion of the remaining enzyme molecules which can escape over the barrier than do aldehyde reactions with high $E_a$ values. Therefore these low $E_a$ value aldehyde reactions produce an initial intensity burst of light unmatched by their high $E_a$ counterparts. However, the population of enzyme molecules complexed with low $E_a$ value aldehydes more quickly diminishes as does the intensity. Thus the low $E_a$ value aldehyde reactions observed for nonanal and decanal are initially bright but are quickly gone, in contrast to the reactions for octanal, undecanal and dodecanal.

Several tests of the model suggest themselves. Luciferases from other bacteria show other patterns of $I_o$, k values for different aldehydes. Are the $E_a$ values consistent with $I_o$, k patterns? Other aldehydes and other conditions need to be tried with *B. harveyi* luciferase. Finally, is an explanation available for tetradecanal behavior?

## REFERENCES


1. Hastings, J. W., J.A. Spudich, G. Malnic, *J. Biol. Chem. 238,* 3100 (1963).
2. Hastings, J. W., Q. H. Gibson, J. Friedland, J. Spudich, *in* "Bioluminescence in Progress" (F. H. Johnson and Y. Haneda, eds.), p 151, Princeton University Press, Princeton, N. J.(1966).
3. Hastings, J. W., K. Weber, J. Friedland, A. Eberhard, G. W. Mitchell, and A. Gunsalus, *Biochem. 8,* 4681 (1969).
4. Ruby, E. G., *these Proceedings*.
5. Nealson, K. H., *in* "Methods in Enzymology" (M.A. DeLuca, ed.) p. 153. Academic Press, New York (1978).
6. Mitchell, G., and J. W. Hastings, *Anal. Biochem. 39,* 243 (1971).
7. Hastings, J. W., and G. Weber, *J. Opt. Soc. Am. 53,* 1410 (1963).
8. Gunsalus-Miguel, A., E.A. Meighen, M. Z. Nicoli, K. H. Nealson, and J. W. Hastings, *J. Biol. Chem. 247,* 398 (1972).

9. Baldwin, T.O., M.Z. Nicoli, J.E. Becvar, and J.W. Hastings, *J. Biol. Chem. 250,* 2763 (1975).
10. Shannon, P., R.P. Presswood, R. Spencer, J.E. Becvar, J.W. Hastings, and C. Walsh, *in* "Mechanisms of Oxidizing Enzymes" (T.P. Singer and R.N. Ondarza, eds.) p. 69 Elsevier/North Holland, New York (1978).
11. Tu, S-C., *Biochem. 18,* 5940 (1979).

# ACTIVE CENTER STUDIES ON BACTERIAL LUCIFERASE: MODIFICATION WITH METHYL METHANETHIOLSULFONATE

Miriam M. Ziegler
Thomas O. Baldwin

Department of Biochemistry
University of Illinois
Urbana, Illinois

## I. INTRODUCTION

The activity of bacterial luciferase from *Beneckea harveyi* has been shown to be very sensitive to thiol-specific reagents, loss of activity being correlated with the alkylation of a single particularly reactive thiol on the $\alpha$ subunit in the tryptic peptide Phe-Gly-Ile-Cys-Arg (1-3). The $FMNH_2$[2] binding affinity of the alkylated enzyme is reduced by >10-fold (1). Both binding of the aldehyde substrate and a cycle of enzymatic oxidation of $FMNH_2$ (as well as binding of the product FMN) can protect the enzyme against inactivation, suggesting that the reactive cysteine is in or near the active center; however, the $pK_a$ of the thiol is 9.4, too high to allow correlation with the pH profiles for $FMNH_2$ binding or catalytic parameters (1).

The reactive thiol is located in a hydrophobic environment (2,4) in a cleft about 17 Å in length (4). Consistent with the proposal that the cysteinyl residue is in or near the active center are the observations that a) the reactivity of

[1]*This work was supported by grants from the National Science Foundation (NSF PCM 79-25335) and the National Institute on Aging (AG-00884).*

[2]*Abbreviations used: FMN and $FMNH_2$, oxidized and reduced riboflavin 5'-phosphate; MMTS, methyl methanethiolsulfonate.*



ISBN 0-12-208820-4

the thiol is altered (5) in mutant enzymes reported to have active center lesions (6), and b) the thiol has recently been shown to reside within the protease-labile region of the α subunit (7).

The present study was undertaken to better evaluate the location of the reactive thiol relative to the active center. To this end, three experiments were performed: 1) the effect of high concentrations (10-100 x $K_d$) of FMN upon reactivity of the thiol was determined, 2) the binding of $FMNH_2$ to luciferase stoichiometrically modified with N-n-octylmaleimide was measured by circular dichroism spectroscopy, and 3) the modification of the enzyme with methyl methanethiolsulfonate, which forms an $-SCH_3$ mixed disulfide with the protein, was analyzed.

## II. MATERIALS AND METHODS

Luciferase was purified from an aldehyde-deficient mutant (M17) of *B. harveyi* and assayed by our published method (8). FMN and iodoacetamide were obtained from Sigma and N-n-octylmaleimide from Pfaltz & Bauer; methyl methanethiolsulfonate was a generous gift from G.L. Kenyon.

Luciferase solutions were dialyzed to remove reducing agent and inactivation (alkylation) reactions were carried out and monitored as previously described (1,2). Solutions for circular dichroism measurements were prepared in 0.1 M Bis-Tris, pH 7, and reduced with 5 mM $Na_2S_2O_4$ under a layer of mineral oil. Circular dichroism spectra were obtained on a Jasco J-40A spectropolarimeter at 25° in a 2 mm pathlength cuvette (9).

## III. RESULTS AND DISCUSSION

A saturating concentration of FMN completely protects luciferase from inactivation by iodoacetamide (Fig. 1). The partial protection reported earlier (1) was presumably due to the low concentration of FMN relative to the $K_d$ of the complex (10). This finding is particularly interesting when compared with our earlier observation (11) that modification of the reactive thiol with N-n-octylmaleimide had little effect on the binding of FMN.

The binding of $FMNH_2$ to luciferase stoichiometrically modified with N-n-octylmaleimide was monitored by circular dichroism spectroscopy essentially as described by Becvar and

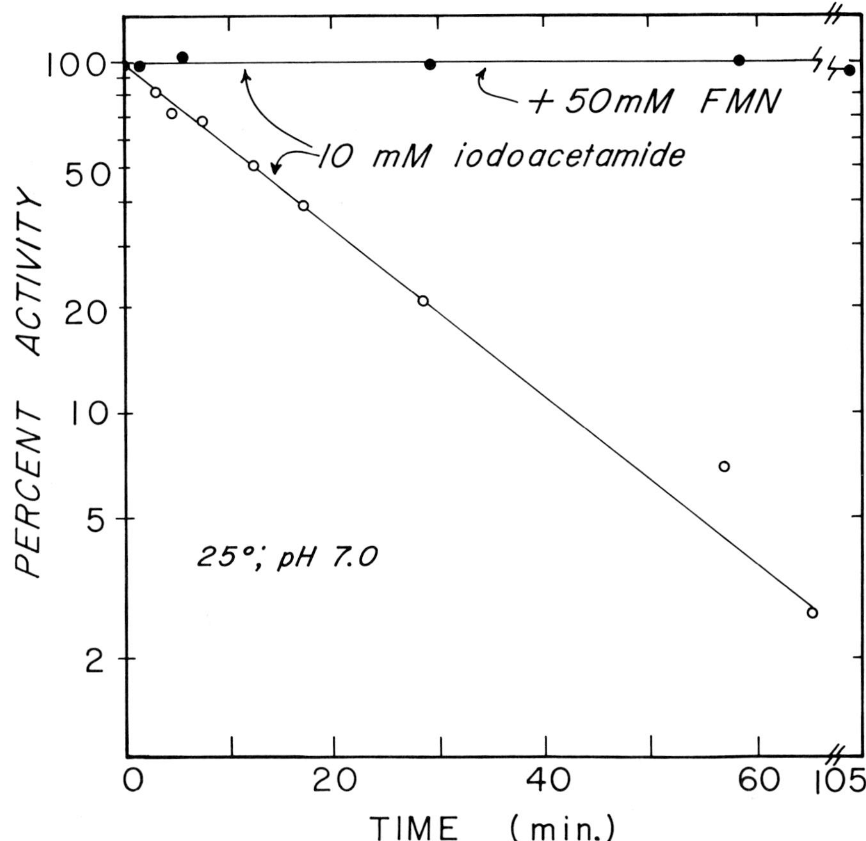


*FIGURE 1. FMN protection of luciferase from inactivation by iodoacetamide. Luciferase (0.1 mM) was incubated with 10 mM iodoacetamide in 20 mM phosphate, pH 7.0, in the presence (closed circles) and absence (open circles) of 50 mM FMN.*

Hastings (9) (Fig. 2). The results of these measurements demonstrate two points. First, $FMNH_2$ does bind to the modified luciferase. Second, the circular dichroism spectrum of $FMNH_2$ bound to the modified enzyme is quite different from the spectrum of $FMNH_2$ bound to the native enzyme. The results of a titration experiment monitored by circular dichroism (data not shown) suggest that $FMNH_2$ binding to the modified enzyme is much weaker than to the native enzyme, consistent with the earlier kinetic determination (1).

If the reactive thiol were in the active center in a site crucial for binding of $FMNH_2$ and/or aldehyde, then the

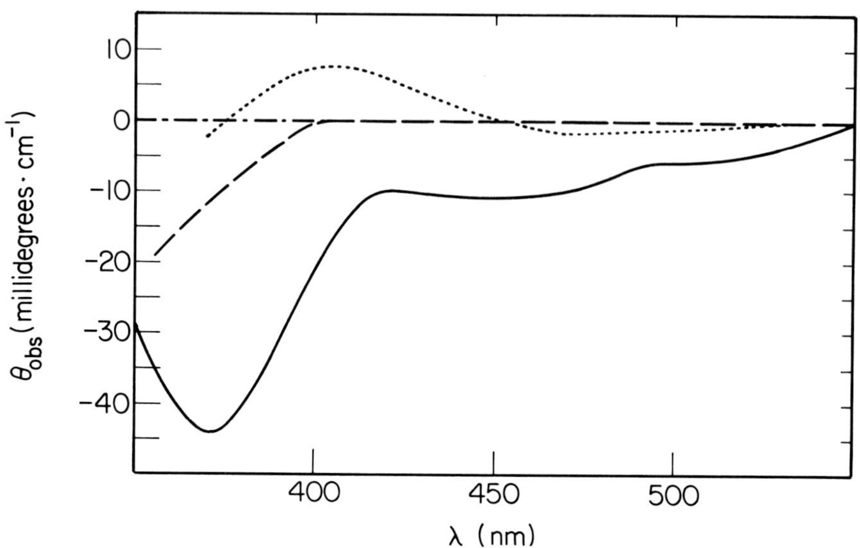


*FIGURE 2. Circular dichroism spectra of luciferase-bound $FMNH_2$. Native luciferase:$FMNH_2$ complex, ca. 0.3 mM (———); 0.55 mM N-n-octylmaleimide-modified luciferase (1.3 moles/mole; 5% activity remaining) + 0.51 mM $FMNH_2$ (·····); 1.0 mM $FMNH_2$ alone (- - - -); native or modified luciferase alone (-·-·-).*

observed inactivation might be due to steric hindrance rather than to modification of a residue directly involved in the mechanism of the enzyme. Such a situation has been reported for creatine kinase (12). To test this hypothesis, luciferase was allowed to react with various molar ratios of methyl methanethiolsulfonate (MMTS) (Fig. 3, open circles). This reagent results in formation of the $-SCH_3$ mixed disulfide with sulfhydryl groups (12,13). The results of this experiment show that MMTS completely inactivates luciferase at an extrapolated ratio of 1:1. To determine with more precision whether the *ca.* 20% activity remaining after addition of 1 molar equivalent of MMTS were due to activity of luciferase with the essential thiol modified or due to reaction of *ca.* 20% of the added reagent with other luciferase thiols, luciferase was first incubated with iodoacetamide in the presence of 50 mM FMN as shown in Figure 1. The reaction was stopped after 100 min by addition of 2-mercaptoethanol; modified enzyme retained essentially complete (>95%) activity. The FMN, 2-mercaptoethanol and its carboxamidomethyl adduct were

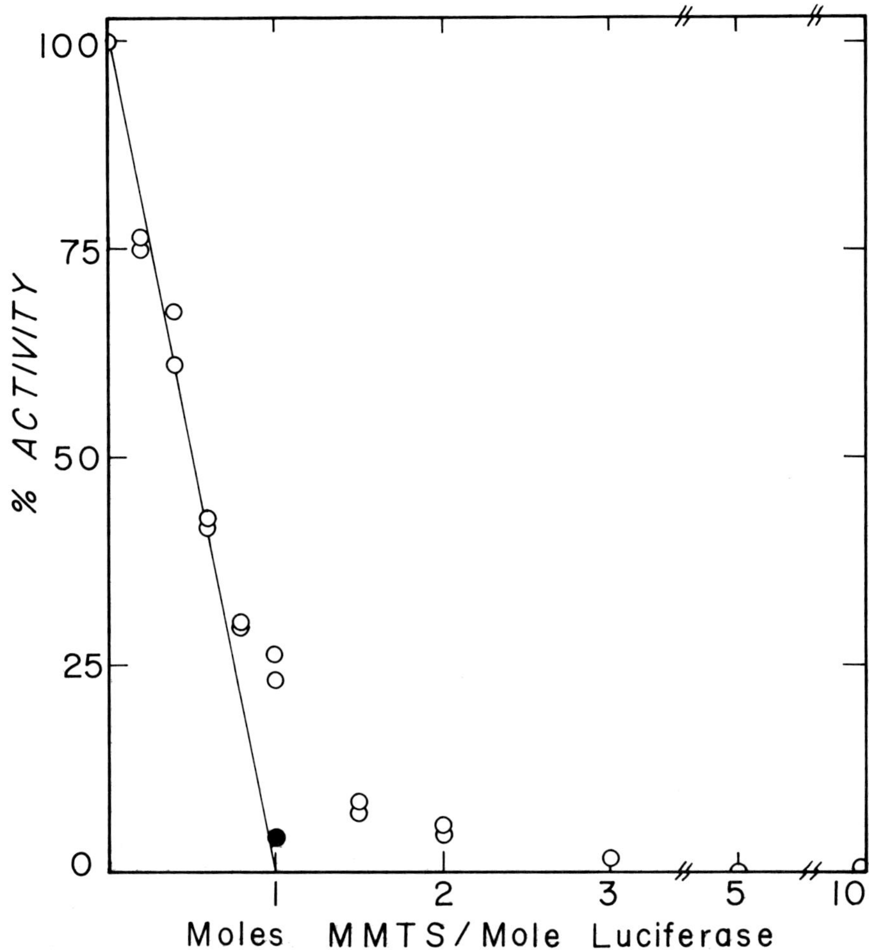


*FIGURE 3. Titration of luciferase with methyl methanethiolsulfonate. Luciferase (5.0 μM) was incubated at 25° with various amounts of MMTS in 50 mM Bis-Tris, pH 7.0. Activity was assayed after 12 min and 25 min (open circles); no change in activity occurred during this time interval. The closed circle shows activity remaining after addition of 1 mole MMTS/mole luciferase after prior alkylation of the enzyme with iodoacetamide in the presence of 50 mM FMN (Fig. 1).*

removed by dialysis. Subsequent addition of 1 mole MMTS per mole luciferase without FMN resulted in loss of 96% of the activity of the enzyme (Fig. 3, solid circle). These results show that other thiols on luciferase do react with iodoacetamide even in the presence of bound FMN. Modification of these thiols with iodoacetamide does not lead to inactivation, but it does result in increased specificity of reaction of MMTS with the essential thiol following removal of FMN. These experiments demonstrate that modification of the reactive thiol of *B. harveyi* luciferase with the small, nonpolar, uncharged $-SCH_3$ group leads to loss of measurable luciferase activity. While this observation certainly does not prove that the thiol is involved with the chemistry of the light-emitting reaction, it does show that the reactive thiol is in a critical region of the active center and that even a minor perturbation has a dramatic effect on the activity of the enzyme.

## REFERENCES


1. Nicoli, M.Z., E.A. Meighen, and J.W. Hastings, *J. Biol. Chem. 249,* 2385-2392 (1974).
2. Nicoli, M.Z., and J.W. Hastings, *J. Biol. Chem. 249,* 2393-2396 (1974).
3. Nicoli, M. Ziegler, Ph.D. Thesis, Harvard University, Cambridge, MA 02138 (1972).
4. Merritt, M.V., and T.O. Baldwin, *Arch. Biochem. Biophys. 202,* 499-506 (1980).
5. Baldwin, T.O., M.S. Currie, M.Z. Nicoli, and T.W. Cline, *Fed. Proc. 34,* 681 (1975).
6. Cline, T.W. and J.W. Hastings, *Biochemistry 11,* 3359-3370 (1972).
7. Baldwin, T.O., S.K. Rausch, J.J. Dougherty, and M.V. Merritt, These *Proceedings.*
8. Hastings, J.W., T.O. Baldwin, and M.Z. Nicoli, *Meth. Enzymol. 57,* 135-152 (1978).
9. Becvar, J.E. and J.W. Hastings, *Proc. Natl. Acad. Sci. USA 72,* 3374-3376 (1975).
10. Baldwin, T.O., M.Z. Nicoli, J.E. Becvar, and J.W. Hastings, *J. Biol. Chem. 250,* 2763-2768 (1975).
11. Nicoli, M.Z., T.O. Baldwin, J.E. Becvar, and J.W. Hastings, *in* "Flavins and Flavoproteins" (T.P. Singer, ed.), pp. 94-100. ASP, Amsterdam (1976).
12. Smith, D.J. and G.L. Kenyon, *J. Biol. Chem. 249,* 3317-3318 (1974).
13. Kenyon, G.L. and T.W. Bruice, *Meth. Enzymol. 47,* 407-430 (1977).

# ISOLATION AND REACTION PRODUCT CHARACTERIZATION OF AN OXYGENATED BACTERIAL LUCIFERASE INTERMEDIATE FORMED WITH A FLAVIN ANALOG[1]

Shiao-Chun Tu

Department of Biophysical Sciences
University of Houston
Houston, Texas

## I. INTRODUCTION

The bioluminescence reaction catalyzed by bacterial luciferase involves an intermediate formed by the reaction of $O_2$ with the enzyme-bound $FMNH_2$[2] (1). This intermediate, designated II, has an unusually long lifetime. It has been isolated at -20°C in 50% ethylene glycol-phosphate buffer, and characterized spectroscopically (2-4) and with respect to reaction products (5). Based on these studies, II was postulated to be a 4a-hydroperoxyFMN-luciferase species. II has also been isolated at 0°C in aqueous phosphate buffer and found to be fully active in reaction with an aldehyde substrate to emit bioluminescence in the absence of free $O_2$ (6). However, detailed chemical and spectral analyses could not be satisfactorily made because II still decayed appreciably at 0°C. Recently it has been found that non-aldehyde aliphatic compounds reversibly complex with and stabilize II (7,8), and II-alcohol complexes have been isolated at 0°C and characterized (8). In the present study we found that long-chain alcohol, e.g. dodecanol, also

[1]*This work was supported by Robert A. Welch Foundation Grant E-738 and National Institute of General Medical Science Grant GM 25953.*

[2]*Abbreviations used: $FMHH_2$, reduced FMN; II and cpII, oxygenated luciferase intermediate formed with $FMHN_2$ and reduced* ω-carboxypentylflavin, respectively; ω-carboxypentylflavin, 7,8-dimethyl-10(ω-carboxypentyl)isoalloxazine.




ISBN 0-12-208820-4

effectively stabilized oxygenated luciferase intermediate formed with flavin analogs. We were able to isolate an oxygenated luciferase-ω-carboxypentylflavin intermediate, designated cpII, as a dodecanol complex and to determine the reaction products of the isolated intermediate in the absence and presence of aldehyde.

## II. MATERIALS AND METHODS

Luciferase was purified from *Beneckea harveyi* cells (9), Luciferase concentrations were determined based on an absorption coefficient of 1.2 (0.1%, 1 cm) at 280 nm (10). Reagent solutions of decanal (Aldrich) and 1-dodecanol (Matheson, Coleman and Bell) were prepared in 95% ethanol. Horseradish peroxidase (type VI), FMN, and riboflavin were obtained from Sigma. All other flavins were generous gifts from Dr. Donald B. McCormick at Emory University.

Luciferase activity was measured in a calibrated photometer (11) by a non-turnover method involving the use of the oxygenated luciferase-flavin intermediate or its dodecanol complex preformed at a designated temperature. To a 0.5-ml phosphate buffer containing 0.25 mg luciferase, 50 μM oxidized flavin, zero or 20 μl of 10 mM dodecanol a few milligrams of sodium dithionite were added to reduce the flavin. An equal volume of air-saturated buffer was immediately introduced to form the intermediate. Aliquots (50μl each) were withdrawn at different times and each injected aerobically into 1 ml 0.05 M phosphate, pH 7, containing 0.26 mM decanal for measurements of bioluminescence activity at 23°C. Stabilities of the intermediate and its dodecanol complex can be determined by following the rates of decrease in the remaining bioluminescence capacity.

The reaction of horseradish peroxidase with $H_2O_2$ is known to associate with spectral changes (12). The $H_2O_2$ concentration of an unknown sample can be determined based on the decrease in $A_{400}$ of a peroxidase solution upon mixing with the sample; corrections should be made with respect to the dilution factor and the background $A_{400}$ of the sample itself.

## III. RESULTS

After the solution of a preformed oxygenated luciferase intermediate or its dodecanol complex was allowed to stand at a designated temperature, the bioluminescence capacity decreased exponentially as a function of time. The half-life of the intermediate species formed with flavin derivatives, with the exception of cpII, were comparable to or even longer than that of II at 0°C (Table I). Interestingly, dodecanol markedly stabilized all the oxygenated luciferase flavin intermediates examined (Table I), thus rendering the intermediate isolation achievable. cpII was prepared at 0°C in 0.4 ml 0.1 M phosphate, pH 7, containing 3 mg luciferase, 0.1 mM ω-carboxypentylflavin, and 20 μl of 10 mM dodecanol by the dithionite reduction method described above. The sample was immeditely applied to a Sephadex G-25 column (1 x 15 cm) preequilibrated and eluted, at 0-2°C under dimmed light, with the same buffer containing saturating level of dodecanol. cpII-dodecanol was obtained in the void volume, and the isolated species exhibited an absorpiton spectrum peaking at 375 nm (Figure I). Upon standing, the absorption spectrum of cpII-dodecanol gradually changed to that of oxidized ω-carboxypentylflavin, with two isosbestic points at 378 and 400 nm.

Reaction products of cpII-dodecanol and II-dodecanol (isolated similarly) were determined, in the absence and presence of decanal, with respect to light output, flavin content, and $H_2O_2$ formation (Table II). Very little light was detected in the decay of II- or cpII-dodecanol in the absence of aldehyde. However, both emitted greatly enhanced bioluminescence upon reaction with decanal; the quantum yields of 0.14 and 0.06, based on the flavin contents, for

*TABLE I. Effect of Dodecanol on the Half-life of Intermediate II Species Formed at 0°C with Flavin Derivatives*[a]

| *Flavin* | *$t_{1/2}$ (min)* Dodecanol None | *$t_{1/2}$ (min)* Dodecanol Added |
|---|---|---|
| *FMN* | *11* | *320* |
| *ω-Carboxypentylflavin* | *2* | *75* |
| *3-Carboxymethyl FMN* | *14* | *138* |
| *2-ThioFMN* | *14* | *312* |
| *Riboflavin* | *17* | *40* |
| *2',3'-Diacetyl FMN* | *25* | *138* |

[a]*Determined in 0.05 M phosphate, pH 7.*

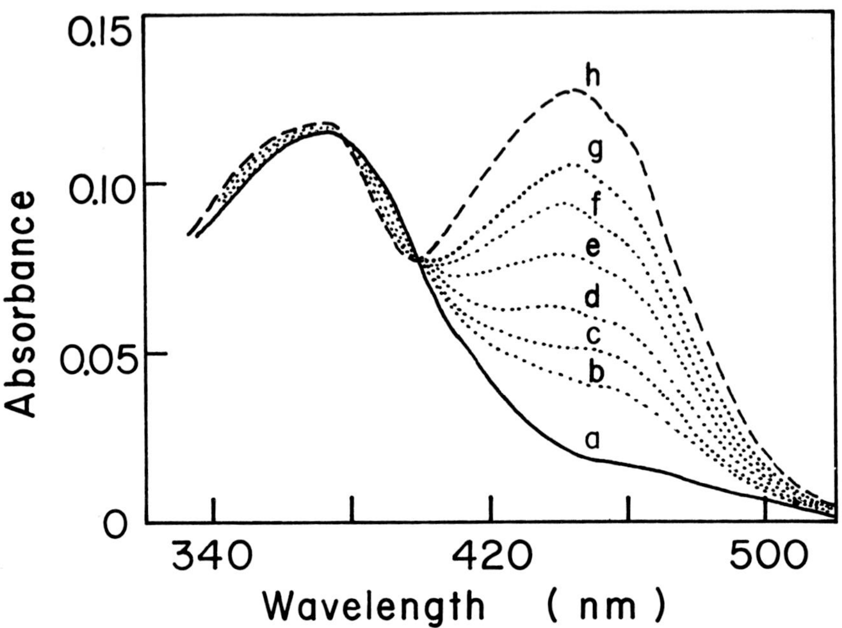


*FIGURE 1. Absorption spectra of cpII-dodecanol and the flavin decay product. The absorption spectrum of cpII-dodecanol (a) was measured at 0°C immediately after its isolation. The sample was quickly warmed up to 10°C and spectra were taken after 10, 20, 30, 50, 75, and 100 min (spectra b through g, respectively). The sample was then warmed up to 25°C for about 15 min and the spectrum of the final flavin decay product (h) was determined. The background absorption due to luciferase alone was subtracted.*

II- and cpII-dodecanol, respectively, are essentially the same as those obtained in the absence of dodecanol (13,14), indicating that the alcohol binding is reversible. In the dark reaction (no aldehyde added), both II- and cpII-dodecanol yielded equal molar quantities of $H_2O_2$ and oxidized flavin. With decanal added (the light reaction), II-dodecanol yielded very little $H_2O_2$. On the other hand, cpII-dodecanol produced about 40% of light output as that of

*TABLE II. Reaction Products of II- and cpII-dodecanol in the Absence and Presence of Decanal*

| Sample[a] | Flavin (n mole) | $H_2O_2$ (n mole) | Quantum yield[b] | $H_2O_2$/Flavin |
|---|---|---|---|---|
| *II-dodecanol* | | | | |
| No aldehyde | 6.4 | 6.2 | 0.008 | 0.97 |
| Decanal added | 7.2 | 1.0 | 0.14 | 0.14 |
| *cpII-dodecanol* | | | | |
| No aldehyde | 5.5 | 5.7 | 0.001 | 1.04 |
| | | 6.0 | | |
| | | 6.1 | | |
| Decanal added | 5.6 | 1.4 | 0.06 | 0.25 |
| | 5.5 | 0.8 | | 0.15 |
| | 5.5 | 1.2 | | 0.22 |
| | 5.6 | | 0.06 | |

*[a]Samples were isolated and kept at 0°C. Aliquots (0.5 ml each) were withdrawn and added to 0.7 ml 0.05 M phosphate, pH 7, containing zero or 0.36 mM decanal for measurements of light emission at 23°C. Subsequently each sample was assayed for flavin content based on $A_{450}$ and $H_2O_2$ by reacting with horseradish peroxidase.*

*[b]Based on the flavin content.*

II-dodecanol in the presence of decanal but the $H_2O_2$ formed was comparable to or only slightly above the amounts obtained from II-dodecanol.

## IV. DISCUSSION

In the present study, we have found that dodecanol markedly stabilizes the oxygenated luciferase intermediate formed with various flavins. Since 4a-hydroperoxyflavins are highly labile in aqueous solution (15), the observed stabilizing effect of dodecanol is likely attributable to an increased hydrophobicity at the luciferase active site resulted from the binding of this long-chain alcohol.

The isolated cpII-dodecanol exhibits an absorption spectrum closely resembling that of II (2) and II-tetradecanol (8). Upon standing, the absorption gradually changes to that of oxidized flavin; the well-defined isosbestic points observed indicate that only two principal forms of the flavin chromophore are present during the time of spectral changes.

The yields of oxidized flavin and $H_2O_2$ from II-dodecanol via both the light and the dark reactions (Table II) are in good accord with those determined for II (5). In the dark reaction, cpII-dodecanol also formed equal molar quantities of oxidized flavin and $H_2O_2$. In the presence of decanal, cpII-dodecanol formed the same amount of oxidized flavin but the $H_2O_2$ produced is only $\leq$ 25% of that of flavin (Table II). It is known that the formation of the excited emitter is obligatorily coupled to the oxidation of aldehyde whereas the $H_2O_2$ is produced only in the dark reaction (5,16). If the low quantum yield of cpII-dodecanol (40% of that of II-dodecanol) in the light reaction is due to a poor yield of the excited state, correspondingly more $H_2O_2$ (60%, or even more if there is an artificial background $H_2O_2$ formation, of that of total flavin) should be generated. Apparently, this is not the case. The present findings suggest that the reduced quantum yield of cpII-dodecanol is, at least in part, due to a lower emitting efficiency of the excited emitter. Many reduced flavin analogs have been tested with bacterial luciferase and all were found to produce low light yields (13, 17, 18). The present findings with cpII indicates that the reduced light yields observed with flavin analogs may not necessarily reflect their true substrate activities in the luciferase-catalyzed monooxygenation of aldehyde.

## V. SUMMARY

Dodecanol stabilizes the oxygenated intermediate formed with several flavin derivatives. The intermediate obtained with reduced 7,8-dimethyl-10-(ω-carboxypentyl)isoalloxazine has been isolated at 0°C in neutral phosphate buffer as a dodecanol complex, and its absorption spectrum was determined. Without aldehyde addition, this isolated intermediate decays to yield equal molar quantities of oxidized flavin and $H_2O_2$ with very little light production. With decanal added, bioluminescence occurs with a quantum yield (0.06) about 40% of that obtained with the authentic flavin substrate; the $H_2O_2$ formed is about $\leq$ 25%, on a molar basis, of the amount of oxidized flavin recovered. The reduced light output of this oxygenated flavin derivative-luciferase intermediate is suggested to be, at least in part, due to a low emitting efficiency of the excited state formed in its reaction with an aldehyde substrate.

REFERENCES


1. Hastings, J.W., and Q. H. Gibson, *J. Biol. Chem.* 238, 2537 (1963).
2. Hastings, J.W., C. Balny, C. LePeuch, and P. Douzou, *Proc. Natl. Acad. Sci. U.S.A.* 70, 3468 (1973).
3. Balny, C., and J. W. Hastings, *Biochemistry* 14, 4719 (1975).
4. Ghisla, S., J. W. Has t ings, V. Favaudon, and J. -M. Lhoste, *Proc. Natl. Acad. Sci. U.S.A.* 75, 5860 (1978).
5. Hastings, J. W., and C. Balny, *J. Biol. Chem.* 250, 7288 (1975).
6. Becvar, J. E., S. -C. Tu, and J. W. Hastings, *Biochemistru* 17, 1807 (1978).
7. Baumstark, A. L., T. W. Cline, and J. W. Hastings, *Arch. Biochem. Biophus.* 193, 449 (1979).
8. Tu, S. -C., *Biochemistry* 18, 5940 (1979).
9. Gunsalus-Miguel, A., E. A. Meighen, M. Z. Nicoli, K. H. Nealson, and J. W. Hastings, *J. Biol. Chem.* 247, 398 (1972).
10. Tu, S. -C., Baldwin, T. O., J. E. Becvar, and J. W. Hastings, *Arch. Biochem. Biophys.* 179, 342 (1977).
11. Hastings, J. W., and G. Weber, *J. Opt. Soc. Am.* 53, 1410 (1963).
12. Noble, R. W., and Q. H. Gibson, *J. Biol. Chem.* 245, 2409 (1970).
13. Meighen, E. A., and R. E. MacKenzie, *Biochemistry* 12, 1482 (1973).
14. Becvar, J. E., and J. W. Hastings, *Proc. Natl. Acad. Sci. U.S.A.* 72,3374 (1975).
15. Kemal, C., and T. C. Bruice, *Proc. Natl. Acad. Sci. U.S.A.* 73, 995 (1976).
16. Shannon, P., Presswood, R. B., R. Spencer, J. E. Becvar, J. W. Hastings, and C. Walsh, *in* "Mechanisms of Oxidizing Enzymes" (T. P. Singer, and R. N. Ondarza, eds.), p. 69. Elsevier, Amsterdam (1978).
17. Mitchell, G., and J. W. Hastings, *J. Biol. Chem.* 244, 2572 (1969).
18. Tu, S. -C., J. W. Hastings, and D. B. McCormick, *Fed. Proc.* 36, 722 (1977).

DISCUSSION

Dr. Ziegler

Dr. Lee, since the lumazine protein from *Photobacterium* is apparently quite clean now, have you attempted to make antibody against it, to give you a handle on quantitating it in crude extracts during growth?

Dr. Lee

No.

Dr. Ziegler

It would be interesting if in fact it were coinduced with the other luminescent system components.

Dr. Lee

Yes, I think the studies that Dr. Meighen has done on the induction of luminescence accessory proteins is a very good line of investigation. But, we are not pursuing it.

Dr. Tu

I would like to ask Dr. Meighen, there are reports in the literature indicating that horse liver alcohol dehydrogenase is a multifunctional enzyme; it reversibly converts alcohol to aldehyde and also converts aldehyde to acid. I noticed in your talk that you have been looking at the activities designated as the aldehyde reductase and the aldehyde dehydrogenase. Have you looked into the possibility that these two activities might be associated with the same protein species?

Dr. Meighen

There are two aldehyde reductases in P. phosphoreum. They are two well resolved peaks, and we have just recently picked up the aldehyde dehydrogenase activity in P. phosphoreum as well, and it separates from both the aldehyde reductases as well as the fatty acid reductase; so they are separate peaks on column chromatography at the present time. But, I do agree that the horse liver alcohol dehydrogenase is reported to have the activity capable of taking aldehyde into acid as well.

Mr. Leisman

I have a question for Dr. Ulitzur. You have reported that tetradecanal is the aldehyde *in vivo* based on assays *in vivo* of dark mutants using tetradecanoic acid and other studies. My data *in vitro* with what may be the complete system of the yellow emitting strain, that is luciferase,




ISBN 0-12-208820-4

Mr. Leisman

oxidoreductase and a fluorescent protein, shows that the color of the bioluminescence is shifted with the chair length of the aldehyde. I am proposing that the identify of the aldehyde *in vivo* can be tested by isolating aldehyde mutants, and then measuring the emission spectra after adding different chain length aldehydes back to these. Dr. Nealson and I have isolated mutants of the yellow emitting strain and tested these with various aldehydes. Additional of dodecanal stimulates an emission even more yellow than the emission of the wild type, while addition of tetradecanal stimulates an emission closest to that of the wild type. My question is, have you looked at the emission spectra of any of the mutants that you have worked with, after adding tetradecanoic acid or any aldehydes?

Dr. Ulitzar

No.

Dr. McElroy

What is your explanation for the change in color with aldehyde?

Dr. Nealson

For the moment I have to stick with the interpretation that was presented in the talk that the most likely reason is changes due to the interactions of these two different proteins. The aldehydes, probably the association, effect changing the energy transfer from one to the other. When the proteins are loosely associated we will get more blue emission, and when tightly associated we get more yellow emission: and until we can disprove this it seems the most likely hypothesis.

Dr. Hastings

With regard to the color of emissions as related to the chain length of the aldehyde, such an effect was also reported in a thesis from my laboratory (K. McClosky, B. A. Thesis Chemistry Department, University of Illinois, Urbana, 1960) using aldehyde dark mutants that had been isolated by Dr. McElroy's group.

Dr. Ward

I would like to direct a question to any of the speakers that discussed the accessory proteins. Could you give us any other information on the stoichiometry between the luciferase and accessory proteins, and is anything known at all about the binding constants? In particular what I am interested in

Dr. Ward

knowing is how soon will we be able to isolate some complexes between these fellows for crystallization purposes?

Dr. Lee

The only information on the stoichiometry with the lumazine protein extracts is what you can conclude from staining on gels, and within the factor of the staining abilities, I think the quantity of lumazine protein is probably less than luciferase but not much less. Protein-protein association, that's a very difficult subject. There is protein-protein association. The kinetic interaction constant is in the micromolar range, but, the direct experiments are still being pursued.

Dr. Ward

Can the complexes be isolated?

Dr. Lee

We have not isolated any complexes.

Dr. Baldwin

I would like to add something to that, if I could. Tom Holzman, a graduate student in my lab, has some rather convincing data that if the cells are lysed gently and he stays away from ammonium sulfate, he can show that the luciferase activity will not pass through a 3000,000 dalton cut-off amicon filter. This suggests that there is, in fact, a complex with which the luciferase is associated. I think that all of us who have worked in the bacterial bioluminescence have felt that this must be the case for quite some time. Dr. Nealson has some very interesting data about the quantum yield of the luminescent system *in vivo* (Karl and Nealson (1980) J. Gen. Microbiol. 117, 357-368). I don't think that those of us who believe in natural selection and evolution can believe that our protein inside a cell is going to have a quantum yield of 0.0. When we are talking about quantum yields with the purified luciferase system, we are talking about the wrong thing, because when those proteins start interacting there is probably going to be a change in the quantum yield, as was shown by the data of Dr. Nealson and Mr. Leisman with their Y-1 protein. Dr. Lee's data also suggest this point very stongly.

Dr. Ornstein

With respect to the *in vitro* results which show that the FMN fluorescent spectra and the bioluminescence emission spectra are different, could you explain why the simple

Dr. Ornstein

minded hypothesis that the enzyme-FMN-peroxide reaction transfers its energy to another enzyme FMNOH which, in fact, is the fluorescent species isn't the simplest explanation? If the species that reacts with the aldehyde transfers the energy to another enzyme complex #2 and that emits, wouldn't that take care of the data on the difference in the spectra?

Dr. Hastings

Would it require the existence of a second enzyme complex for light emission?

Dr. Ornstein

No. I am saying you have a sea of molecules of E-FMNOOH; the first one that reacts with the aldehyde passes its energy back to one that hasn't reacted yet which is the one that fluoresces but does not decompose, and then that one will later react with aldehyde.

Dr. Hastings

If I understand you correctly, an experiment pertaining to your proposal was done by Dr. Becvar, showing that only a single flavin molecule is required for light emission. He found that the quantum yield, based on the limiting reactant, either $FMNH_2$ or luciferase, was independent of the ratio of the two reactants over a range of $10^6$ or more (PNAS 72: 3374, 1975). Stated otherwise, at increasing dilutions the probability of the energy transfer that you speak of would become increasingly low and the yield would go down. It did not.

Dr. Ornstein

In those dilution experiments, have careful spectral measurements been made to show that a great dilutions the spectra are not effected?

Dr. Hastings

I don't believe so, at least not over the full range studied by Dr. Becvar. But the reaction of the purified system is well characterized and the spectrum is apparently not dependent on dilution.

Dr. Ornstein

My second question - in your diagrams you show that the reaction of the enzyme-FMN-peroxide complex can go in one direction when reacting with aldehyde to produce the chemiluminescent reaction, but, it also can decompose to FMN and peroxide and enzyme; but you show that, at least, being partially reversible - are there ways of driving the system

Dr. Ornstein
from the peroxide and so that you have enzyme, FMN and excess of peroxide and aldehyde to drive the reaction efficiently?

Dr. Hastings
I don't think efficiently, at least, in anything that we have done on that system. The Japanese group has gotten much higher efficiencies with photobacterium phosphoreum. Dr. Watanabi is here, I believe, and maybe he can add to it as to what the efficiency was or maybe Dr. Bevcar remembers. Their efficiency was higher, wasn't it?

Dr. Bevcar
Yes.

Dr. Seliger
I would like to ask a general question of the panel in relation to this very exciting observation by Dr. Ulitzur. Is it the case with the aldehyde dark mutants that the induction of the luciferase follows the same time course as with the wild type, and that presumably these accessory enzymes that Dr. Meighen has found are related to the production of the aldehyde which is required for the luminescence? Now, the time course that Dr. Ulitzur indicated for the amazing increase in bioluminescence observed with the dark mutant - is that consistent with time course for induction that is observed for either the wild type or the dark mutants for *de novo* synthesis of luciferase?

Dr. Nealson
It probably is, Dr. Seliger, because we are dealing with cells that are already at a density and in a condition where they could be induced, but they are super repressed mutants if I understand Dr. Ulitzur's mutants correctly. So that you are simply depressing them by adding these agents and a rapid derepression can occur on the order of minutes. So it is a different class of mutants than the aldehyde mutants that exhibit normal induction. What seems to be missing in Dr. Ulitzur's data right now is the evidence that luciferase itself or any single set of proteins is induced; there are only light measurements, so it isn't known if there is only one thing missing or if the whole operon is missing at moment. But, it certainly is time-wise consistent with the induction phenomenon.

Dr. Seliger
The question is might you expect complete *de novo* synthesis of an enzyme within seven minutes.

Dr. Hastings

The time course of the light intensity increase that one sees in normal cells growing in liquid culture reflects a combination of growth, production of inducer, and response to inducer. If you take non-induced cells and add autoinducer, then induction occurs promptly.

Dr. Nealson

Certainly we can see induction within ten minutes after addition of the purified autoinducer.

Dr. McElroy

One thing that is interesting -- the earlier aldehyde mutants that we obtained -- the luciferase was actually higher by a factor of 2 compared to the wild type.

Dr. Ulitzur

The DNA interelating agents induct the *de novo* systhesis of luciferase. This appears about 6 to 7 minutes (under optimal conditions) after the addition of these agents to the dark variant culture. However, in these mutant cells there is quite a lot of unexpressed luciferase. It is about 100th of what you really find *in vivo* conditions. Therefore, upon addition of acridine dyes or other DNA interculating agents to the dark varient we observe first fast increase in the *in vivo* luminescence that exceeds the increase in the *in vitro* activity. This is due to the fact that at the beginning we have high synthesis of aldehyde that allows the expression of the cellular luciferase; only then at the end of this period the *in vivo* and *in vitro* activity are exactly the same.

Dr. Nealson

I would just add that, if you don't do some enzyme assays, you can't prove that. It would be nice to know the levels of the various enzymes during the rapid increase in luminescence.

Dr. Ulitzur

The luciferase synthesis under optimal conditions - it takes 6.2 minutes to get the maximum, however, in the dark mutant there is quite a lot of unexpressed luciferase. It is about 100th of what you really find in *in vivo* conditions. Therefore, upon addition of acridine dyes or other DNA interculating agents to the dark varient, we have first higher increasing amounts of *in vivo* luminescence than in the *in vitro* activity. This is due to the fact that at the beginning we have high systhesis of aldehyde that express the unexpressed luciferase; only then at the end of this period

Dr. Ulitzur

the *in vivo* and *in vitro* are exactly the same.

Dr. Nealson

I would just add that, if you don't do some enzyme assays, you can't prove that. It would be nice to know the level of these various enzymes during that rapid increase. Do you have luciferase assays showing that there is unexpressed luciferase in the cells prior to the run?

Dr. Ulitzur

Yes.

Dr. Hastings

Yes. We have made similar observations with several of the different types of aldehyde dark mutants isolated in my laboratory. In part, the higher level of luciferase in dark aldehyde mutants may be due to the existence of a suicide pathway in the light reaction which leaves the wild type cell with an accumulation of inactive inhibited luciferase. It is not strictly product inhibition, but perhaps "false" product inhibition, the inhibitor probably being a non-covalently bound 4a substituted flavin. The material has been separated from luciferase and shown to have an interesting property: upon flash irradiation it emits characteristic bioluminescence dependent upon added aldehyde and oxygen but not flavin. We have called this "light inducible luciferase". Dark aldehyde mutants lack it, but if such mutants are grown with exogenously added aldehyde it is formed.

# IV
# FIREFLY BIOLUMINESCENCE

FRONT ROW- left to right: J.R. Blinks, T. Goto, M. Kamen, M. Ringrose, and K. Van Dyke

SECOND ROW- left to right: W. D. McElroy and J. Wampler

# THE CHEMISTRY AND APPLICATIONS OF FIREFLY LUMINESCENCE

W. D. McElroy and Marlene DeLuca

Department of Biology and Department of Chemistry
University of California, San Diego

## I. INTRODUCTION

Since the crystallization of firefly luciferase (1) and luciferin (2) and the synthesis of the latter (3), great progress has been made in our understanding of the organic and enzymatic mechanisms leading to light emission. In the present abstract we will review briefly the information on the firefly system with an emphasis on the kinetics of the reaction and how this influences the use of firefly luciferase in the detection of ATP in bioassays. For details the reader is referred to the following reviews (4) and (5).

## II. REACTIONS CATALYZED BY FIREFLY LUCIFERASE

Firefly luciferase catalyzes the following three reactions:

$$E + LH_2 + Mg\ ATP \longrightarrow E\ LH_2\text{-}AMP + PP \quad (1)$$

$$E\ LH_2\text{- } AMP + O_2 \longrightarrow oxyluciferin + CO_2 + AMP + light \quad (2)$$

$$E + L + Mg\ ATP \rightleftarrows E\ L\text{-}AMP + PP \quad (3)$$

In the first reaction, the carboxyl group of luciferin ($LH_2$) is the key site for the formation of an anhydride with adenylic acid from ATP accompanied by the formation of inorganic pyrophosphate (PP). This activation step is analogous to the fatty-acid and amino acid activation reactions.

In the second reaction, oxygen combines with the enzyme bound luciferyl adenylate which leads to a decarboylation of




ISBN 0-12-208820-4

the luciferin and the formation of an excited produce (oxyluciferin) that subsequently decays to the ground state with light emission and the liberation of free adenylic acid. The data indicate that the enzyme must extract a proton from the 4 carbon atom before oxygen is added. The addition of oxygen leads to the formation of a four-membered peroxide ring after the loss of AMP. The later remains bound to the enzyme and influences the nature of the excited state.

If synthetic $LH_2$-AMP is added to luciferase, light emission occurs without requiring ATP. For each $LH_2$ used, one quantum of light is emitted, one mole of oxygen is used, and one mole of $CO_2$ is released.

The peak emission for bioluminescence of *Photinus pyralis* is 562 mμ. However, other fireflies have different peak emission even with the same luciferin, indicating that the differences must be due to a change in the enzyme structure. Since the relative hydrophobicity of the solvent is known to effect the fluorescence properties of $LH_2$ and the product, a change at the $LH_2$ binding site could explain the color differences. However, the stereochemistry of the nucleotide attachment is also of importance in determining the color of the light. If 3-iso ATP is used instead of ATP, a significant amount of the light emitted is red even at pH 7.5. Although Σ-ATP is not active in reaction 1, synthetic $LH_2$-Σ-AMP does function in reaction 2 and the light is red. The results from the Σ-AMP and iso-ATP demonstrate that the binding of the adenylate to the enzyme induces changes in the enzyme structure that must be sustained during the subsequent decarboxylation that leads to the enzyme-product excited state. Studies by White *et al.* (6) indicate that in chemiluminescence in organic solutions the monanion of the decarboxylated, 4 keto derivative of $LH_2$ is the red emitter. In the presence of excess base the red chemiluminescence shifts to a yellow-green emission. These results suggest that proton abstraction at carbon 5 is essential and that the dianion is the light emitter in the yellow-green chemiluminescence.

## III. KINETICS OF THE LIGHT EMISSION

When ATP is rapidly mixed with a luciferin-luciferase mixture at 25°C there is lag of 25 msec before any light is emitted. This is followed by a slow rise of light intensity that requires approximately 0.3 seconds to reach the peak light emission. The results indicate that at least two slow overall and independent steps occur prior to light emission. The slow steps must occur after the synthesis of $LH_2$-AMP since the rates with the latter substrate are no different from those starting with ATP and $LH_2$. When one prepares E-$LH_2$-AMP under anaerobic conditions and then injects oxygen, the 25 msec lag is eliminated and the time to reach max rate is changed from 0.3 sec to 60 msec (7).

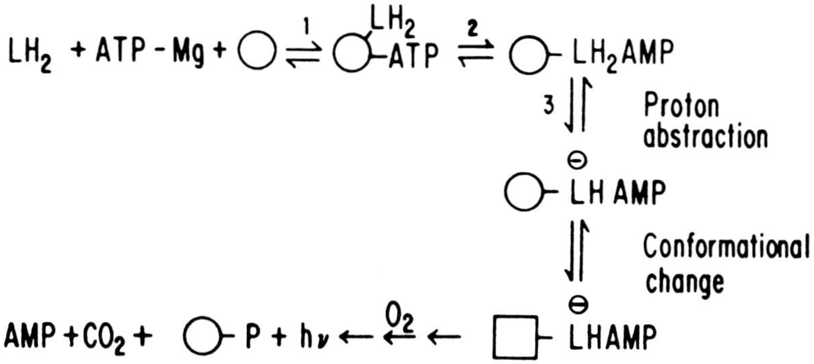

$LH_2$ + ATP-Mg + ○ ⇌(1) ○-$LH_2$-ATP ⇌(2) ○-$LH_2$AMP

⇅ 3 Proton abstraction

○-$LH^{\ominus}$AMP

⇅ Conformational change

AMP + $CO_2$ + ○-P + hν ← $O_2$ ← □-$LH^{\ominus}$AMP

*FIGURE 1. Postulated sequence of events occurring during luciferase-catalyzed reactions. In reaction 1, the binding of the substrates and in reaction 2 synthesis of $LH_2$-AMP occur rapidly relative to later reactions. The enzyme-bound $LH_2$-AMP after proton abstraction (step 3) undergoes a slow conformational change so that it is in a form to which oxygen adds rapidly. The experiment in which the enzyme was incubated with the substrates in the absence of oxygen would proceed through step 3 and the conformational change.*

The scheme presented in Fig. 1 suggests the most likely explanation. It depends on the nature of the $LH_2$-AMP binding to the luciferase. From previous studies it is known that large conformational changes occur in the enzyme when all substrates are added (8). The lag phase could be due to such changes in order to bring a specific proton acceptor into the proper steric position with respect to the substrate. Other data indicates that there must be large conformational changes even after proton abstraction and this could explain the slow rise time. There are many examples of hysteretic enzymes that exhibit a slow response to its substrates; however, luciferase appears to be unique in that there are two distinct slow steps involving conformational changes.

When ATP concentrations are varied from 0.2 pmol to 2μ mol there are large changes in the kinetics of the overall light

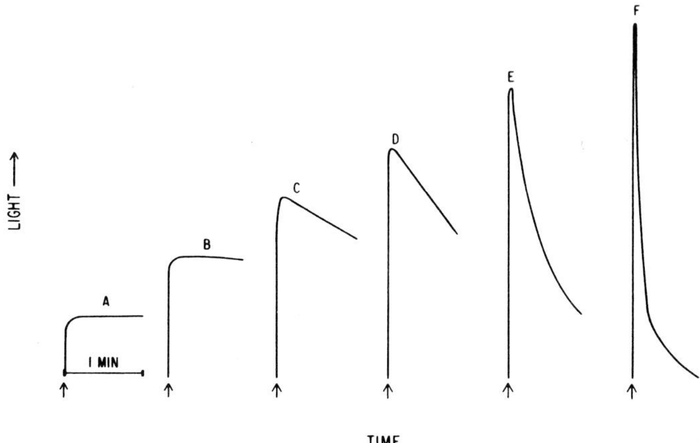


*FIGURE 2. Time course of light emission with increasing concentrations of ATP. Final concentration of ATP was A = 0.2 pmol; B = 2 pmol; C = 20 pmol; D - 200 pmol; E = 2 nmol; F = 2 μmol. The reaction contained 0.2 μg of luciferase. Reaction conditions are given under Materials and Methods. The increase in light intensity indicated on the ordinate is not quantitatively related to the ATP concentration but reflects only a qualitative presentation.*

emission (9). The results are shown in Fig. 2 for both crude and crystalline luciferase.

At low ATP concentrations, light emission rises to a maximum and remains constant for a minute or longer. At higher ATP concentrations, the light intensity increases (greater product production) but the decay rate also increases due to product inhibition. High concentrations of sodium arsenate, sodium sulfate, magnesium sulfate as well as other salts inhibit the peak light emission and thus prevents the decay (Fig. 3). However, by using these high salts the sensitivity of the ATP assay using luciferase is greatly reduced. Dilution of the ATP unknown could achieve a similar result and at the same time dilute possible inhibitors in the preparation.

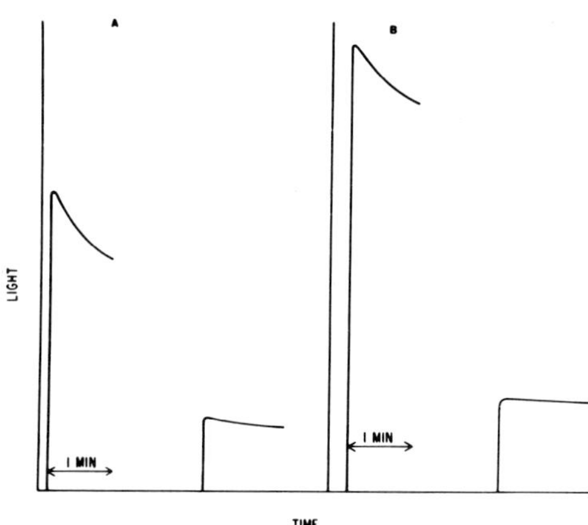


*FIGURE 3. Effect of sodium arsenate on the kinetics of light emission from purified luciferase (A) and crude luciferase (B). The reaction was initiated by the injection of 0.1 nmol of ATP. (A) Forty micrograms of crystalline luciferase in standard reaction mix (first flash) and in the presence of 0.05 M sodium arsenate (second flash). (B) 20λ of Sigma crude enzyme in standard reaction mix (first flash) and in the presence of 0.05 M sodium arsenate (second flash).*

## IV. USE OF FIREFLY LUCIFERASE IN BIOASSAYS

Firefly luciferase has been used to measure ATP in a variety of systems. These include coupled reactions in which ATP is generated (photosynthesis and oxidative phosphorylation) or when ATP is used (ATPase). Creatine kinase isoenzymes can be readily determined using this system as well as bacterial content in fluids. These and other assays using the firefly luciferase have been extensively reviewed in two recent publications (10, 4).

As noted previously, the assay is specific for ATP, however, precursors to ATP will produce light provided the appropriate enzymes for ATP formation are present. Crude enzyme extracts, for example, contain a number of enzymes that catalyze the formation of ATP from ADP. Under these conditions flash height measurements are essential in order to prevent errors from these transphosphorylation reactions. Purified crystalline firefly luciferase is readily available from a number of commercial firms at a reasonable price. Research workers should use this material rather than crude extracts.

The light emitting reaction is sensitive to high salt concentrations, thus it is essential to dilute the unknown in some cases to lower the final salt concentration (to approximately 0.1 mM or less). Heavy metals change the color of the light from yellow-green to red. Thus, if a phototube with low red sensitivity is used an apparent inhibition can be observed due to the shift in the color of the light. Reagents that interact with sulfhydryl groups will inhibit the luciferase reaction. Most of these inhibitory effects can be detected and corrected by using internal controls. The pH of the reaction should be run at 7.5 - 7.8.

## V. IMMOBILIZED LUMINESCENCE IMMUNOASSAY

Firefly luciferase has been attached to glass beads and covalently linked to Sepharose (11, 12). The immobilized enzymes retains a high percentage of its enzymatic activity but more importantly it is extremely stable and can be used repeatedly in bioassays. Recently this immobilized technique has been used to develop a luminescent immunoassay for various drugs. The initial experiments were concerned with the determination of methotrexate (Wannlund *et al.*). Methotrexate was covalently linked to firefly luciferase which resulted in a product that contained two methotrexates per mol of luciferase. Approximately 60% of the catalytic activity of luciferase was retained. The luciferase-methotrexate was readily bound by antibody raised against methotrexate-hemocyanin. Using antibody immobilized on Sepharose mixed with the luciferase-methotrexate, it was possible to demonstrate that the enzyme conjugate bound was inversely proportional to the concentration of free methotrexate (Fig. 4). Using this procedure, it is possible to detect 2.5 pmoles of methotrexate. With a double antibody procedure, it is possible to detect 0.5 picomoles.

Using this luminescent antibody procedure it is possible to detect a number of drugs and other compounds of biological interest at a sensitivity level equivalent to radioimmunoassays.

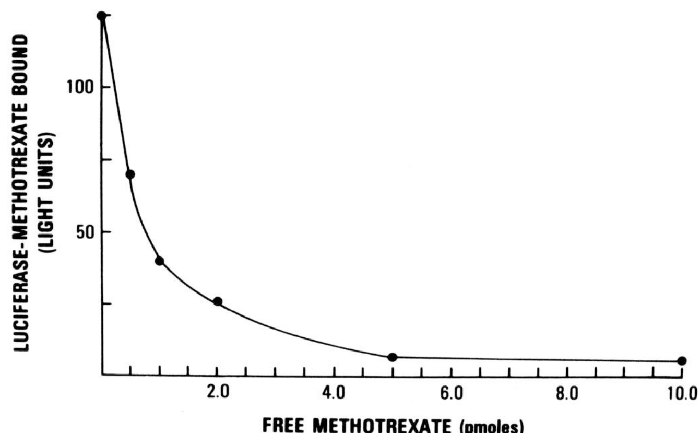


*FIGURE 4. Competitive binding curve of free methotrexate with luciferase-methotrexate to Sepharose-antibody as measured by bioluminescence.*

REFERENCES

1. Green, A. A., and W. D. McElroy, *Biochem. Biophys. Act.* 20, 1970 (1956).
2. Bitler, B., and W. D. McElroy, *Arch. Biochem. Biophys.* 72, 358 (1957).
3. White, E. H., R. McCapra, and G. F. Field, *J. Am. Chem. Soc.* 85, 337 (1963).
4. DeLuca, M. and W. D. McElroy, *in* "Methods in Enzymology LVII," M. DeLuca (ed.). Academic Press, New York, p. 3 (1978).
5. DeLuca, M., *Adv. Enzymol.* 44, 37 (1976).
6. White, E. H., E. Rapaport, H. H. Seliger, and T. A. Hopkins, *Bioorganic Chem.* 1, 92 (1971).
7. Deluca, M., and W. D. McElroy, *Biochemistry* 13, 921 (1974).

8. DeLuca, M., and M. Marsh, *Arch. Biochem. Biophys.* 121, 233, (1967).
9. DeLuca, M., J. Wannlund, and W. D. McElroy, *Anal. Biochem.* 95, (1979).
10. International Symposium on Analytical Applications of Bioluminescence and Chemiluminescence. (1979), Schram, E., and P. E. Stanley, (eds.). State Printing and Publishing Company, Westlake Village, California.
11. Lee, Y., E. Jablonski, and M. DeLuca, *Anal. Biochem.* 80, 496 (1977).
12. Unpublished observations.

# APPLICATIONS OF FIREFLY LUMINESCENCE[1]

Arne Lundin

Bioluminescence Centre
LKB-Produkter AB
Bromma, Sweden

## I. INTRODUCTION

The firefly reaction has been used for analytical purposes since 1947 when McElroy discovered that the reaction required ATP (1). It was soon realized that the firefly reaction could not only be used for sensitive assays of ATP but for assays of all metabolites and enzymes participating in ATP converting reactions (2). A large number of applications based on the firefly assay have been published in biochemistry, biology, immunology and medicine. Although used extensively in some laboratories for special applications the firefly assay has not gained a general acceptance in routine laboratories in e.g., clinical chemistry or clinical microbiology.

A serious obstacle in a more widespread use of the firefly assay has been the lack of highly purified and standardized reagents. A particularly troublesome property of previously available reagents has been that the light was emitted as a brief flash after addition of ATP. However, with purified reagents a stable light emission proportional to the ATP concentration can be obtained under appropriate reaction conditions (3). This does not only facilitate the assay of ATP in biological extracts but also makes it possible to monitor the ATP concentration in ATP converting systems simply by continuously measuring the light emission. Several

[1] *This work was partly supported by the Swedish Natural Science Research Council*




ISBN 0-12-208820-4

applications of this ATP monitoring technique have been published in assays of enzymes and metabolites and in monitoring of electron transport linked phosphorylation and lysis of cells.

The applications of firefly luminescence have recently been reviewed in detail by Lundin (4). The present paper is limited to an overview of the most important types of applications emphasizing the analytical improvements represented by the ATP monitoring concept.

## II. ANALYTICAL CONSIDERATIONS

Instrumentation for analytical luminescence has recently been reviewed by Stanley (5). However, a few points of special importance in connection with the firefly assay and ATP monitoring will be mentioned. The firefly assay as any other enzymatic assay should be performed at a constant temperature for maximum accuracy. Several luminometers having temperature control of the measuring cuvette are available on the market. Some of them are also equipped with built-in light standards for instrument control and/or stirring facilities to allow low volume additions of reagents during measurements. In high sensitivity work the most important parameter to compare between instruments is the signal to noise ratio at the wavelength obtained in the firefly reaction. The photomultipliers used in the luminometers are less sensitive to the red light obtained at pH<7 or in the presence of certain cations than to the yellow-green light obtained at pH 7.75, i.e., the optimum of the firefly reaction. Thus a correct comparison between different instruments can only be made measuring the signal using the same firefly reagent and the same ATP concentration under the appropriate reaction conditions. Besides sensitivity it is also an advantage if the dynamic range of the instrument covers the entire linear range of the firefly assay, i.e., from the detection limit to a few micromolar ATP concentration.

A detailed comparison of firefly reagents has recently been published (6). Luciferase activity is highly variable between reagents from different manufacturers and cannot be estimated from data given by the manufacturers since no internationally accepted light standard or internationally recommended assay is available. The light emission is dependent on the concentrations of luciferase, luciferin and ATP. If the reagent contains saturating amounts of luciferin the analytical results will be less influenced by minor pipetting errors. The sensitivity of the assays is normally determined by the reagent backgrounds, i.e., levels of contaminating ATP or other metabolites or enzymes. A firefly reagent suitable for ATP monitoring is now commercially available (LKB-Wallac, Turku, Finland). The luciferase in this reagent is prepared by ammonium sulphate precipitation and isoelectric focusing and appears as a single band both in analytical isoelectric focusing and SDS gel electrophoresis (7). The purity is similar to the crystalline enzyme prepared according to Green and McElroy (8).

In choosing reaction conditions one should be particularly careful when high sensitivity is required or when ATP monitoring conditions should be obtained. The pH optimum of the firefly reaction is pH 7.75, although the entire interval pH 6-8 can be used if a decreased sensitivity is accepted (3). Luciferase is inhibited by high ionic strength and by a specific anion inhibition increasing in the series acetate <chloride<bromide<nitrate<iodide<perchlorate (4). Tris acetate, glycylglycine acetate or imidazole acetate are often suitable buffers. In ATP monitoring the rate of the luciferase reaction should be constant throughout the measuring time. Thus there should be no activation or inactivation of luciferase or luciferin during the assays and the termperature should be constant. The optimum temperature of the reaction is generally 25 centigrades (4). The linear range of the assay of ATP is determined by the Km value, which is somewhat influenced by reaction conditions (4). Generally there is some deviation from linearity above a few micromoles per liter.

Sample pretreatment to avoid analytical interference from the sample can, due to the sensitivity of the method, often be done simply by diluting the sample. In e.g., estimation of bacterial biomass by the firefly assay it is necessary to use strongly denaturing extraction agents to obtain a good yield of adenine nucleotides and complete and rapid inactivation of all ATP converting systems in the bacterial cells. Such agents cause inhibition of luciferase if extracts cannot be diluted. With ATP monitoring reagent this problem can be overcome by measuring the light emission before and after addition of a known amount of ATP (internal standard technique). A comparison of 10 different methods used for extraction of bacteria has been published (9). Under appropriate storage conditions stability of ATP in biological extracts is not a problem provided that ATP converting reactions are inhibited by EDTA included in the extracting agent (4,9). For example dilute extracts of whole blood prepared with triton X-100 and EDTA can be stored at room temperature for 2 days without measurable decrease of the ATP level (4).

## III. APPLICATIONS

The various applications of the firefly assay are overviewed in Tables I and II. Specific examples and references can be found in a recent review (4) or obtained from the author.

### *A. Determination of Adenine Nucleotides*

The assay of ATP can be performed with any firefly reagent, although crude reagents require certain precautions in the evaluation of their light signal. Thus peak light intensity should be used rather than integration of the light signal or the intensity after a specified time (10). With ATP monitoring reagent a stable light level is obtained. This means that reagent and sample can be mixed without special equipment and can be measured or integrated for any period of time. If inhibition of the luciferase reaction from the samples is expected the stable light level makes it possible to measure the light intensity before and after addition of a known amount of ATP and thus to obtain an individual ATP standard value for each sample.

With ATP monitoring reagent end-point determination of all three adenine nucleotides can be made in the same reaction mixture. After the assay of ATP the pyruvate kinase reaction is allowed to convert ADP to ATP and then the combined pyruvate and adenylate kinase reactions to convert AMP to ATP (3).

The use of ATP as an index of biomass is based on the sensitivity of the firefly assay and the constant level of ATP in all organisms. The firefly assay has been used for estimation of biomass in oceanography for a long time (11).

Using selectively lyzing agents it is possible to differentiate between different types of cells. Screening for bacteriuria can e.g., be done after degradation of non-bacterial ATP in the urine by pretreatment with triton x-100 and apyrase (12).

In monitoring of fermentation and activated sludge processes (sewage treatment plants) the ATP level can be used not only as an index of biomass but also of biological activity, since the ATP level is affected by growth conditions, metabolic inhibitors, etc. In measurements on vitamin and antibiotic effects on bacterial cells the ATP level is used as a growth index. Alternatively specific effects on intra- or extracellular ATP levels can be utilized (13). As compared to conventional microbiological techniques the

TABLE I. *Overview of Analytical Applications of the Firefly Assay*

---

*Microbiology (general and applied)*

*Environmental studies (determination of biomass in marine/fresh water, marine/lake sediments, soil, etc.).*

*Quality control (determination of biomass in drinking water and production control in dairies, breweries and other food industries).*

*Fermentation monitoring (process control and growth optimization).*

*Sewage treatment (biomass/biological activity in activated sludge and waste water).*

*Biochemistry*

*Energy metabolism studies (determination of nucleotides, intermediates in glycolytic pathway, etc.)*

*Oxidative phosphorylation (monitoring of formation and degradation of ATP in mitochondria and submitochondrial particles).*

*Photophosphorylation (monitoring of formation and degradation of ATP in chloroplasts and bacterial chromatophores).*

*Membrane studies (monitoring of lytic reactions).*

*Determination of enzymes and metabolites (see Table II).*

*Immunology*

*Determination of antigens and antibodies (labeling with enzymes and metabolites participating in firefly or other ATP converting reaction).*

*Complement fixation (monitoring of membrane damage).*

*Medicine*

*Clinical bacteriology (bacteriuria screening, determination of vitamins and antibiotics, antibiotic susceptibility testing).*

*Hematology (erythrocyte disorders and viability, platelet aggregation).*

*Clinical chemistry (determination of enzymes and metabolites in serum, cerebrospinal fluid, urine and biopsy specimens).*

---

bioluminescent methods for determination of vitamins or antibiotics and for antibiotic susceptibility testing are more rapid and more suited for automation.

In hematology the firefly assay can be used to study changes in the erythrocyte ATP level appearing in certain pathological situations. The use of ATP for monitoring of the viability of erythrocytes in blood banks may turn out to be an important application. The firefly assay has been used for measuring ATP and ADP in connection with platelet aggregation.

### *B. Monitoring of Electron Transport Linked Phosphorylation*

Photophosphorylation and oxidative phosphorylation as well as ATPase activity can be conveniently studied by ATP monitoring. Using this technique it has e.g., been possible to determine the amount of ATP formed after a single 5 usec flash causing only one turn-over of the cyclic electron flow in chromatophores from the photosynthetic bacterium *Rhodospirillum rubrum*. At high ATP levels product inhibition of luciferase makes ATP monitoring with a stable light level impossible (3). However, Lemasters and Hackenbrock (14) have shown that this problem can be mathematically compensated for by multiple additions of internal ATP standards. This technique was used by the authors for studying oxidative phosphorylation.

### *C. Monitoring of Lytic Events*

Monitoring of lytic phenomena should be an important future application of ATP monitoring, since several important biological reactions involve the lysis of cells. The leakage not only of ATP but, under appropriate reaction conditions, of all metabolites and enzymes participating in ATP converting reactions can be followed. Leakage of ATP as a result of antibiotic action has been used for determination of gentamicin concentration (13). Complement fixation, platelet aggregation and action of bacterial cytolysins (15) are other phenomena that can be followed by ATP monitoring.

### *D. Luminescent Immunoassays*

In immunoassays ATP, luciferin and luciferase or other metabolites and enzymes participating in ATP converting reactions are of potential interests as markers of antibodies or antigens. Carrico et al. (16) have shown that specific protein binding reactions can be monitored with an ATP labeled ligand and firefly luciferase. The $P^1$-{2-[N-(2,4-dinitrophenyl) amino]ethyl}- $P^4$ -(5′ - adenosine) tetraphosphate conjugate was reactive (after degradation to liberate ATP) in the firefly reaction. After reaction with antibody specific to the dinitrophenyl residue the conjugate lost its reactivity; which could be restored competitively by addition of N-di-nitrophenyl-β- alanine. Further work will tell how the firefly system is best utilized in immunoassays.

### *E. Immobilized Systems*

Firefly luciferase can be immobilized on glass beads or Sepharose, although the yield has been fairly low (17). The economical advantages of the immobilized assay of ATP is doubtful, since luciferin, which is the most expensive component in the firefly assay, cannot be reused. However, in more complex systems with several enzyme systems or for specialized purposes immobilized systems may turn out to be highly interesting.

### *F. Determination of Enzymes and Metabolites*

Table II summarizes the enzymes and metabolites that have been determined using the firefly assay. In principle all enzymes and metabolites participating in ATP converting reactions or enzymatically linked to such reactions can be determined by the firefly assay. Whenever reaction conditions are compatible with the firefly reaction, the most convenient and accurate way to perform such determinations is ATP monitoring.

Metabolites can be assayed either by end-point determination or kinetically. The end-point determination of ATP, ADP and AMP in a single reaction mixture has already been mentioned and by adding phosphodiesterase also cAMP can be assayed in this mixture (3). Similarly a large number of enzymes can be kinetically determined using ATP monitoring. The most thoroughly studied examples are serum creatine kinase and creatine kinase B-subunit activity (the latter being assayed after immunoinhibition of M-subunit activity). Optimized versions of these assays are commercially available as kits for diagnosis of acute myocardial infarction (LKB-Wallac) and are likely to be the first bioluminescent assays to be accepted as routine methods in the clinical chemistry laboratories. In the future the bioluminescent methods

TABLE II. Determination of metabolites and enzymes by firefly assay[a]

| Metabolite | Enzyme |
|---|---|
| ATP | Luciferase (direct reaction) |
| | Apyrase and ATPase (degradation) |
| ADP | Pyruvate kinase |
| AMP | Adenlyate Kinase |
| cAMP | cAMP phosphodiesterase |
| Adenosine phosphosulphate | ATP sulphurylase |
| Adenosine tetraphosphate, | Various enzymes present in |
| GTP, GDP, GMP, CTP, UTP | crude luciferase reagents. |
| cGMP | cGMP phosphodiesterase and guanylate kinase. |
| Coenzyme A | Luciferase (release of product inhibition). |
| Creatine phosphate | Creative kinase |
| 1, 3-Diphosphoglycerate | Phosphoglycerokinase |
| Glycerol, triglycerides | Glycerol kinase |
| Glucose | Hexokinase |
| Phosphoenolpyruvate | Pyruvate kinase |
| Pyrophosphate | ATP sulphurylase |
| Phosphoribosyl ATP | ATP phosphoribosyl transferase |

[a]Assay procedures have been described for all metabolites given in left column and, with a few exceptions, also for all corresponding enzymes given in right column.

should be particularly suitable when only small specimens can be obtained e.g. serum from newborns or biopsy samples.

## IV. CONCLUSIONS

A large number of applications of the firefly assay have been published. The introduction of the ATP monitoring concept increases the applicability of the firefly system both by making new types of assays possible and by making those already suggested more convenient and reliable. Much work is presently devoted to the development of various assays based on ATP monitoring such as determination of various metabolites and enzymes, monitoring of electron transport linked phosphorylation and monitoring of lytic reactions. Other lines of development also involving the firefly system are concerned with immobilized enzyme systems and luminescent immunoassays. However, the most important contribution in the next few years will have to be standardization of reagents, instrumentation and methods. As soon as international recommendations on how to perform assays and express results obtained with the firefly system have been agreed upon the analytical potential of the ATP monitoring technique may make it an analytical tool as important as the spectrophotometric 340 nm system based on NAD(P)H converting reactions.

## REFERENCES

1. McElroy, W. D., *Proc. Nat. Acad. Sci. USA 33,* 342 (1947).
2. Strehler, B. L., and J. R. Totter, *Arch. Biochem. Biophys. 40,* 28 (1952).
3. Lundin, A., A. Rickardsson, and A. Thore, *Anal. Biochem. 75,* 611 (1976).
4. Lundin, A., *in* "Clinical and Biochemical Applications of Luminescence" (L. J. Kricka and T. J. N. Carter, eds.), Marcel Dekker, Inc. (in press).
5. Stanley, P., *in* "Clinical and Biochemical Applications of Luminescence" (L. J. Cricka and T. J. N. Carter, eds.) Marcel Dekker, Inc. (in press).
6. Webster, J. J., J. C. Chang, J. L. Howard, and F. R. Leach, *J. Appl. Biochem.* (in press).
7. Lundin, A., A. Myhrman, and G. Linfors, in this book.
8. Green, A. A., and W. D. McElroy, *Biochem. Biophys. Acta 20,* 170 (1956).
9. Lundin, A. and A. Thore, *Appl. Microbiol. 30,* 713 (1975).

10. Lundin, A. and A. Thore, *Anal. Biochem. 66,* 47 (1975).
11. Holm-Hansen, O. and C. R. Booth, *Limnol. Oceanogr. 11,* 510 (1966).
12. Thore, A., S. Ansehn, A. Lundin, and S. Bergman, *J. Clin. Microbiol. 1,* 1 (1975).
13. Nilsson, L., *Antimicrob. Agents Chemother. 14,* 812 (1979).
14. Lemasters, J.J., and C. R. Hackenbrock, *in* "Methods in Enzymology," vol. 57 (M. Deluca, ed.), p. 36, Academic Press, New York (1978).
15. Fehrenbach, F-J, H. Huser, and C. Jaschinski, *FEMS Microbiol. Lett. 7,* 285 (1980).
16. Carrico, R. J., K-K Yeung, H. R. Schroeder, R. C. Boguslaski, R. T. Buckler, and J. S. E. Christner, *Anal. Biochem. 76,* 95 (1976).
17. Lee, Y., I. Jablonski, and M. DeLuca, *Anal. Biochem. 80,* 496 (1977).

# DYNAMIC MEASUREMENT OF ATP WITH FIREFLY LUCIFERASE BIOLUMINESCENCE[1]

John J. Lemasters

Laboratories for Cell Biology
Department of Anatomy
School of Medicine
University of North Carolina at Chapel Hill
Chapel Hill, North Carolina

Dynamic measurement of ATP employing the continuous luminescence signal of firefly luciferase has been applied to a number of biological systems. The first quantitative application was for the measurement of oxidative phosphorylation by mitochondria and submitochondrial particles (1,2). A partial list of other applications includes measurement of photophosphorylation by bacterial chromatophores (3), assay of various enzymes such as creatine phosphokinase (4,5), and analysis of adenine nucleotide mixtures by coupled enzyme assay (5). In general, these applications may be considered in terms of the concentrations of ATP employed - low (<1μM), intermediate (between 1μM and 25μM), and high (>25μM).

## I. ATP LESS THAN 1μM

The major complicating factor for the continuous measurement of ATP by luminescence is inhibition of light production by a product of the luciferase reaction, presumably oxyluciferin (1,6). However, as emphasized by Lundin and coworkers (5), at concentrations of ATP below about 1μM, product inhibition of luminescence is negligible in many

[1]*This work was supported by a Grant-in Aid from the American Heart Association with funds contributed in part by the North Carolina Heart Association.*




ISBN 0-12-208820-4

cases, and luminescence becomes a direct and constant measure of ATP.

Product inhibition increases with the absolute rate and duration of the light producing reaction. Thus, product inhibition can be minimized by employing small amounts of luciferin and luciferase, by partially inhibiting the luciferase reaction as with high salt, or by conducting experiments over a brief time course (2,5,7).

## II. ATP BETWEEN 1μM AND 25μM

The constraint that ATP concentration be held to less than 1μM is often inconvenient. The Michaelis constant of most ATP utilizing enzymes, for example, is greater than 1μM making kinetic analysis of such enzymes impossible if ATP concentrations greater than 1μM cannot be employed. Moreover, it may be desirable to make simultaneous spectrophotometric or electrode measurements, and such techniques may be less sensitive than luminescence.

Fortunately, product inhibition follows simple non-competitive kinetics which permits the use of internal standards to quantitate the product inhibited luminescence signal (1,6). At concentrations below 25μM, ATP can be calculated by a simple linear extrapolation since enzyme reaction rates are linear with substrate at substrate concentrations less than one tenth of Km. In Figure 1, the magnitude of this correction is illustrated for endogenous ATP production by rat liver mitochondria. The dashed line is luminescence, and the solid line is actual ATP concentration as determined from several parallel experiments in which ATP standard was added after various intervals. As expected, the correction becomes greater as ATP concentration and the length of incubation increase.

## III. ATP MORE THAN 25μM

At ATP concentrations greater than 25μM, enzyme saturation effects become significant and the luminescence response to ATP is no longer linear (Figure 2). Since the kinetics of product inhibition are understood, accurate quantitation after addition of ATP standard can be made using equations described previously (1,6,8). Such calculations can be rapidly made with the aid of a programable calculator and are based on the assumption that Km remains constant during product inhibition.

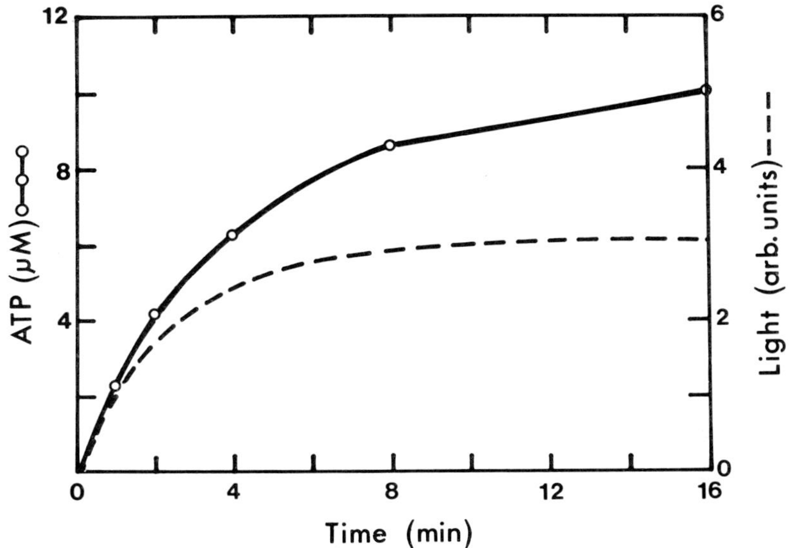


*Fig. 1. Luminescence (---) and extramitochondrial ATP concentration (O—O) during respiratory State 4 incubation. From Reference (12).*

## IV. INHIBITION OF LUMINESCENCE AND THE IMPORTANCE OF INTERNAL STANDARDS

Firefly luciferase luminescence is subject to inhibition by a long list of substances including anesthetics (9), ions generally-especially chloride (7,10), and the fluorescent probes 1,5-anilinonaphthalenesulfonate (ANS) and 2,6-toluidinonaphthalene sulfonate (TNS) (11). In addition, several structurally unrelated uncouplers of oxidative phosphorylation such as dinitrophenol, dicoumarol, and carbonylcyanide p-trifluoromethoxyphenyl hydrazone (FCCP) inhibit luminescence. For the uncoupler carbonyl cyanide m-chlorophenyl hydrazone (CCCP), this inhibition is competitive with respect to luciferin, and Ki is about 1μM (Figure 3).

Since luminescence is subject to inhibition by so many different compounds, it is essential to use internal ATP standards even when there is no interference by product inhibition. The same consideration applies for ATP measurement by routine flash height assay. The use of internal

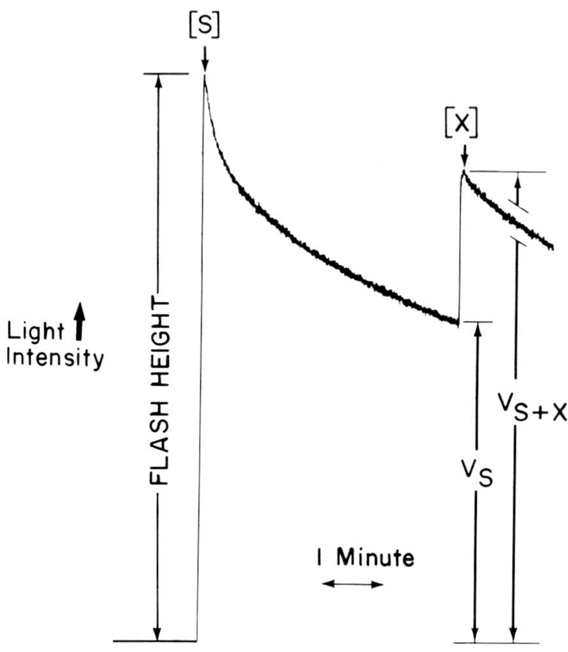


*Fig. 2. Product inhibition of firefly luciferase luminescence. Aliquots of 239μM ATP were added at [S] and [X] to a medium containing 10 units/ml of luciferase and 7.1μM luciferin. From Reference (1).*

standards during a continuous recording requires that the reaction medium be constantly stirred. In this regard, it is unfortunate that most commercially available luminometers make no provision for continuous stirring.

## V. SUMMARY

Firefly luciferase bioluminescence can be utilized to monitor ATP sensitively and quantitatively on a continuous basis. At ATP concentrations below 1μM, there is generally no need to correct for product inhibition of the luminescence reaction. Above 1μM, internal standardization is required, and between 1μM and 25μM this can be done by

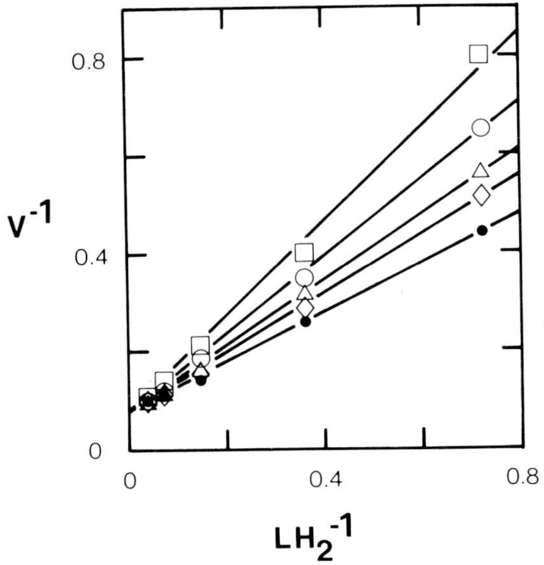


*Fig. 3. Inhibition of firefly luciferase luminescence by carbonylcyanide m-chlorophenyl hydrazone (CCCP). Flash height was measured after addition of 500μM ATP to a reaction medium containing 5 units/ml of firefly luciferase, 5mM $MgCl_2$, 10mM NaPi buffer, pH 7.4, 1.4μM to 28μM luciferin ($LH_2$) and 0 (●—●), 0.2μM (◇—◇), 0.4μM (△—△), 1μM (○—○) or 2μM (     ) CCCP, 23°C. The inverse of flash height (V) in arbitrary units is plotted versus the inverse of luciferin ($LH_2$) in units of $\mu M^{-1}$.*

linear extrapolation. Above 25μM saturation effects become evident, and a hyperbolic extrapolation is necessary. It is prudent, however, to use internal standards at all times since so many substances inhibit the luciferase reaction. Uncouplers of oxidative phosphorylation, in particular, are competitive inhibitors with respect to luciferin. Finally, it is emphasized that luminometers should be equipped for continuous stirring so that internal standards may be employed.

REFERENCES

1. Lemasters, J. J., and Hackenbrock, C. R., *Biochem. Biophys. Res. Commun. 55*, 1262-1270 (1973).
2. Lemasters, J. J., and Hackenbrock, C. R., *Eur. J. Biochem. 67*, 1-10 (1976).
3. Lundin, A., Thore, A., and Baltscheffsky, M., *FEBS Lett. 79*, 73-76 (1977).
4. Witteven, S.A.G.J., Sobel, B. E., and DeLuca, M., *Proc. Natl. Acad. Sci. USA 71*, 1384-1387 (1974).
5. Lundin, A., Rickardsson, A., and Thore, A., *Anal. Biochem. 75*, 611-620 (1976).
6. Lemasters, J. J., and Hackenbrock, C. R., *Biochemistry 16*, 445-447 (1977).
7. DeLuca, M., Wannlund, J., and McElroy, W. D., *Anal. Biochem. 95*, 194-198 (1979).
8. Lemasters, J. J., and Hackenbrock, C. R., *Meth. Enzymol. 57*, 36-50 (1978).
9. Ueda, I., Kamaya, H., and Eyring, H., *Proc. Natl. Acad. Sci. 73*, 481-485 (1976).
10. Denburg, J. L., and McElroy, W. D., *Arch. Biochem. Biophys. 141*, 668-675 (1970).
11. DeLuca, M., *Biochemistry 8*, 160-166 (1969).
12. Lemasters, J. J., and Hackenbrock, C. R., *J. Biol. Chem. 255*, 5674-5680 (1980).

# PRELIMINARY REPORT ON THE PINK-COLORED *CYPRIDINA* LUCIFERASE, A NATURAL MODEL OF THE LUCIFERIN-LUCIFERASE COMPLEX

Toshio Goto
Hideshi Nakamura

Department of Agricultural Chemistry
Nagoya University
Nagoya, Japan

## I. INTRODUCTION

About 60 years ago Harvey (1) observed that a test tube of clear solution of crude *Cypridina* luciferase, although it may give off no light at first when shaken with air, after standing a day or so emits quite a bright light if disturbed, and later found that reducing agents such as sodium hydrosulfite are effective in restoration of the luminescence by reduction of "oxyluciferin" present in the luciferase solution. Anderson (2) established that the oxidation with light production in presence of luciferase gives an oxidation product which cannot be reduced, but oxidation without light production, taking place spontaneously or with oxidants like ferricyanide, is reversible. During the spontaneous oxidation of luciferin the uv maximum changes from 435 nm to 465 nm and finally becomes colorless (3). The first oxidation product is the "reversibly oxidized luciferin" [we named it luciferin-R (4)], which gives a dim light with luciferase (4,5). Addition of hydrosulfite or sodium borohydride to this solution gave a flush of light. We suggested luciferin-R to be a dimer of *Cypridina* luciferin (4,6).

It has long been known that concentrated solution of luciferase sometimes has a slightly pink color (525 nm) which is not related to the luciferase activity and disappears by addition of hydrosulfite (7).




ISBN 0-12-208820-4

## II. PINK-COLORED *CYPRIDINA* LUCIFERASE

We could isolate the pink-colored luciferase which is supposedly a complex of luciferase and pink-colored luciferin. Curiously enough, this pink-colored luciferin is not identical with the luciferin-R obtained by chemical oxidation (and probably by spontaneous oxidation) of luciferin without luciferase. The pink-colored luciferase is fairly stable in aqueous solutions. Its uv spectrum indicates about 1:1 ratio of luciferase and pink-colored luciferin by assuming the ε value of luciferase at 280 nm to be 53400 (8) and oxidized luciferin at 520 nm to be 10000 (4) (Fig. 1).

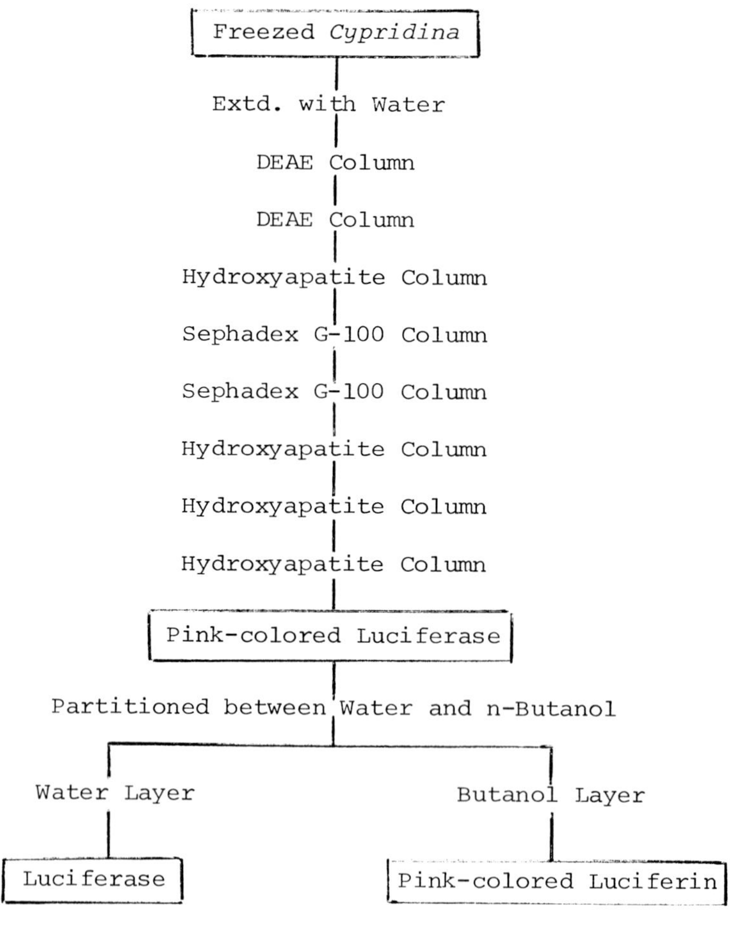

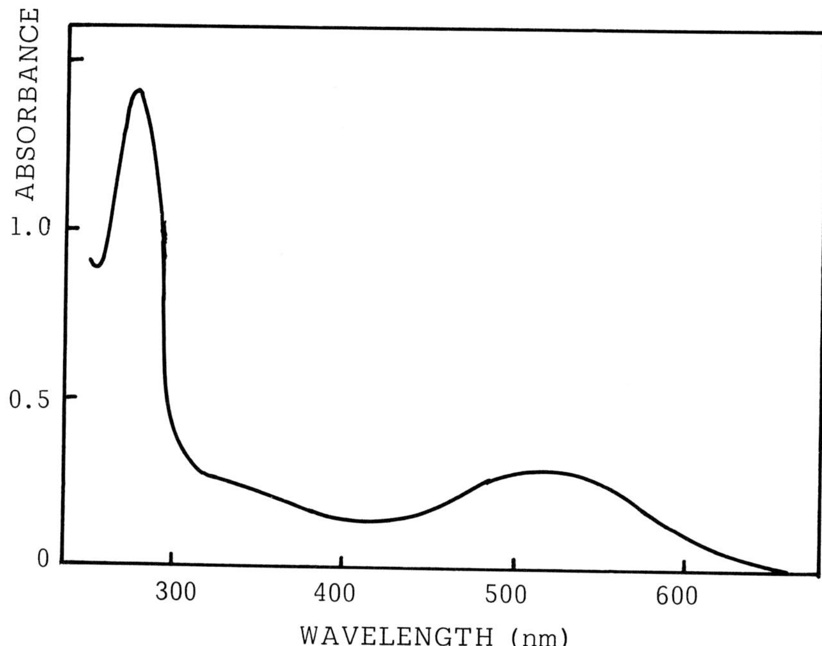


*FIGURE 1. Uv spectrum of the pink-colored luciferase in 0.1M phosphate buffer (pH 6.8) containing 0.2M NaCl.*

The pink-colored luciferin in the complex is easily extracted with butanol. The extracted pigment, however, cannot be introduced again in the luciferase. A large bathochromic shift of the pink-colored luciferase (520nm) compared with the pink-colored luciferin in phosphate buffer (pH 6.8) (473 nm) and in dimethyl sulfoxide (463 nm) may indicate a heterogeneous environment around the pigment in the luciferase. The pink-colored luciferase and the pink-colored luciferin show very large circular dichroism (CD) (Fig. 2), whereas *Cypridina* luciferin and the chemically oxidized luciferin-R gave almost no CD, suggesting that the pink-colored luciferin was formed in the chiral environment of the luciferase.

Interestingly, sodium borohydride reduction of the pink-colored luciferase did not proceed completely at once. Almost half of the pink-colored luciferin is reduced at once and successive addition of sodium borohydride is necessary after each luminescence ceased (Fig. 3). As early as 1944 Johnson and Eyring (9) reported that a portion of the

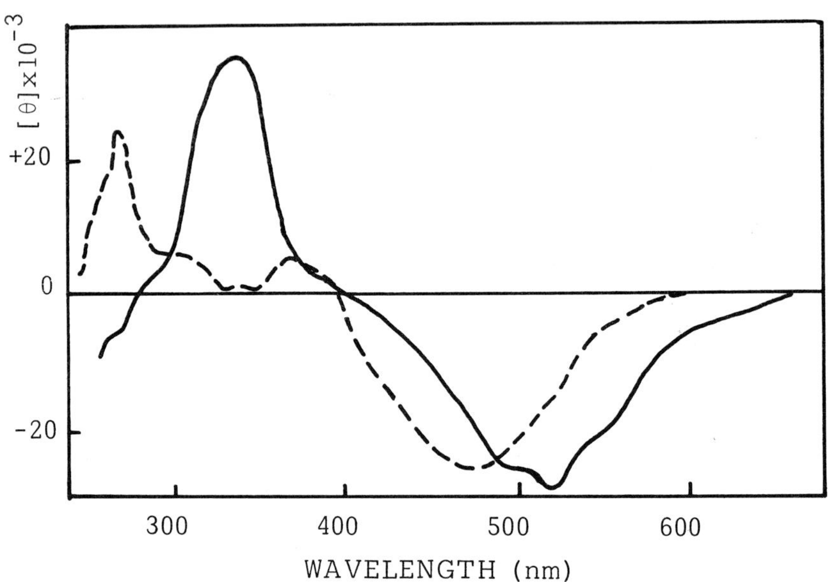


*FIGURE 2. Circular dichroism spectra of the pink-colored luciferase in 0.1M phosphate buffer (pH 6.8) containing 0.2M NaCl (——), and the pink-colored luciferin in n-butanol (----).*

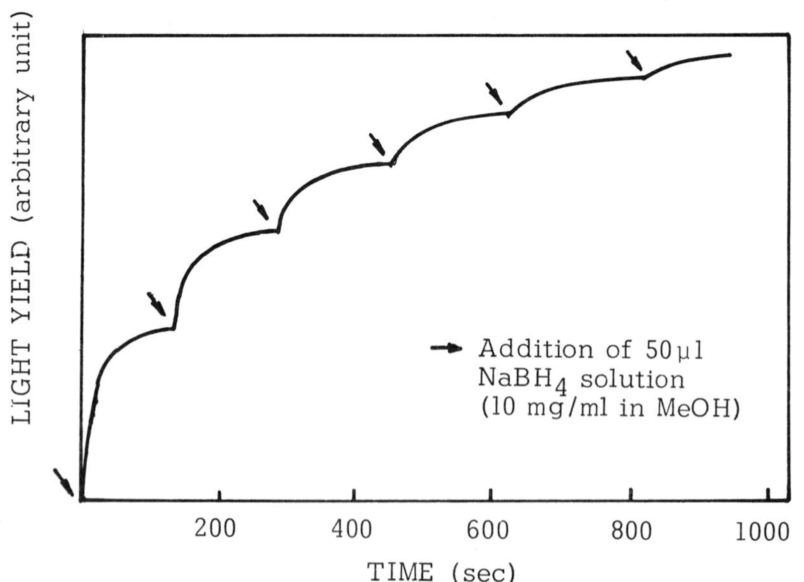


*FIGURE 3. Light emission of the pink-colored luciferase by addition of sodium borohydride in 0.1M phosphate buffer (pH 6.8) containing 0.2M NaCl.*

luciferase solution gave luminescence repeatedly on successive addition of an excess of hydrosulfite then oxygen, but finally ceased. This observation indicates the presence of the pink-colored luciferase. The reason for this phenomenon is unknown.

This pink-colored luciferase may be a model of luciferin-luciferase complex and worthwhile to study further.

REFERENCES

1. Harvey, E. N., "Bioluminescence," p. 307-309. Academic Press, New York (1952).
2. Anderson, R. S., *J. Cell. Comp. Physiol 8,* 261 (1936).
3. Chase, A. M., *J. Biol. Chem. 150,* 433 (1943).
4. Goto, T., *Pure Appl. Chem. 17,* 421 (1968).
5. Chakravorty, P. N., and R. Ballentine, *J. Am. Chem. Soc. 63,* 2030 (1941); Chase, A. M., *J. Cell. Comp. Physiol. 33,* 113 (1949).
6. Goto, T., I. Kubota, N. Suzuki, Y. Kishi, and S. Inoue, *in* "Chemiluminescence and Bioluminescence," (M. J. Cormier, D. M. Hercules, and J. Lee, eds.), p. 325. Plenum Press, New York (1973).
7. Shimomura, O., F. H. Johnson, and T. Masugi, *Science 164,* 1299 (1969).
8. Shimomura, O., F. H. Johnson, and Y. Saiga, *J. Cell. Comp. Physiol. 58,* 113 (1961); Tsuji, F. I., and R. Sowinski, *J. Cell. Comp. Physiol. 58,* 152 (1961).
9. Johnson, F. H., and H. Eyring, *J. Am. Chem. Soc. 66,* 848 (1944).

# STANDARD ASSAY FOR TOTAL CREATINE KINASE AND THE MB-ISOENZYME IN HUMAN SERUM WITH FIREFLY LUCIFERASE

Karl Wulff
Fritz Stähler
Wolfgang Gruber

Boehringer Mannheim GmbH
Research-Center Tutzing
Tutzing, FRG

## I. INTRODUCTION

The enzyme creatine kinase (EC 2.7.3.2) catalyzes the formation of ATP and creatine from creatine phosphate and ADP:

$$\text{creatine phosphate} + \text{ADP} = \text{creatine} + \text{ATP}$$

Human creatine kinase is a dimer formed by two different kinds of subunits, the M-subunit and the B-subunit. Three possible dimer structures are found in human serum: the MM-form, which is muscle-specific, the BB-isoenzyme being brain-specific, and the heart-specific MB-isoenzyme. In acute myocardial infarction the level of the heart-specific isoenzymes increased significantly in human blood serum. The precise assay for the activity of creatine kinase CK-MB is of great value for the diagnosis and control of acute myocardial infarction (1).

There have been many attempts being published in the literature to assay creatine kinase by following the formation of ATP by this enzyme with the firefly bioluminescence system (2-4). All of them have more or less severe disadvantages in being not standardized. The aim of this investigation is to present standard conditions for a bioluminescence assay of creatine kinase.



ISBN 0-12-208820-4

## II. RESULTS AND DISCUSSION

In 1975 Lundin and Thore (5) found that under special conditions using a very pure firefly luciferase it is possible to get a constant light output at a given ATP concentration. These results can be reproduced using twice crystallized luciferase from Photinus pyralis. The decrease in light intensity is less than 5 per cent over a period of 20 minutes and parallels the decrease in ATP concentration. Since the luciferase is a very slow enzyme the decrease in ATP concentration due to the bioluminescence reaction is negligible. In a coupled bioluminescence assay for creatine kinase the light intensity (I) is proportional to the ATP concentration and the slope of a linear increase in light intensity gives the activity of the creatine kinase ($A_{CK}$) according to the following equations:

$$I = \frac{d\ (h \cdot v)}{dt} = const \cdot C_{ATP}$$

$$A_{CK} = \frac{dC_{ATP}}{dt} \propto \frac{dI}{dt}$$

A great problem in any creatine kinase assay in human blood serum is a possible interference by adenylate kinases present in the sample. There are two kinds of adenylate kinase, one coming from erythrocytes and the other coming from the liver. The latter can be inhibited by AMP and the erythrocyte enzyme is inhibited by diadenosine pentaphosphate. Based on these results for the photometric assay the German Society for Clinical Chemistry recommended standard conditions for the concentrations of creatine kinase substrates, buffers, and concentrations of the inhibitors of the adenylate kinases, which are shown in Table I.

In a firefly bioluminescence assay these adenylate kinase inhibitors cause problems. It has been demonstrated by Momsen (7) that diadenosine pentaphosphate shows a light generating reaction with firefly luciferase which is not due to a contamination by ATP and which simulates an ATP content of 0.8 mol per cent. We have been able to reproduce his results. AMP on the other hand is a well known inhibitor of luciferase, which inhibits at pH 7.75 at the recommended concentration the firefly luciferase quantitatively. From this many investigators concluded that creatine kinase assays cannot be performed with bioluminescence in the presence of AMP (2-4)

TABLE I. *Standard Conditions for Creatine Kinase Assay (5)*

| Solute | Concentration[a] |
|---|---|
| Imidazole-acetate buffer, pH 6.7 | 100 mmol/ℓ |
| Creatine phosphate | 30 mmol/ℓ |
| ADP | 2 mmol/ℓ |
| Magnesium acetate | 10 mmol/ℓ |
| N-Acetylcysteine | 20 mmol/ℓ |
| AMP | 5 mmol/ℓ |
| Diadenosine pentaphosphate | 10 µmol/ℓ |

[a]All concentrations are final concentrations in the assay.

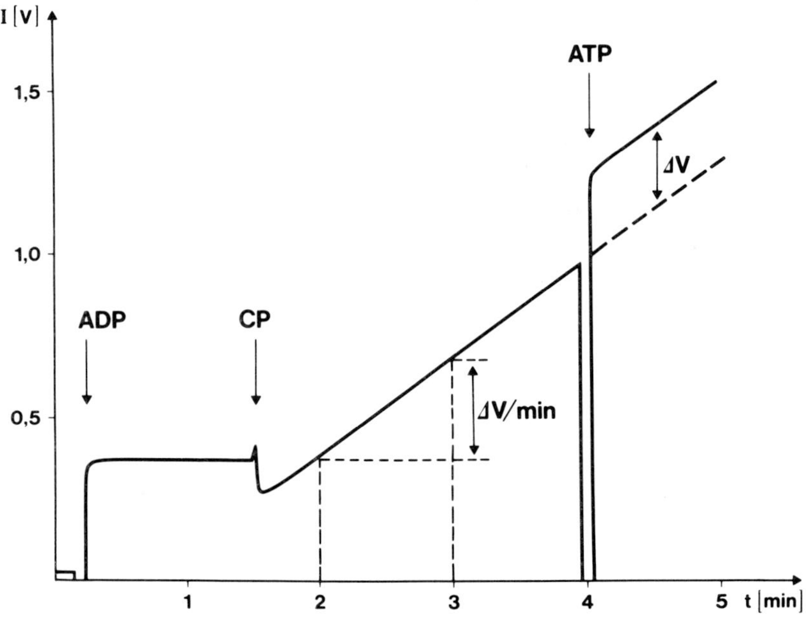


FIGURE 1. *Time course of a typical bioluminescence creatine kinase assay.*

But, most of them overlooked that the effect of AMP on the firefly luciferase is strictly pH dependent, as has been worked out by McElroy's group 10 years ago (8). From their results it follows that at pH 6.7 an AMP concentration of 5 mmol/ℓ will only cause an 80 per cent inhibition of the luciferase. The rest activity is sufficiently high to do a coupled assay for creatine kinase activity with the sensitivity required.

In Fig. 1 a typical experiment for a creatine kinase assay is shown. AFter a 5 minute period of preincubation for reactivation of creatine kinase the reaction is prestarted by the addition of ADP to check for an uninhibited rest activity of adenylate kinase. Then creatine phosphate is added to the reaction mixture causing a low inhibition of the firefly luciferase which may be due to so far unknown contaminants in the creatine phosphate. Within a few seconds an essentially linear increase in light intensity starts being due to the increase in ATP concentration. For internal calibration a small amount of an ATP solution is added causing a well defined step like increase in ATP concentration. Using this step the slope of the curve can be calculated in the dimension "µmol per minute" giving the creatine kinase activity in International Units.

Theoretically the sensitivity of this assay is only limited by the purity of the reagents. The main interference will be expected by the ATP contamination present in ADP. Using commercially available reagents without further purification we routinely reach a sensitivity of 0.2 U/ℓ creatine kinase in the sample by using a one by 10 dilution of the serum. The great advantage of the bioluminescence creatine kinase assay is its superior precision compared with the photometric assay. The total creatine kinase in a serum sample diluted with a 6% serum albumin solution to give a final activity of 1.3 U/ℓ sample was assayed with a CV of 4.3%. As has been shown earlier, this bioluminescence creatine kinase assay correlates well with the photometric one (9).

The most convenient assay for the MB-isoenzyme of creatine kinase to perform is the one using specificly inhibiting antibodies against the M-subunit (10). Using this assay with the bioluminescence reaction will cause two main problems:

1. If the serum contains extremely high levels of the MM-isoenzyme there will be a background of up to 5 U/ℓ of uninhibited MM-isoenzyme because the antibody does not completely inhibit the enzyme.

2. The presence of very high levels of adenylate kinase will cause similar problems, because both of the inhibitors for adenylate kinases do not inhibit the enzyme completely.

Due to these possible interferences in some extreme cases the increase in sensitivity by the bioluminescence reaction is compensated by the uninhibited background activities.

These problems are expected to be less severe using the chromatographic separation (11) of the creatine kinase isoenzymes on DEAE-sephadex A50 under conditions where the MM-isoenzyme as well as the adenylate kinases are washed through the column and only the MB- and the BB-isoenzymes are retained. After washing the column both the B-subunit containing isoenzymes are eluted using a different buffer. The activity of both these isoenzymes is then assayed in the eluent.

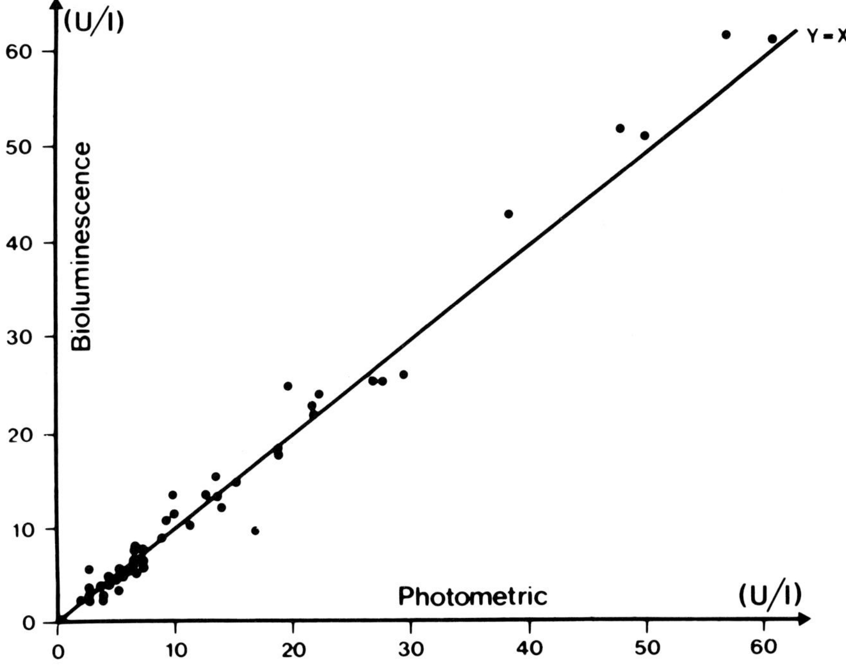


*FIGURE 2. Comparison of bioluminescence and photometric creatine kinase MB assay after chromatographic separation.*

With this chromatographic assay we compared by using 51 human sera from myocardial infarction patients the classical photometric method with our bioluminescence assay. The results are shown in Fig. 2. It can be seen that there is an excellent agreement. The correlation coefficient calculated from these data is r = 0.991.

REFERENCES


1. Haux, P., et al., *J. Clin. Chem. Clin. Biochem.* 17, 747 (1979).
2. Lundin, A. and I. Styrelius, *Clin. Chim. Acta* 87, 199 (1978).
3. Witteveen, S.A.G.J., et al., *Proc. Nat. Acad. Sci.* 71, 1384 (1979).
4. Tarkkanen, P., et al., *Clin. Chem.* 25, 1644 (1979).
5. Lundin. A. and A. Thore, *Anal. Biochem.* 66, 47 (1975).
6. Anonymus, *J. Clin. Chem. Clin. Biochem.* 15, 225 (1977).
7. Momsen, G., *Biochem. Biophys. Res. Commun.* 84, 816 (1978).
8. Lee, R. T., et al., *Arch. Biochem. Biophys.* 141, 38 (1970).
9. Wulff, K., et al., *Fresenius Z. Anal. Chem.* 301, 173 (1980).
10. Würzburg, U., et al., *Klin. Wschr.* 54, 357 (1976).
11. Mercer, D. W., *Clin. Chem.* 20, 36 (1974).

## DISCUSSION

Dr. Tsuji

I would like to ask Dr. Goto if he knows how the pink colored luciferin might be functioning, if it is bound to the catalytic site?

Dr. Goto

These are very preliminary results; so, we don't know it actually. Pink colored luciferase can catalyze bioluminescence with usual luciferin; pink colored luciferin does not inhibit any bioluminescence so this pink colored luciferin can be replaced very easily with luciferin or may be attached to a different active site of the luciferase; we don't know yet.

Dr. Keith Ward

I'd like to ask a question about the stability of firefly luciferase. In any field as specialized as this, there always arise myths and folklore about how one should handle various preparations, and since this afternoon we have in this one room so many experts on handling firefly luciferase, it would be a great help to a neophyte like myself, and perhaps other people in the audience, if you could tell me whether a lot of things I have heard about the preparations are just myth or whether they are founded on fact. First of all, apparently the cleaner you get these preparations the less stable they are. It is not clear to me whether that is true or not. Dr. McElroy referred to the fact that in 10% ammonium sulfate at pH 7, I believe, the preps are stable indefinitely. That hasn't been the case in our hands, and I don't know if it is true in other people's laboratories or not. The other thing is: is it necessary or helpful to use sulphhydryl protecting reagents when you store the enzyme?

Dr. McElroy

When you are storing the firefly enzyme in ammonium sulfate you do have to be extremely careful to keep the tube covered otherwise ammonia will distill off at that pH (7.5) and the pH goes down. When that happens, you will get inactivation.

Dr. Ward

But you don't buffer the preparations to prevent that from happening?




ISBN 0-12-208820-4

Dr. McElroy

Even with buffer the loss of ammonia still leads to a decrease in pH. The enzyme can be protected by adding albumen.

Dr. Hastings

Many preparations are supplied with additives. Maybe Dr. Lundin would like to make a comment?

Dr. Lundin

Most commercial preparations contain a lot of other proteins. For example, Bovine serum albumin. I must say that firefly luciferase is one of the most stable proteins that I have ever experienced. You can keep it in room temperature at least for a day with loss of 10% activity - something like that.

Dr. Hastings

Crystalline firefly luciferase is supplied by a couple of firms. I think Dr. Wulff might comment on the crystalline material with which he has experience, which does not have additives, I believe.

Dr. Wulff

Yes, we are producing the crystallized enzyme then lyophilizing it. At this state at room temperature or even elevated temperature, it is very, very stable.

Dr. Lundin

I would like to confirm that both the crystalline enzyme and preparations with Bovine serum albumin - I would not recommend the addition of SH compounds, as you suggest, because you get some activation initially and then it goes up to fairly high activity and is stable there, for say, a week or so in the refrigerator; then suddenly all of the activity disappears.

Dr. Ward

Is the enzyme light sensitive?

Dr. McElroy

Not if you have it really pure. If you have luciferin or dehydroluciferin bound to it, then light will affect it.

Dr. Hastings

Due to photosensitized oxidation, perhaps.

Dr. Ward

I had heard, in fact, that one could not lyophilize the material, but, I noticed that Dr. Wulff does.

Dr. McElroy

Actually, you also have to be careful of your luciferin because that is photosensitive.

Dr. DeLuca

I wanted to ask Dr. Lundin: When you add pyrophosphate as well as D-luciferin to your preparations, how much sensitivity do you lose?

Dr. Lundin

By adding pyrophosphate, I don't lose sensitivity, since I add very low concentrations, I think it is $10^{-5}$ molar. By adding L-luciferin, I lose the activity partly, but it's not important since the most expensive component in the firefly assay is not luciferase. It is luciferin, and by using this mixture of L- and D-luciferin you get a decrease of the apparent Km value, and thus we can have an optimized assay at a lower level of luciferase as is the case if one had just pure D-luciferin.

Dr. McElroy

Well, why do you want to inhibit your enzyme? Why don't you just lower your ATP concentration, if you are worried about the product inhibition?

Dr. Lundin

There are situations when I cannot do it. For example, in the CK assay the formation of ATP up to 10.6 liter is very rapid. It may take you less than seconds, if you have a high serum. I want to really extend the linear range of the assay up to a maximum, and that is also where I want to have the constant light; and that only be obtained using pyrophosphate, and some inhibitor.

Dr. McElroy

Well, then you could do what Strehler and Totter did - add arsenate.

Dr. Lundin

Yes, but then I would not get the advantage of having an optimized assay, with respect to luciferin. That would mean that I would have additional pipetting errors. Effects of pipetting errors in the assay; this is a way to get an analytically favorable situation up to $10^{-6}$ moles per liter.

Dr. Lemasters

Could you just use less luciferase to begin with?

Dr. Lundin

I cannot do that because the same ration of the enzyme will be inhibited if I use a lower concentration. I mean it's not formation of oxyluciferin in the solution. We get this product complex, and that is the reason why some part of the enzyme is taken away from the reaction.

Dr. McElroy

The magnesium pyrophosphate is a competitive inhibitor of ATP and the L enters into the activation reaction which inhibits the luciferase, and so, basically what you are saying is that you are just decreasing your active sites for ATP utilization.

Dr. Lundin

Yes.

Dr. Totter

I wanted to comment about the arsenate business. I am not sure the whole story has appeared about that. The difference between the work we did and Dr. McElroy was our use of very crude enzyme always, in contrast to that of Dr. McElroy which was mostly with purified enzyme. What he says about the instrumentation being difficult for us is correct. Although in the latter part of the experiments we had a photon counter. We used a fluorimeter which had been developed by Oliver Lowry, which was by far the most sensitive instrument available in the early fifties. But, the arsenate was not in the central part of it. The holding of the light fairly constant really depended on two things: 1) a high magnesium concentration, 2) a high phosphate buffer concentration; and it had to go so high that the magnesium phosphate precipitated. But, we didn't find that trouble with arsenate, and that was the sole reason for the use of arsenate.

Dr. Presswood

I'd like to mention some experiments which Dr. Wulff, Dr. Hastings, and I recently performed. The question we wished to address was "what is the turnover number of the firefly luciferase?" Our mode of experimentation was to use the stop-flow apparatus, quenching a reaction containing a low level of ATP which exhibited a constant light output over a long period of time. We quenched the reaction with apyrase to remove ATP or EDTA to remove magnesium. Our

Dr. Presswood

results were not compatible with a turnover number of ten per minute. I'd like to solicit from Dr. Wulff some comments as to what these experiments might imply.

Dr. Wulff

When discussing the firefly luciferase, we always speak of product inhibition. Product inhibition results from this reaction: enzyme + oxyluciferin which I will call "p" is in equilibrium with the E P complex for which the equilibrium constant is known - it was published by Dr. Goto and co-workers in 1972. It is not extremely high. In an experiment where you have initially a fast decrease in light, followed by a slower decrease; if you integrate the light, you can, by using a calibrated instrument and the published quantum yield for this reaction correlate the emitted photons to the molecules of product formed, and you will end up with a number which is extremely low. We did this experiment. I have a manuscript which is not completely finished. Our results are in contradiction to a product inhibition. Also, if you have a product inhibition you will have no light increase if you add an additional amount of luciferase. What you observe, however, is that you get a new cycle. This has been published in this famous paper by McElroy, et. al. 27 years ago; this old experiment is in contradiction to the assumption of product inhibition. What we did was to follow the reaction and extrapolate to zero time. Then we found out that the specific activity of the luciferase is smaller or equal to 0.1 micromoles per minute per milligram at T=0. From this number, we calculated the turnover number of about 10 or greater. This means that one cycle of the reaction will take about 6 seconds. Then, in the stopped flow experiment, we took away the magnesium, and then followed the rate of the light decay. We did the same with ATP and ended up, in both cases, with a half life time of about 120 milliseconds. Then we took away the oxygen by using glucose oxidase and glucose and we ended up with a half life time of about 170 milliseconds. So this is in some disagreement. The overall reaction should last approximately 6 seconds. But, these major steps are very much faster; so, at the moment, we have no explanation for this. The dissociation constant for this product complex is so small, that you get a constant signal if your amount of enzyme is about orders of magnitude above the amount of product complex formed in the given time interval. In the experiment, I showed on my slide, in 20 minutes we calculated a decrease in ATP concentration calculated from the light output of 0.05%; but a decrease in enzyme concentration of 2.5%. This is well in agreement with the small decrease in intensity.

Dr. McElroy

In the luminescent reaction, when the enzyme ends up inhibited, you have two product molecules per enzyme. I think that's going to change your kinetics.

Dr. Wulff

Yes, it may be. It makes a factor of two difference, but it is not in complete disagreement.

Dr. McElroy

As you know, it is a slow turnover after the initial flash.

Dr. Wulff

Yes, of course, it is a slow turnover, and you can destroy this complex with pyrophosphate, as has been published 27 years ago.

Dr. McElroy

Pyrophosphate only has a secondary effect, when you have dehydroluciferin in your luciferin. Then you get a stimulation.

Dr. Hastings

What is the pyrophosphate effect in the absence of dehydroluciferin?

Dr. McElroy

Pyrophosphate is just a potent inhibitor; it competes with ATP. The secondary stimulation depends entirely upon the formation of free enzyme from E-LAMP.

Dr. Lundin

How, then, do you explain the experiments of Gates and DeLuca some years ago in which they showed that you could isolate the enzyme product complex only with both oxyluciferin and AMP and the enzyme only in the presence of pyrophosphatase?

Dr. McElroy

Pyrophosphatase pulls the reaction so that all enzyme would be converted to E-LAMP. Pyrophosphate will reverse that reaction. If you take a crude enzyme that has a bound dehydroluciferin, the secondary addition of pyrophosphate stimulates it.

Dr. Lundin

I suppose that this preparation was completely clean.

Dr. Hastings

Dehydroluciferin is not normally added to the reaction, is it?

Dr. McElroy

No.

Dr. Hastings

Therefore, the question is - what is the pyrophosphate effect in a normal luciferase assay?

Dr. McElroy

If you have a mixture of dehydroluciferin and luciferin and you add pyrophosphate, you will get a stimulation because you are freeing enzyme by breaking down the E-LAMP.

Dr. Hastings

Is there dehydroluciferin in the normal luciferin?

Dr. McElroy

Most luciferin has a little dehydroluciferin in it, if you are not careful, because you do get photo-oxidation of luciferin.

Dr. Hastings

Did you want to comment on the nature of the structure of the inhibitory compound, the EP - 2 complex?

Dr. McElroy

Howard Seliger and others have shown that when two $LH_2$-AMP reacts with the enzyme, you get 100% inhibition of the enzyme. If you isolate the product inhibited enzyme, one of the product molecules readily comes off and the enzyme is active again.

Dr. Hastings

Do both of those products have the same structure as confirmed by thin layer chromotography or some other method?

Dr. McElroy

Both products are the same as that made by Dr. Goto earlier.

Dr. Hastings

And the enzyme might be a dimer and it might be a monomer?

Dr. McElroy?

We don't know.

Dr. LeMasters

Some of these reactions are carried out at low concentrations of ATP. The fact that net turnover is less than one does not necessarily imply that inhibition is irreversible. It is a statistical phenomenon; some enzyme molecules are not reacting at all, and some turnover more than once. Certainly, at published values for the dissociation constant of oxyluciferin, perhaps Dr. Goto can comment on this, you would have significant rates of dissociation of inhibitor from the enzyme.

Dr. McElroy

You need ATP and free luciferin to obtain the slow turnover. If you start with $LH_2$-AMP and no ATP, then you immediately get 100% inhibition after you've reacted the enzyme with two $LH_2$-AMP's.

# V
# RENILLA, AEQUOREA, EARTHWORM, AND LAND SNAIL BIOLUMINESCENCE

FRONT ROW- left to right: H. Seliger, M. DeLuca, Y. Haneda, and Mrs. Haneda

SECOND ROW- left to right: E. Engel, F. Tsuji, M. Cormier, and A. Lundin

# RENILLA AND AEQUOREA BIOLUMINESCENCE

Milton J. Cormier

Department of Biochemistry
University of Georgia
Athens, Georgia

## I. INTRODUCTION

A number of physiological studies have shown that the luminescence of *Renilla* and other anthozoan coelenterates is under the control of a nerve net (1-7) and that the bioluminescent proteins of both anthozoans and hydrozoans are located intracellularly in specialized cells, termed photocytes (1). Biochemical investigations on coelenterate bioluminescence suggest that their nerve-linked luminescence is controlled by $Ca^{2+}$ (8-11). What follows is a brief summary of the chemistry and enzymology of *Renilla* and *Aequorea* bioluminescence.

## II. RENILLA BIOLUMINESCENCE

Evidence, based on *in vitro* studies, suggest that three proteins are directly involved in the *in vivo* bioluminescence of *Renilla*. All three have been purified to homogeneity and characterized (11-13). Only one of these is catalytic while the second protein controls the color of light emission and the third protein apparently couples nerve excitation to bioluminescence through $Ca^{2+}$.

### A. *Renilla Luciferase*

The catalytic protein, *Renilla* luciferase (EC 1.13.12.5), is active as a single polypeptide chain monomer of molecular weight 35,000. Luciferase catalyzes the reaction shown in




ISBN 0-12-208820-4

Fig. 1. The enzyme contains a single luciferin binding site with a dissociation constant of $3 \times 10^{-8}$M (14). Oxygen is apparently incorporated into the product, oxyluciferin while $CO_2$ is derived from carbon 3 of luciferin. The initial product of the reaction is an electronic excited state of a luciferase-oxyluciferin monoanion complex which relaxes to the ground state with the production of blue light ($\lambda_{max}$ = 480 nm). The quantum yield ($\phi_B$) for the reaction is 6.9% (15). A careful examination of the emission spectra reveals one major and one minor emission band. Less than 1% of the light is

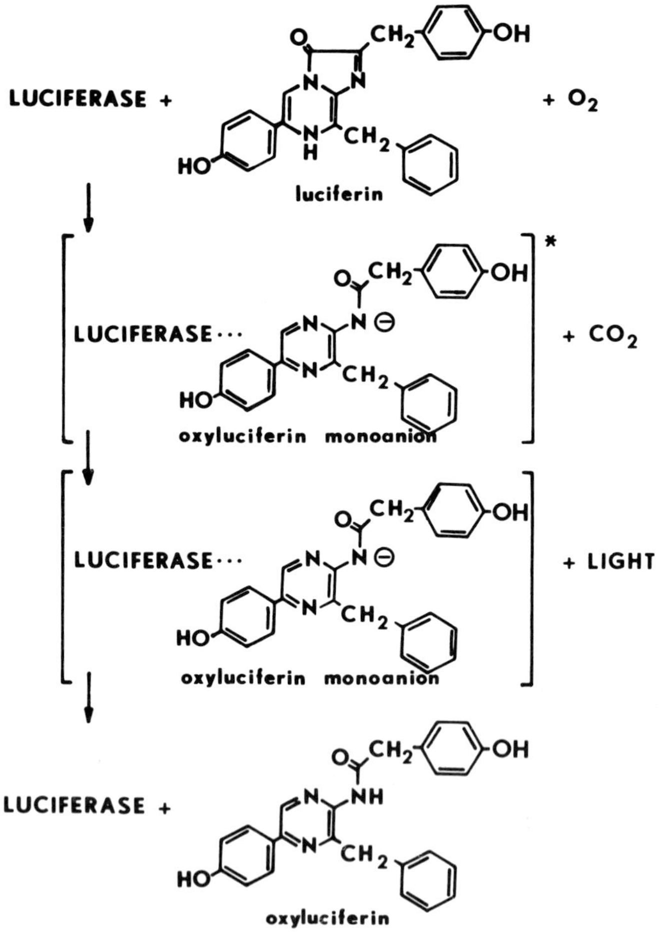

LUCIFERASE + luciferin + $O_2$

↓

[LUCIFERASE··· oxyluciferin monoanion]* + $CO_2$

↓

[LUCIFERASE··· oxyluciferin monoanion] + LIGHT

↓

LUCIFERASE + oxyluciferin

*FIGURE 1. Reaction catalyzed by* <u>*Renilla*</u> *luciferase.*

emitted in the near ultraviolet ($\lambda_{max}$ = 395 nm) and is derived from a neutral species excited state of the luciferase-oxyluciferin complex (13-15). Most of the light (>99%) occurs at 480 nm.

### B. *Green Fluorescent Protein (GFP)*

Although the luciferin-luciferase reaction described above produces blue light an examination of *Renilla* bioluminescence *in vivo* reveals a green emission ($\lambda_{max}$ = 509 nm) which is characteristically narrow and structured (16, 17). The

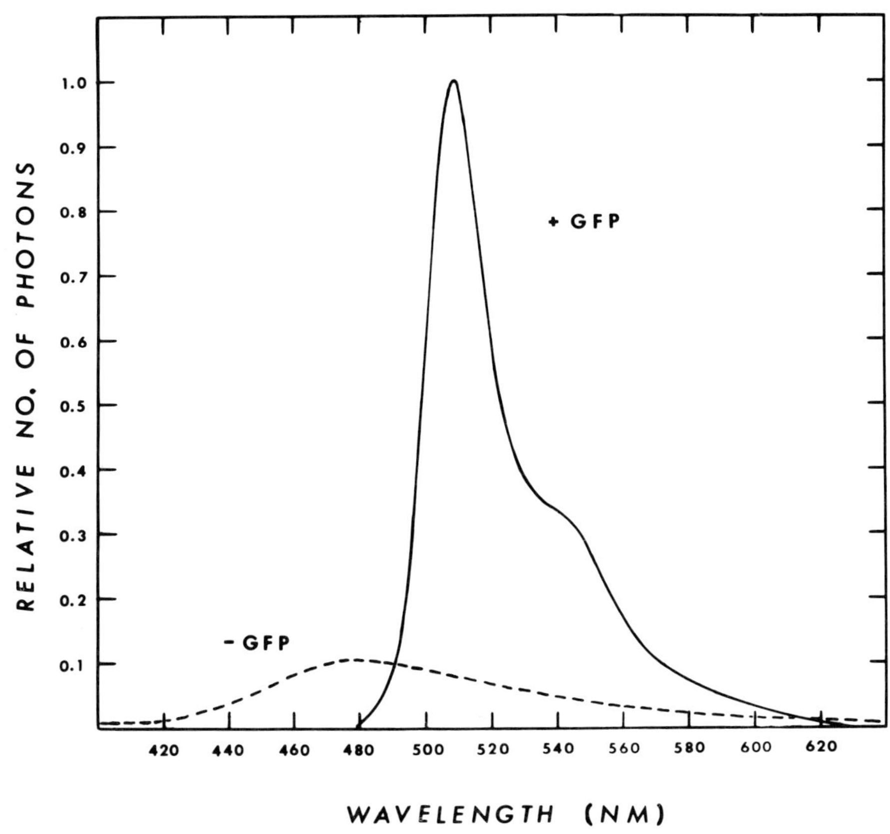


*FIGURE 2. Effect of green fluorescent protein (GFP) on the bioluminescence emission of a Renilla luciferase-luciferin reaction mixture.*

difference between the *in vivo* and *in vitro* emissions is due to the fact that bioluminescence in *Renilla* is a sensitized process. The sensitizer is a protein, termed the green fluorescent protein (GFP), which has been isolated and characterized (12). GFP, whose fluorescence quantum yield is 80%, is a dimer (MW = 54,000) of identical subunits. Each subunit contains at least one covalently linked chromophore of unknown structure which is responsible for its characteristic spectral properties.

The fluorescence of GFP is identical to the green *in vivo* luminescence of *Renilla* (16). Furthermore, the green *in vivo* emission can be duplicated *in vitro* by the addition of GFP ($10^{-7}$ M, final concentration) to a luciferase-luciferin reaction mixture. A shift in spectral distribution occurs, from the broad blue to a narrow green emission as shown in Fig. 2. This spectral shift occurs apparently due to a non-radiative energy transfer process. Evidence for this explanation is substantial. Firstly, the radiative quantum yield for luciferin oxidation increases about 2-fold. Using certain luciferin analogues this quantum yield increase can be as high as 200-fold (15). Secondly, the process involves a highly specific interaction between luciferase and GFP to form a 1:1 complex (12, 18).

Studies with luciferin analogues have shown that the luciferase-bound oxyluciferin monoanion singlet excited state can be quenched by solvent and/or protein functional groups. Furthermore, energy transfer to GFP can favorable compete with this quenching process. We also find that, under energy transfer conditions, emission from the neutral species excited state of oxyluciferin remains unchanged whereas emission from the oxyluciferin monoanion species excited state is eliminated upon the addition of GFP (15). These results suggest that energy transfer to GFP occurs from the monoanion species excited state.

### C. *Luciferin Binding Protein (BP-$LH_2$)*

A protein, which couples *Renilla* bioluminescence to a nerve impulse through $Ca^{++}$, has been isolated and characterized (11). It is a single polypeptide chain protein of molecular weight 18,500 and contains two high affinity $Ca^{2+}$ binding sites (Kd = $1.4 \times 10^{-7}$ M). The native protein also contains one mole of luciferin (Fig. 1) non-covalently bound to it (19) and is generally referred to as luciferin binding protein (BP-$LH_2$).

The probable *in vivo* function of BP-$LH_2$ may be understood by an examination of the interaction of the isolated components of the *Renilla* bioluminescence system as illustrated in Fig. 3. In the absence of $Ca^{2+}$, the addition of BP-LH to luciferase does not produce light because its bound luciferin is unavailable. When $Ca^{2+}$ is added, the $Ca^{2+}$ binding sites on BP-$LH_2$ are filled. This is followed by a conformational change in BP-$LH_2$ which makes its luciferin available for reaction with luciferase. Thus an *in vitro* mixture of BP-$LH_2$, GFP, and luciferase will produce a flash of green light upon $Ca^{2+}$ addition. The emission characteristics of the light are identical to that observed *in vivo*.

A single live *Renilla* can be made to emit most of its light within seconds upon $K^+$ depolarization. This amount of light, corrected for *in vivo* quantum yield, is approximately equal to the amount of BP-LH obtained per animal upon extraction. We have also found that the *in vivo* molar ratio of BP-$LH_2$:luciferase:GFP is approximately 10:3:1. Thus it appears that BP-$LH_2$, luciferase, and GFP are prepackaged and ready for triggering by the nerve net of *Renilla*. There is

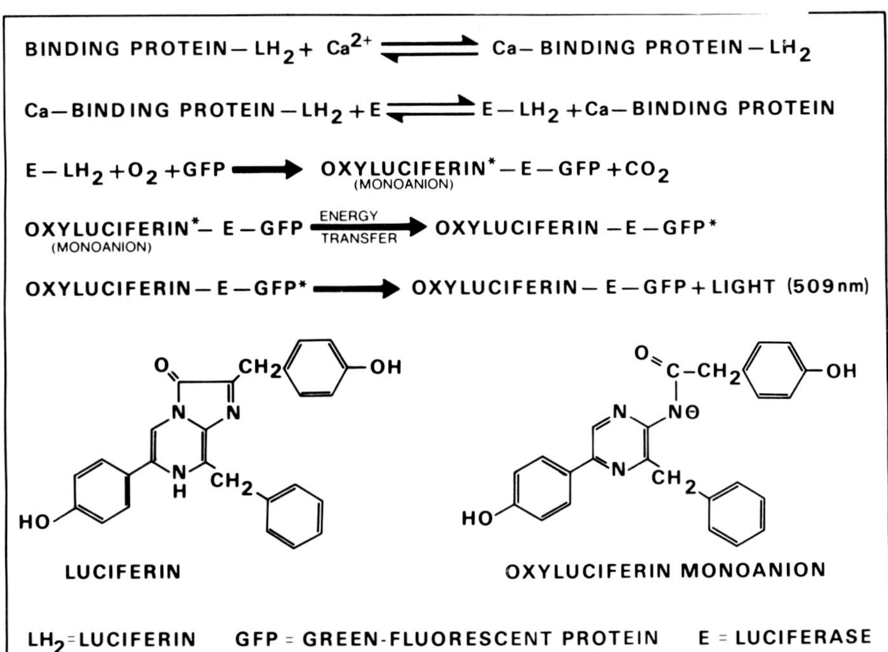


*FIGURE 3. Reaction pathway of the $Ca^{2+}$-triggered green in vitro luminescence of Renilla.*

additional evidence for this view. Firstly, luminescence from whole *Renilla* polyps, induced by $K^+$ depolarization, is dependent upon $Ca^{2+}$ in the external medium. Secondly, membrane-bounded vesicles (lumisomes) have been isolated from *Renilla* and found to contain all three proteins involved in bioluminescence-luciferase, GFP and BP-$LH_2$ (8). These vesicles have the ability to take up $Ca^{2+}$ when the $Na^+/K^+$ ratio is high on the inside and low on the outside (20-22). Under these conditions, the addition of $Ca^{2+}$ to a lumisome suspension results in a flash of green light ($\lambda_{max}$ = 509 nm).

The observations suggest a plausible model which could explain the coupling of a nerve impulse to bioluminescence. A nerve impulse could arrive at the photocyte membrane through the nerve net and initiate $Ca^{2+}$ uptake by depolarization of the photocyte membrane. An intracellular $Ca^{2+}$ transient would result followed by $Ca^{2+}$ binding to BP-$LH_2$. This would allow the luciferin on BP-$LH_2$ to react with luciferase thus triggering the onset of bioluminescence. A rapid sequestration or removal of free $Ca^{2+}$ from the photocyte could terminate light emission.

## III. AEQUOREA BIOLUMINESCENCE

### A. *Aequorin*

A protein from the luminous jellyfish *Aequorea*, which produces blue light ($\lambda_{max}$ = 469 nm) in the presence of $Ca^{2+}$, was discovered and isolated by Shimomura *et al.*(9). They termed this protein a photoprotein and referred to it specifically as aequorin. Photoproteins have also been examined in other hydrozoans and ctenophores (23-27).

*Auquorin* has several properties in common with the *Renilla* luciferin binding protein (BP-$LH_2$). For example, *aequorin* is a single polypeptide chain protein of molecular weight 20,000. It contains approximately 3 high affinity $Ca^{2+}$ binding sites (10, 28-30). In addition, *aequorin* contains approximately one mol of coelenterate-type luciferin (Fig. 1) non-covalently bound to the protein (31).

A comparison of *aequorin* and the *Renilla* luciferase-luciferin luminescence systems indicates that the chemistry is the same. As indicated above the luciferins are identical. Furthermore the products of the $Ca^{2+}$ triggered luminescence of *aequorin* are $CO_2$ and oxyluciferin (32, 33; see Fig. 1). An interesting difference is the lack of a dissolved oxygen-requirement for *aequorin* luminescence (9). However, native *aequorin* can be regenerated, in the absence of $Ca^{2+}$, by incubating the apoprotein with 2-mercaptoethanol, luciferin or luciferin analogues, and dissolved oxygen (8, 34).

Taken together the above data suggest that oxygen is incorporated into a site on apoaequorin (not luciferin) to form an oxygenated species, possibly a hydroperoxide. The binding of $Ca^{2+}$ to *aequorin* presumably induces a conformational change in the protein resulting in the transfer of the oxygenated species to luciferin to form a luciferin hydroperoxide intermediate. Fig. 4 illustrates a model for *aequorin* luminescence and recharging based on the above interpretation of the data. According to the model photoproteins are oxygenases which are most unique since, in the absence of $Ca^{2+}$, they exist as oxygen-containing enzyme-substrate intermediates that are remarkably stable.

### B. *Green Fluorescent Protein (GFP)*

Like *Renilla*, *Aequorea* contains a GFP which functions as an energy transfer acceptor (35, 36). Unlike *Renilla* the mechanism of energy transfer, based on quantum yields, appears to be primarily a radiative transfer (37). *Aequorea* GFP exists in the native state as a monomer of molecular weight 30,000 (38). Although the absorption spectrum of *Aequorea* GFP

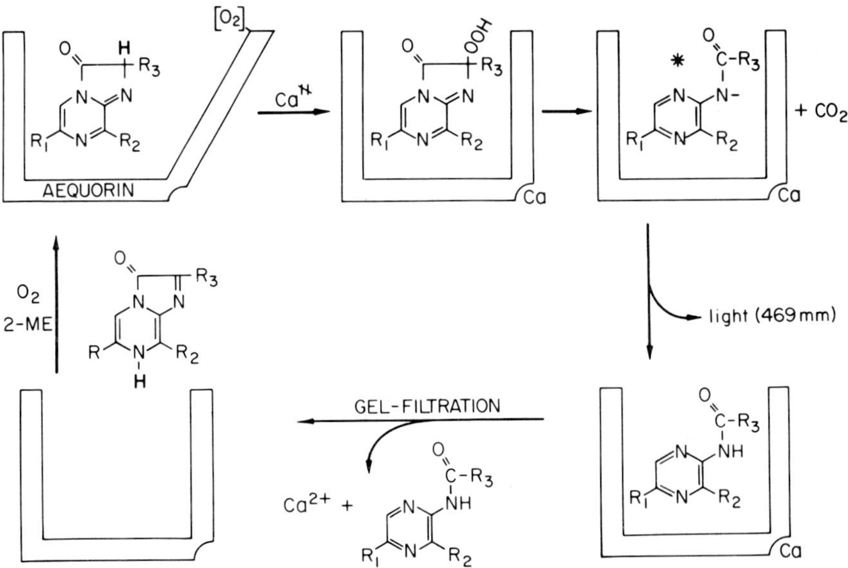


*FIGURE 4. Model for $Ca^{2+}$-triggered light emission from aequorin and its recharging.*

differs from that of *Renilla* their fluorescence emission spectra are identical with essentially identical fluorescence quantum yields (39). Recent spectrophotometric and fluorimetric data on denatured *Aequorea* and *Renilla* GFP suggest that the chromaphores are identical (39). A structure for the chromaphore of *Aequorea* GFP has recently been proposed by Shimomura (40).

## REFERENCES


1. Morin, J. G., *in* "Coelenterate Biology: Reviews and Perspectives" (Muscatine, L. and Lenhoff, H. M., eds.) pp. 397-438, Academic Press, New York (1974).
2. Parker, G. H., *J. Exp. Zool.* 31, 475 (1920).
3. Nicol, J. A. C., *J. Exp. Biol.* 32, 299 (1955).
4. Buck, J. B., *Biol. Bull.* 144, 19 (1973).
5. Anderson, P. A. V. and J. F. Case, *Biol. Bull.* 149, 80 (1975).
6. Anderson, P. A. V. *in* "Coelenterate Ecology and Behavior" (Mackie, G. O., ed.) pp. 609-18. Plenum Press, New York (1976).
7. Satterlie, R. A., P. A. V. Anderson, and J. F. Case *in* "Coelenterate Ecology and Behavior" (Mackie, G. O., ed.) pp. 619-27. Plenum Press, New York (1976).
8. Cormier, M. J. *in* "Bioluminescence in Action" (Herring, P. J., ed.) pp. 75-108. Academic Press, London (1978).
9. Shimomura, O., F. H. Johnson, and Y. J. Saiga, *Cell. Comp. Physiol.* 59, 223 (1962).
10. Blinks, J. R., F. F. Prendergast, and D. G. Allen, *Pharmacol.Rev.* 28, 1 (1976).
11. Charbonneau, H. and M. J. Cormier, *J. Biol. Chem.* 254, 769 (1979).
12. Ward, W. W. and M. J. Cormier, *J. Biol. Chem.* 254, 781 (1979).
13. Matthews, J. C., K. Hori, and M. J. Cormier, *Biochem.* 16, 85 (1977).
14. Matthews, J. C., K. Hori, and M. J. Cormier, *Biochem.* 16, 5217 (1977).
15. Hart, R. C., J. C. Matthews, K. Hori, and M. J. Cormier, *Biochem.* 18, 2204 (1979).
16. Wampler, J. E., K. Hori, J. Lee, and M. J. Cormier, *Biochem.* 10, 2903 (1971).
17. Morin, J. G. and J. W. Hastings, *J. Cell Physiol.* 77, 313 (1971).
18. Ward, W. W. and M. J. Cormier, *Photochem. Photobiol.* 27 389 (1978).

19. Hori, K., H. Charbonneau, R. C. Hart, and M. J. Cormier, *Proc. Nat'l Acad. Sci. USA* 74, 4285 (1977).
20. Anderson, J. M. and M. J. Cormier, *Biochem. Biophys. Res. Commun.* 68, 1234 (1976).
21. Henry, J. P., *Biochem. Biophys. Res. Commun.* 62, 253 (1975).
22. Anderson, J. M. and M. J. Cormier, *Biochem. Biophys. Res. Commun.* 81, 114 (1978).
23. Shimomura, O., F. H. Johnson, and Y. Saiga, *J. Cell. Comp. Physiol.* 62, 9 (1963).
24. Morin, J. G. and J. W. Hastings, *J. Cell. Physiol.* 77, 305 (1971).
25. Campbell, A. K., *Biochem. J.* 143, 411 (1974).
26. Ward, W. W. and H. H. Seliger, *Biochem.* 13, 1500 (1974).
27. Ward, W. W. and H. H. Seliger, *Biochem.* 13 1491 (1974).
28. Shimomura, O., F. H. Johnson, and Y. Saiga, *J. Cell. Comp. Physiol.* 62, 1 (1963).
29. Shimomura, O. and F. H. Johnson, *Nature, Lond.* 227, 1356 (1970).
30. Allen, D. G., J. R. Blinks, and F. G. Prendergast, *Science* 195, 996 (1977).
31. Ward, W. W. and M. J. Cormier, *Proc. Nat'l. Acad. Sci. USA* 72, 2530 (1975).
32. Shimomura, O. and F. H. Johnson, *Tetrahedron Lett.* 31 2963 (1973).
33. Shimomura, O. and F. H. Johnson, *Proc. Nat'l. Acad. Sci. USA,* 75, 2611 (1978).
34. Shimomura, O. and F. H. Johnson, *Nature* 256, 236 (1975).
35. Johnson, F. H., O. Shimomura, Y. Saiga, G. Gershman, G. T. Reynolds, and J. R. Waters, *J. Cell. Comp. Physiol.* 60, 85 (1962).
36. Morise, H., O. Shimomura, F. H. Johnson, and J. Winart, *Biochem.* 13, 2656 (1974).
37. Ward, W. W., *Photochem. Photobiol. Rev.* 4, 1 (1979).
38. Prendergast, F. G. and K. G. Mann, *Biochem.* 17, 3448 (1978).
39. Ward, W. W., C. W. Cody, R. C. Hart, and M. J. Cormier, *Photochem. Photobiol.* 31, 611 (1980).
40. Shimomura, O., *FEBS Lett.* 104, 220 (1979).

# PROPERTIES OF THE COELENTERATE GREEN-FLUORESCENT PROTEINS

William W. Ward

Department of Biochemistry & Microbiology
Rutgers University, Cook College
New Brunswick, New Jersey

## I. INTRODUCTION

Green-fluorescent proteins (GFP) are accessory molecules in the bioluminescence reactions of certain of the coelenterates (1-4). Species that contain GFP produce green bioluminescence via energy transfer from excited state oxyluciferin (5, 6). The GFP's that have received the most attention are those isolated from the sea pansy *Renilla reniformis* and the jellyfish *Aequorea aequorea* (5, 7). These proteins, abbreviated R-GFP and A-GFP, are acidic, globular proteins with similar amino acid compositions and monomer molecular weights of 27,000-30,000 (5, 7, 8). They differ, however, in a variety of other properties including absorption and excitation spectra and mechansim of energy transfer (9).

### A. *Purification and Characterization of GFP*

Methods for purifying *Renilla* (7) and *Aequorea* (5) GFP's to apparent homogeneity are published. The electrophoretically pure proteins are reported to have absorption ratios of 1.0 for A-GFP (395 nm/280 nm) and 5.6 for R-GFP (498 nm/280 nm). A partial list of the physical properties of these proteins appears in Table I. With slight modifications in the purification methods, we have recently achieved ratios of 1.2 and 6.3, respectively, indicating that a significantly higher degree of purification is possible (9, 10).



ISBN 0-12-208820-4

*TABLE I. Selected Physical Properties of the GFP's*

| Property | *Renilla* GFP[a] | *Aequorea* GFP |
|---|---|---|
| Isoelectric point | 5.34 ± 0.07 | ~4.8[b] |
| Molar extinction Coefficients | 133,000 (498 nm) 21,000 (280 nm) | 24,000[c] (393 nm) 25,000[c] (280 nm) |
| Fluorescence QY | 80 ± 2% | 72 - 78% |
| Molecular weight | 27,000 (monomer) 54,000 (dimer) | 30,000 |

[a]From reference 7, [b]Approximated from ion exchange elution [c]From reference 5, using 30,000 MW, [d]From reference 8.

### B. Energy Transfer Mechanisms

GFP is separated from luciferase (or photoprotein) during purification (5,7). The *in vitro* reaction between GFP-free luciferase and luciferin (or pure photoprotein and Ca++) generates blue light (5, 11). Addition of GFP under proper conditions produces green luminescence spectrally identical to the live animal flash (5, 6). With *Aequorea*, co-immobilization of GFP and photoprotein appears necessary for green *in vitro* luminescence (5). In the sea pansy, green light is generated from dilute solutions of GFP and luciferase (6), suggesting that these proteins form a complex prior to light emission. Excitation energy from luciferase-bound oxyluciferin must be transferred to GFP while the proteins are complexed (12). Evidence for protein-protein interaction during energy transfer (ET) in *Renilla* includes: (a) full ET at 0.5 μM GFP, (b) disruption of ET by increasing μ to 0.1 M, (c) or by modifying GFP lysines, (d) luciferase-GFP cross reactions between closely related species only, and (e) luciferase-GFP complex formation on a gel filtration column pre-equilibrated with GFP (12).

## II. THE GFP CHROMOPHORE

Spectral characteristics of the GFP's result from a covalently-bound chromophore (13, 14). Such a chromophore, released from A-GFP by papain digestion, has recently been characterized (13). The proposed chromophore is a cyclic tripeptide apparently derived from the primary sequence of A-GFP. The R-GFP chromophore has not yet been chemically

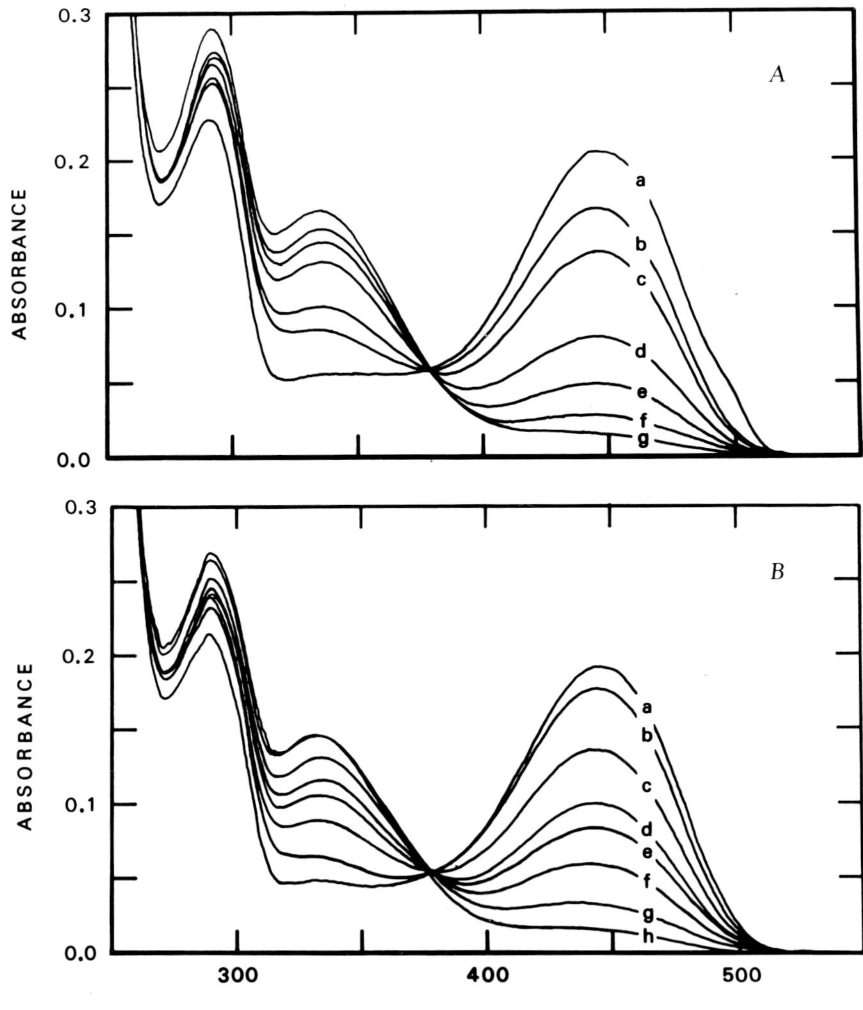


*FIGURE 1A. Degradation of R-GFP with time at pH 13.5 Native R-GFP was denatured by warming to 60°C for 1 min at pH 12.5 then dialyzed 12 hr. at pH 12.5. The pH was raised to 13.5 and spectra were taken at 0, 5.5, 33, 110, 186, 320, and 596 min. intervals (curves a-g, respectively). Note the small undenatured component in curve (a) absorbing at 498 nm.*

*FIGURE 1B. A-GFP. Spectra measured at 0, 5.5, 45, 88, 120, 188, 377, and 1092 min. intervals (curves a-h, respectively).*

identified; however, we recently showed that the chromophore of R-GFP may be identical with that of A-GFP (14). Spectral characteristics of guanidine-denatured and pronase-digested GFP's, for example, are identical. Spectrophotometric titration in guanidine shows a common isosbestic point at 405 nm and a pK of 8.1 for ionization of the chromophore. Below pH 8.1, a form absorbing at 383-4 nm predominates and above pH 8.1 the major form of the chromophore absorbs at 447-8 nm (14). Treatment of either protein with 0.3*M* sodium hydroxide solition irreversibly alters the 447-8 nm absorbing species, generating a new chromophore that absorbs maximally at 333-335 nm. The transition requires several hours at room temperature, going through a sharp isosbestic point with time at 378 nm (Figs. 1A and 1B). The large spectral differences between *native* proteins (see Table I) appear to result from differences in non-covalent interactions between the chromophore and other regions of the protein rather than from structural differences between chromophores per se.

A chromophore/protein ratio has not been precisely determined, although 1:1 stoichiometry for A-GFP can be inferred (13). We have determined extinction coefficients for acid-denatured (pH $\simeq$ 2) GFP both in acid and in base (see Table II). Corresponding values for denatured A-GFP ield an extinction coefficient for native A=GFP of 27,600 at 393 nm. This compares favorably with the independently based extinction coefficient of 24,000 (Table I) determined by Morise *et al.*(5). These data clearly indicate that both GFP species contain the same chromophore/protein ratio.

## III. PHYSICAL PROPERTIES

### A. *Fluorescence*

The spectral characteristics of native *Renilla* GFP (absorption, excitation and emission) are virtually unaffected by prolonged treatment under strongly denaturing conditions (14) such as 8*M* urea, 6*M* guanidine hydrochloride, or 1% sodium dodecyl sulfate (with and without reducing agents and at temperatures up to about 45-50°C). A-GFP is somewhat more labile, slowly bleaching under these conditions at room temperature or below. Both proteins retain their fluorescence characteristics in dilute aqueous buffer solution at temperature as high as 65°C. However, there is a dramatic difference between the two GFP's as pH is varied at room temperature (Fig. 2). The fluorescence of R-GFP is constant from pH 6 to pH 12.5, followed by an almost total loss above pH 13. A-GFP, on the

*TABLE II. Molar Extinction Coefficients of Denatured GFP*

| *Range of pH* | *Wavelength* | *Renilla GFP* | *Aequorea GFP* |
|---|---|---|---|
| <6 | 383-4 | 30,000 | 29,400 |
| >10 | 447-8 | 44,000 | 44,100 |
| 1-13 | 405 | 20,000 | 20,400 |

other hand, appears to increase in fluorescence more than 2-fold between pH 9 and 12 before also dropping off at pH 13. This apparent fluorescence enhancement results from a reversible shift in the excitation maximum (but not the emission maximum) from 393-5 nm to about 470 nm. Corresponding rever-

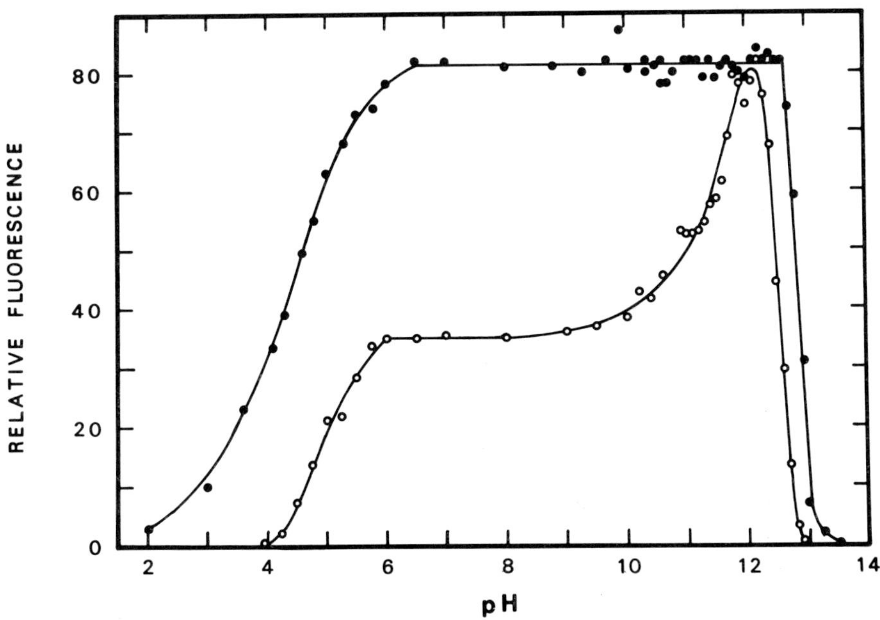


*FIGURE 2. GFP fluorescence vs. pH. Solid circles for R-GFP, open circles for A-GFP. Samples incubated 5 ± 0.5 min at 22 ± 2°C before fluorescence was measured on a Turner 110 fluorimeter with blue lamp, Ditric FITC excitation filter, and Corning 3-70 emission filter. Buffers were: 0.05M each of glycine·HCl, sodium phosphate, sodium citrate (pH 2-11), 0.025M $Na_2HPO_4$ (pH 11-11.9), 0.05M KCl·NaOH buffer (pH 12-13), and 0.1 to 1.0M NaOH (pH 13-14).*

sible changes in the absorption spectra of A-GFP with pH are shown in Fig. 3. These spectral shifts may account for deviations from Beer's law upon dilution as previously reported (5). It is interesting that R-GFP undergoes no such spectral shift under any perturbing conditions that have been investigated.

### *B. Conformational Analysis by CD*

Circular dichroism (CD) is a sensitive method for detecting conformational changes in proteins. In this study the CD signal was measured while temperature was continuously varied from 25° to 80°C. Measurements were made at 205 nm where an optimum difference was observed between CD spectra of native and boiled GFP (greater differences were observed at shorter wavelengths, but at the sacrifice of signal/noise

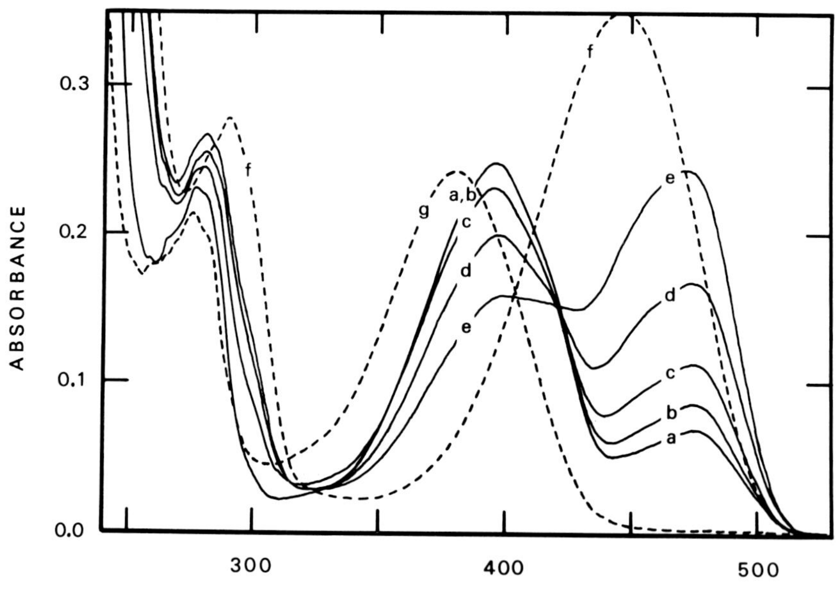


*FIGURE 3. A-GFP absorption spectra at various pH's. Samples incubated 30 min. at 22 ± 2°C at pH: 5.46 (a), 8.08 (b), 10.22 (c), 11.07 (d), 11.55 (e), 13.0 (f) and 1.0 (g). For curves (a-e) the buffer contained 0.01M each of sodium citrate, sodium phosphate and glycine. Sample (f) was in 0.1M NaOH and sample (g) was in 0.1M HCl. Note isosbestic points at 422 nm and 405 nm.*

ratio). The results of these experiments (Fig. 4) indicate that loss of fluorescence closely parallels disruption of secondary structure. Markedly varying the temperature gradient, prolonging heating at 80°C, or increasing temperature to 85°C had no effect on the CD signal except for a small change that correlated with the appearance of turbidity.

It would appear that rapid and extensive unfolding of the protein chain during heat denaturation is responsible for loss of GFP fluorescence, thus supporting our hypothesis that integrity of protein conformation is essential for the maintenance of a fluorescent form of the chromophore (14).

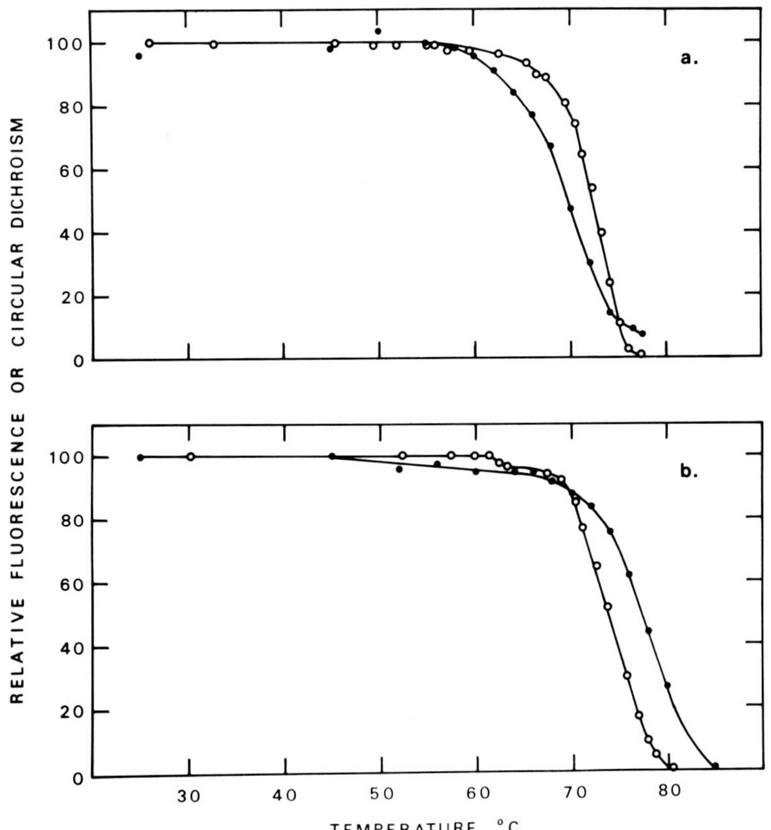


*FIGURE 4. Circular dichroism (open circles) and fluorescence melting curves for Renilla (a) and Aequorea (b) GFP. CD sample was continuously heated at a rate of 20°C per hour. Fluorescence read at 22 ± 2°C after heating at the same rate. All samples in 0.01M sodium phosphate pH 8.0 buffer.*

Preliminary correlations between fluorescence and CD, using guanidine·HCl, acid, or base as the perturbing agent, also support this conclusion.

REFERENCES


1. Johnson, F. H., O. Shimomura, Y. Saiga, L. C. Gershman, G. T. Reynolds, and J. R. Waters, *J. Cell. Comp. Physiol.* 60, 85 (1962).
2. Morin, J. G., and J. W. Hastings, *J. Cell. Physiol.* 77, 313 (1971).
3. Wampler, J. E., K. Hori, J. W. Lee, and M. J. Cormier, *Biochem.* 10, 2903 (1971).
4. Wampler, J. E., Y. D. Karkhanis, J. G. Morin, and M. J. Cormier, *Biochem. Biophys. Acta,* 314, 104 (1973).
5. Morise, H., O. Shimomura, F. H. Johnson, and J. Winant, *Biochem.* 13, 2656 (1974).
6. Ward, W. W. and M. J. Cormier, *J. Phys. Chem.* 80, 2289 (1976).
7. Ward, W. W. and M. J. Cormier, *J. Biol. Chem.* 254, 781 (1979).
8. Prendergast, F. G. and K. G. Mann, *Biochem.* 17, 3448 (1978).
9. Ward, W. W., *in* "Photochemical and PHotobiological Reviews" (K. Smith, ed.) Vol. 4, p. 1. Plenum Press, New York, (1979).
10. Roth, A. F. and Ward, W. W., unpublished.
11. Hori, K., H. Charbonneau, R. C. Hart, and M. J. Cormier, *J. Proc. Nat. Acad. Sci. USA* 74, 4285 (1977).
12. Ward, W. W. and M. J. Cormier, *Photochem. Photobiol.* 27 389 (1978).
13. Shimomura, O., *FEBS Lett.* 104, 220 (1979).
14. Ward, W. W., C. W. Cody, R. C. Hart, and M. J. Cormier, *Photochem. Photobiol.* 31, 611 (1980).

# APPLICATIONS OF AEQUORIN

John R. Blinks

Department of Pharmacology
Mayo Foundation
Rochester, Minnesota

## I. PROPERTIES OF AEQUORIN

Calcium-activated photoproteins isolated from marine coelenterates have properties that make them extremely useful as tools in experimental biology. Aequorin, the photoprotein extracted from the hydromedusan *Aequorea forskalea* (1) is the best-known and most readily available of these substances. It is a self-contained bioluminescent system consisting of a single polypeptide chain (M.W. 20,000) to which an imidazolopyrazinone chromophore (luciferin) is tightly but not covalently bound. Oxygen is also bound to the complex in some as yet undetermined way. The binding of calcium ions to the protein facilitates the interaction of these two ligands, with the oxidation of the chromophore, the generation of an excited state, and the emission of a photon ($\lambda_{max}$ 469 nm). (For further details see refs. 2,3,4).

The luminescent reaction proceeds at a very low rate in the absence of $Ca^{++}$, but increases by a factor of more than $10^6$ in the presence of saturating concentrations of $Ca^{++}$. A log-log plot of the relation between $[Ca^{++}]$ and luminescent intensity is shown in Fig. 1. The slope of this curve (2.5) indicates that the luminescent reaction of the aequorin molecule involves the binding of at least 3 calcium ions.




ISBN 0-12-208820-4

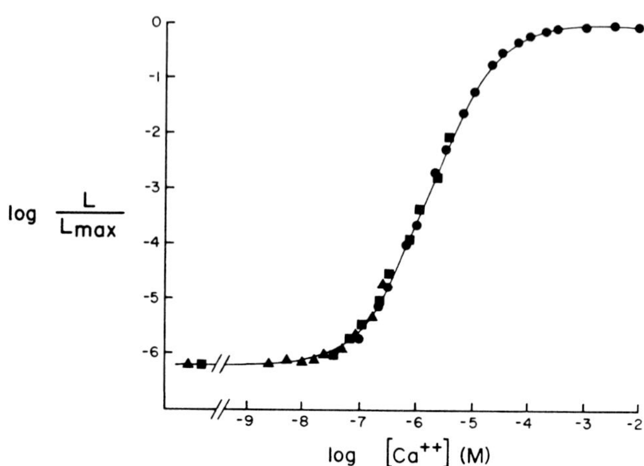


*FIGURE 1. Relation between [$Ca^{++}$] and aequorin luminescence. Log-log plot of results obtained with simple dilutions of $CaCl_2$ and two different $Ca^{++}$-buffer systems. All solutions contained 150 mM KCl and 5 mM PIPES (piperazine-N,N'-bis (2-ethanesulfonic acid)) and were at pH 7.0, 22°C. $L/L_{max}$ indicates peak light intensity as a fraction of that obtained in saturating [$Ca^{++}$]. $Ca^{++}$ concentrations were established as follows: by dilution of $CaCl_2$ in solutions previously freed of contaminating $Ca^{++}$ by passage through columns of Chelex-100 resin (Bio-Rad); with 1 mM EGTA (ethyleneglycol bis(β-aminoethylether)-N,N'-tetraacetic acid) buffer, log apparent $Ca^{++}$ binding constant (K') 6.45; with 1 mM CDTA [(1,2-cyclohexylenedinitrilo)tetraacetic acid] buffer, log K' = 7.60. The two points on the far left (undefined abscissae) were obtained with the chelators alone (no added $Ca^{++}$). The curved line was calculated for a two-state model with three $Ca^{++}$ binding sites per aequorin molecule. (Data replotted from ref. 5 which should be consulted for further details. This figure reproduced from ref. 6 by permission of the publisher.)*

The position of the calcium concentration-effect curve is sensitive to monovalent salt concentration, and the influence of $Ca^{++}$ is apparently specifically antagonized by $Mg^{++}$. $Sr^{++}$ and $Ba^{++}$ are weak (partial) agonists; the lanthanides are more potent than $Ca^{++}$ (Fig. 2).

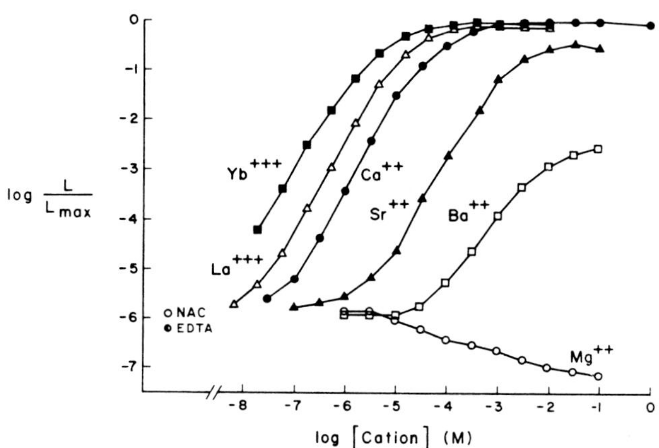


*FIGURE 2. Effects of various cations on aequorin luminescence. Concentration-effect curves were determined at 21°C by diluting salts of the various ions in a Chelex-treated solution of 150 mM KCl, 5 mM PIPES, pH 6.5. The point marked NAC indicates the level of luminescence in this solution with no additional cation: the point marked EDTA indicates the level after the addition of 1 mM EDTA. (Reproduced from ref. 6 with permission of the publisher. For further details see the original paper.)*

Aequorin exhibits heterogeneity, and at least twelve separate luminescent species can be distinguished on isoelectric focusing. All of these components migrate together on gel filtration and on SDS gels. The isoaequorins have been partially resolved by preparative gel electrophoresis, and it is now clear that they differ somewhat with respect to calcium sensitivity and to the level of $Ca^{++}$-independent luminescence (I. R. Neering, W. G. Wier, and J. R. Blinks, in preparation). Different batches of aequorin differ somewhat in these characteristics in the relative proportions of the isoaequorins present.

## II. APPLICATIONS OF AEQUORIN

### *A. As a $Ca^{++}$-indicator*

Aequorin has been used as a $Ca^{++}$-indicator both *in vitro* and in living systems. Its novelty has led to its use *in vitro* in circumstances where calcium-sensitive electrodes might have served as well or better, and the recent development of highly sensitive $Ca^{++}$-selective microelectrodes (7) has probably eliminated most advantages that aequorin had for such applications except when [$Ca^{++}$] changes too rapidly to be followed by the rather slowly responding electrodes. The speed of response of aequorin, though not rapid enough for all purposes (8) is very much greater than that of any ion-selective electrode.

Aequorin is particularly well-suited for use as an intracellular $Ca^{++}$-indicator. In contrast to that of extracellular fluids, the range of [$Ca^{++}$] likely to be encountered inside cells is well matched to the $Ca^{++}$-sensitivity of aequorin. Aequorin also has the advantage of relative specificity: $Ca^{++}$ is the only ion likely to occur naturally in biological systems in concentrations sufficient to trigger the luminescent reaction. Aequorin usually tolerates the intracellular environment well -- the calcium concentrations of most cells at rest are low enough so that the photoprotein is not consumed at a troublesome rate. Aequorin has proved to be relatively non-toxic in a large variety of cells. It has now been used successfully as an intracellular $Ca^{++}$-indicator in more than 40 different types of cells ranging from protozoa to human muscle cells (9,10). Aequorin has proved particularly useful in muscle cells because alternative techniques (metallochromic dyes or $Ca^{++}$-sensitive microelectrodes) cannot readily be used in tissues that are subject to movement. Figure 3 shows an example of the aequorin signal recorded during the twitch of a frog skeletal muscle fiber.

With the technique of microscopic image intensification (11) aequorin makes it possible not only to detect changes in intracellular calcium concentration, but also to localize them within the cell. This marriage of techniques has only begun to be exploited (e.g., 12,13,14) but it appears to be a powerful one and one that will give information that cannot be obtained in any other foreseeable way.

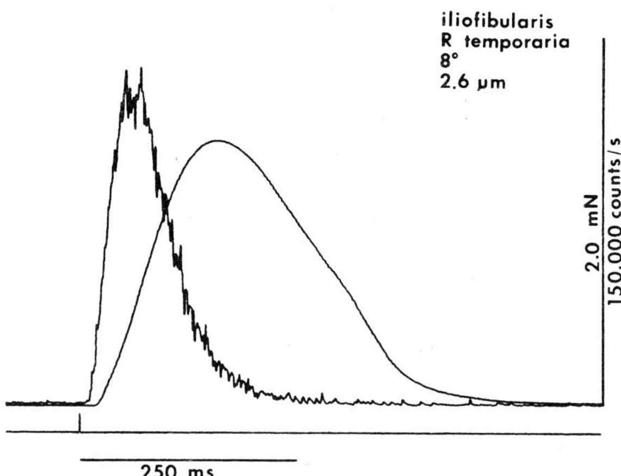


*FIGURE 3. Aequorin signal and force developed by a single frog skeletal muscle fiber in an isometric twitch. Lower trace = stimulus; noisy trace = light record.*

## *B. As a Model System*

Apart from the intrinsic interest attached to its unusual mechanism of luminescence (see previous paper), aequorin offers interesting possibilities as a model system for the study of more general properties of proteins and of protein-ligand interactions. The importance of calcium-binding proteins as regulators of cell function is becoming more and more apparent, and aequorin may have important properties in common with other members of the group. Allen et al.(5) have recently proposed a two-state model for the aequorin reaction that can account for the shape of the $Ca^{++}$ concentration-effect curve. If this model continues to appear appropriate, aequorin may prove useful for the study of two-state mechanisms as such. The similarity between certain properties of aequorin and those of drug receptors is striking (6). Influences of a variety of local and general anesthetics agents on the aequorin reaction have been described in two recent papers (15,16). Although there are still unexplained discrepancies between the results reported in the two papers, these findings have generated speculation that studies on photoproteins could prove useful in efforts to understand the molecular basis of anesthesia.

REFERENCES

1. Shimomura, O., F. H. Johnson, and Y. Saiga, *J. Cell. Comp. Physiol. 59*, 223 (1962).
2. Blinks, J. R., F. G. Prendergast, and D. G. Allen, *Pharmacol. Rev. 28*, 1 (1976).
3. Cormier, M. J., *in* "Bioluminescence in Action" (P. J. Herring, ed.), p. 75. Academic Press, London (1973).
4. Shimomura, O., and F. H. Johnson, *in* "Detection and Measurement of Free $Ca^{2+}$ in Cells" (C. C. Ashley and A. K. Campbell, eds.), p. 73. Elsevier/North Holland, Amsterdam (1979).
5. Allen, D. G., J. R. Blinks, and F. G. Prendergast, *Science 195*, 996 (1977).
6. Blinks, J. R., D. G. Allen, F. G. Prendergast, and G. C. Harrer, *Life Sciences 22*, 1237 (1978).
7. Ammann, D., P. C. Meir, and W. Simon, *in* "Detection and Measurement of Free $Ca^{2+}$ in Cells" (C. C. Ashley and A. K. Campbell, eds.), p. 117. Elsevier/North Holland, Amsterdam (1979).
8. Hastings, J. W., G. Mitchell, P. H. Mattingly, J. R. Blinks, and M. van Leeuwen, *Nature (Lond.) 222*, 1047 (1969).
9. Blinks, J. R., *Photochem. Photobiol. 27*, 423 (1978).
10. Blinks, J. R., *in* "Techniques in Cellular Physiology" (P. F. Baker, ed.). Elsevier/North Holland, Amsterdam (1981).
11. Reynolds, G. T., *in* "Detection and Measurement of Free $Ca^{2+}$ in Cells" (C. C. Ashley and A. K. Campbell, eds.), p. 227. Elsevier/North Holland, Amsterdam (1979).
12. Rose, B., and W. R. Loewenstein, *J. Membrane Biol. 28*, 87 (1976).
13. Gilkey, J. C., L. F. Jaffee, E. B. Ridgway, and G. T. Reynolds, *J. Cell Biol. 76*, 448 (1978).
14. Taylor, D. L., J. R. Blinks, and G. T. Reynolds, *J. Cell Biol. 86*, in press (1980).
15. Kamaya, H., I. Ueda, and H. Eyring, *Proc. Natl. Acad. Sci. U.S.A. 74*, 5534 (1977).
16. Baker, P. F., and A.H.U. Schapira, *Nature 284*, 168 (1980).

# EARTHWORM BIOLUMINESCENCE

John E. Wampler

Bioluminescence Laboratory
Department of Biochemistry
University of Georgia
Athens, Georgia

## I. OVERVIEW

While bioluminescence has not been recognized as a general feature of earthworm biology, the number of known luminous species has grown in recent years. When these observations are correlated with distribution data, the phenomenon appears much more common than one might at first expect. Previous reviews (1-3) and reports of new species (4,7) show that there are at least 33 species in 16 genera which exhibit bioluminescence. These species are classified into three families of oligochaetes (4). More importantly, native bioluminescent species are found on five continents and several very broadly distributed species are bioluminescent. The map of Fig. 1 shows the distribution of some of the luminous species. This is not a complete record nor has any attempt been made to exhaustively plot all of the reported localities of these species, but it does serve to illustrate the wide distribution of bioluminescent earthworms. In addition, the ease with which we have recently discovered luminescence in species not previously known to be luminous (6, 7) suggests that other common species may also be bioluminescent.

The physiology of earthworm bioluminescence has been confused by some casual, not too careful, observations in the early literature. The bulk of the literature, however, along with recent comparative studies indicates that all luminous species exhibit similar physiology. The luminescence originates in a viscous fluid exuded by the worms following stimulation. In many species this luminescence is associated with

[1]*The studies of earthworm bioluminescence have been supported by the Nat'l. Science Foundation (Grant No.PCM 76-15842).*




ISBN 0-12-208820-4

large cells suspended in the fluid (for review see reference 7). The source of the fluid and the cells is the coelomic cavity and, at least in *Diplocardia longa*, the cells are degenerated, free chloragogen cells (8). In a few species, notably *Pontodrilus matsushimensis* (Lynch, unpublished), *Pontodrilus bermudensis* (7), and *Microscolex phosphoreus* (9), a cellular source for the luminescence has been questioned. These same species are also lacking the dorsal pores through which the other worms exude the luminescent fluid. However, most studies still indicate that the exudate of these worms is coelomic fluid (7, 10) and that there may be a particulate source of the bioluminescence (10 and Jamieson and Wampler, unpublished observations).

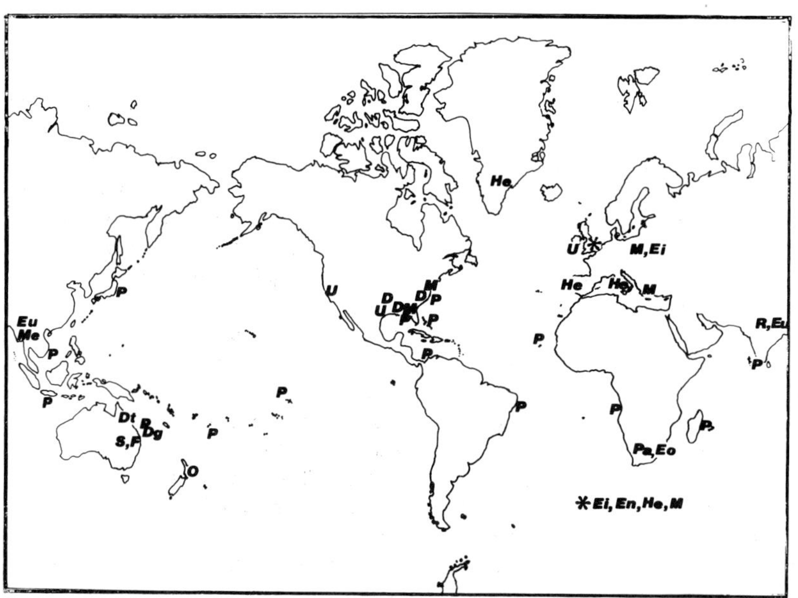


*FIGURE 1. Distribution of bioluminescent earthworms. Luminous species in each of the following genera have been found in the localities indicated by the letter abbreviations: D - Diplocardia (3 species), Dg - Digaster, Dt - Diplotrema, Ei - Eisenia, En- Enchytraeus, Eo - Eodrilus (3 species), Eu - Eutyphdeus (6 species), F - Fletcherodrilus (2 species), He - Henlea, M - Microscolex, Me - Megascolex, O - Octochaetus, P - Pontodrilus (2 species), Pa - Parachilota (4 species), R - Ramiella, S - Spenceriella (4 species), U - unidentified.*

Until quite recently biochemical studies had been carried out on only three species, *Eisenia submontana* (11), *Octochaetus multiporous* (12) and *D. longa* (13-18). Only in the case of *D. longa* has a luciferin and luciferase both been isolated and purified (14-16). We have recently extended these studies with identification of the chemical structure of *D. longa* luciferin (17) and with chemical and physical characterization of a high specific activity form of *D. longa* luciferase (18). Failures to isolate luciferin and luciferase activities in earthworms in the past(1, 9, 19) were probably due to the assumption that, like most other bioluminescence, the biochemistry depended on molecular oxygen. This view was reinforced by observations of *in vivo* oxygen dependence in some species, although Komarek (11) noted that *E. submontana* luminescence could occur in a pure hydrogen atmosphere. Cormier and coworkers (13) first reported that the *D. longa* reaction is stimulated by hydrogen peroxide and is not dependent on oxygen *in vitro*. Our recent studies of the comparative biochemistry of other species show that hydrogen peroxide stimulates bioluminescence in extracts of their coelomic fluid in all cases and that *D. longa* luciferin and/or luciferase also stimulate emission from such extracts. Table I summarizes the comparative data for a wide variety of luminescent species.

## II. THE BIOLUMINESCENCE OF *D. LONGA*

*D. longa* is a large acanthodrile, a member of one of the few native North American genera. Two other species in this genus are known to be luminous (7), but three non-luminous species have also been examined (Wampler, unpublished). *D. longa* is found in the southern states and advertises its presence by a relatively distinct mound of castings. It is large (some speciments over a foot long) and exudes copious amounts of sticky coelomic fluid when excited by chemical, tactile or electrical stimuli. The luminescence is bright (around $10^{12}$ photons per second, total peak flux) and blue-green in color ($\lambda_{max}$ = 500 nm).

Close examination of the exuded fluid shows that it is replete with large cells and small granules. The cells appear to be the source of the granules since they are packed full of several types of granular particles but contain few other subcellular organelles (8). Through dissection, tests for bioluminescence in dissected tissues and electron microscopic comparison of cellular morphology, these cells have been identified as free chloragogen cells which have

*TABLE I. Comparative Physiology and Biochemistry of Luminescent Earthworms.*

| *Species* | *Localities* | *Luminescent exudate* | *Coelomocytes Observed* | *In Vitro*[a] *Stimulation* | *Color* (λ *max*) | *References* |
|---|---|---|---|---|---|---|
| *Digaster keasti* | *Fraser Island, Australia* | *yes* | *yes* | *H* | *whitish* | *4* |
| *Diplocardia longa* | *Southern U.S.* | *yes* | *yes* | *H,La,Li* | *blue-green (500)* | *13-18* |
| *Diplotrema heteropora* | *Townsville, Australia* | *yes* | *yes* | *H,La* | *green (545)* | *5,7* |
| *Eisenia submontana* | *Poland* | *yes* | *yes* | - | *green-blue* | *11* |
| *Euthyphoeus peguanus* | *Rangoon* | *yes* | - | - | *whitish* | *22* |
| *Fletcherodrilus fasciatus* | *South Queensland, Australia* | *yes* | *yes* | *H,La* | *whitish* | *6,7* |
| *Microscolex phosphoreus* | *Europe, Asia, etc* | *yes* | *yes* | *H,La,Li* | *green (538)* | *10*[b] |
| *Octochaetus multiporous* | *North Island, New Zealand* | *yes* | *yes* | *H,La,Li* | *yellow (570)* | *7,12,23* |
| *Pontodrilus bermudensis* | *U.S., Australia, Africa, etc.* | *yes* | *yes* | *H,La* | *whitish (550)* | *6,7* |
| *Spenceriella curtisi* | *South Queensland, Australia* | *yes* | *yes* | *H,La,Li* | *green (535)* | *6,7* |

[a]*Stimulated by H - hydrogen peroxide, La- D. longa luciferase, Li- D. longa luciferin.*
[b]*Wampler, unpublished results.*

degenerated following their release from the chloragogen tissue. The multifunctional tissue surrounds the dorsal blood vessel and is part of the lining of the coelomic cavity. With the help of Dr. George Reynolds' image intensified microscope system (20), we were able to localize the luminescence within these cells (Fig. 2).

Purification and characterization of *D. longa* luciferin and luciferase have required the efforts of several people (13-18). Luciferin is a simple aliphatic aldehyde, N-isovaleryl-3-aminopropanal (I), which is easily synthesized (17, 21).

$$\begin{matrix} CH_3 \\ & CH{-}CH_2{-}C(=O){-}\underset{\displaystyle H}{\underset{|}{N}}{-}CH_2{-}CH_2{-}C(=O)H \\ CH_3 \end{matrix} \qquad (I)$$

*D. longa* luciferase is a large (300,000 daltons), asymmetric ($f/f_0 = 1.9$) protein which contains lipid (2%), carbohydrate (6%), unusual amino acids (5% proline and 6% hydroxyproline) and EPR silent copper in its active form (18).

We have recently shown through studies that examined the order of addition kinetics, the continuous variation kinetics, the effects of catalase on the reaction, and the conditions required for luciferase turnover that the true substrate of luciferase in the bioluminescence reaction is N-isovaleryl-3-amino-1-hydroxypropane hydroperoxide (18). The hydrogen peroxide reaction with luciferin to form this compound appears to be an uncatalyzed, spontaneous event which is similar to the hydration reaction of aldehydes. Under the assay conditions previously used where luciferase is in the presence of significant concentrations of free peroxide no turnover occurs and

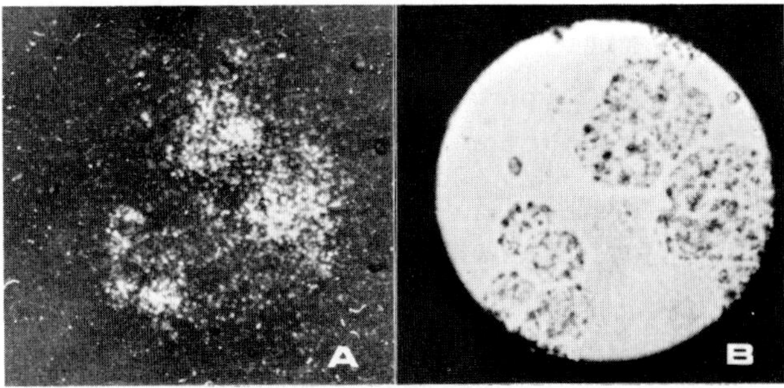


*FIGURE 2. A. Bioluminescence originates from within isolated D. longa coelomocytes. B. The cells of part A with substage illumination. From Rudie and Wampler, reference 8.*

the quantum yield for luciferase is 0.2% (15). However, if the peroxide titer is kept low using excess luciferin to form the hydroperoxide, luciferase exhibits its catalytic function (i.e., 63% quantum yield for luciferase while the hydroperoxide quantum yield is only 3%).

Neither the nature of the emitter nor the products of the bioluminescence reaction have been identified. The spectrum produced appears to be determined by the luciferase molecule (see Fig. 3), since addition of *D. longa* luciferin to an exhausted *in vitro* reaction from an extract of *P. bermudensis* give no spectral change, but addition of *D. longa* luciferase to the same type of sample causes a shift of the spectrum toward that of *D. longa* (7). Similar results were obtained with an extract from *O. multiporous* (7). This suggests that the emitter is either a transient, protein bound species or an energy transfer acceptor which accompanies the luciferase. However, luciferase purified from *D. longa* is not fluorescent (15, 18) and neither is luciferin or any of its likely degradation products. This suggests that the emitter is a transient species in the *D. longa* reaction. However, since fluorescent components have been detected in the coelomic

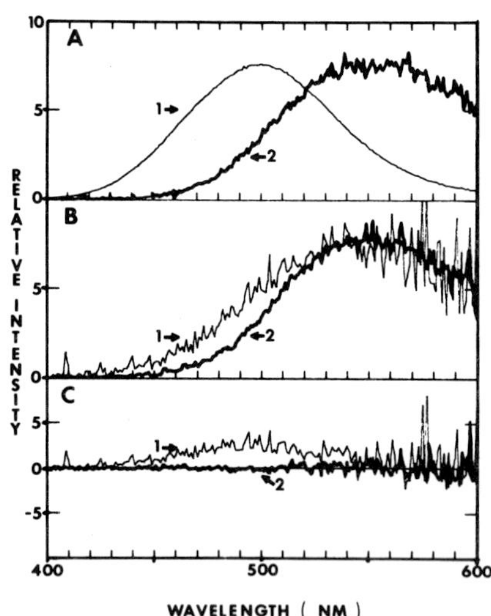


*FIGURE 3. A. In vitro bioluminescence of D. longa (1) and P. bermudensis (2) extracts. B. Following decay of the emission from a P. bermudensis extract, bioluminescence was reinitiated by addition of D. longa luciferase (1) or luciferin (2). C. Difference spectra of data from panels A and B.*

fluid of other species (9, 11, 12), a role for energy transfer can not be ruled out entirely. This is particularly true considering the range of emission maxima for the various species (Table I).

## III. ANALYTICAL APPLICATIONS FOR EARTHWORM BIOLUMINESCENCE

Earthworm bioluminescence is a sensitive analytical system for detecting hydrogen peroxide (sensitivity of around 10 picomoles per assay) and for components of biochemical reactions which produce peroxide (21, 22). With its broad pH optimum (>50% of maximal activity from pH 7 to 8.5), its insensitivity to extraneous solute species (protein, metal ions, enzymes, etc.), and with kinetic profiles which are diagnostic of the presence of those substances which do interfere, the *D. longa* reaction can be easily coupled to detection of low concentrations of oxidases and their substrates (22). This use of the *D. longa* system has recently been demonstrated in the detection of glucose, galactose, and putrescine and for each of the corresponding oxidases (22). The only limitation to broad use of earthworm bioluminescence as an analytical tool is the availability of *D. longa*. This points out the need for further studies of the biochemistry of the other bioluminescent species, particularly those with broad distribution.

## IV. ACKNOWLEDGMENTS

Many people have contributed to our developing knowledge of this phenomenon, but special thanks are due to Mr. Roy Holland who has operated the *D. longa* collection program, to Dr. Barrie Jamieson for many helpful discussions and insights, and to Dr. Milton Cormier for his encouragement.

## REFERENCES

1. Harvey, E. N., *in* "Bioluminescence", p. 233 ff, Academic Press, New York (1952).
2. Cormier, M. J., *in* "Chemical Zoology" (M. Florkin and B. Scheer, eds.) Vol. IV, p. 467. Academic Press, New York (1969).
3. Herring, P. J., *in* "Bioluminescence in Action " (P. Herring, ed.) p. 215. Academic Press, New York (1978).

4. Jamieson, B. G. M., *Proc. Roy. Soc. Qd.* 88, 83 (1977).
5. Dyne, G. R., *Mem. Qld. Mus.* 19, 373 (1979).
6. Jamieson, B. G. M., and J. E. Wampler, *Aust. J. Zool.* 27, 637 (1979).
7. Wampler, J. E., and B. G. M. Jamieson, *Comp. Biochem. Physiol.* 66B, 43 (1980).
8. Rudie, N. G., and J. E. Wampler, *Comp. Biochem. Physiol.* 59A, 1 (1978).
9. Bersis, D. S., *Fol. Biochim. Biol. Graeca* 14, 5 (1977).
10. Skowron, S., *Biol. Bull.* 45, 191 (1928).
11. Komarek, J., *Bull. Int. Acad. Sci. Boheme* 1934, 1 (1934).
12. Johnson, F. H., O. Shimomura, and Y. Haneda, *in* "Bioluminescence in Progress" (F. Johnson and Y. Haneda, eds.) p. 385. Princeton University Press, Princeton, New Jersey (1966).
13. Cormier, M. J., P. Kreiss, and P. M. Prichard, *in* "Bioluminescence in Progress" ref. cit., p. 363 (1966).
14. Bellisario, R., and M. J. Cormier, *Biochem. Biophys. Res. Comm.* 43, 800 (1971).
15. Bellisario, R., T. E. Spencer and M. J. Cormier, *Biochem.* 11, 2256 (1972).
16. Rudie, N. G., H. Ohtsuka, and J. E. Wampler, *Photochem. Photobiol.* 23, 71 (1976).
17. Ohtsuka, H., N. G. Rudie, and J. E. Wampler, *Biochem.* 15, 1001 (1976).
18. Rudie, N. G., M. Mulkerrin, and J. E. Wampler, *Biochem.*, in press (1980).
19. Gilchrist, J. D. F., *Trans. R. Soc. South Africa* 7, 203 (1918-19).
20. Reynolds, G. T., *Q. Rev. Biophys.* 5, 295 (1972).
21. Mulkerrin, M. G., and J. E. Wampler, *Methods in Enzymology* 57, 375 (1978).
22. Gates, G. E., *Rec. Indian Mus.* 27, 471 (1925).
23. Benham, W. B., *Nature* 60, 591 (1899).

# LUMINESCENCE ACTIVITY OF THE LAND SNAIL *QUANTULA STRIATA*

Yata Haneda

Yokosuka City Museum
Yokosuka
Kanagawa 238, Japan

Until our discovery of *Dyakia striata* in 1942, luminescence was not known in the terrestrial gastropods. The luminescence of the snail was discovered by Mr. Kumazawa, an entomologist, and I while working at the Shonan Museum (Raffles Museum) in Singapore. The snail was found on the lawn of the Goodwood Park Hotel, Scottroad, Singapore. We were both surprised to find it to be luminous. A report was prepared and sent to Tokyo for publication, but the manuscript and all copies of it were destroyed by fire during the war.

On being repatriated from Singapore, I published a short note in 1946 entitled "A luminous land snail *Dyakia Striata* found in Malaya" which was based on recollections of the earlier observations. The same findings were also reported at the bioluminescence conference in Asilomar, Pacific Grove, California, in 1954.

With assistance from the National Science Foundation, I visited Singapore in April 1960, and collected more live specimens of the snail and returned them to Japan for study, and made further observations on the snail (Haneda, 1969). In addition to the main luminous organ located in the mucous fold beneath the mouth, I found the surface of the mantle and the foot of the young snail to be continuously luminous, a fact which I had overlooked in my 1946 report.

If the snail retracts its body into the shell the dim, continuous luminosity of the mantle (Fig. 1) may still be observed through the translucent shell with the dark-adapted eye. Only 2 or 3 specimens out of 10 show this



ISBN 0-12-208820-4

condition, which appears to be an individual characteristic.

According to Dr. Horikoshi, a malacologist, the correct name of the snail is *Quantula Striata*. *Quantula* is commonly found during the rainy season in Singapore, Malay Peninsula, and on Batam and Kundure Island, Rhio Archipelago. It is nearly always on the lawn and in grass along the roadside, and also hibiscus fence, but never on tree trunks and leaves.

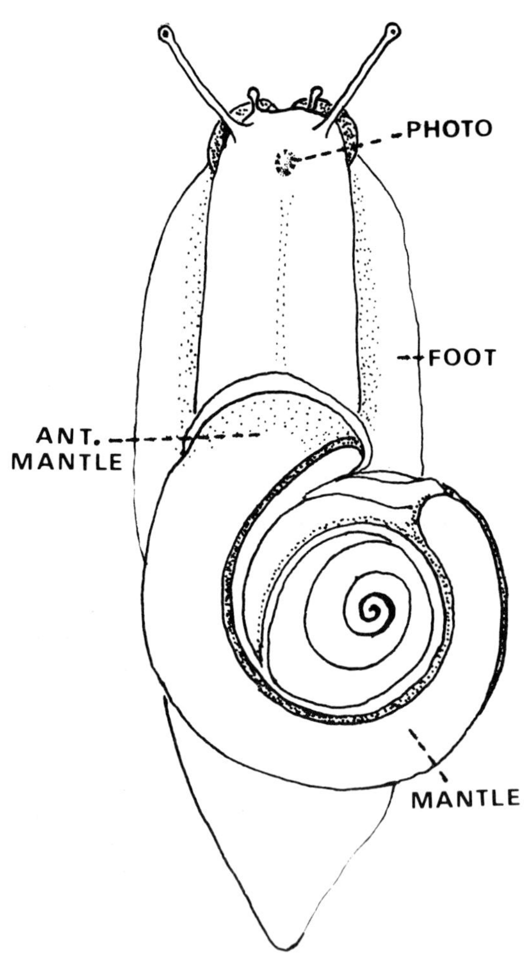


*FIGURE 1. Diagram of the main luminous organ (PHOT) and dim continuous luminous part, showing mantle and foot, with shell removed, of Quantula striata.*

The diameter of the shell is approximately 20-25 mm. and the color of the shell is a mixture of brown and pale purple. The flicker light is observed at the anterior region of the head foot (Fig. 2) and invisible from the exterior when the snail moving with the foot well extended. The luminescence is visible from all directions through the translucent muscle of the body. The time interval of the flash was found to depend on the stage of development of the snail (Haneda, 1963).

In 1968 I collected more specimens in Singapore and cultivated them in a terrarium at 20-25°C., the terrarium on a bed of soil. The snails were maintained on a diet of cucumber, eggplant and other vegetables. After two months,

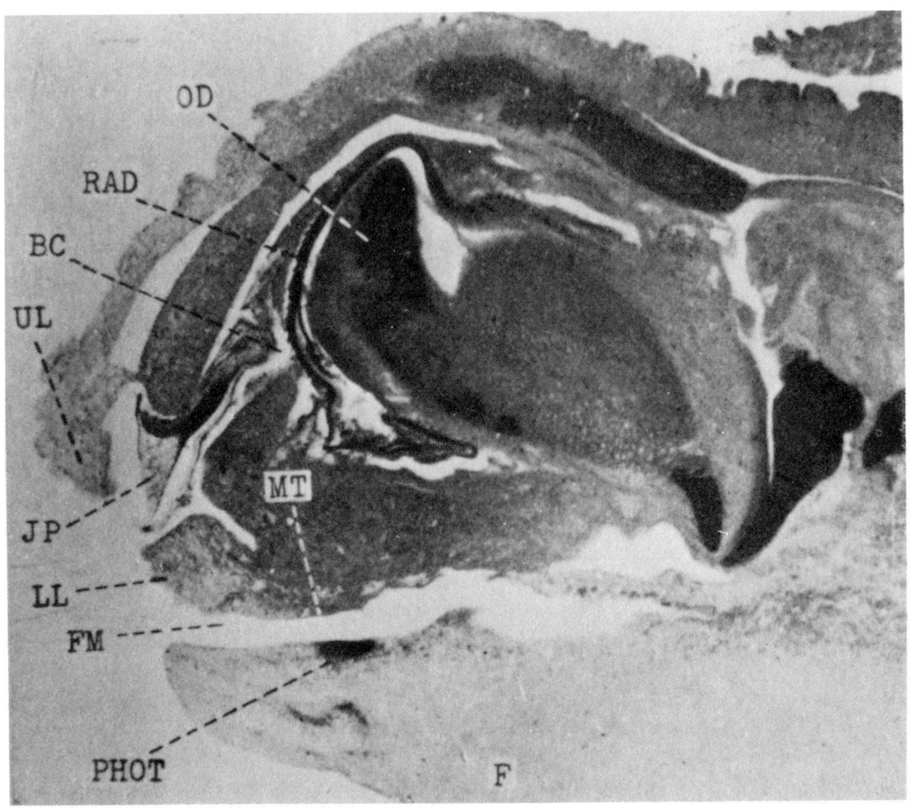


*FIGURE 2. Longitudinal section of the head of snail, showing main luminous organ (PHOT), situates on the lower wall of mucous hold (FM) under the mouth; UL, upper lip; LL, lower lip, JP, Jaw Plate, RAD, Radula, MT, mucous tissue.*

many larval snails were found in the terrarium. They recognized from white eggs spawned in the soil. The eggs appeared ellipsoid in shape; major and minor axes measured approximately 1.0 mm. and 0.8 mm. respectively. About 10 days after spawning, each eggs contained an embryo which emitted a continuous weak light that was visible through transparent shell. A week later, a small larval snail was hatched which grew into an adult in about a month. The adult snail was practically non-luminous and whatever luminescence was observed appeared to be an individual characteristic.

In specimens 20.0 mm. in shell diameter, the main luminous organ measured 1.8 mm x 1.5 mm. As the snail matured, the flicker rate diminished and in the case of some individuals, no luminescence was observed. However, a dim continuous light was observed from the mantle of adult specimens.

The time interval of the flash was found to depend on the stage of development of the snail. Immediately after hatching, when the shell diameter was about 1.0 mm., the snail emitted a weak light which appeared to be continuous over the entire surface of the foot. However, when observed under low magnification, the diffuse glow was observed to be due to many small flashes scattered over the entire foot. The spontaneous flashes of light from the main luminous organ whose duration varied between 2 to 3 seconds. After 2 - 4 flashes, there was a pause which lasted 30 - 40 seconds, after which another cycle started.

Concerning more detailed observations and its life history were reported in 1969 as a cooperative work with Dr. F. I. Tsuji. In 1968 Bassot and Martoja collected specimens of the snail in a jungle of Cambodia and reported them as *Hemiplecta Weinafliana*. The snail was subsequently identified as *Quantula* (Bassot, personal communication, 1969). Therefore these *Quantula* still remain the only known luminous species in the Pulmonata.

In 1975 Parmentier and Barnes of the University of California, Santa Barbara, collected specimens of the snail in the garden of Singapore and obtained chart recordings of the flashes of the snail. But they did not observe dim light of the mantle and the foot of the young snail. In August 1979 and February 1980 I collected more live specimens of this snail from the same locality of Singapore.

Attempts were also made to preserve the luminous system for biochemical studies, but the results were unsuccessful.

Luminescence was intracellular and no luminous slime was secreted by all the luminous cells. In animals preserved in formalin, the oval shaped main luminous organ below the mucous fold at the head exhibited a pale yellow fluorescence was observed under ultraviolet ray.

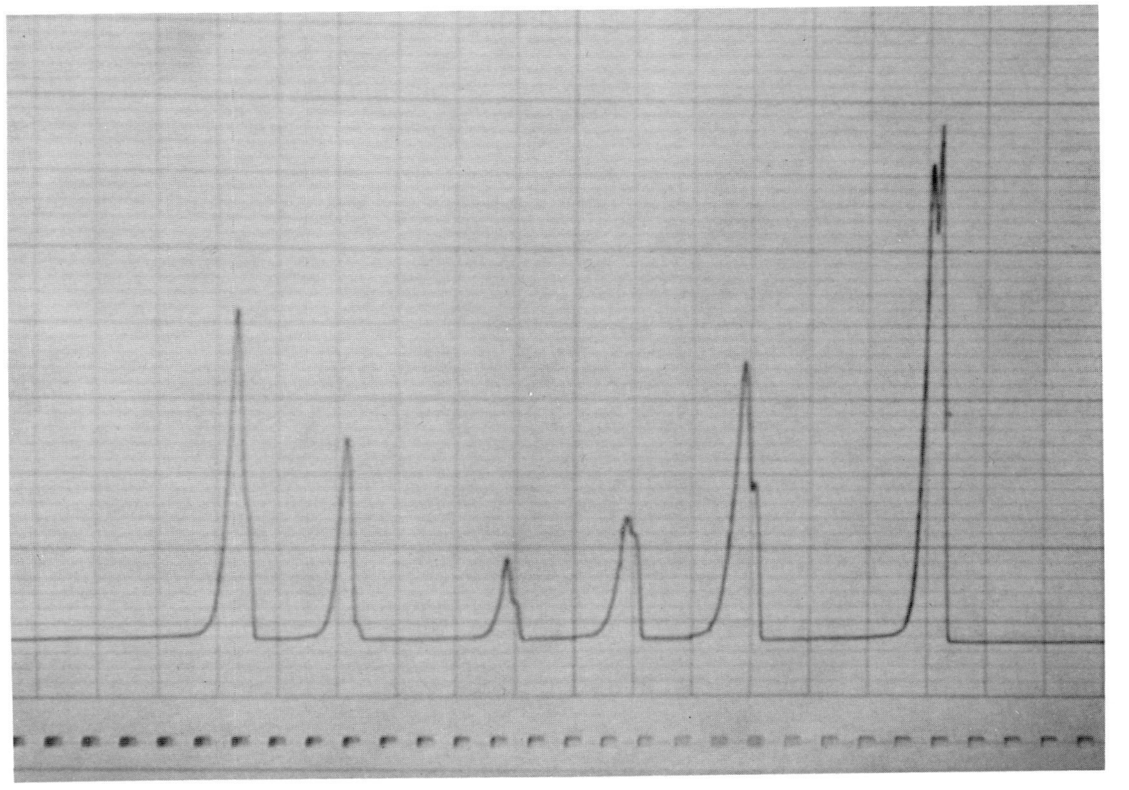

*FIGURE 3. Chart record of the flash light of main luminous organ of adult snail. 60 mm./min. 1 frame:10 sec.*

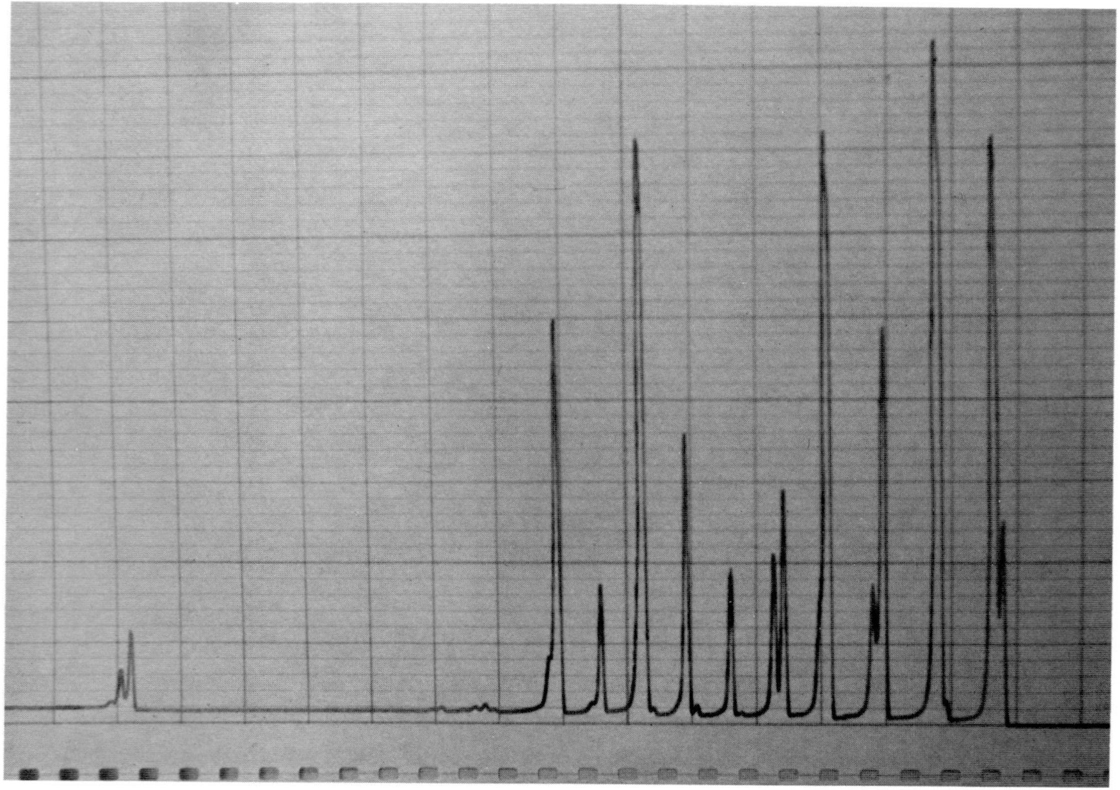

*FIGURE 4. Chart record of flash light of main luminous organ of young snail. 60mm./min.*

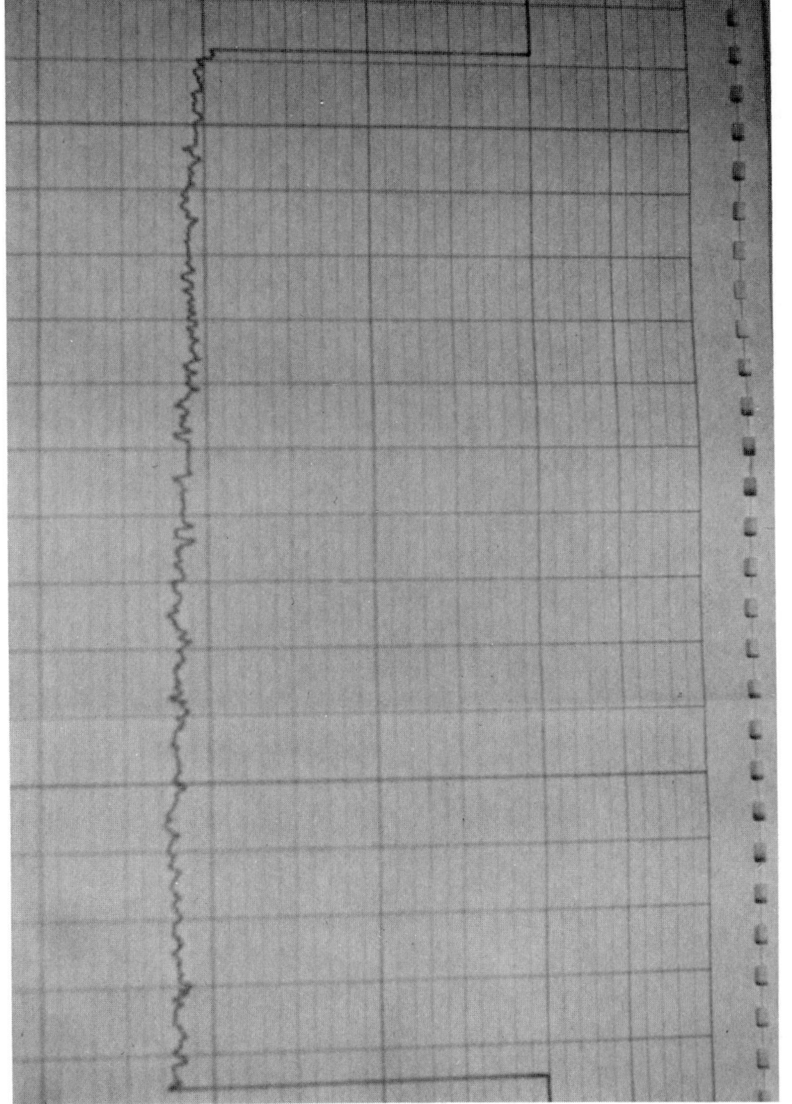

FIGURE 5. *Chart record of continuous dim light of the mantle of adult snail.*

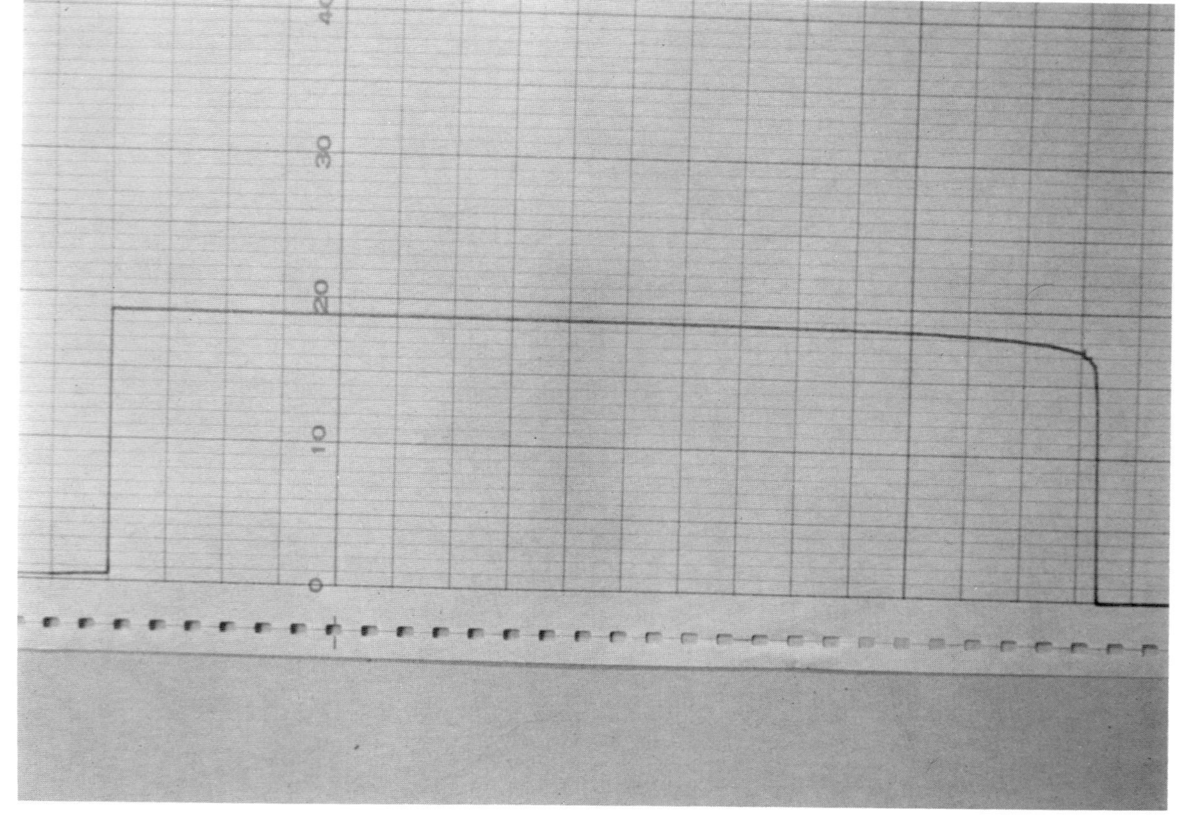


*FIGURE 6. Chart record of light of a colony of luminous bacteria (Photobacterium).*

Hot and cold water extracts of crushed animals when mixed gave a negative luciferin and luciferase reaction. Specimens frozen in dry ice did not reappear luminescence when thawed.

Trace recordings of the luminescence activity of the main luminous organ and the mantle and foot were made with a photomultiplier photometer and chart recorder.

Comparisons were then made with recordings of the light from a culture of luminous bacteria. The luminous bacteria emitted light of a constant intensity, whereas the light of the mantle and the foot of the young snail flickered continuously (Haneda, 1979). The snail did not respond to either electrical or mechanical stimulation and bodily injury had no effect on luminescence. Crushing the animal resulted in immediate disappearance of the luminescence. Flash light from the main luminous organ always recognized only the snail in natural condition and the foot well extended.

In the case of nearly all luminous organisms, the frequency or intensity of luminescence is usually increased by stimulations, but with *Quantula*, no such relationship was observed while one would expect the luminescence system to be under nervous control, the present findings suggest that *Quantula* possesses a unique system which only future research will unravel.

## REFERENCES

Bassot, M. J. -M. and M. M. Matrtoja (1968). "Presence dum organe Lumineux transiteire chez le Gasteropode Pulmone Hemiplecta weinkauffiana," *C.R.S. Acad. Sci.*, Paris, 226, 1045-1049.

Haneda, Y. (1946). "A Luminous Land Snail, *Dyakia Striata* found in Malaya (in Japanese) Seibutsu (Living Organisms) 1 (5-6), 244-298.

Haneda, Y. (1963). "Further studies on a luminous land snail *Quantula Striata,* in Malaya," *Sc. Rt. Yokosuka Ct. Mus. 8,* 1-9.

Haneda, Y. (1979). "Flash patterns of the light on luminous land snail. *Quantula Striata*(Gray) from Singapore," *Sci. Rt. Yokosuka Ct. Mus. 26,* 31-33.

Parmentier, J. and A. T. Barnes (1975). "Observations on the luminescence produced by the Malayan gastropod *Dyakia Striata," Malay Nat. J. 28,* 173-180.

## DISCUSSION

Dr. McElroy

In your beautiful slide, Milt, you didn't show the kinase, and its possible function in that whole system. I have two questions there. Is the luciferyl sulfate the primary storage for luciferin, and does the luciferin-binding protein have any influence on the conversion of luciferyl sulfate to luciferin?

Dr. Cormier

Luciferyl sulfate does appear to be the primary storage form of free (unbound) luciferin. The luciferin-binding protein has no effect on the conversion of luciferyl sulfate to luciferin. However, a related calcium-binding protein, such as calmodulin, just might, and we are looking at that.

Dr. Campbell

I would like to ask Dr. Cormier whether he has looked at any calcium-blocking agents with renilla. I have looked quite extensively at cells isolated from obelia, and have demonstrated that potassium-stimulated luminescence requires extracellular calcium and is blocked by D-600. Luminescence is also stimulated but much more weakly by lithium ions, which of course, would be expected if there was a sodium-dependent calcium efflux from these cells. I would also like to draw his attention to some work which Peter Herring and I did on RRS Discovery last September, where we looked at a number of other coelenterate medusae, and also some luminescent radiolarians. The luminescent radiolarians, for examply, Thalassicola, seem to have a calcium-activated photoprotein which is virtually identical to that in the coelenterates and to our knowledge, this is the first time its been demonstrated outside either ctenophores orcnidarians. You can reactivate the photoprotein Thalassicolin with the same prosthetic group with which one can reactivate aequorin or obelin. This was a prosthetic group synthesized for us by Russell Hart. Also, in some of the Scyphozoans such as atolla, if you homogenize these in a medium that you would normally use for extracting photoproteins, one gets a resting luminescence which is not stimulated by calcium. We are just wondering if it might be worth looking at some of the other scyphozoans to see if they might have a renilla-type system rather than a true photoprotein system there. Finally, I'd like to make a technical point about the use of photoproteins at high temperature. Because I am of course most interested in the role of calcium as a regulator, mainly in mammalian cells, many of my experiments are done at 37 degrees centigrade. When you do kinetic studies, many people inject a small proportion



ISBN 0-12-208820-4

Dr. Campbell

of photoprotein into a solution which is already at the temperature they want to study. Unless you have a protein carrier present, you get an irreversible denaturation of the protein, with a very low quantum yield reaction. And so, to do kinetics at 37 degrees centigrade, it is essential to start with a protein at 37°C, and have a protein carrier present (approx. 0.1-1 mg/ml). Any carrier protein will do; for example, bovine serum albumin. Otherwise, you see two phenomena. You see the calcium-dependent luminescence and you also see the denaturation caused by the temperature effect. I was wondering whether Dr. Blinks or Dr. Cormier had observed this effect themselves with the photoproteins they have studied.

Dr. Blinks

We have never used a carrier such as albumin. We have done stop-flow studies at a variety of temperatures, including 37 degrees, in which the solutions are prewarmed just before the reaction is initiated. It's certainly true that the photoproteins are less stable at warmer temperatures, but I'm not entirely sure I understand what you are referring to when you speak of denaturation. The quantum yield also goes down with temperature; Shimomura and Johnson showed that many years ago.

Dr. Campbell

The point is that some reports in the literature show that at higher temperatures you get very fast saturating rate constants with calcium. I think this is possibly because you're seeing a combination of the normal saturating rate constant with calcium together with a 'denaturation'. We can't say whether it's true protein denaturation. In EGTA at 37 degrees, with low protein concentration, you don't get much light output. When you take the protein back to room temperature, you can no longer get a calcium-activated luminescence.

Dr. Blinks

We have not investigated that phenomenon because we have not observed it in work with aequorin. Nor have we ever seen such high rate constants for the rise of luminescence as have been reported by Loschen and Chance. In our hands the rate constant for the rise of luminescence follows a linear Arrhenius plot up to 40°C at least. I see no evidence there to suggest the operation of more than a single process.

Dr. W. Ward

Well I know there are rather large differences in the temperature stability of photoproteins, at least the ctenophore photoproteins are much more sensitive to temperature than are the medusae photoproteins. So I've observed the same thing that you're talking about, I believe; the more rapid rise and extremely faster decay at high calcium concentrations with ctenophore photoproteins at high temperature, The melting temperature of mnemiopsin is something on the order of 25 degrees, whereas it's quite a bit higher for aequorin.

Dr. K. Ward

A question for Dr. Cormier. Is there any circular dichroic evidence for there being tertiary structural changes in the luciferin-binding protein when calcium replaces luciferin?

Dr. Cormier

That experiment is on the agenda for this fall. We have observed that when calcium binds to the protein, you see a several-fold enhancement of tyrosine fluorescence. You also see dramatic changes in the spectral properties of luciferin, as well as in the fluorescence properties of the luciferin.

Dr. K. Ward

Another question for Dr. Ward. In your experiments where you get the complex between the green fluorescent protein and luciferase, what is it that you think holds those together and what makes them fall apart, and how strongly are they held together--under what conditions?

Dr. W. Ward

We have no evidence for a tight complex under any conditions. The proteins are readily dissociable under all conditions that we have tried. The trick that's used in Hummel-Dreyer chromatography is that the green-fluorescent protein is present in a very high concentration continuously throughout the chromatography, so the column is essentially turned green before you even start putting the luciferase on. The GFP concentration that we used was $10^{-4}$ molar. Under those conditions, the luciferase forms a complex with GFP, simply because it's always in the presence of GFP. The interpretation that we made, and Milt described that earlier, is that of a rapid equilibrium. In other words, a very fast $k_1$ and a very fast $k_2$ that would indicate that these proteins are flickering back and forth. So there really is no evidence that they ever exist as a tight complex

Dr. W. Ward

with a high dissociation constant. The evidence that an association must occur includes the Hummel-Dreyer experiment plus some of the others that I mentioned. If you modify lysine substituents, you can demonstrate that these proteins simply will not interact, and there's no evidence for energy transfer at all with the lysines modified.

Dr. Hastings

I would like to ask Milt Cormier if he could review for us the evidence that oxygen is bound to the protein in the complex prior to the addition of calcium,and to correct me if I'm wrong, is it not true that the same postulate has been included in the Shimomura and Johnson model?

Dr. Cormier

There is a difference in the two models. In the latest model of Shimomura and Johnson, they have oxygen bound to the 2-position of luciferin in the form of a peroxide, which is then covalently linked to the protein. That's different from saying that the chromophore, the luciferin, is free, and with oxygen being attached at some other point. Initially, a great deal of effort went into trying to explain the 454 nanometer absorption of aequorin, as you might recall. When Ward and Seliger described their work on the photoproteins mnemiopsin and berovin, they found that the visible absorption characteristics of those proteins differed considerably from that of aequorin. And in model studies on luciferin analoques, we showed that we could mimic the absorption characteristics, including the molar extinction coefficient characteristics, of all of those photoproteins, by placing the luciferin analoques or luciferin itself, in fact, in different environments. The appropriate perturbation of luciferin by the protein would thus explain the absorption characteristics of that particular photoprotein. We also have shown (Bill Ward did this in our lab), that under very mild conditions you can quantitatively extract the luciferin from the photoproteins mnemiopsin and berovin. That is, 98% yields were observed. Furthermore, the extracted luciferin, as far as we can tell by all criteria, is identical to the luciferin found in aequorin. The yield of aequorin luciferin in our hands is about 50%, but I think putting all of the evidence together would suggest that since one can extract the luciferin of various photoproteins rather quantitatively under very mild conditions--by that I mean pH 6 buffers containing β-mercaptoethanol--I think that that argues against having luciverin in the form of a peroxide to begin with. In order to extract luciferin from aequorin one must also

Dr. Cormier

include $NaHSO_3$ in the buffer. Under these conditions one could argue (as Shimomura and Johnson have) that an oxidized luciferin species is being reduced to luciferin. However, such an argument would make aequorin unique among photoproteins. I think it is simpler to argue that all photoproteins are basically similar but that their absorption and bioluminescent emission characteristics may differ because of photoprotein-dependent differences in the luciferin environment.

# VI
# APPLICATIONS OF BIOLUMINESCENCE

## OVERVIEW OF APPLICATIONS OF BIOLUMINESCENCE

Philip E. Stanley

Department of Clinical Pharmacology
The Queen Elizabeth Hospital
Woodville, South Australia

### I. INTRODUCTION

The analytical uses of bioluminescence are found among such widely dispersed disciplines as soil science, clinical chemistry, marine biology and the dairy industry. Inevitably it is therefore difficult for workers to be aware of the advances that may be relevant to their endeavours but which have been published in a section of journals they never consult. This meeting is thus important in bringing together those disciplines as was the previous symposium held in Brussels two years ago (1) and the ATP-Methodology meetings of 1975 and 1977 (2,3). The publication of a whole volume dealing with the topic (4) and another one in press (5) will serve to improve the situation even further.

The reaction which has been exploited to the greatest extent is that of firefly luciferase with ATP and this has been used to measure the latter. As little as $10^{-15}$ moles may be measured and thus the method is capable of the assay of ATP in single cells (6). Bacterial luciferase (from *Photobacterium* or *Beneckea*) together with a NAD(P)H:FMN oxido-reductase (7) can effectively be used to measure NADH and/or NADPH, dehydrogenase enzymes and their substrates (7-10). The intact bacterium may also serve as a useful analytical tool (11-13). Mention, albeit brief, must be made of the photoprotein aequorin which provides an important micro-assay for ionic calcium and may be used at the intracellular level (14).




ISBN 0-12-208820-4

Until recently users of the technique were generally obliged to prepare and purify their own enzymes and in my opinion this inhibited many workers who might have otherwise used the techniques. I am glad to say the situation has changed markedly in the last three years or so and now around ten companies market purified enzymes and reagents (see Appendix). Similarly there has been a surge in the range of equipment designed for use in this technique. This has boosted activity in biomass assays where many samples are to be analyzed, e.g. water monitoring, dairy industry, bacteruria, since such features as data processing and automation have been introduced (see Appendix). This author has recently reviewed the topic of instrumentation as it pertains to the clinical field (15).

Because of space limitations, I have in the following sections endeavoured to cover briefly the main or novel areas and concentrate on a number of more recent applications with examples drawn from a variety of disciplines. Previous reviews should be consulted for a more expansive and detailed account (6,8,10,16-19).

## II. ASSAY PROCEDURES

### A. *The Assay of ATP with Firefly Luciferase*

There appears to be a move away from the use of peak or flash height measurements towards the integration of sample signal over a fixed time interval. This may be due to the introduction of commercial preparations which produce a light output rising to a maximum in a second or so and thence decreasing at only a few percent per minute. Such preparations may be very effectively used to not only measure stat levels of ATP but also to monitor changing concentrations. Good examples of the latter include the measurement of creatine phosphokinase subunits (for myocardial infarction) (20,21) and electron transport linked ATP formation in chloroplasts and mitochondria (22,23).

Much has been written concerning the buffer and pH needed to produce optimal analytical conditions and a detailed study of this problem will appear shortly (24). The presence of an appropriate pyrophosphate concentration is also important in the maintenance of luminescence (25). It is worthwhile noting that pyrophosphate may occur as a contaminant in phosphate buffers.

The use of the ATP assay may be broadly divided into two classes (a) measurement of cell numbers (b) assay of enzymes and substrates, cell viability and function. Consider firstly the measurement of cell numbers. This procedure is useful in screening procedures, e.g. bacteria in urine (26,27) and bacteria and somatic cells in milk (28,29) where a grading of a sample is needed or a positive or negative number is required concerning a threshold or set value of cell concentration. In others, biomass measurements may be valuable to measure an on-going process, e.g. fermentation of beer (30), activated sludge (31), disinfection control (32), microorganisms in soil (33) and sea water (34), or as a tool in microbial ecology (35).

The firefly system may also be employed to measure enzymes and their substrates. Such coupling techniques may be used for creatine phosphokinase (17,18,20,21,36), ATP-sulphurylase (37) and substrates such as creatine phosphate (38). ATP level may also be used as a measure of cell viability in for example spermatazoa (39,40) and erythrocytes (41). The assay has also been used to study cell membrane properties of erythrocytes (42). Immobilized firefly luciferease has also been successfully exploited in measuring enzymes and substrates (43).

### *B. The Assay of NAD(P)H with Bacterial Luciferase and Oxidoreductase*

These assays are based on the following enzyme system (simplified):

$$\text{NAD(P)H} + \text{H} + \text{FMN} \xrightarrow[\text{oxidoreductase}]{\text{NADH:FMN}} \text{NAD(P)} + \text{FMNH}_2$$

$$\text{FMNH}_2 + \text{O}_2 \xrightarrow[\text{long chain aldehyde}]{\text{luciferase}} \text{FMN} + \text{H}_2\text{O} + \text{Light}$$

The specificity for adenine nucleotide depends on the specificity of the oxidoreductase (7) and in principle the assay may be used to measure dehydrogenase enzymes and their substrates (8,10) providing that favourable equilibrium conditions can be established. Examples would include malate dehydrogenase, malate and oxaloacetate (44) and pyruvate and NAD (45).

Sensitivity for reduced nucleotide is in the femtomole range and the time to peak light output is of the order half a minute for analytical conditions. Light output declines slowly thereafter.

Immobilized bacterial luciferase/oxidoreductase have been used with effect to measure enzymes and substrates and perhaps will be useful in the clinical chemistry laboratory in the future (46).

Whole bacterial cells can be used in the bioassay of lipase and phospholipases (11) for mutagenic substances (12) and to monitor water quality (13).

## III. CONCLUSION

Bioluminescence assays by virtue of their extreme sensitivity and specificity are being increasingly used in a wide range of disciplines. Many of the practical hindrances are being removed by the availability of quality controlled reagents for the assay of ATP and NAD(P)H and the introduction of flexible and sensitive instrumentation.

## REFERENCES

(1) Schram, E., and P. Stanley, eds. "Proceedings of the International Symposium on Analytical Applications of Bioluminescence and Chemiluminescence" (Brussels, 1978) State Printing and Publishing, Inc., Westlake Village, (1979).

(2) Borun, G.A., ed. "ATP Methodology Seminar" SAI Technology Co., San Diego (1975).

(3) Borun, G.A., ed "2nd Bi-Annual ATP Methodology Seminar" SAI Technology Co., San Diego (1977).

(4) DeLuca, M., ed. *Methods in Enzymology* 57, (1978).

(5) Kricka, L.J., and T.J.N. Carter, eds. "Clinical and Biochemical Applications of Luminescence" Marcel Dekker Inc. (in press).

(6) Brolin, S.E., G. Wettermark, and H. Hammar, *Strahlentherapie* 153, 124 (1977).

(7) Gerlo, E. and J. Charlier, *Eur. J. Biochem.* 57, 461 (1977).

(8) Stanley, P.E. in "Liquid Scintillation Counting" (M.A. Crook and P. Johnson, eds.), Vol. 3, p. 253, Heyden, London, (1974).

(9) Wulff, K., R. Pagel, and F. Staehler, *Fresenius Z. Anal. Chem.* 301, 163 (1980).

(10) Thore, A., *Annals Clin. Biochem.* 16, 359 (1979.

(11) Ulitzur, S. in reference 1, p. 135.

(12) Ulitzur, S., I. Weiser, and S. Yannai, *Mutation Research* 74, 113 (1980).
(13) Bulich, A.A., and M.W. Greene, in reference 1, p. 193.
(14) Blinks, J.R., *Annals N.Y. Acad. Sci.* 307, 71 (1978).
(15) Stanley, P.E., in reference 5.
(16) Stanley, P.E., in reference 1, p. 89.
(17) Gorus, F., and E. Schram, *Clin. Chem.* 25, 512 (1979).
(18) Whitehead, T.P., L.J. Kricka, T.J.N. Carter, and G.H.G. Thorpe, *Clin. Chem.* 25, 1531 (1979).
(19) Stanley, P.E. in "Liquid Scintillation - Science and Technology " (A.A. Noujaim, C. Ediss, and L. Weibe, eds.) Academic, New York (1976).
(20) Lundin, A., and I. Styrelius, *Clin. Chim, Acta* 87, 199 (1978).
(21) LKB-Wallac Test kit for CK (see Appendix).
(22) Lundin, A., M. Baltscheffsky, and B. Höjer, in reference 1, p. 339.
(23) Lemasters, J.J. and C.R. Hackenbrock, in reference 4, p. 36.
(24) Webster, J.J., J.C. Chang, E.R. Manley, H.O. Spivey, and F.R. Leach, *Anal. Biochem.* 105,(1980) (in press).
(25) Ahmad, M., E. Moreels, and E. Schram, *Arch. Intern. Physiol. Biochim.* 88, 217 (1980).
(26) Curtis, G.D.W., and H.H. Johnston, in reference 1, p. 448.
(27) Lumac, B.V., Test kit for bacteriuria (see Appendix).
(28) Lumac, B.V. Test kit for bacteria in milk (see Appendix).
(29) Lumac, B.V., Test for mastitis. Applications Bulletin no. 503B (see Appendix).
(30) Hysert, D.W., F. Kovecses, and N.M. Morrison, *J. Amer. Soc. Brew. Chem.* 34, 145 (1976).
(31) Statham, M., and D. Langton, *Process Biochemistry* 10, 25 (1975).
(32) Tifft, E.C. and S.J. Spiegel, *Environm. Sci. Technol.* 10, 1268 (1976).
(33) Jenkinson, D.S., and J.M. Oades, *Soil Biol. Biochem.* 11, 193 (1979).
(34) Azam, F., and R.E. Hodson in reference 3, p. 109.
(35) Karl, D.M., *Microbiological Reviews* (in press).
(36) Bostick, W.D., M.S. Denton, and S.R. Dinsmore, *Clin. Chem.* 26, 712 (1980).
(37) Balharry, G.J.E., and D.J.D. Nicholas, *Anal. Biochem.* 40, 1 (1971).
(38) Jabs, C.M., W.J. Ferrell, and H.J. Robb, *Clin. Chem.* 23, 2254 (1977).
(39) Brooks, D.E., *J. Reproduct. Fert.* 23, 525 (1970).
(40) Lumac, B.V. Medical Applications 509 (see Appendix).

(41) Kirkpatrick, F.H., D.G. Hillman, and P.L. La Celle, *Experientia* 31, 653 (1975).
(42) Wettermark, G., S.E. Brolin, and L. Juhlin, *J. Coll. Interf. Sci.* 73, 287 (1980).
(43) Lee, Y., I. Jablonski, and M. DeLuca, *Anal. Biochem.* 80, 496 (1977).
(44) Stanley, P.E., in reference 4, p. 181.
(45) Agren, A., C. Berne, and S.E. Brolin, *Anal. Biochem.* 78, 229 (1977).
(46) Haggarty, C., E. Jablonski, L. Stav, and M. DeLuca, *Anal. Biochem.* 88, 162 (1978).

## APPENDIX

The following appendix comprises suppliers and manufacturers of instruments, reagents and kits used in analytical bioluminescence which are known to the author as of July 1980. While it may not be entirely complete I hope it will be of value to the reader seeking commercial products.

*Appendix*

| *Supplier or manufacturer* | *Product Code*† |
|---|---|
| American Instrument Co.<br>Silver Spring, MD 20910, U.S.A. | A |
| Analytical Luminescence Laboratory<br>Westlake Village, CA 91361, U.S.A. | A,C,D,E,F,G |
| Antonik Laboratories<br>Elk Grove Village, IL 60007, U.S.A. | E |
| Berthold Laboratorium<br>Wildbad, 1, W. Germany | A |
| Boehringer Mannheim GmbH.<br>Mannheim 31, W. Germany | E,F |
| Calbiochem-Behring Corp.<br>La Jolla, CA 92037, U.S.A. | E |
| Du Pont De Nemours<br>Wilmington, DE 19898, U.S.A. | A,E |
| LKB-Wallac<br>Turku, Finland | A,E,H |

| | | |
|---|---|---|
| Lumac B.V.<br>Schaesberg, Holland | Lumac Systems Inc.<br>Titusville, FL 32780<br>U.S.A. | A,B,C,D,E,<br>F,G,H |
| Marwell International AB<br>Solna, Sweden | | A,C,D |
| Millipore Corporation<br>Freehold, NJ 07728, U.S.A. | | F |
| New Brunswick Scientific Co. Inc.<br>Edison, NJ 08817, U.S.A. | | A,E |
| Packard Instrument Co. Inc.<br>Downers Grove, IL 60515, U.S.A. | | A,D,E,G |
| SAI Technology Inc.<br>San Diego, CA 92121, U.S.A. | | A,E |
| Sigma Chemical Co.<br>St. Louis, MO 63178, U.S.A. | | E,F |
| Skan AG<br>Basel, Switzerland | | A |
| Turner Designs<br>Mountain View, CA 94043, U.S.A. | | A |
| Vitatect Corp.<br>Alexandria, VG 22308, U.S.A. | | A |

†A) Instruments; B) Portable units; C) Automatic units; D) Data processing capability; E) Firefly luciferase; F) Bacterial luciferase; G) Extractants and reagents; H) Kits.

# IMMUNOASSAYS MONITORED BY CHEMILUMIGENIC LABELS

R. C. Boguslaski
H. R. Schroeder

Immunochemistry Laboratory
Ames Research and Development Department
Ames Division
Miles Laboratories, Inc.
Elkhart, Indiana

## I. INTRODUCTION

Numerous compounds present at low levels in biological fluids are currently determined by various types of immunoassays which use radiolabels as tracers. To avoid the inconvenience and stability problems associated with radiolabels, we have used chemilumigenic compounds to monitor these assays. The new labels can be detected at picomolar (pM) levels in a few seconds by luminescence and can be used in various homogeneous and heterogeneous immunoassay formats. This paper will briefly describe some of the results we have obtained from assays which employ chemiluminngenic labels.

## II. HOMOGENEOUS ASSAYS

The basic difference between a homogeneous and heterogeneous assay is that the former procedure does not require any separation steps, while phase separation is essential in the latter. We have developed a homogeneous assay for biotin (1) using a conjugate of biotin and isoluminol (6-amino-2,3-dihydrophthalazine-1 ,4-dione). Where the conjugate was treated with an oxidizing agent, such as potassium superoxide or hydrogen peroxide lactoperoxidase, light production was linear with conjugate concentration with the




ISBN 0-12-208820-4

lowest detectable level of conjugate being 5 nM. When a constant level of the conjugate was treated with various levels of avidin, the natural binding protein for biotin, and the mixture subjected to oxidation, light production increased up to ten-fold as avidin levels rose to 14 μg/ml. Light production then declined at higher avidin levels, probably due to general protein inhibition of the chemiluminescent reaction. Various control experiments demonstrated that enhancement of light production was associated with the specific binding reaction between avidin and the biotin portion of the conjugate (1).

The enhancement phenomenon served as the basis for a homogeneous assay. Avidin and the biotin-isoluminol conjugate at levels which gave near optimum light enhancement were combined with various levels of biotin. Competition between the two forms of biotin for binding sites on avidin occurred and was reflected in the light producing ability of the reaction mixture. As the levels of biotin increased the analyte was able to compete more successfully than the conjugate for binding sites on avidin which resulted in a concomitant decrease in light production. The range of the assay was 50 to 400 nM biotin.

A similar assay was developed for progesterone (2). Here an isoluminol derivative was covalently linked to progesterone-11α-hemisuccinate to form a chemilumigenic conjugate, which could be detected at levels as low as 0.14 pmol/tube. When this conjugate was treated with increasing levels of antiserum to progesterone, the total light yields of the reaction mixture increased to a maximum level of four-fold. When progesterone was allowed to compete with the conjugate for binding sites on the antibody, the enhancement in light emission was diminished. The assay covered a range of 25-400 pg/assay tube. The plasma progesterone levels in 15 samples from non-pregnant women determined by the chemiluminescence immunoassay (CIA) were compared with results from the same samples obtained by a radioimmunoassay (RIA) procedure. The methods agreed satisfactorily with the equation for the regression line being $y = 0.89x + 0.48$; $r = 0.974$.

## III. HETEROGENEOUS ASSAYS

A heterogeneous competitive binding immunossay was developed for the measurement of thyroxine ($T_4$) in serum (3). A thyroxine-isoluminol conjugate was synthesized. The compound emitted light upon oxidation by $H_2O_2$ in the presence of microperoxidase. Peak light intensity was linearily related to conjugate concentration to a lower limit of 0.1 nM. The cross reactivity of the thyroxine conjugate with antibody raised against thyroxine was 35%.

The $T_4$ conjugate was adsorbed to a small 1 ml column of Sephadex G-25 and then serum was diluted in 0.1 M NaOH and applied to the column. The column extracted and adsorbed both thyroxine species. Serum interferences were removed with a 75 mM barbital buffer wash, pH 8.6, and $T_4$ antibody was added followed by a 1 hour incubation. The antibody bound fraction was eluted with an aliquot of the barbital buffer and the aliquot analyzed for the level of chemilumigenic label. Thyroxine from serum diminished the amount of $T_4$ conjugate bound to antibody and consequently decreased the level of light produced in this fraction.

The assay covered the clinically significant range of 25-200 nM, had an intra-assay precision of ±5% (CV) and generated results which compared favorably with those determined by a reference radioassay ($y = 0.95x + 5.9$, $r = 0.98$; $n = 28$).

We have also developed several solid phase immunoassays for proteins monitored by chemilumigenic labels. The heterogeneous methods are, in general, much more sensitive than the homogeneous approaches, since the former procedures are able to eliminate interferences which limit the sensitivity of the latter. The heterogeneous methods are discussed by Schroeder elsewhere in this volume.

## IV. CONCLUSIONS

Chemilumigenic labels are viable alternatives to radiolabels for monitoring immunoassays. Assays for haptens and proteins, using either homogeneous or heterogeneous formats, have been developed with these labels. The heterogeneous assays are best able to take advantage of the sensitivity inherent in the chemilumigenic labels and produce assays which are equivalent in sensitivity to radioassays. An important, but less well-recognized advantage of using these labels to monitor immunoassay arises from the fact that the monitoring reaction (light emission) takes only a few seconds. Thus, these labels may reduce assay time substantially, especially in mass screening applications.

REFERENCES

1. Schroeder, H. R., P. O. Vogelhut, R. J. Carrico, R. C. Boguslaski, and R. T. Buckler, *Anal. Chem.* 48, 1933 (1976).
2. Kohen, F., M. Pazzagli, J. B. Kim, H. R. Lindner, and R. C. Boguslaski, *FEBS Letters* 104, 201 (1979).
3. Schroeder, H. R., F. M. Yeager, R. C. Boguslaski, and P. O. Vogelhut, *J. Immunol. Meth.* 25, 275 (1979).

# ASPECTS ON THE POTENTIALITIES OF BIOLUMINESCENCE ASSAY IN CELL BIOLOGY AND MICROPHYSIOLOGY

Sven E. Brolin
Gunnar Wettermark

Department of Medical Cell Biology
University of Uppsala
Uppsala, Sweden

Department of Physical Chemistry
The Royal Institute of Technology
Stockholm, Sweden

In the rapidly expanding field of general cell biology much research concerns the biochemical properties of cells and their components. For elucidation of these properties bioluminescence analysis is a powerful tool. There are many lines of development possible for such analyses. We have selected three directions which appear to possess particularly great potentialities. They may be labelled micro analyses, *in situ* tests and direct monitoring.

## I. MICROANALYSES

In order to gain desired biochemical information, analyses are generally performed using large amounts of the same kind of cells. In many situations, however, and particularly in medical cell biology, cells which only can be isolated in small amounts may attract our interest when they exert fundamental physiological functions. Such cells may also be involved in the development of diseases and particularly so when the generation of neural and hormonal signals is disturbed by metabolic disorders. Studies of small explants composed of such cells are of considerable




ISBN 0-12-208820-4

interest in medical cell biology. Sensitive methods are available for evaluating their function by analyzing the discharge of signal substances. The limited size of the samples as determined by the access to the cells concerned is in reality of advantage in many experiments in which as far as a favourable surface-mass ratio is essential for exchange with a surrounding substrate medium. When we now wish to relate the physiological function of cells to their biochemistry the work in the microscale requires highly sensitive analytical methods. Among these bioluminescence analysis possesses notable potentialities not only in sensitivity but also in convenience of the analytical performance. Thus several assays can be carried out in a single step by addition of the sample to a light yielding solution composed of a luciferase supplemented with properly selected additives. In addition to the now well known micro assays of ATP, $FMNH_2$, NADH and NADPH (1-3), it is also possible to determine NAD+ and NADP+ in a single step (4-6). The direct assay of the oxidized pyridine nucleotides is accomplished by continuous reduction within the light yielding solutions.

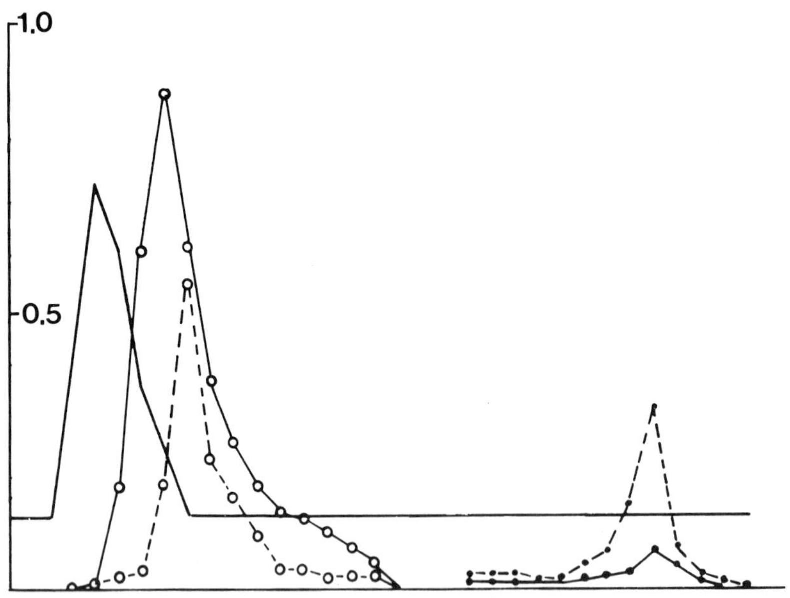


*FIGURE 1. Selective removal of dehydrogenases from a bacterial luciferase extract through column chromatography on Cibachron blue F3G-A; ——— Ultraviolet absorption, —o—o—o Luciferase, --o---o--Oxidoreductase, ---●---● Lactate dehydrogenase, ●———● Malate dehydrogenase.*

Certain substrates can also be directly assayed via the reduced nucleotides (7-8). Furthermore, enzymes are determinable (1,9) using the same principle, but dehydrogenases contaminating bacterial luciferases constitute a limitation. Through chromatographic purification of the luciferase, however, satisfactory preparations for micro assay of enzymes can be obtained, see Fig. 1. The elution of luciferase can be easily followed by electrochemical determinations (Fig. 2). The single step procedure permits a large number of data to be collected in a short time and allows data processing techniques to be used for improving the analytical situation (9).

## II. *IN SITU* TESTS

Cells in many locations are difficult to isolate for analyses. Others may be seriously affected by removal procedures. The difficulties of access are met for instance with the innermost cell layers of tubular organs. The ureter represents this situation. By successive infusion and biochemiluminescence analyses of the outflow an extraction spectrum can be obtained, characterizing the biological status

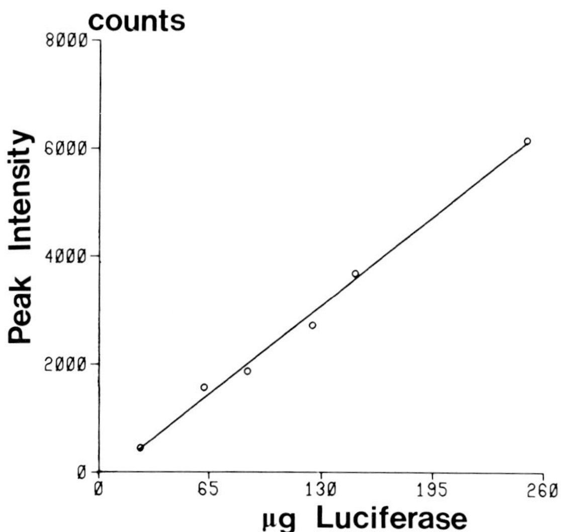


*FIGURE 2. Calibration curves for electroluminescence assay of bacterial luciferase, cf. procedure described in ref. 10.*

of the tissue, see Fig. 3. It is also possible to check the condition of homogeneous tissue by extraction from an artificial cavity. This may be of value in cases where the integrity of the tissue should be maintained. Metabolic changes can be followed *in situ*. Devitalization of organs upon removal can be easily assessed, see Fig. 4.

## III. DIRECT MONITORING

Bioluminescence can be applied for directly monitoring the outflow of substances from cells. To this effect luciferase and appropriate reactants are added to their physiological environment. The outflow may be followed by adding

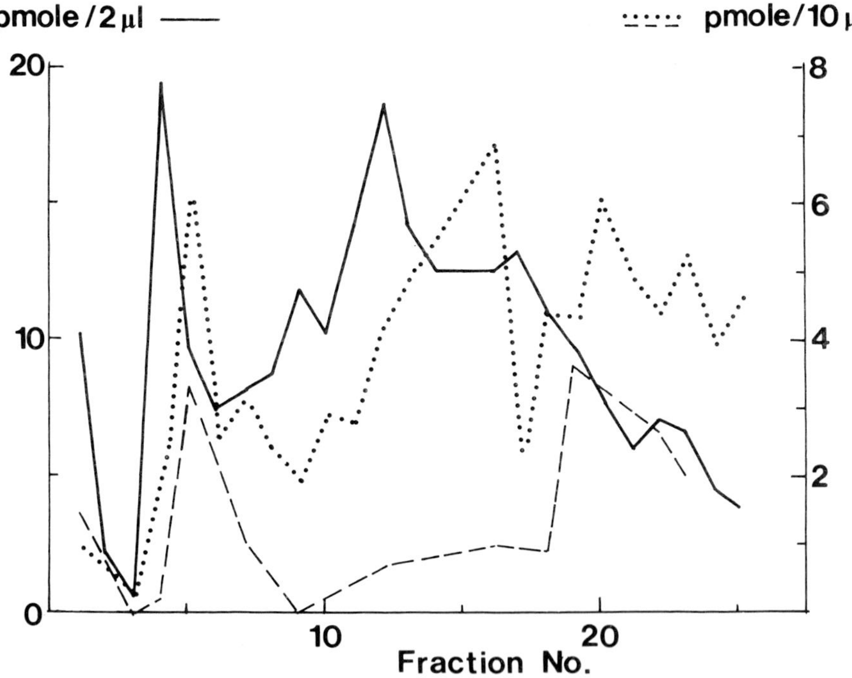


*FIGURE 3. Extractions from ureter of a rat with 0.05% saponin. The first three fractions represent washing with saline. Thereafter the fractions represent successive eluates from perfusion each with 40 µl of saponin solution during 40 sec. ——Total adenylate (ATP+ADP+AMP); ···· NAD$^+$; ----Malate dehydrogenase.*

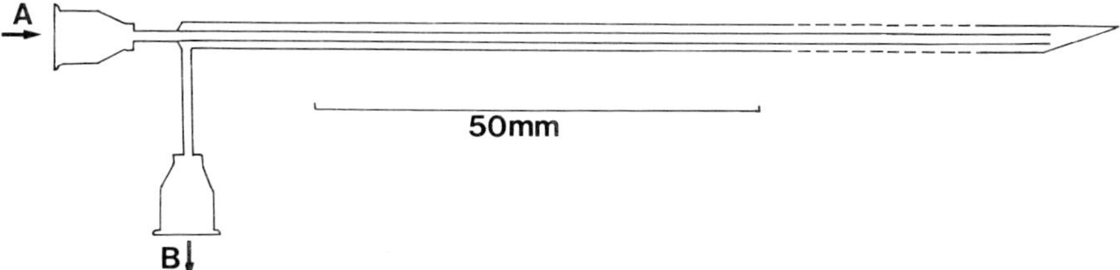


*FIGURE 4a. Double walled needle used for perfusion studies*

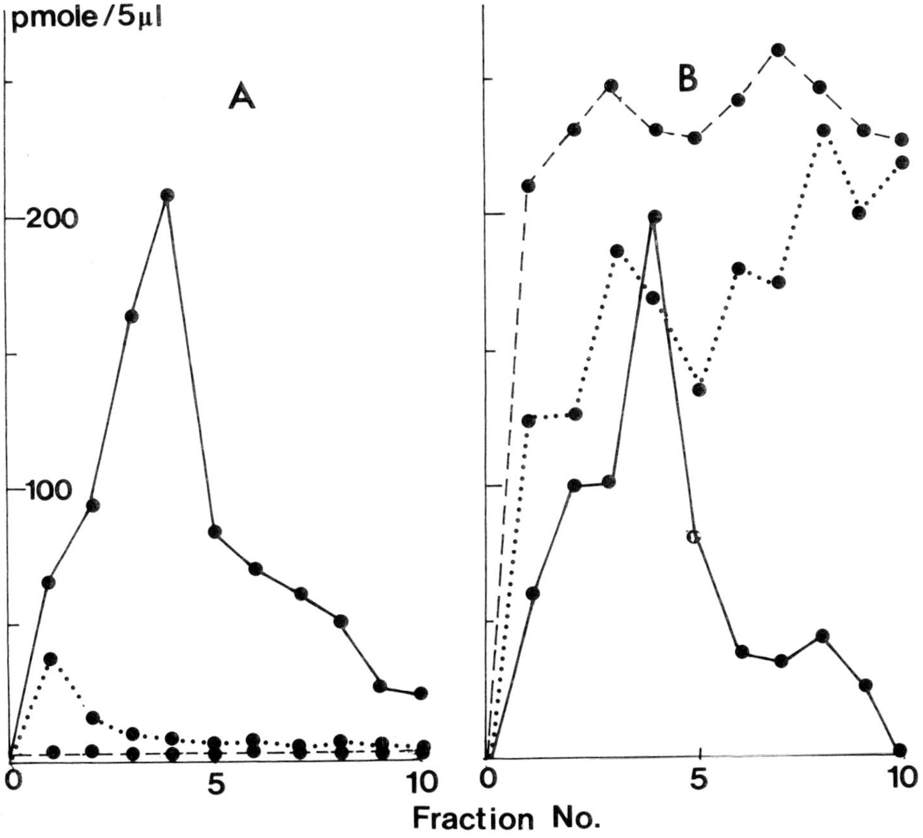

*FIGURE 4b. Perfusion studies on pig muscle using the needle shown in Fig. 4a. ——— living animal, ····· one hour after killing of the pig, ---- after three days storage at 4°C. Curve A: ATP content. Curve B: Total adenylate (ATP+ ADP+AMP). Each fraction represents the result of injecting 1 ml distilled water into the muscle and withdrawing the extract after 5 min.*

various agents. Continuously increasing the concentration of such an agent and recording of the light emission may yield luminescence spectra, Fig. 5.

## IV. SUMMARY

The convenience of a single step performance is well recognized in assay of reduced pyridine nucleotides by means of bacterial luciferase. A continuous reduction within the luciferase solution can be arranged and makes it possible to analyze directly also the oxidized nucleotides as well as certain substrates and enzymes.

New techniques have been designed to accomplish bioluminescence assay of cells *in situ*, which is of particular advantage when they are difficult to isolate or very dependent on their surrounding.

Monitoring outflow of metabolites from cells can be arranged so that luminescence spectra are obtained. These may be utilized to characterize the response to different agents. The luminescence spectra may be employed for recognization of cells and for evaluation of their condition.

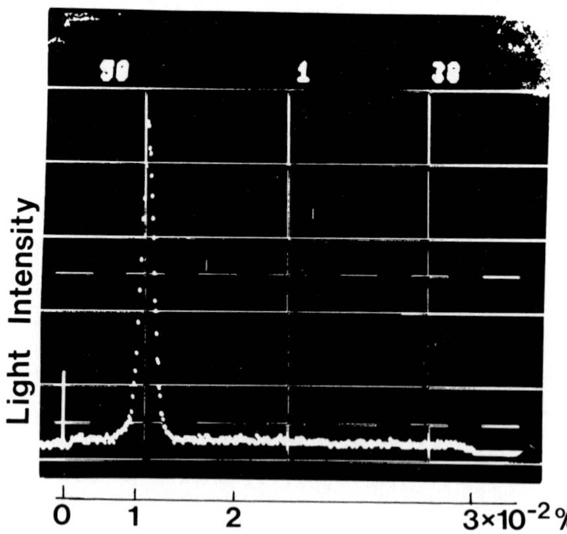


*FIGURE 5. "Luminescence spectrum" obtained by subjecting erythrocytes to lysis through continuous addition of saponin to a suspension of erythrocytes in a fire-fly extract (11).*

REFERENCES


1. Methods in Enzymology Vol. LVII: "Bioluminescence and Chemiluminescence" (M. A. DeLuca, ed.), Academic Press, New York (1978).
2. Stanley, P. E., *Anal. Biochem.* *39*, 441 (1971).
3. Brolin, S. E., E. Borglund, L. Tegnér, and G. Wettermark, *Anal. Biochem.* *42*, 124 (1971).
4. Brolin, S. E., and S. Hjertén, *Molec. & Cellul. Biochem.* *17*, 61 (1977).
5. Ågren, A., S. E. Brolin, and S. Hjertén, *Biochem. Biophys. Acta* *500*, 103 (1977).
6. Hutton, J. C., A. Sener, and W. J. Malaise, *in* "Proceedings International Symposium on Analytical Applications of Bioluminescence and Chemiluminescence" (E. Schram and P. Stanley, eds.), p. 166. State Printing & Publishing, Inc., Westlake Village, Calif. (1979).
7. Brolin, S. E., *Bioelectrochem. Bioenerg.* *4*, 257 (1977).
8. Stanley, P. E., *in* Methods in Enzymology Vol. LVII: "Bioluminescence and Chemiluminescence" (M. A. DeLuca, ed.), p. 181, Academic Press, New York (1978).
9. Brolin, S. E., to be published.
10. Wettermark, G. and S. E. Brolin, *in* "Proceedings International Symposium on Analytical Applications of Bioluminescence and Chemiluminescence" (E. Schram and P. Stanley, eds.), p. 212. State Printing & Publishing, Inc., Westlake Village, Calif. (1979).
11. Wettermark, G., S. E. Brolin, and L. Juhlin, *J. Colloid Surface Chem.* *73*, 287 (1980).

## DISCUSSION

Dr. Antonik

In your address, Dr. Stanley, you stated that you were not aware of an FDA-approved product for the clinical laboratory using bioluminescence. There is a CK bioluminescent product available from Antonik Laboratories that has FDA approval since 1975.

Dr. Stanley

Thank you.

Dr. DeLuca

From your brief presentation on the chemiluminescent assays as potential replacements for the radioimmuno assays, I gather that you feel there is a future in this, and that it will be possible to work out some of the problems of serum interferences and background light adequately.

Dr. Boguslaski

The only way I think I can answer that honestly is with a real firm maybe. When you talk about them serving as replacements for radioimmunoassays, that question has at least two components. It has a technical component: Are these assays going to be technically competent? And I can answer that,"yes", I think you can work out the serum interference problems, etc., and have valid assays. But the more nebulous part of the question revolves around marketability, and that itself is composed of a whole host of parameters and other questions. That's where the question really lies. The only way these assays are really going to replace RIAs is if they show substantial advantages. The whole name of the game is advantages. And they just can't be marginal advantages, they can't just be replacement of A with B for replacement's sake.

Dr. Stanley

I would like to echo that. Running a clinical laboratory, you really have to have a very good reason for changing from a well tried method which works in your hands to another method which has only marginal advantages.

Dr. Lindner

I would like to sound a more optimistic note on the possibility of chemiluminescent monitored assays as a viable alternative to radioimmunoassay. Our group has been heavily involved in establishing radioimmunoassay for steroids. When we first talked about this ten or eleven years ago, there was the same degree of skepticism whether this would ever be feasible, and could replace the existing methods that were



ISBN 0-12-208820-4

Dr. Lindner

firmly established in clinical laboratories. People pointed out, quite correctly, that studies at Columbia University, in which antisera to steroids were generated, had shown that these lacked adequate specificity. However, within a few years it was possible to show in our laboratory, and independently by others (e.g. Niswender and Midgley), that with proper attention to antigen design one can develop radioimmunoassays for all the important steroid hormones and their metabolites with considerable specificity, and these assay systems are now widely used in clinical and research laboratories. Now, as Dr. Boguslasky rightly pointed out, this very success may impede further development, since it is difficult to establish that a new technology is superior when the existing one is fairly satisfactory. However, I do not believe that there can be an ultimate methodology, and radio-immunoassay, too, has its foot of clay. In some countries, the problem of disposing of radioactive waste presents difficulties. I won't assert that there is a real danger here, but the issue has become at least a major nuisance. In Britain, the Health and Safety Act for workers has exacerbated - or created greater awareness of - this problem and here in the U. S., I am told, similar problems have turned up. In France, you have to have a degree in medical physics to run a radio-immunoassay laboratory. Other difficulties exist in the developing countries: a student from Korea assured me that there are only two scintillation counters in his country, and he would have no access to these. Clearly, then, in the developing world a simpler technology and cheaper instrumentation would be welcome. A further advantage of chemiluminescence-based immunoassays is that the labeled ligands are stable and have a much longer shelf-life than radio-labeled steroids. Moreover, unlike conventional R. I. A. systems, some of the prototype chemiluminescence-assays presented at this meeting require no phase-separation step and thus are quicker and more amenable to automation. I realize that we are still at the beginning of this development and many teething troubles remain to be overcome; but as you may have seen in the poster sessions, we now have workable chemiluminescence immunoassays for at least three plasma steroids - estriol, progesterone and cortisol - and for several conjugated urinary steroid metabolites. These can tell the physician whether a woman has ovulated, whether she is in the follicular or luteal phase of her cycle, and when pregnant, whether placental endocrine function is developing normally; or whether a patient is eucorticoid, addisonian or cushingoid. These are the kind of questions a clinician is interested in and he can get his answer possibly in 1 - 2 hours.

Dr. Lindner

It would usually take rather longer by radioimmunoassay. In sensitivity and specificity these assays now approach the performance of radioimmunoassay, and trials conducted in conjunction with the King's College Medical School, London (G. Barnard & W.Collins) and The Endocrinology Unit, University of Florence (M. Pazzagli & M. Serio) suggest that the results of the two assays when applied to plasma are closely correlated. Now, it seems important that our colleagues in industry also take a more sanguine view of this, and take an active part in developing optimal instrumentation and procuring suitable reagents so that this new technology can be tested out in a number of leading reference centers and clinical laboratories to establish whether it offers significant advantage. The results of our pilot studies are sufficiently encouraging to make me believe that assays of this type may well find a place in clinical laboratory practice.

Dr. Campbell

Perhaps I could just draw your attention to Steve Simpson's poster this afternoon. He has shown that not only is it possible to establish assays in the clinically-useful range for substances of biological and medical interest using chemiluminescent labels as alternatives to 125I, with at least the equivalent sensitivity which we hope eventually to improve on, but also that they have one very clear technical advantage over any of the 125I labels, and that is their stability on storage. The labels that he has produced so far have been tested over a period of two years, retaining biological activity and chemiluminescent activity. Of course people in this audience familiar with radioimmunoassay will be familiar with the fact that 125I labels lose their radioactivity and biological activity very quickly and that you have to keep remaking your labels. This means that with chemiluminescent compounds one could make one very large bulk preparation which could be used by a very large number of laboratories over a considerable period of time.

I would also like to direct a question at Dr. Schram relating to the temperature sensitivity of photomultiplier tubes. In my hands, the photomultiplier tubes are used with cell holders in the range of 8 degrees to 37 degrees. Although the background, i.e. the dark current, can be fairly sensitive to temperature, the absolute sensitivity doesn't seem to vary very much. These tubes are bialkali tubes. We do actually have a system where we can cool the photo tube to -30 degrees. I wondered whether your data is either a reflection of the photomultiplier tube itself, for example differences between bialkali and trialkali photomultipliers where the trialkali ones are much more temperature-sensitive, or whether it was because of the wavelength of the firefly system. Of course, most of the wavelengths that we work at with photoproteins and chemiluminescent labels are nearly always below 500 nm, although with energy transfer or with the photoproteins, we can get up to about 510 nm. The firefly system emits much more towards the yellow, and I wonder whether this is a region where the photomultiplier tube temperature sensitivity is worse than it is with ours.

Dr. Schram

As I mentioned, the background of the photomultiplier will usually not be limiting as far as one is using good-quality photomultipliers, of course. As a matter of fact, decreasing the temperature will decrease the background of the photomultiplier. But I don't think there is need for that in most counters. What I want to stress is not the emission spectrum of firefly luciferase, is on the edge

Dr. Schram

of the sensitivity spectrum of most photomultipliers used in commercial counters now. We are using RCA 8850. In that case, a slight shift in the emission spectrum of luciferase will have a dramatic influence on the measured luminescence, because the sensitivity of the photomultiplier decreases.

Dr. LeMasters

One of the problems, I think, with using luminescence versus absorbance, fluorescence or radioactivity is the liability of the signal--the fact that it may be susceptible to a variety of types of inhibition. In yesterday's session with firefly enzyme, some of the inhibitors or extraneous factors that can quench luminescence were mentioned. I wonder if the panel could comment on common or unsuspected sources of inhibition or quenching of luminescence in these clinical assays which one needs to be careful of.

Dr. Stanley

Changes in pH are what come to mind immediately. If you are looking at, for example, a spectrum of urine samples, color and turbidity are also very variable in urine samples. Also, the presence of compounds that may have built up, say if the patient is in renal failure. You've got a whole spectrum of drugs and their metabolites, which may inhibit.

# VII
# SUMMARY AND FUTURE APPLICATIONS OF LUMINESCENCE

# OVERVIEW

T. P. Whitehead

Wolfson Research Laboratories
Department of Clinical Chemistry
Queen Elizabeth Medical Centre
Edgbaston
Birmingham

I think the statistics of the presentations at this International Symposium are of interest. In the auditorium there were some forty presentations by lecturers and discussants. I calculate that six or perhaps seven were on applications the rest were in the main concerned with the basic chemistry and biochemistry of luminescence. We were priviledged to hear the leaders in the field and their colleagues discuss the latest developments in explaining the phenomenon of luminescence. It gave the impression that since the first Symposium there have been steps forward but there remained many unsolved problems and there was little that I could detect that improved the situation regarding analytical applications of luminescence but perhaps I am not perceptive enough.

The situation in the poster sessions was different. There were seventy-six poster titles and about one half were analytical applications and the majority of those were in the clinical field. Of the 27 titles concerned with Chemical applications

6 were concerned with Phagocytosis




ISBN 0-12-208820-4

3 were concerned with Bacteria uria
5 were concerned with Enzymatic Activity Analyses
8 were concerned with Immunoassay
5 were Miscellaneous

I should have welcomed some of these being presented as papers in the auditorium in front of those scientists whose way of life is luminescence so that they could comment and contribute. I particularly regret that Marlene DeLuca's excellent work on coupling methotrexate to luciferase only merited one slide and two minutes comment in the auditorium. The possible analytical applications since the first Symposium have not increased, but there has been some consolidation of proposed applications.

I am not a haematologist or microbiologist but it is obvious that in haematology the potential areas are in the study of phagocytocis, platelet and erythrocyte viability. In microbiology the detection of ATP has obvious applications in the field of detection of bacteria. The detection of antigens and virus will be by immunoassay and the role of luminescence in such assays is dealt with later.

With regard to clinical chemistry, the modern routine clinical laboratory in developed countries is inevitably a conservative organisation, reluctant to change because it is frequently bounded by external quality assessment, controlled by various types of Government organisations, increasingly cost conscious and solutions to problems once solved are difficult to change.

A large number of clinical chemistry laboratories can now perform the common analyses required to provide a clinical service by methods which are cheap, easy to perform, precise and reasonably accurate. They are also fast and can be performed by automation. Twelve of the most common determinations in my own laboratory can, as a biochemical profile, be performed for a total sum of three dollars including direct and indirect costs.

For these reasons it is not easy for luminescent methods to replace those in current use for the commonly used tests if there is no guarantee of the supply and quality of reagents. Equally, if effective automation is not readily available and the cost of analyses increases without a guarantee of equivalent precision and accuracy this is an additional problem.

Luminescence at the present time can be described as a technique looking for a role in the clinical laboratory. However, I am convinced that this is a temporary phase and my task is to try to identify that role for luminescence which I am sure will occur in the next few years.

The main types of analyses performed in a clinical chemistry laboratory may be conveniently divided for purposes of this paper into:

- ions
- metabolic substrates
- drugs
- vitamins
- enzymes
- hormones
- proteins

*A. Ions*

The role of luminescence in this area of work is at present confined to the determination of ionised calcium in single cells by Aequorin. However, this is at the present time a technique confined to the research laboratory.

*B. Metabolic Substrates*

For many routine assays within this group of substances there are at the present time convenient, cheap, fast, accurate and precise techniques which can be automated. If luminescence techniques are to be introduced in this area of analyses, then the commercial availability of automated luminescent analysers is essential. The analyses are normally performed in large numbers.

Increasingly these substances are being assayed by specific enzymes or groups of enzymes (eg., glucose, cholestrol, urea, uric acid, triglycerides, lactate, ethanol). For many laboratories the costs of some of these enzymes are prohibitively high. Many of these reactions can be conveniently linked to luminescent reactions and with increased sensitivity beyond that normally associated with colorimetric assay this would reduce the unit cost of an analyses without sacrificing speed and maintaining accuracy and precision.

*C. Drugs*

Enzyme immunoassay techniques (EMIT) for the assay of drugs are becoming established in many clinical laboratories. They offer speed, ease of use, specificity along with acceptable precision. The end-point may be detected using BL. The outstanding advantage in adopting such an approach is the savings in costly reagents. In our own laboratory we have, without difficulty, carried out twenty drug assays using EMIT combined with BL with the amount of reagents normally used for one colorimetric assay.

### *D. Vitamins*

There is almost a complete paucity of good methods for assaying vitamins in the clinical chemistry laboratory. It is frequently stated that vitamin deficiency is not an important clinical problem in developed countries. So often technological development is needed before such statements can be challenged. Luminescence reactions, because of the sensitivity when coupled with enzymes used as reagents to chemically alter the vitamin and therefore give specificity, have considerable potential in this important area of nutrition. There are obvious uses of ATP assay in microbiological assay of vitamins.

### *E. Enzymes*

The papers from Sweden in the last Symposium showed how enzyme activity could be measured with luminescence. The technique of continuous ATP monitoring is certainly, for some enzymes, as reproducible, rapid and convenient as the spectrophotometric assay, but has greater sensitivity. It will however need considerable persuasion to introduce such techniques into the present clinical chemistry laboratories for activity measurements of enzymes at present in clinical use. It would be regarded by some as frustrating to present efforts on standardization. Yet, in the future, our ability to detect enzyme activity in biological fluids at levels undetectable by conventional spectrophotometric methods must surely be the result of more widespread use of luminescence in the research enzyme laboratory.

### *F. Hormones*

Many of these substances are now being assayed by RIA whether they are protein or steroids. The remarks made below regarding proteins are thus applicable to hormones.

### *G. Proteins*

This is the greatest actual and potential area of

expansion in clinical chemistry and luminescence should have an important part in this expansion. It is frequently difficult for my colleagues and I to think outside "proteins" when we are considering the development of our subject in the next ten years. Inevitably such thoughts are based upon and stimulated by the developments in immunoassay and there is, in our minds, no doubt that this is where much of the future of luminescence assay lies in the clinical chemistry laboratory.

The use of luminescence has advantages over conventional immunoassays in terms of sensitivity, safety, stability of reagents, potentially cheaper reagents and equipment and the possibility of avoiding separation steps.

Such assays may involve CL or BL labelling of the antigen or antibody or the use of CL or BL to detect the end point of an enzymeimmunoassay.

The following is a convenient classification of the types of immunoassay available and is taken from "Clinical and Biochemical Applications of Luminescence" edited by L. J. Kricka and T. J. N. Carter and published by Marcel Dekker, New York (in press).

## *H. Classification*

- Luminescent Immunoassay (LIA)
- Luminescent Enzyme Immunoassay (LEIA)
- Luminescent Enzyme Multiplied Immunoassay Techniques (LEMIT)
- Luminescent Cofactor Immunoassay (LCIA)

The routine application of labels and luminescent reactions to the monitoring of immunological reactions is an exciting prospect for the future. It is, however, only one amongst a number of possible replacements for radioactive labels. Luminescent immunoassay is still at a very early stage of development and its future success will largely depend upon improvements in the quality of luminescent reagents, labelling techniques and the availability of automatic luminometers.

My colleagues and I look forward to seeing you at the next Symposium in England in 1982. The next two years will be crucial for this subject and there is no doubt that the theoretical basic science debate on chemi- and bioluminescence will continue. There is no doubt that luminescence will remain an important tool in various research activities. If the present research activities do not lead to the routine application of analytical procedures by 1982 I think the subject may well go into recession with regard to commercial and applied science interest.

# VIII
# ABSTRACTS
## A. Chemiluminescence and Related Assays

# A PRELIMINARY STUDY OF THE MEASUREMENT OF URINARY PREGNANEDIOL-3α-GLUCURONIDE BY A SOLID-PHASE CHEMILUMINESCENCE IMMUNOASSAY

G. Barnard
W. P. Collins

Department of Obstetrics and Gynaecology
King's College Hospital Medical School
Denmark Hill, London SE5 8RX

F. Kohen
H. R. Lindner

Department of Hormone Research
The Weizmann Institute of Science
Rehovot, Israel

## I. INTRODUCTION

There is still the need for the development of simple, reliable methods to assess ovarian function in women, particularly over long periods of time. One approach to this objective has been to measure the concentration of estrone-3-glucuronide and pregnanediol-3α-glucuronide in serial samples of early morning urine (1). The results of this study indicated that the levels of the two steroid metabolites could be used to monitor follicular development and luteal function.

This abstract describes an immunoassay for urinary pregnanediol-3α-glucuronide using a chemiluminescent derivative of the steroid metabolite. Separation of bound and free hormone is effected by the use of a solid-phase system, the specific IgG being adsorbed onto the surface of the polystyrene assay tubes. The aspiration of the assay buffer and subsequent washing leads to the removal of all interfering




ISBN 0-12-208820-4

compounds with the concomitant significant reduction of background chemiluminescence. Thus the method can be used to assay urine samples without prior purification. The assay is comparable in specificity and sensitivity to the established radioimmunoassay.

## II. MATERIALS AND METHODS

### *A. Steroid-isoluminol conjugate*

Pregnanediol-3α-glucuronide was covalently linked to 6(N-aminohexyl)-N-ethyl-amino-2,3-dihydrophthalazine-1, 4-dione (AHEI) to yield pregnanediol-3α-glucuronide-AHEI conjugate by a procedure previously described (2). The proposed structure of this derivative is shown in Figure 1.

### *B. Antibody-coated tubes*

Antiserum to pregnanediol-3α-glucuronide was kindly donated by Dr. W. F. Coulson (Courtauld Institute of Biochemistry, London). Anti-pregnanediol-3α-glucuronide IgG was prepared by chromatography on Sepharose-Protein A. Disposable polystyrene tubes (Luckham LP3) were coated with antibody by overnight incubation at 4°C with each tube containing 200μl of suitably diluted IgG in a barbitol buffer

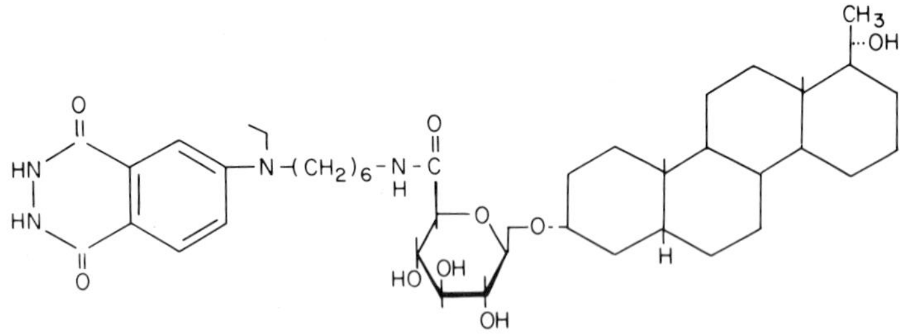

*FIGURE 1. The proposed structure of pregnanediol-3α-glucuronide aminohexyl ethyl isoluminol (AHEI) conjugate.*

(0.07M, pH9.5). Prior to use, the excess antibody was removed by aspiration and the addition of 200μl of phosphate buffer (0.1M, pH6.9 containing 0.15M NaCl and 200mg/l BSA). After a further two hours incubation at 4°C, this buffer was also aspirated.

### *C. Immunoassay*

Stock solutions of the conjugate and steroids were prepared in ethanol and stored at 4°C. When required the compounds were diluted in assay buffer (phosphate, pH6.9). 100μl of standard or sample together with 100μl of pregnanediol-3α-glucuronide-AHEI (100pg) were added to the antibody-coated tubes and incubated overnight at 4°C. Subsequently, the contents of the tubes were aspirated. 1ml wash buffer (0.1M phosphate, pH8 containing 6g/l NaCl) was added to each tube and aspirated.

### *D. Light emission and measurement*

Microperoxidase was dissolved at 1mg/ml in phosphate buffer (pH8) and this stock solution kept at 4°C. The working solution was obtained by diluting in buffer to 1.3μM enzyme. The oxidant solution was prepared by adding 30μl of 30% $H_2O_2$ solution to 5ml double distilled water.

Light emission was measured using a Luminometer 1250 (LKB-Wallac) together with an automatic injection system (Hook and Tucker Microspenser). Prior to the initiation of chemiluminescence by the injection of 100μl of $H_2O_2$, 200μl of microperoxidase solution was added to the assay tube. The light emission was monitored by the trace of a flat bed recorder and the peak heights (mV) subsequently measured.

## III. RESULTS

### *A. Standard dose-response curve*

Using the assay conditions described, a standard dose-response curve was obtained with a range of 0.156 to 5ng/tube (Fig. 2).

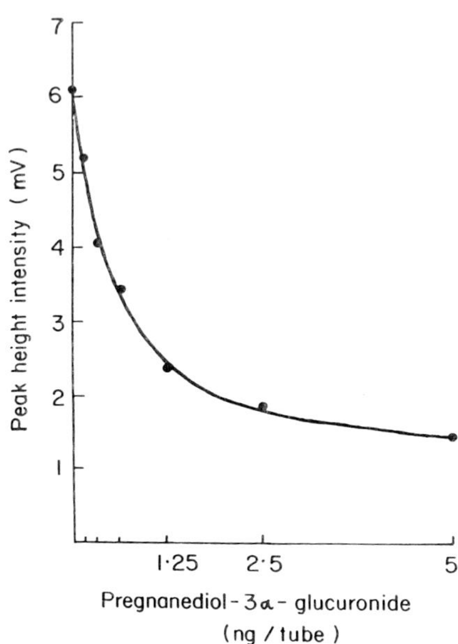


*FIGURE 2. Representative dose-response curve for pregnanediol-3α-glucuronide*

*B. The measurement of pregnanediol-3α-glucuronide in serial samples of early morning urine collected throughout two normal menstrual cycles*

Early morning urine specimens were collected daily by two normal healthy female volunteers throughout their entire menstrual cycles. Aliquots of these specimens were suitably diluted with assay buffer (x200) and 100μl of diluted urine (equivalent to 0.5μl urine) was added to the antibody-coated tubes. The concentration of pregnanediol-3α-glucuronide in the samples was measured by both conventional radioimmunoassay and solid-phase chemiluminescence immunoassay (Figs. 3 and 4).

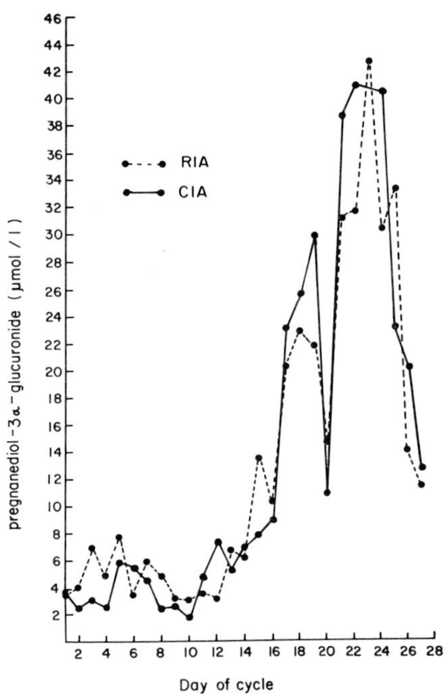


*FIGURE 3. The concentration of pregnanediol-3α-glucuronide in cycle No. 1.*

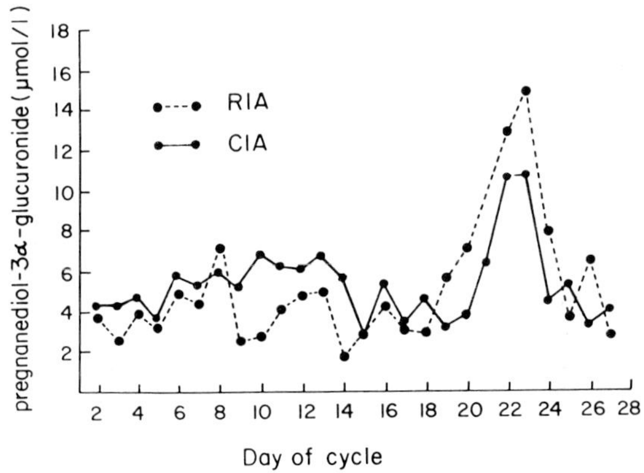


*FIGURE 4. The concentration of pregnanediol-3α-glucuronide in cycle 2.*

*C. The correlation between the measurement of pregnanediol-3α-glucuronide by radioimmunoassay and solid-phase chemiluminescence immunoassay (Fig. 5).*

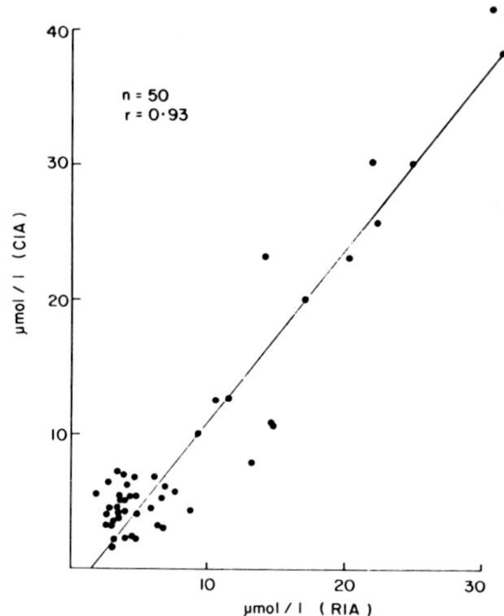


*FIGURE 5. The correlation between the two methods*

## IV. DISCUSSION

In recent years increasing interest has been shown in the use of chemiluminescent markers as the basis of an alternative approach to immunoassay. Several methods have been developed using the principles of antibody-enhanced chemiluminescence that have not required a phase-separation step (2-8). These homogeneous assays have shown great promise and further developmental work is progressing. Nevertheless, as with all homogeneous assays, great care needs to be exercised to avoid interference from substances in biological samples.

In this abstract we have described a simple, reliable heterogeneous chemiluminescent immunoassay for the measurement of urinary pregnanediol-3α-glucuronide which obviates the need for prior purification of the urine and which has a sensitivity and specificity similar to the conventional radioimmunoassay. The antibody-coated tubes are simple to prepare and separation of bound and free hormone is effected by aspiration. Thus the assay can be performed with the minimum financial outlay on equipment. The LKB-Wallac Luminometer 1250 used in these experiments is one of the cheaper of presently available luminescence photometers. Although an integration facility would be of advantage, the simple measurement of peak height is sufficient.

We believe that the method described represents a simple non-invasive approach to monitor one facit of ovarian function in women. The application of the same principles to the measurement of related compounds will provide simple, non-isotopic procedures to assess different aspects of endocrine function.

## ACKNOWLEDGEMENTS

This work has been supported by grants to HRL from WHO, Ford Foundation and Rockefeller. HRL is an International Fogarty Fellow at the National Institute of Health. We thank LKB-Wallac for the use of the instrument.

## REFERENCES

1. Collins, W. P., P. O. Collins, M. J. Kilpatrick, P. A. Manning, J. M. Pike, and J.P.P. Tyler, *Acta Endocr.* 90, 336 (1979).
2. Kohen, F., M. Pazzagli, J. B. Kim, H. R. Lindner, and R. C. Boguslaski, *FEBS Letts.* 104, 201 (1979).
3. Schroeder, H. R., P. O. Vogelhut, R. J. Carrico, R. C. Boguslaski, and R. T. Buckler, *Anal. Chem.* 48, 1933 (1976).
4. Schroeder, H. R., R. C. Boguslaski, R. J. Carrico, and R. T. Buckler, *Methods Enzymol.* 57, 424 (1978).
5. Kohen, F., J. B. Kim, and H. R. Lindner, *in* "Proceedings of the Second International Symposium on Luminescence," this volume, Academic Press, New York, in press.
6. Kohen, F., J. B. Kim, G. Barnard, and H. R. Lindner, *in* Proceedings of the Second International Symposium on Luminescence," this volume, Academic Press, New York, in press.
7. Kohen, F., J. B. Kim, G. Barnard, and H. R. Lindner, *Steroids,* in press (1980).
8. Kohen, F., M. Pazzagli, J. B. Kim, and H. R. Lindner, *Steroids,* in press (1980).

# CHEMILUMINESCENT REACTIONS OF ALCOHOLS AND ALDEHYDES

Ralph J. Bushnell

Department of Biology
Sonoma State University
Rohnert Park, California

## I. INTRODUCTION

Chemiluminescence is commonly encountered in preparations for liquid scintillation counting of readioactive isotopes. This problem is especially prevalent in labelled materials derived from plants. Attempts to bleach the chlorophyll or other pigments in such preparations that are also likely to contain aldehyde and alcoholic substances lead to extremely high background counts. I had noted that the highest backgrounds occur when aldehyde and alkali bases were in the same preparations. Because of the well-known involvement of aldehydes and α-peroxylactones in bioluminescent systems, it was decided to investigate further the aldehyde chemiluminescence phenomena. Aldehydes are best known in bacteria for their involvement in bioluminescent reactions (1,2,3). They have also been investigated for Annelids (4). α-peroxylactones recently have been considered for fireflies (5).

## II. PROCEDURES

Procedures have been modified as information accumulated.

### A. *Reaction Preparations*

The first system consisted simply of dehydrated ethanol saturated with NaOH or KOH in 20 cc scintillation vials. These were placed in the counting position of a Beckman 150 Scintil-



ISBN 0-12-208820-4

lation Counter and left there for several days with frequent readouts of the accumulated counts. The counts were recorded with a teletype printer as the per minute rate.

After several hours lag period the count rate per minute began to increase gradually to reach a peak. After the peak there began a gradual decline in count rate. Variation in duration of lag period from some absolute ethanol samples led to discovery that this is sometimes due to the presence of acetaldehyde.

The second procedure consisted of addition of various aldehydes to alcohols before injecting over to alkali hydroxide. This method for two or three carbon alcohols shortened the lag period and decreased decay time to a few minutes and increased the peak light output.

The third procedure consisted simply of injecting aldehyde in milliliter amounts over the dry alkali pellet. A rapid light yielding reaction usually followed such injections and yellow coloration developed on the surface of the alkali masses. This deposit is similar to the formose reaction mixture resulting from the formose reaction (6).

Attempts are being made to separate the solid reaction products but this is proving as difficult as it is for those of the formose reaction (7).

The operation of the scintillation counter was checked before and after each experiment with calibrated carbon-14 sources.

### *B. Background Counts*

Potassium hydroxide has 0.012 per cent of its K as isotopic $^{40}K$. The beta decay from this amount shows up as Cerenkov radiation from the solid state and also in solution. This radiation is recorded in the liquid scintillation counter as part of the background. For several grams the counts per minute may amount to several thousand. An average count for twenty 7 gram samples, dry in the scintillation vials, is 2600 CPM, while for 15 grams it is 3900 CPM. The released betas behave as electrons in the pellet matrices. The probably account for differences between KOH and NaOH in the progress of the reactions of such substrates with alcohols and aldehydes. The CPM background count for 7 to 15 grams of NaOH usually lies between 20 and 50 CPM.

*TABLE I. Chemiluminescent Response of Alcohols to Alkali Hydroxides*

| Alcohol | Base | $t_1$[a] | Peak[b] | $t_2$(½ decay)[c] |
|---|---|---|---|---|
| Ethyl | KOH | 200-400 | $3.0 \times 10^4$ to $1.0 \times 10^5$ | 200-500 |
| Ethyl (in hexane) | KOH | 150 | $1.5 \times 10^4$ | 500 |
| Ethyl | NaOH | 150+ | $2.4 \times 10^4$ | still gaining |
| Propyl | KOH | 240 | $3.0 \times 10^4$ | 500 |
| Propyl | NaOH | 180+ | $3.0 \times 10^5$ | still gaining |
| Allyl | KOH | 5 | $4.5 \times 10^5$ | 5 |
| Allyl | NaOH | 68 | $1.3 \times 10^5$ | 14 |
| Isopropyl | KOH | 108-465 | $2.0 \times 10^4$ | long |
| Isopropyl | NaOH | 1414+ | $2.2 \times 10^3$ | still gaining |
| Butyl | KOH | 1800 | $4.3 \times 10^5$ | long |

[a] $t_1$ = time in minutes for the system to attain highest per minute count.

[b] highest per minute count attained.

[c] $t_2$ = time in minutes from time of peak photon output for decay to a value one-half that of the peak.

TABLE II. Chemiluminescent Response of Aldehydes With Alkali Hydroxides

| Aldehyde | Base | $t_1$[a] | Peak light[b] | $t_2$[c] |
|---|---|---|---|---|
| Acetaldehyde | KOH | 2 | $4.5 \times 10^4$ | 2½ |
| Acetaldehyde | NaOH | 2 | $1.3 \times 10^4$ | 15 |
| Propanal | KOH | 3 | $3.0 \times 10^5$ | 13 |
| Propanal | NaOH | 35 | $3.9 \times 10^5$ | 123 |
| Butanal | KOH | 4 | $3.0 \times 10^5$ | 12 |
| Butanal | NaOH | 18 | $4.2 \times 10^5$ | 88 |
| Crotonaldehyde | KOH | 2½ | $4.2 \times 10^5$ | 8 |
| Crotonaldehyde | NaOH | 5 | $2.0 \times 10^5$ | 12 |
| Valeraldehyde | KOH | 32 | $1.0 \times 10^5$ | 63 |
| Hexanal | KOH | 40-80 | $1.0 \times 10^5$ | >150 |
| Decanal | KOH | 5 | $1.1 \times 10^4$ | 21 |
| Decanal (In ETOH) | KOH | 14 | $2.5 \times 10^5$ | 45 |
| Dodecanal | KOH | 8 | $3.0 \times 10^5$ | long |
| Dodecanal (In ETOH) | KOH | 80 | $6.4 \times 10^4$ | long |

[a] $t_1$ = time in minutes for the system to attain highest per minute count

[b] Peak height = light output per minute at the highest point

[c] $t_2$ = time in minutes to decay to a rate equal to one-half that per minute at the peak rate

## III. RESULTS

### *A. Chemiluminescent Response of Alcohols*

In response to presence of alkali hydroxides onset of chemiluminescence of dehydrated alcohols exhibit variations among samples (Table I). Presence of traces of aldehydes apparently affects earliness of response. Usually a period of several hours is required for a peak light output to develop if aldehydes are not present at the start.

Isopropyl and butyl alcohols exhibit longer lag periods than ethanol and propanol. Some samples are affected strongly by the fluorescent laboratory lights. This environmental factor must be controlled in order to determine comparative response times.

NaOH is slower than KOH in inducing the light reaction but allows the process to continue for longer periods.

The unsaturated allyl alcohol reacted rapidly with KOH, reaching its peak at the end of five minutes. However, allyl alcohol and NaOH reacted only after greater time lapse.

### *B. Chemiluminescent Response of Aldehydes*

Aldehydes respond to the presence of alkali hydroxides by chemiluminescence faster than do their corresponding alcohols. Acetaldehyde produces a light peak within two minutes, and decays in two minutes as a response to KOH. The response to NaOH of acetaldehyde is slower than to KOH (Table II).

The longer carbon chain aldehydes tend to respond by photon production less rapidly when exposed to alkali hydroxide. Thus time factors involved in the responses are matters of minutes for aldehydes as compared to hours for all alcohols except allyl alcohol.

## IV. DISCUSSION

Primary reactions occur at the surface of the alkali hydroxide matrix (pellet or flake). This is where the several yellow reaction products accumulate.

The alkali hydroxides as solids apparently are complex systems containing water (15 per cent in KOH) and absorbed gases. Carbon dioxide may be ionized in the matrix. Oxygen is difficult to eliminate and may be present as the ionized superoxide $O_2$. In fact it is possible to develop a trace of the superoxide color on the surface of the moist pellets by passing oxygen over them for extended periods. It is possible also that alkali peroxides may be present in traces. So it is a complex system that the aldehydes encounter on reaching the surface of such matrices.

The yellow reaction products in the various systems probably include glyoxal, biacetyl, crotonic acids, and perhaps alkali superoxide. It is possible that a thermally unstable yellow dioxetane is formed. This could serve as the source of the excited volatile species responsible for the chemiluminescence.

Alkali metals and hydroxides are well known agents of aldehyde polymerizations (8).

The aldol reactions:

(1) $2CH_3CHO \rightleftarrows CH_3CH(OH)CH_2CHO$

(2) $CH_3CH(OH)CH_2CHO \rightleftarrows CH_3CH{=}CHCHO$

The product of the second reaction, crotonaldehyde, is one of the possible reactants with oxygen to yield a dioxetane:

```
CH3HC ——— CHCHO
  |         |
  C  ———    O
```

This dioxetane would probably rapidly decompose to acetaldehyde and glyoxal. The dioxetane literature has been reviewed (9).

Alkali hydroxides also induce enolization of carbonyls so that vinyl species could possibly be the origin of dioxetanes (10).

(3) $CH_3CHO \underset{}{\overset{-OH}{\rightleftharpoons}} H_2C{=}CHO$

## V. CONCLUSIONS

The contact between an aldehyde and the surface of a mass of alkali peroxide results in a complex set of reactions.

One or more of the reactions result in a chemiluminescent phenomenon.

The reaction residue left of the surface of the hydroxide mass is a yellow substance.

Within the yellow substance is the efficient initiator for light reactions of alcohols or unsaturated lipids, should contact between the latter two and the initiator eventually develop.

Saturated aqueous alkali hydorxide and aldehydes luminesce much less efficiently than do pellets covered by hexane.

At least one of the intermediates formed during the course of the light reactions is very sensitive to the fluorescent laboratory lights. Depending on the stage of the reaction the light output may be reduced or accelerated as a result of light exposure.

Within the solid residue adherent to the unreacted KOH resides a species that is light sensitive and capable of inducing a rapid chemiluminescent response of alcohols subsequently added to the mixture. It is thus the initiator of the next round of photon emissions.

The unsaturated alcohols and unsaturated aldehydes react very rapidly with alkali hydroxides as measured by the rate of light emission. The unsaturated bond seems to represent a clue for determination of the mechanisms involved in this chemiluminescence.

## REFERENCES

(1) Hastings, J.W., J. Spudich, and G. Malic, *J. Biol. Chem.* 238, 3100-3105 (1963).

(2) Meighen, E.A., *Bioch. Biophys. Res. Commun.* 87, 1080-86 (1979).

(3) Hastings, J.W. and K.H. Nealson, *Ann. Rev. Microbiol.* 31, 549-95 (1977).

(4) Rudie, N.G. and J.E. Wampler, *Photochem. and Photobiol.* 29, 171-174 (1979).

(5) Adam, W. and O. Cueto, *J. Am. Chem. Soc.* 101, 6511-15 (1979).

(6) Matsuura, T., Y. Shigemasa, and C. Sakazawa, *Chem. Lett.* 713-714 (1974).

(7) Weiss, A.H. and T. John, *J. of Catal.* 32, 216-229 (1974).

(8) Guthrie, J.P., *Can. J. Chem.* 52, 2037-40 (1974).

(9) Bartlett, P.D. and M.E. Landis, In "Singlet Oxygen," (Wasserman, H.H. and R.W. Murray) pp. 243-83, Academic Press, New York (1979).

(10) Blank, B., A. Henne, G.P. Laroff and H. Fischer, *Pure and Appl, Chem* 41, 475-94 (1975).

# DEFECTS ASSOCIATED WITH THE TRANSMEMBRANE POTENTIAL OF GRANULOCYTES IN CHRONIC GRANULOMATOUS DISEASE

Vincent Castranova
George S. Jones
Ruth M. Phillips
David Peden
Knox VanDyke

Appalachian Laboratory for
Occupational Safety and Health
and
Department of Physiology, Pediatrics,
and Pharmacology and Toxicology
West Virginia University
Morgantown, West Virginia

## I. INTRODUCTION

Chronic granulomatous disease is a disorder characterized by a high susceptibility to bacterial infection (1, 2). This disease has been traced to a defect in granulocyte function. These cells fail to kill bacteria even though there in no apparent decrease in bacterial ingestion (3). Evidence indicated that this disease is characterized by an absence of the "respiratory burst" associated with phagocytosis by granulocytes (4). Indeed, granulocytes from patients with chronic granulomatous disease fail to produce antibacterial substances, such as, superoxide anion or hydrogen peroxide, in response to particle-exposure (5, 6). In addition, these cells fail to generate chemiluminescence or increase hexose monophosphate shunt activity in response to bacteria (7).

Recent data indicate that stimulation of granulocytes with chemotactic factors, phorbol myristate acetate, or concanavalin A results in a rapid depolarization followed by a prolonged hyperpolarization of the transmembrane potential (8, 9). In contrast, stimulation with A23187, i. e., a calcium ionophore,



ISBN 0-12-208820-4

results in a rapid and prolonged depolarization (9). Evidence indicates that these shifts in transmembrane potential precede the release of superoxide anion or the generation of chemiluminescence induced by these stimulants (8, 9, 10, 11). Therefore, it has been suggested that depolarization of the membrane potential acts as a signal which triggers the "respiratory burst".

The objective of this investigation is to determine if chronic granulomatous disease alters the transmembrane potential response of granulocytes to stimulants. The stimulants tested were n-formyl-methionyl-leucyl-phenylalanine, phorbol myristate acetate, concanavalin A, and A23187.

## II. METHODS

Human granulocytes were isolated by dextran sedimentation followed by centrifugal elutriation as described previously (12). Purity of this granulocyte preparation was about 95% with a cell viability of greater than 95%.

The transmembrane potential of human granulocytes was measured at 22°C using a fluorescent probe, Di-S-$C_3$ (5) (9, 12). Excitation and emission wavelengths were set at 622 and 665 nm, respectively. All samples contained $2.3 \times 10^7$ cells suspended in 3 ml of HEPES medium [145 mM NaCl, 5 mM KCl, 10 mM Na HEPES, 5 mM glucose, and 1 mM $CaCl_2$ (pH = 7.4)] plus 0.66 µg/ml Di-S-$C_3$ (5).

Generation of chemiluminescence was measured at 37°C using a liquid scintillation counter as described previously (13). Each sample contained $1 \times 10^6$ cells in 5 ml of HEPES medium plus $10^{-8}$ M luminol.

Chronic granulomatous disease was diagnosed by the history of recurrent infections, and absence of particle-stimulated reduction of nitroblue tetrazolium or generation of chemiluminescence.

## III. RESULTS

We have reported that treatment of normal granulocytes with n-formyl-methionyl-leucyl-phenylalanine (a chemotactic factor), phorbol myristate acetate, concanavalin A (a lectin), or A23187 ( a calcium ionophore) stimulates superoxide release and the generation of chemiluminescence (8, 9). The data also indicate that shifts in the transmembrane potential precede this "respiratory burst" by granulocytes (8, 9).

N-formyl-methionyl-leucyl-phenylalanine, phorbol myristate acetate, and concanavalin A produce a biphasic shift in transmembrane potential, i. e., a rapid depolarization followed by a prolonged hyperpolarization. In contrast, the response to A23187 is monophasic, i.e., a rapid and prolonged depolarization.

Normal shifts in granulocyte transmembrane potential in response to n-formyl-methionyl-leucyl-phenylalanine are shown in Fig. 1. Note that an increase in the fluorescence of Di-S-C (5) indicates depolarization while a decrease in fluorescence indicates hyperpolarization. Granulocytes from healthy donors exhibit similar transmembrane potential shifts in response to phorbol myristate acetate and concanavalin A (8, 9). In contrast, n-formyl-methionyl-leucyl-phenylalanine treatment of granulocytes from a patient with chronic granulomatous disease does not result in a transmembrane potential shift (Fig. 1). Concanavalin A treatment of diseased granulocytes also fails to induce changes in transmembrane potential while phorbol myristate acetate causes only slight depolarization.

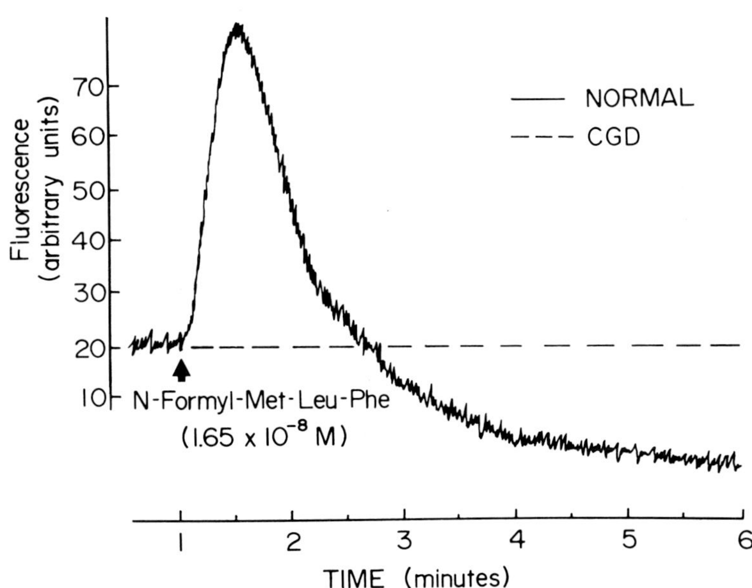


*FIGURE 1. Effect of n-formyl-methionyl-leucyl-phenylalanine on the potential of normal and diseased (CGD) granulocytes.*

It has been suggested that membrane depolarization triggers the release of superoxide anion from granulocytes (8, 9). Since these stimulants fail to induce potential shifts in diseased granulocytes, they should also fail to induce the "respiratory burst". A normal chemiluminescence response to n-formyl-methionyl-leucyl-phenylalanine treatment is shown in Fig. 2. Note that chemiluminescence peaks well after membrane depolarization is induced, i.e., very little chemiluminescence occurs during the first minute after the addition of stimulant. Note also that treatment of diseased granulocytes with this chemotactic factor results in almost no chemiluminescence. In addition, phorbol myristate acetate also fails to induce superoxide release from granulocytes in chronic granulomatous disease (14).

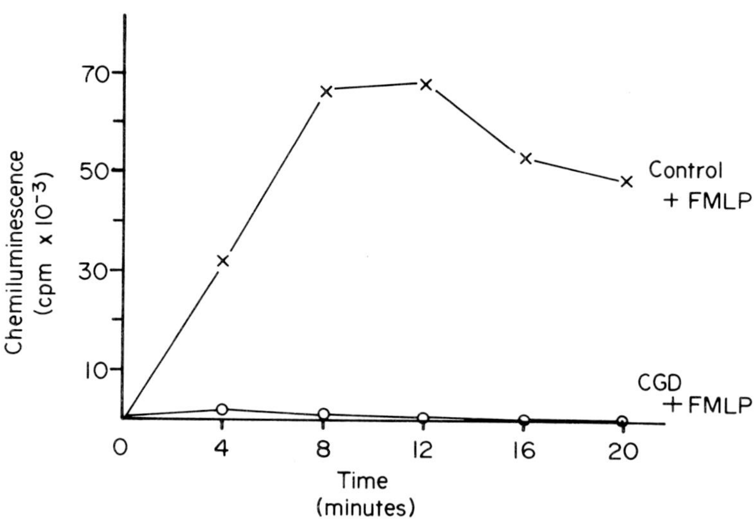


*FIGURE 2. Effect of n-formyl-methionyl-leucyl-phenylalanine (FMLP) on the generation of chemiluminescence with granulocytes from a healthy donor (normal) and from a patient with chronic granulomatous disease (CGD).*

The effects of A23187 on the transmembrane potential of normal and diseased granulocytes are shown in Fig. 3. Note that A23187, a calcium ionophore, induces a similar monophasic shift (depolarization) in the potential of both normal and diseased cells. In contrast, A23187 induces chemiluminescence with normal cells but not with granulocytes from patients with chronic granulomatous disease.

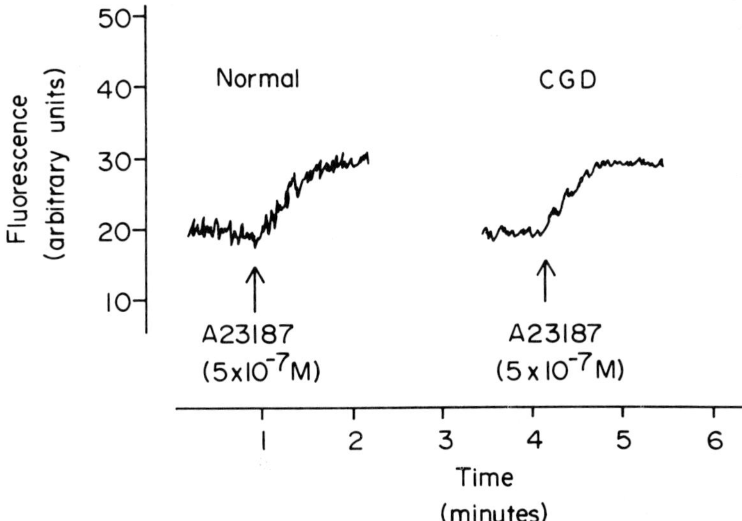


*FIGURE 3. Effect of the calcium ionophore (A23187) on the transmembrane potential of granulocytes from a healthy donor (normal) and from a patient with chronic granulomatous disease (CGD).*

## IV. DISCUSSION

Granulocytes from patients with chronic granulomatous disease do not exhibit a "respiratory burst" in response to particle-stimulation (4). This failure to release active forms of oxygen would explain the inability of these diseased cells to kill bacteria (15). It has, therefore, been concluded that the defect in chronic granulomatous disease resides in the oxidase (7).

Recent evidence indicates that stimulants induce shifts in the transmembrane potential of granulocytes which precede and presumably trigger the "respiratory burst" (8, 9, 10, 11). Reports indicate phorbol myristate acetate fails to change the membrane potential of diseased cells (10). Therefore, it was suggested that the major defect in chronic granulomatous disease is due to the inability of stimulant to alter the membrane permeability to ions (10).

Our data indicate that stimulants do not induce chemiluminescence with diseased cells, i.e., there is no "respiratory burst". Three of the stimulants also fail to perturb the transmembrane potential of diseased granulocytes, i.e., n-formyl-methionyl-leucyl-phenylalanine, concanavalin A, and phorbol myristate acetate fail to induce the normal biphasic shift in potential. In contrast, depolarization occurs with diseased cells treated with a calcium ionophore (A23187), yet chemiluminescence is not generated. We conclude that there are at least two defects associated with chronic granulomatous disease: 1) a defect which prevents many stimulants from affecting ionic permeabilities and, thus, the membrane potential, and 2) a defect which prevents depolarization from triggering superoxide release.

## REFERENCES

1. Johnson, R. B. Jr., and R. L. Baehner, *Pediatrics*. 48, 730 (1971).
2. Quie, P. G., E. L. Kaplan, A. R. Page, F. L. Gruskay, and S. E. Malawista, *N. Engl. J. Med.* 278, 976 (1968).
3. Quie, P. G., J. G. White, B. Holmes, and R. A. Good, *J. Clin. Invest.* 46, 668 (1967).
4. Homes, B., A. R. Page, and R. A. Good, *J. Clin. Invest.* 46, 1422 (1967).
5. Curnutte, J. T., D. M. Whitten, and B. M. Babior, *N. Engl. J. Med.* 290, 593 (1974).
6. Paul, B., and A. J. Sbarra, *Biochim. Biophys. Acta* 156, 168 (1968).
7. Allen, R. C., R. L. Stjernholm, M. A. Reed, T. B. Harper, S. Gupta, R. H. Steele, and W. W. Waring, *J. Infect. Dis.* 136, 510 (1977).
8. Jones, G. S., M. E. Wilson, K. VanDyke, and V. Castranova, *Fed. Proc.* 39, 2613 (1980).
9. Jones, G. S., K. VanDyke, and V. Castranova, *J. Cell. Physiol.*, in press (1980).
10. Whitin, J. C., C. E. Chapman, E. R. Simons, M. E. Chovaniec, and H. J. Cohen, *J. Bio. Chem.* 255, 1874 (1980).
11. Seligmann, B. E., E. K. Gallin, D. L. Martin, W. Shain, and J. I. Gallin, *J. Mem. Biol.* 52, 257 (1980).

12. Jones, G. S., K. VanDyke, and V. Castranova, *J. Cell. Physiol.*, in press (1980).
13. Trush, M. A., M. E. Wilson, and K. VanDyke, *Meth. Eng.*, *LVII*, 462 (1978).
14. Newburger, P. E., H. J. Cohen, S. B. Rothchild, J. C. Hobbins, S. E. Malawista, and M. J. Mahoney, *N. Engl. J. Med.* *300*, 178 (1979).
15. Babior, B. M., *N. Engl. J. Med.* *298*, 659 (1978).

# ROOM TEMPERATURE PHOSPHORESCENCE OF SOME PHARMACEUTICAL IMPORTANT IMIDAZOLES

F. Abdel Fattah
W. Baeyens
P. De Moerloose

Department of Pharmaceutical Chemistry and
Drug Quality Control
State University of Ghent, Belgium

## I. ABSTRACT

Room temperature phosphorescence spectrometry has been developed into a sensitive spectrochemical method of analysis for a number of organic compounds when absorbed on a suitable solid support.

In the present communication, spectral data, limits of detection, decay times and analytical possibilities for some imidazoles of pharmaceutical interest are reported under various experimental conditions using room temperature phosphorimetry on filter paper.

The present work is not only to be seen as an extension of the methods of analysis of these imidazole derivatives but may as well be considered as a contribution to the rapidly increasing room temperature phosphorescence techniques.

## II. INTRODUCTION

Phosphorescence emission of organic compounds is a phenomenon that has been observed in the gas phase and, more frequently, in rigid media such as plastic films and boric acid glasses, or at cryogenic temperatures (1,2) in the glasses of degassed solutions.



ISBN 0-12-208820-4

Room temperature phosphorimetry (RTP) has recently attained considerable interest as this relatively new analytical technique is based on the phosphorescence emitted at room temperature by organic compounds absorbed on various solid substrates such as filter paper, silica gel, asbestos, and sodium acetate (3-29). Original papers simply reported these phosphorescence phenomena at room temperature and no analytical applications were dealt with. Later on, the observed phenomena were developed towards analytical usefulness. RTP appears to offer a fast, economical, and convenient method of analyzing a variety of molecules, many of biological interest (6). As the external heavy atom effect on the 77K phosphorescence of many organic compounds was well-known (30-32), the use of external heavy atom perturbers (e.g. sodium iodide, lead or thallium salts) in the room temperature phosphorescence method was investigated for a wide variety of compounds of biological and pharmaceutical interest by various authors (10, 16, 24, 26-28, 33). The effects of moisture, oxygen and the nature of the support-phospher interaction are very critical in these experiments (21). Lloyd (34) described a packed flow-through cell technique for the measurement of the room temperature phosphorescence spectra of absorbed compounds.

The imidazoles under investigation include mebendazole (I) and flubendazole (II), both broad-spectrum anthelmintics; econazole nitrate (III), miconazole nitrate (IV) and imazalil (V), all potent fungicides; carnidazole (VI), a trichomonacide, metomidate hydrochloride (VII) and etomidate (VIII), both hypnotics, and the anthelmintics levamisole hydrochloride (IX) (with growing importance in the treatment of cancer and viral diseases), dexamisole hydrochloride (X) (antidepressant psychoenergizer) and tetramisole hydrochloride (XI).

## III. EXPERIMENTAL

### *A. Reagents and Materials*

All imidazoles used in this section were generous gifts from Janssen Pharmaceutica, Beerse, Belgium, and could be used without further purification. The pharmaceutical formulations were obtained from the same manufacturer.

De-ionized water was used throughout; the quality was controlled by luminescence scanning at the highest instrumental sensitivity settings. Only in exceptional cases bi-distilled water had to be used.

All reagents were of analytical grade (Merck, U.C.B.,

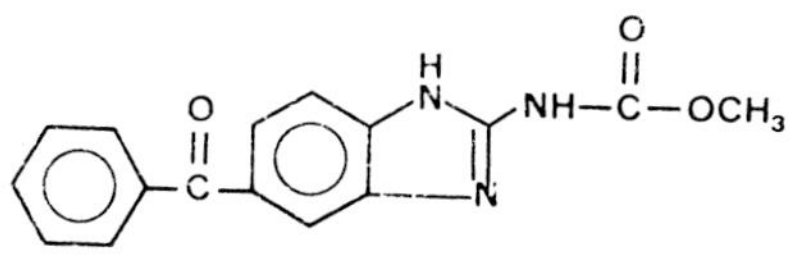

(I)

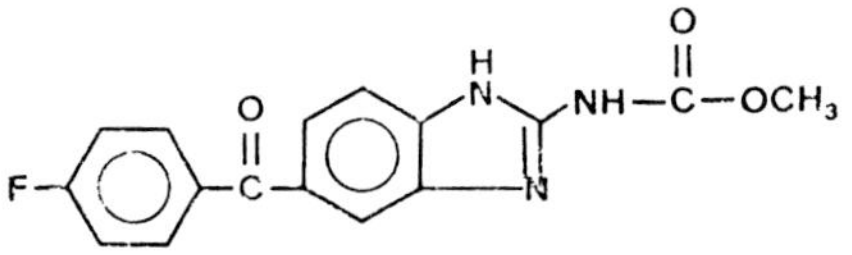

(II)

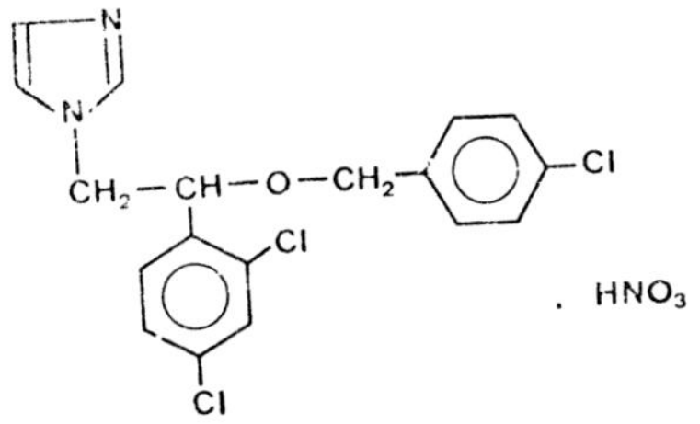

(III) · $HNO_3$

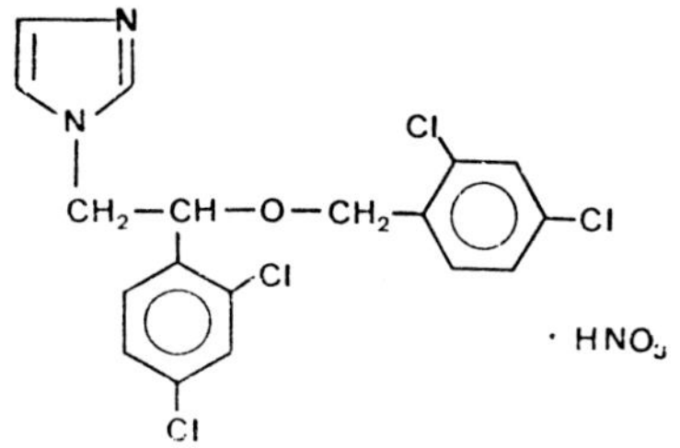

(IV) · $HNO_3$

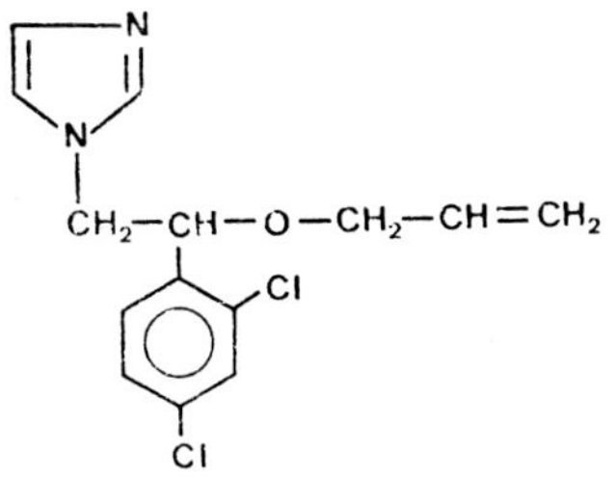

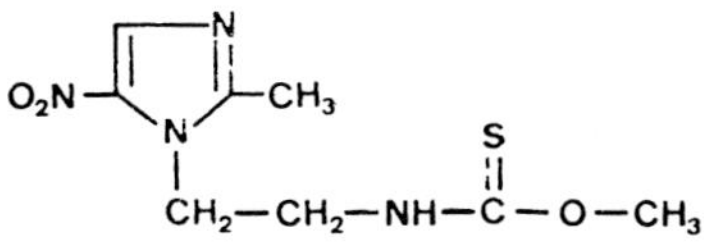

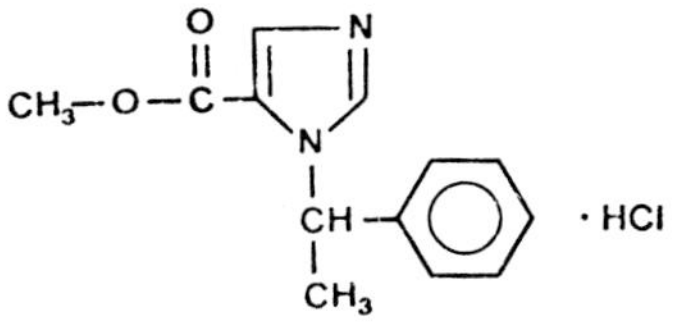

(V)

(VI)

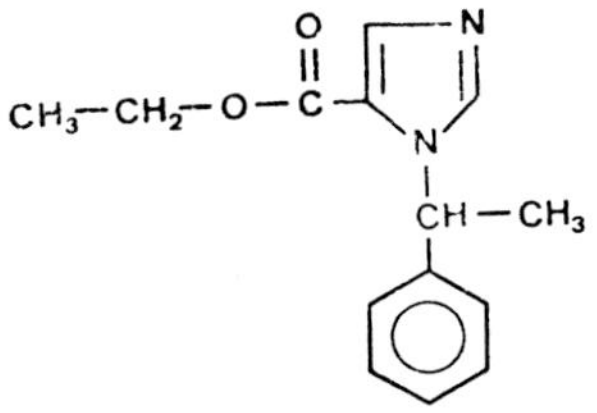

(VII) · HCl

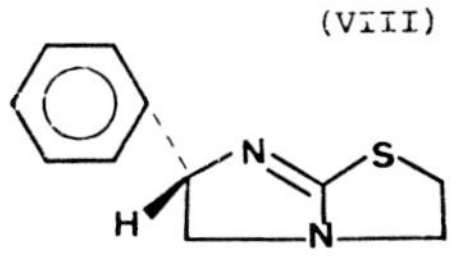

(VIII) (R)-(+)

(IX) (S)-(-).HCl

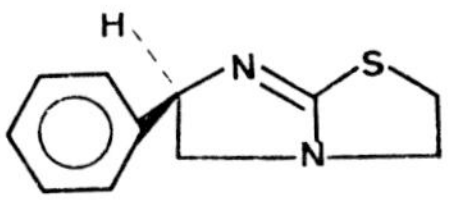

(X) (R)-(+).HCl

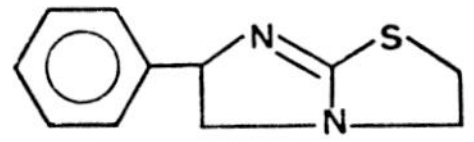

(XI) (±)-HCl

J.T. Baker), except for thallium (I) acetate and lead tetra-acetate (purim quality, Fluka) that were found satisfactory.

### B. *Instruments*

Luminescence spectra and measurements (fluorescence, room temperature and low temperature phosphorescence) were taken on an Aminco-Bowman spectrophotofluorometer (American Instrument Co., Silver Spring, Md.), equipped with the Aminco-Keirs phosphoroscope attachment (rotating can type, cat. c 558-62140).

The rotating sample-cell assembly as described by Hollifield and Winefordner (35), and modified by Zweidinger and Winefordner (36), was used as well.

RTP measurements were performed with a self-constructed sample holder (Fig. 1) that was designed for use with samples on paper at room temperature replacing the standard quartz Dewar assembly normally used at 77K. This holder is described elsewhere (37).

### C. *Procedure*

RTP on silica plates did not yield useful results.

Phosphorescent background intensities of a variety of filter papers were investigated, Whatman 42 (11.0 cm ashless filter paper circles) yielded satisfactory results and was thus used throughout.

The compounds investigated were dissolved in de-ionized water, if they were water-soluble, or preferably in ethanol if not readily soluble in water, or in any other volatile organic solvent as the application of volatile solvents is much more practical for quantitative RTP estimations. The poor solubility of mebendazole and flubendazole in most common solvents caused extra problems for the analytical and preparative determinations as basic or acid solutions had to be employed. As drying of the samples is essential, moisture causing RTP quenching, the spotting and measuring procedures require special attention (37).

### D. *Results and Discussion*

All imidazoles under investigation except for mebendazole and flubendazole exhibit very weak phosphorescence when absorbed on filter paper (Whatman 42) at room temperature. However, good 77K phosphorescence signals were obtained from mebendazole, flubendazole, cardinazole, metomidate and etomidate (37).

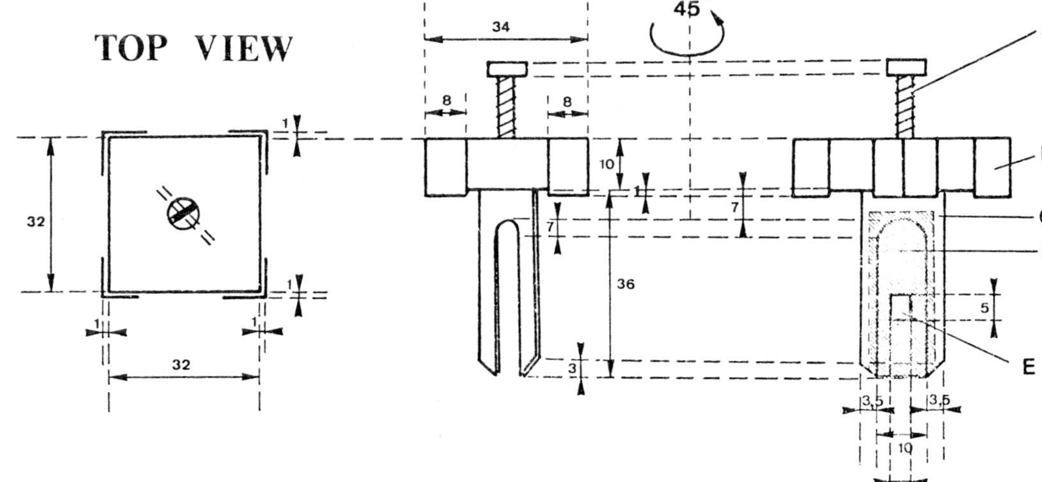


*Fig. 1. Schematic diagram of a new sample holder system for room temperature phosphorescence studies. All dimensions are in mm. A = screw to facilitate manipulations of the holder; B = plastic cap (Perflex) with aluminium supports at the corners; C = aluminium two-bladed holder; D = inserted filter paper with marked window (E) from the 4 mm slit. This sample holder unit fits into the standard Aminco phosphoroscope accessory instead of the required cooling devices. By spotting in the centre of the filter paper window, the whole amount of analyte is excited.*

Mebendazole and flubendazole yielded good phosphorescence signals. For the remaining imidazoles no remarkable differences from reagent blank signals could be registered, except for econazole, metomidate and etomidate which showed a slight RTP activity when treated with thallium. Table 1 illustrates the qualitative and semi-quantitative aspects of the mebendazole RTP characteristics in alkaline medium and the effect of various ions.

From Table 1 it is clear that from all heavy metals used as reagents for enhancing RTP yield of alkaline mebendazole samples taken as an example, lead (IV) and thallium (I) give good results, lead (IV) tetraacetate producing highest emission intensities while mercuric nitrate brings along phosphorescence quenching. All heavy metals seem to cause a hypsochromic shift in the excitation spectra of mebendazole. Only

*TABLE 1*

*Effect of various ions on the room temperature phosphorescence characteristics of mebendazole[a] in sodium hydroxide.*

| Reagent (in $H_2O$ unless stated otherwise) | $\lambda_{EXC.}$ (nm) | $\lambda_{EM.}$ (nm) | Relative intensity (arbitrary units) | $\tau_{RTP}$ (s) |
|---|---|---|---|---|
| 1M NaOH | (285sh),360 | 498 | 58 | 0,61 |
| 1M NaI | (285sh),360 | 494 | 61 | 0,57 |
| 1M NaI in 1M NaOH | (275sh),365 | 510 | 51 | 0,57 |
| 0.2M $AgNO_2$ | (285sh),330 | 481 | 49 | 0,58 |
| 0,02M $Pb(CH_3COO)_4$ | 332 | 475 | 100 | 0,55 |
| 0,2M $CH_3COOOTl$ | (300sh),346 | 505 | 74 | 0,56 |
| 0.2M $Hg(NO_3)_2$ | 295 | 502 | 17 | 0,52 |

[a]Whatman 42 filter paper + 3.0 µl mebendazole (1 µg/µl 1M NaOH, gentle heating in water bath) + directly 3.0 µl reagent, all being spotted in the centre of the window, 10 min. IR 65°C drying followed by reading in the rotating can (average speed) while dry nitrogen passes over the sample. Blank spectra were run simultaneously. Reported data are averages of at least 6 determinations. sh = shoulder.

silver (I) and lead (IV) produces a significant blue-shift of the phosphorescence wavelength. Resolution techniques (38) can minimize metal ion background interferences.

All heavy metals influence the decay time of alkaline mebendazole on filter paper; in each case a slight decrease of this value can be observed. All reagents still worked in the abundance of free oxygen, the RTP phenomenon probably is no more sensitive to collisional quenching by oxygen.

Sodium iodide as such has only a slight increasing effect on the phosphorescence intensity although the decreasing influence on $\tau$ is similar to the heavy metal influence. The combined use of NaI-NaOH leads to an analogous effect on $\tau$ but results in a lower emission value; a negative influence of the total $[Na^+]$ concentration being probably at the origin of this phenomenon.

Vo-Dinh et al. (10) state that when drying solutions containing an excess of sodium hydroxide in the presence of $CO_2$ (air), the sample may become incorporated into a NaOH-$Na_2CO_3$ matrix, $Na_2CO_3$ being an extra parameter to be taken into account in these investigations.

In the external heavy atom effect, environmental atoms of high atomic numbers enhance molecular spin-forbidden transitions via a spin-orbital coupling mechanism (26), an effect that has been employed to enhance phosphorescence intensity in both low temperature and room temperature studies of some organic compounds.

In addition to Table 1, the excitation and phosphorescence wavelength maxima for mebendazole 77K are $\lambda_{EXC.}$ = (256), 320nm $\lambda_{EM.}$ = 450 nm (absolute ethanol), for flubendazole 77K $\lambda_{EXC.}$= (258), 320 nm; $\lambda_{EM.}$ = 450 nm (absolute ethanol), given for comparison (37).

The strong phosphorescence characteristics exhibited by mebendazole and flubendazole when absorbed on filter paper were worked out qualitatively and quantitatively.

In Figure 2 the influence of IR heating time at 65° is shown. Ten minutes proved to be a good heating period; higher temperatures were not tried as these might cause simple decomposition. Figure 3 represents the effect of blowing time with dried nitrogen on the RTP emission. Five minutes gave reproducible results and could be used furthermore.

The concentration curve for mebendazole was established in this way using the optimum conditions. Limit of linearity: 100 ng/2 μl spot.

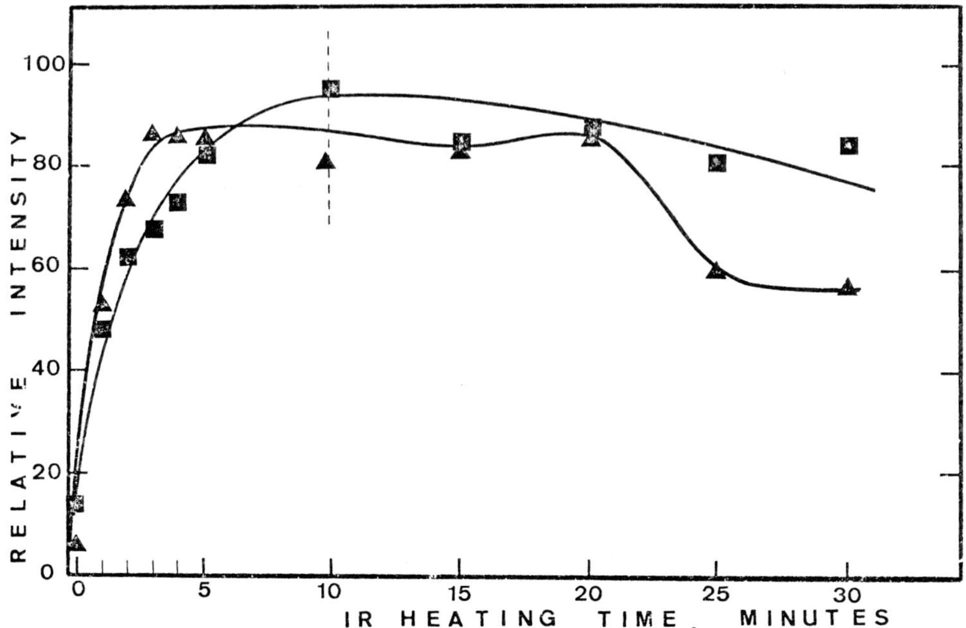


*Figure 2. Influence of sample heating time at 65° by infra-red heating on room temperature phosphorescence intensity of mebendazole (■) and flubendazole (▲) (2,0 µl 0,05 µg/µl in DMSO = 0.1 µg/spot) on Whatman 42 filter paper.*

*($\lambda_{EXC._{MEB.}}$ = 320 nm, $\lambda_{EM._{MEB.}}$ = 480 nm;*

*$\lambda_{EXC._{FLU.}}$ = 316 nm, $\lambda_{EM._{FLU.}}$ = 473 nm).*

*Each point is the average of three independent determinations.*

Detection limit: 6 ng/2 µl spot. The decay time (lifetime of the excited state) $\tau_{RTP}$ measured at $\lambda_{EXC.}$ = 320 nm, $\lambda_{EM.}$ = 480 nm using the previous conditions = 0,46s.

Establishment of the optimum parameters for the RTP quantisations of flubendazole was accomplished starting from the mebendazole conditions. Similar results were obtained, Fig. 2 and Fig. 3 indicating the influence of infra-red sample heating and of treating the sample with dried nitrogen, respectively. Table 2 outlines the important RTP characteristics for both imidazoles. As was expected, both compounds behave in a similar way.

The phosphorescence characteristics of mebendazole and flubendazole originate from the aromatic carbonyl moiety present in both molecules. It is not very clear to ascertain which type of triplet is responsible for the observed room temperature phosphorescence as the heavy atom effect on intensity and lifetime is not too strong; most probably the lowest triplet of these compound is $(n,\pi^*)$.

Fig. 4 illustrates the experimental analytical curves for mebendazole and flubendazole applied to the solid support in DMSO solution.

The RTP method for mebendazole was applied to the analytical determination in Vermox® 100 mg tablets.

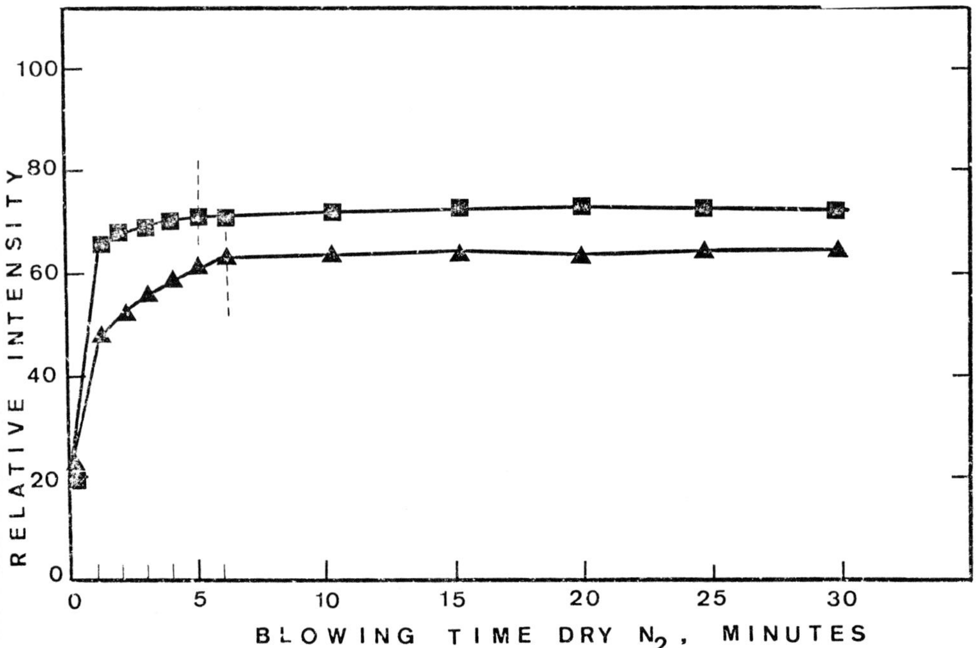


*Figure 3. Effect of blowing time with dry nitrogen over the sample in the cell compartment upon room temperature phosphorescence intensity of mebendazole (■) and flubendazole (▲). Same conditions as for Fig. 2.*

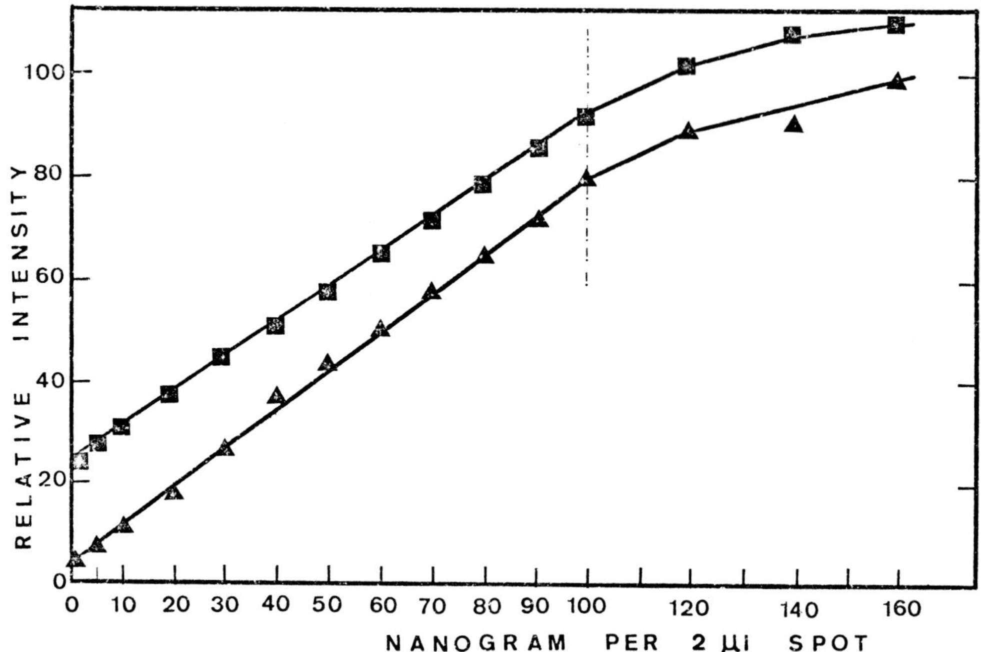


*Figure 4. RTP intensity vs. concentration of mebendazole (■) and flubendazole (▲) spotted from DMSO solutions on Whatman 42 filter paper. For wavelength maxima see Table 2.*

A recovery of 97.6 % against the appropriate standard (average of 14 experiments with individual standard solutions) and a coefficient of variation of 3.2 % were obtained.

Room temperature phosphorescence is a relatively simple and economical technique offering good selectivity and sensitivity and its importance for the specific analysis of low concentrations of various drugs is growing rapidly.

## REFERENCES

1. Zander, M., *in* "Phosphorimetry", Academic Press, New York, 1968, p. 117.
2. Lower, S. K., and M. A. El-Sayed, *Chem. Rev.*, 1966, *66*, 199.
3. Roth, M., *J. Chromatogr.*, 1967, *30*, 276.
4. Schulman, E. M., and C. Walling, *Science*, 1972, *178*, 53.
5. Schulman, E. M., and C. Walling, *J. Phys. Chem.*, 1973, *77*, 902.

6. Paynter, R. A., S. L. Wellons, and J. D. Winefordner, *Anal. Chem.*, 1974, *46*, 736
7. Wellons, S. L., R. A. Paynter, and J. D. Winefordner, *Spectrochem. Acta*, Part A, 1974, *30*, 2133.
8. Seybold, P. G., and W. White, *Anal. Chem.*, 1975, *47*, 1199.
9. White, W., and Seybold, P. J., *J. Phys. Chem.*, 1977, *81*, 2035.
10. Vo-Dinh, T., E. L. Yen, and J. D. Winefordner, *Anal. Chem.*, 1976, *48*, 1186.
11. Vo-Dinh, T., E. L. Yen, and J. D. Winefordner, *Talanta*, 1977, *24*, 146.
12. Vo-Dinh, T., G. L. Walden, and J. D. Winefordner, *Anal. Chem.*, 1977, *49*, 1126.
13. Vo-Dinh, T., G. L. Walden, and J. D. Winefordner, *Appl. Spectrosc. Rev.*, 1977, *13*, 261.
14. Vo-Dinh, T., R. B. Gammage, A. R. Hawthorne, and J. H. Thomgate, *Environ. Sci. Technol.*, 1978, *12*, 1297.
15. Vo-Dinh, T., and R. B. Gammage, *Anal. Chem.*, 1978, *50*, 2054.
16. Jakovljevic, I. M., *Anal. Chem.*, 1977, *49*, 2048.
17. von Wandruszka, R. M. A., and R. J. Hurtubise, *Anal. Chem.*, 1976, *48*, 1784.
18. von Wandruszka, R. M. A., and R. J. Hurtubise, *Anal. Chim. Acta*, 1977, *93*, 331.
19. von Wandruszka, R. M. A., and R. J. Hurtubise, *Anal. Chem.*, 1977, *49*, 2164.
20. Ford, C. D., and R. J. Hurtubise, *Anal. Chem.*, 1978, *50*, 610.
21. Schulman, E. M., and R. T. Parker, *J. Phys. Chem.*, 1977, *81*, 1932.
22. de Lima, C. G., and de M. Nicola, E. M., *Anal. Chem.*, 1978, *50*, 1685.
23. Ford, C. D., and R. J. Hurtubise, *Anal. Chem.*, 1979, *51*, 659.
24. Vo-Dinh, T., and J. R. Hooyman, *Anal. Chem.*, 1979, *51*, 1915.
25. Parker, R. T., R. S. Freelander, E. M. Schulman, and R. B. Dunlap, *Anal. Chem.*, 1979, *51*, 1921.
26. Meyers, M. L., and P. G. Seybold, *Anal. Chem.*, 1979, *51*, 1609.
27. Lue-Yen Bower, E., and J. D. Winefordner, *Anal. Chim. Acta*, 1978, *101*, 319.
28. Lue-Yen Bower, E., and J. D. Winefordner, *Anal. Chim. Acta*, 1978, *102*, 1.
29. Lloyd, J. B. F., and J. N. Miller, *Talanta*, 1979, *26*, 180.

30. Guilbault, G. G., *in* "Practical Fluorescence, Theory, Methods and Techniques", Marcel Dekker, New York, 1973, pp. 103, 165, 187.
31. Boutilier, G. D., and J. D. Winefordner, *Anal. Chem.*, 1979, *51,* 1384.
32. Boutilier, G. D., and J. D. Winefordner, *Anal. Chem.*, 1979, *51,* 1391.
33. Niday, G. J., and P. G. Seybold, *Anal. Chem.,* 1978, *50,* 1577.
34. Lloyd, J. B. F., *Analyst,* 1978, *103,* 775.
35. Hollifield, H. C., and J. D. Winefordner , *Anal. Chem.*, 1968, *40,* 1759.
36. Zweidinger, R., and J. D. Winefordner, *Anal. Chem.*, 1970, *42,* 639.
37. Abdel Fattah, F., W. Baeyens, and P. De Moerloose, unpublished data.

38. Winefordner, J. D., S. G. Schulman, and T. C. O'Haver, *in* "Luminescence Spectrometry in Analytical Chemistry", Wiley-Interscience, New York, 1972, pp. 289-90.

# AN OXYGEN PROBE BASED ON TETRAKIS ALKYL AMINOETHYLENE CHEMILUMINESCENCE[1]

Thompson M. Freeman
W. Rudolf Seitz

Department of Chemistry
University of New Hampshire
Durham, New Hampshire

Tetrakis alkyl amino ethylenes react with molecular oxygen yielding chemiluminescence (CL). The reaction is shown below for the tetrakis dimethylamino derivative:

$$[(CH_3)_2N]_2C=C[N(CH_3)_2]_2 + O_2 \longrightarrow 2\ [(CH_3)_2N]_2C=O$$

Some of the substituted urea product is electronically excited and transfers its energy back to the amino-ethylene starting material which serves as the emitter.

There has been considerable interest in this reaction for possible lighting applications (1-3). It has the advantage that CL is observed at high reagent concentrations or even from pure reagent. However, due to low efficiency and product quenching it has not been applied in practice.

We have been evaluating the possibility of oxygen analysis based on tetrakis alkylamino ethylene CL. The compound we have chosen to use is 1,1',3,3'tetraethyl-2,2-bi(imidazolidine) (EIA), because it can be readily synthesized (4). Unlike the tetrakis dimethyl derivative, EIA is a solid at room temperature and cannot be used pure.

[1] *Partial Financial Support for this work was provided by NSF Grant CHE-7825192.*




ISBN 0-12-208820-4

The analytical device consists of a glass reagent chamber with a volume of about 5 mls. The reagent chamber is separated from the analyte by a 1 mil thick FEP Teflon membrane (area - 10 $cm^2$). The membrane eliminates the possibility of non-volatile interferences. CL is detected by a photomultiplier positioned behind the reagent chamber. When the membrane contacts an oxygen-containing sample, CL is observed as $O_2$ diffuses through the membrane to react with EIA. It takes 15 to 30 seconds to establish steady-state. The steady-state CL intensity is proportional to oxygen partial pressure.

Using 10% EIA in hexane in the reagent chamber, we observe that CL intensity decreases to 10-15% of its initial value in 12 hours if the membrane is exposed to pure $O_2$. By analyzing the shape of the intensity decay curve, we conclude that EIA consumption is the primary process responsible for the decrease in intensity with time. It should be possible to reduce the rate at which intensity decays by increasing the volume of the reagent solution and/or by switching to another aminoethylene such as the N-dimethyl derivative which is a liquid at room temperature. The rate at which EIA is consumed can also be reduced by changing to a less permeable membrane and/or by reducing exposed membrane area, however, this would also reduce sensitivity. Our device had an estimated detection limit on the order of 1 part-per-million (volume/volume) $O_2$ in the gas phase. This is lower than detection limits typically obtained with oxygen electrodes.

The slope of the analytical curve increases considerably with temperature. Higher temperatures facilitate the diffusion of oxygen across the membrane as well as accelerating the rate of CL reaction. Routine use would require a correction factor. The slope of the analytical curve for our device is greater for $O_2$ in air than for $O_2$ in water. We attribute this to slow mass transfer to the membrane surface in water. There was no change in response to $O_2$ in the gas phase in the presence of water vapor.

Although we have not yet carried out an extensive investigation of possible interferences, we anticipate that the CL $O_2$ probe should not be subject to interference. To interfere a compound would have to diffuse across the membrane and react with the reagent to yield CL. There are no compounds known that will do this besides oxygen.

We believe that the CL oxygen probe may be developed into a superior alternative to the oxygen electrode for oxygen measurements. A more complete account of our results will be appearing shortly (5).

REFERENCES


1. Urry, W.H. and J. Sheeto, *Photochem. Photobiol. 4,* 1067 (1965).
2. Paris, J.P., *Photochem. Photobiol. 4,* 1059 (1965).
3. Fletcher, A.N. and C.A. Heller, *J. Phys. Chem. 71*, 1507 (1967).
4. Winberg, H.E., U.S. Patent 3,239,519, March 8, 1966.
5. Freeman, T.M. and W.R. Seitz, submitted to *Anal. Chem.*

# AN IMMUNOASSAY FOR URINARY ESTRIOL-16α-GLUCURONIDE BASED ON ANTIBODY-ENHANCED CHEMILUMINESCENCE

F. Kohen
J.B. Kim
H.R. Lindner

Department of Hormone Research
The Weizmann Institute of Science
Rehovot, Israel

G. Barnard

King's College Medical School
London, England

## I. INTRODUCTION

We have developed an immunoassay for urinary estriol-16α-glucuronide based on antibody-enhanced chemiluminescence. The assay is comparable to radioimmunoassay in sensitivity and specificity and does not require a phase-separation step. The method described here follows an approach used earlier for biotin (1), for thyroxine (2) and for progesterone (3, 4).

Briefly, the carboxy group of the steroid-glucuronide was attached covalently to the free amino group of the chemiluminescent marker aminobutyl ethyl isoluminol. The resulting steroid glucuronide-chemiluminescent marker conjugate emits light upon oxidation with microperoxidase and $H_2O_2$. When the steroid glucuronide chemiluminescent marker conjugate is bound to specific binding protein, the total light production of the conjugate is enhanced. This binding and the consequent enhancement of light emission is inhibited in a competitive manner by the addition of unaltered steroid glucuronide. The



ISBN 0-12-208820-4

results of an immunoassay for urinary estriol-16α-glucuronide based on the above principles are reported here.

## II. MATERIALS AND METHODS

Estriol-16α-glucuronide was covalently linked to 6[N-aminobutyl)-N-ethyl]-amino-2,3-dihydrophthalazine-1,4-dione to yield estriol-16α-glucuronide-ABEI conjugate (Fig. 1) by a procedure previously described (3). Antiserum to estriol-16α-glucuronide was obtained from Dr. W.F. Coulson (The Middlesex Hospital, London). Anti-estriol-16α-glucuronide IgG was prepared by chromatography on Sepharose-Protein A. Extracts from second and third trimester human pregnancy urine were purified by adsorbing diluted urine on Amberlite XAD-2 resin, followed by methanolic elution of the steroid glucuronide (5). All other reagents were as previously described (3-6).

Light emission was measured with a Luminometer Model 2080 (Lumac Systems, Basel), using the automatic injection and integration modes of the instrument. When the Luminometer was used in the automatic injection mode, readings on the Luminometer started two seconds after initiation of the light reaction.

The assay buffer in the chemiluminescent reactions was 0.05 M phosphate (pH 8.0) containing 6 g NaCl/liter. Stock solutions of steroids and chemiluminescent compounds were prepared in ethanol and in water, respectively. They were stored at 4°C and diluted to the desired concentration in assay buffer when required. Microperoxidase was dissolved at 1 mg/ml in 0.01 Tris-HCl (pH 7.4); this stock solution was kept at 4°C. The working solution was obtained by diluting the stock solution in assay buffer to 2.6 μM enzyme. The oxidant solution was prepared by adding 30 μl of 30% $H_2O_2$ solution to 5 ml borate buffer (pH 8.6; 0.06 M).

## III. RESULTS

### A. *A Chemiluminescence Immunoassay for Estriol-16α-Glucuronide*

Using the estriol-16α-glucuronide conjugate shown in Fig. 1 and conditions described in Legend to Fig. 2, an assay curve for estriol-16α-glucuronide was obtained with a linear range of 10 to 100 pg steroid/tube (Fig. 2).

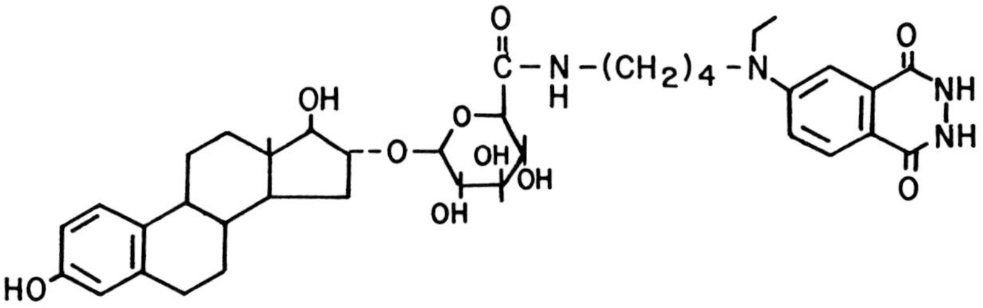

**estriol-16α-glucuronide-ABEI conjugate**

*Fig. 1. Proposed structure for estriol-16α-glucuronide aminobutyl ethyl isoluminol (ABEI) conjugate.*

## *B. Validation of the Estriol-16α-Glucuronide Immunoassay*

*1. Specificity of the Assay.* The specificity of the immunoassay using anti-estriol-16α-glucuronide IgG and chemiluminescence was similar to that observed when using the same antiserum in a radioimmunoassay: light emission was not significantly affected by addition of cortisol, progesterone, testosterone, estrone-3-glucuronide or estradiol-3-glucuronide; cross-reaction, as defined by Thorneycroft *et al.* (7), was 20% for estrone, 25% for estradiol and 48% for estriol.

*2. Precision.* The precision of the assay is shown in Table 1 with regard to intra-assay variation and Table 2 with regard to between assay variation.

## *C. Comparison of Chemiluminescence Immunoassay for Urinary Estriol-16α-Glucuronide with Radioimmunoassay.*

When extracts from second and third trimester human pregnancy urine, purified by adsorbing diluted urine on Amberlite XAD-2 resin, were assayed for estriol-16α-glucuronide by radioimmunoassay and chemiluminescence immunoassay using the same antiserum, the results of the two methods agreed well ($r=0.98$; $y=1.15x - 2.94$, where y are the values determined by radioimmunoassay and x values determined by chemiluminescence immunoassay).

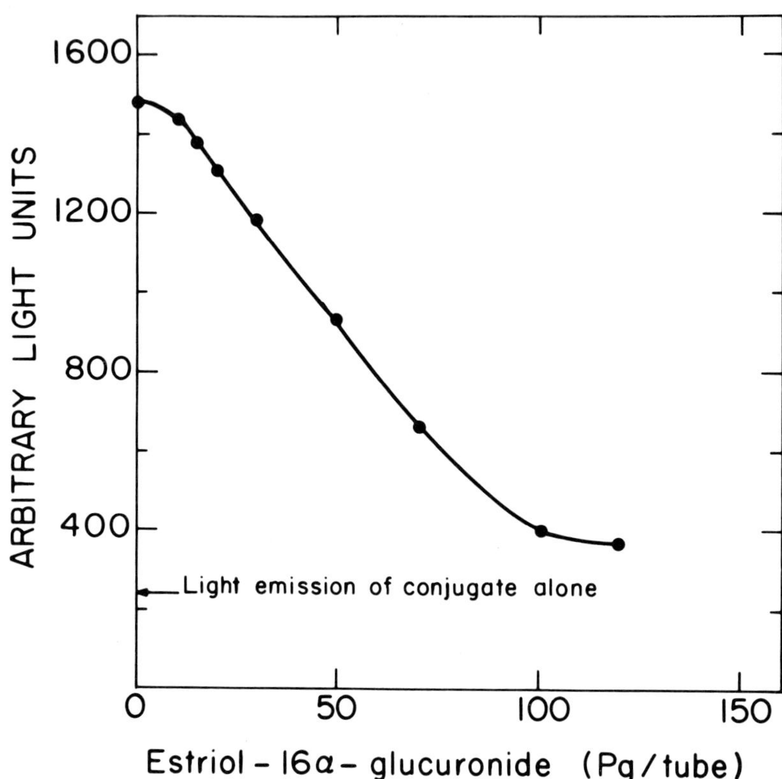


*Fig. 2. Representative dose-response curve for estriol-16α-glucuronide measured by chemiluminescence immunoassay. Varying amounts of estriol-16α-glucuronide were incubated with 0.12 pmol of specific anti-estriol-16α-glucuronide IgG in a total volume of 200 μl of assay buffer for 30 min at 4°C. The estriol-16α-glucuronide-ABEI conjugate (20 pg in 100 μl of assay buffer) was then added, and the incubation was continued for another 30 min at 4°C. The enzyme solution (100 μl) was added, and the oxidant was injected in the dark. Light emission was integrated over the period 3-12 sec after adding the last reagent and is expressed in arbitrary units.*

*TABLE I. Intra-assay Variation of Urinary Estriol-16α-Glucuronide Assay in Pregnancy*

| Sample No. | Mean ± S.D. (μg/ml) | Coeff. of variation (%) | n | Dilution of urine extract used in the assay |
|---|---|---|---|---|
| 33 | 10.97 ± 0.18 | 1.67 | 6 | 1:5000 |
| 38 | 142.27 ± 2.88 | 2.02 | 5 | 1:4000 |
| | 149.96 ± 10.87 | 7.25 | 4 | 1:6000 |

## IV. DISCUSSION

This paper describes the development of an immunoassay for urinary estriol-16α-glucuronide based on antibody-enhanced chemiluminescence. This technique does not require physical separation of bound and free ligand, or counting of radioactivity. The sensitivity achieved (10 pg/tube) is comparable to that obtained by radioimmunoassay. Furthermore, the steroid-chemiluminescent marker conjugate is stable and can be stored at 4°C for at least a year without loss of activity. The entire assay, including the extraction step, is readily performed in two working hours, and the end point determination requires only 10 sec. However, it should be pointed out that the new methodology still requires validation using a wider range of biological material.

*TABLE II. Between-assay Variation for the Urinary-Estriol-16α-Glucuronide Assay in Pregnancy*

| Sample No. | Number of assays | Mean ± S.D. (μg/ml) | Coefficient of variation (%) |
|---|---|---|---|
| 21 | 3 | 32.2 ± 5.14 | 15.9 |
| 22 | 3 | 22.06 ± 1.79 | 8.1 |
| 24 | 3 | 13.5 ± 0.43 | 3.2 |
| 35 | 3 | 33.46 ± 1.74 | 5.2 |
| 40 | 3 | 5.26 ± 1.06 | 20.3 |

## ACKNOWLEDGEMENTS

This work was supported by grants to HRL from WHO, the Ford and Rockefeller Foundations. HRL is a Scholar-in-Residence at the Fogarty International Center of the National Institutes of Health.

## REFERENCES

1. Schroeder, H.R., P.O. Vogelhut, R.J. Carrico, R.C. Boguslaski, and R.T. Buckler, *Anal. Chem. 48,* 1933 (1976).
2. Schroeder, H.R., R.C. Boguslaski, R.J. Carrico, and R.T. Buckler, *Methods Enzymol. 57,* 424 (1978).
3. Kohen, F., M. Pazzagli, J.B. Kim, H.R. Lindner, and R.C. Boguslaski, *FEBS Letts. 104,* 201 (1979).
4. Kohen, F., J.B. Kim, and H.R. Lindner *in* "Proceedings of the Second International Symposium on Luminescence", this volume, Academic Press, New York, in press.
5. Kohen, F., J.B. Kim, G. Barnard, and H.R. Lindner, *Steroids* (in press) (1980).
6. Kohen, F., M. Pazzagli, J.B. Kim, and H.R. Lindner, *Steroids* (in press) (1980).

# ASSAY OF GONADAL STEROIDS BASED ON ANTIBODY-ENHANCED CHEMILUMINESCENCE

F. Kohen
J. B. Kim
H. R. Lindner

Department of Hormone Research
The Weizmann Institute of Science
Rehovot, Israel

## I. INTRODUCTION

Existing radioimmunoassay (RIA) procedures for steroids have the advantage of high sensitivity and specificity. However, RIA involves a time-consuming phase-separation step (1) and the use of a radioactive label may pose waste disposal problems. To avoid these drawbacks while retaining the specificity of an immunoassay, the use of fluorescent (2), enzyme (3) and chemiluminescent (4-7) labels has been explored. We have developed chemiluminescence-based immunoassays for progesterone and estriol that are comparable to RIA in sensitivity and specificity and do not require physical separation of bound and free steroid.

An amino derivative of the chemiluminescent compound isoluminol is covalently linked through an alkyl chain of varying length to progesterone at position 11 or to estriol at position 6. The resulting steroid-chemiluminescent marker conjugate emits light upon oxidation with hematin compounds and $H_2O_2$. When the steroid-chemiluminescent marker conjugate is bound to specific antibody, total light production by the conjugate is enhanced. This binding and the consequent enhancement of light emission are inhibited by the addition of unaltered steroid in a competitive manner. The results of immunoassays for plasma progesterone and estriol based on the monitoring of chemiluminescence are reported here.



ISBN 0-12-208820-4

## II. MATERIALS AND METHODS

11α-Hydroxy-4-pregnene-3,20-dione 11-hemisuccinate was linked covalently to the appropriate derivative of isoluminol (6-amino-2,3-dihydrophthalazine-1,4-dione) to yield the conjugates P-ABEI, P-APEI and P-AHEI (Fig. 1), respectively, essentially as previously described (7). Estriol-6(0)-carboxymethyl oxime was attached to aminopentyl ethyl isoluminol in like manner to yield the conjugate $E_3$-APEI shown in Fig. 5. All other reagents and assay procedures were as previously described (7-9), unless specified otherwise.

Light emission was measured with a Luminometer Model 2080 (Lumac Systems, Basel).

## III. RESULTS

### *A. Refinement of the Progesterone Assay: Influence of Length of Spacer Between Steroid and Isoluminol*

In an attempt to optimize this assay, we examined the effect of shortening the alkyl chain linking the steroid to the marker. It was found that the light yield upon oxidation of P-ABEI (Fig. 1, n=4) and P-APEI (n=5) was considerably lower (Fig. 2) than that of P-AHEI (n=6). However, the increment in light yield caused by binding of the steroid

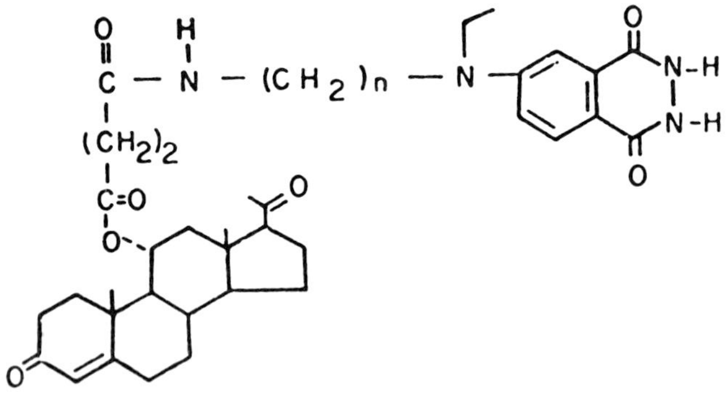

*FIGURE 1. Structures for progesterone-chemiluminescent marker conjugates.*

*n=4: P-ABEI; n=5: P-APEI; n=6: P-AHEI.*

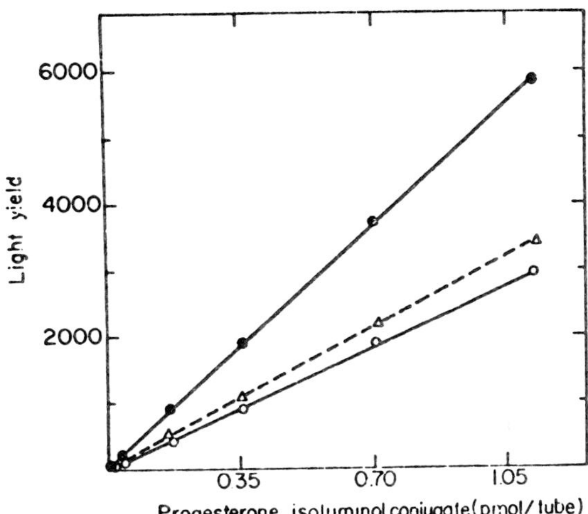


*FIGURE 2. Light yield of progesterone-chemiluminescent marker conjugates upon oxidation*

*Varying amounts of three different progesterone chemiluminescent marker conjugates were dispensed at 4°C into Lumacuvettes in a total volume of 300 µl of assay buffer. Microperoxidase (100 µl of a 2.6 µM solution) was added. The reaction tube was then introduced into the Luminometer and 100 µl of the oxidant (0.19% of $H_2O_2$ solution in borate buffer, pH 8.6, 0.06M) was injected. Light emission was integrated over the period 3-12 sec after adding the last reagent and is expressed in arbitrary units. Legend: o——o, P-ABEI; Δ----Δ, P-APEI; ●——●, P-AHEI.*

conjugate to anti-progesterone-γ-globulin before oxidation was far greater when using P-ABEI than in the case of either of the two conjugates of longer linker-chain length (Fig. 3). An added advantage of P-ABEI was that the assay system was less sensitive to variation in the time interval between mixing the steroid with antibody and marker conjugate, and the addition of the oxidizing system. This interval had to be kept within 30 sec in the assay system previously described (7), but can be extended to 10-30 minutes when using P-ABEI conjugate. In addition, the P-ABEI system gave a more favorable assay curve (Fig. 4), with a wider linear range.

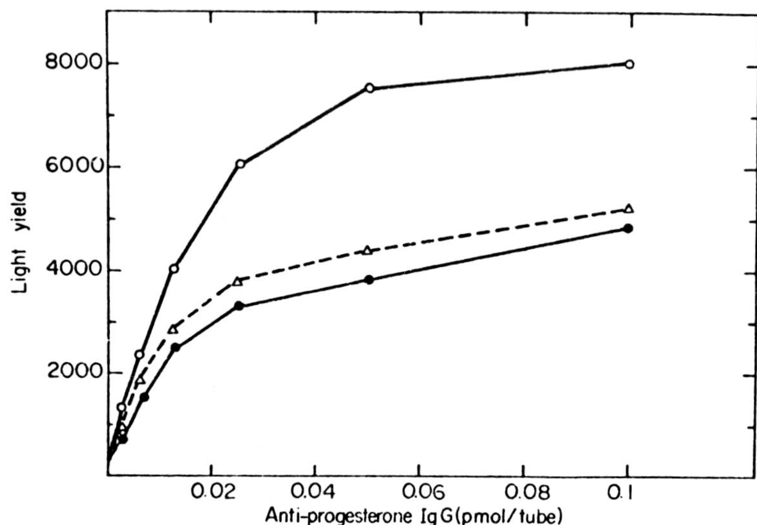


*FIGURE 3. Effect of anti-progesterone IgG concentration on the light yield of progesterone-chemiluminescent marker conjugates upon oxidation.*

*Varying amounts of anti-progesterone IgG were incubated at 4°C for 1 h with 0.06 pmol of each progesterone-chemiluminescent marker conjugate in a total volume of 300 μl of assay buffer. Further treatment as described in Legend to Fig. 2.* ○——○, *P-ABEI;* △---△ , *P-APEI;* ●——● , *P-AHEI.*

*B. A Chemiluminescence immunoassay for Estriol.*

Using the estriol-APEI conjugate (Fig. 5) and conditions described in Legend to Fig. 4, an assay curve for estriol was obtained with a linear segment between 10-100 pg estriol/tube.

*C. Reliability of the Progesterone and Estriol Assay and Application to Biological Material.*

*1. Specificity.* Augmentation of the chemiluminescent light yield was dependent on the use of the homologous antiserum: anti-cortisol IgG did not enhance the light yield of P-ABEI upon oxidation and anti-progesterone-γ-globulin was without effect in the $E_3$-APEI system. The antibody induced increment in light output by P-ABEI was not inhibited by

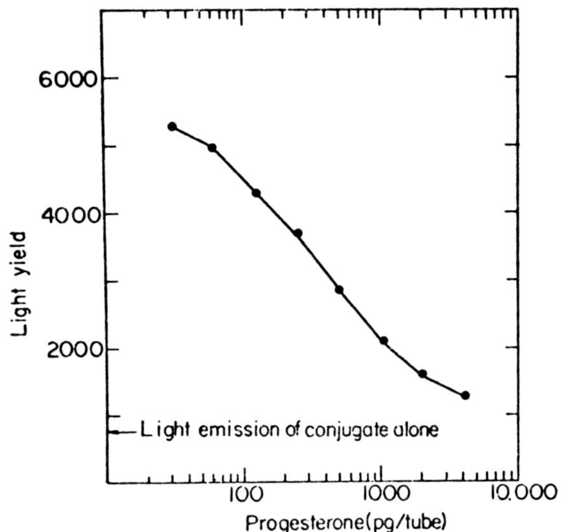


*FIGURE 4. Representative dose-response curve for progesterone measured by chemiluminescence immunoassay.*

*Varying amounts of progesterone were incubated with 0.034 pmol of anti-progesterone IgG in a total volume of 200 µl of assay buffer for 45 min at 4°C. The enzyme solution (100 µl) was added, and the oxidant (100 µl) was injected in the dark. The ordinate shows the light yield recorded by the Luminometer in arbitrary units.*

added cortisol or estradiol; minor cross-reaction was observed with 17-hydroxy-progesterone (1.5%), testosterone (2%), 11α-hydroxyprogesterone (17%) and 11β-hydroxyprogesterone (23%). Likewise, the specificity of the estriol assay paralleled that observed in an RIA system using the same antibody. Thus, using a monoclonal antibody (10) generated with 17β-estradiol-6 (O)-carboxymethyloxime - bovine serum albumin with a high cross-reaction for $E_3$ (50%) but insignificant cross-reaction (<0.1%) for estrone or testosterone as judged by RIA, a similar pattern of cross-reactivity was observed in the chemiluminescence immunoassay.

*2.* The *precision* of the two assays is shown in Tables 1A with regard to intra-assay variation and in Table 1B with regard to inter-assay variation. While there is scope for

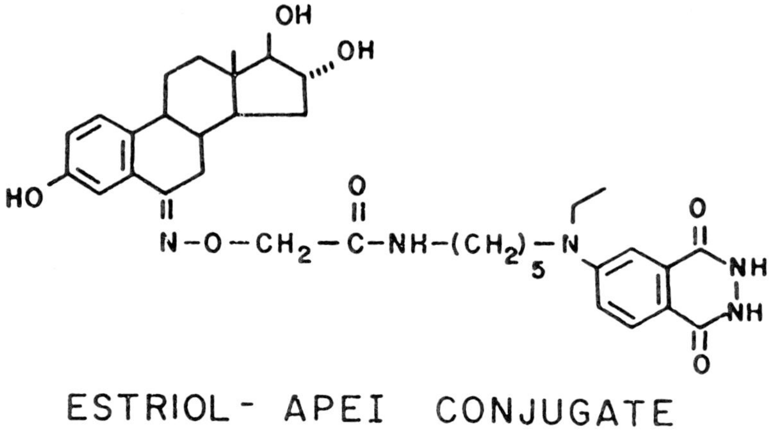

ESTRIOL- APEI CONJUGATE

*FIGURE 5. Proposed structure for estriol-APEI conjugate*

*TABLE I. A. Within-assay Precision*
*Replicate samples (n) of the same plasma pool were assayed simultaneously.*

| | Mean ± S.D. (ng/ml) | Coeff. of variation (%) | n |
|---|---|---|---|
| Progesterone (luteal phase) | 9.4 ± 0.7 | 7.4 | 20 |
| Estriol (3rd trimester pregnancy) | 32 ± 2.9 | 9 | 7 |

*B. Between-assay Precision*
*Replicate samples of the same plasma pool were assayed on different occasions.*

| | Mean ± S.D. (ng/ml) | Coeff. of variation (%) | n |
|---|---|---|---|
| Progesterone (luteal phase) | 9.6 ± 1.38 | 15 | 10 |
| Estriol (late gestation) | 21.6 ± 1.48 | 6.8 | 5 |

improvement, the precision is not greatly inferior to that obtained with common RIA procedures.

*3. Progesterone Concentration in Human Plasma.* Progesterone concentration in samples taken from women during the follicular or mid-luteal phase or from normal human males as determined by the chemiluminescence assay is shown in Table 2. The results are in accordance with those obtained by established methodologies. Likewise, treatment of six previously anovulatory women with human menopausal gonadotropins, followed by human chorionic gonadotropin, led to the expected rise in plasma progesterone level as measured by chemiluminescence from <1 ng/ml to 6.1-15.3 ng/ml.

*TABLE II. Human Plasma Progesterone Levels Measured by Chemiluminescence Immunoassay*

| *Source* | *Progesterone concentration (ng/ml; mean value ± S.D.)* |
|---|---|
| *Female* | |
| *Follicular phase* | *0.65 ± 0.365 (n = 25)* |
| *Mid-luteal phase* | *8.67 ± 4.3 (n = 30)* |
| *Male* | *0.28 ± 0.09 (n = 8)* |

## IV. CONCLUSION

The results of these pilot trials of a chemiluminescence-based immunoassay for progesterone and estriol, combined with similar results for chemiluminescence assay systems for urinary estriol-16α-glucuronide (8) and plasma cortisol (9) presented elsewhere at this meeting, indicate that this novel approach to steroid immunoassay is sufficiently promising to warrant more extensive exploration. Potential advantages over RIA could be the elimination of a phase-separation step and consequent speed and ease of automation, and the lack of radioactive waste. However, these assays should be further validated by application to a wider range of normal and pathological biological material.

ACKNOWLEDGEMENT

This work was supported by grants to HRL from WHO, the Ford and the Rockefeller Foundations. HRL is a Scholar-in-Residence at the Fogarty International Center of the National Institutes of Health.

REFERENCES

1. Kohen, F., Bauminger, S., and H. R. Lindner, *in* "Steroid Immunoassay" (E.H.D. Cameron, S. G. Hillier, and K. Griffiths, eds.), pp. 11-32. Alpha Omega Publishing Ltd., Cardiff, Wales, U.K. (1975).
2. Burd, J. F., R. C. Wong, J. E. Feeney, R. J. Carrico, and R. C. Boguslaski, *Clin. Chem. 23,* 1402 (1977).
3. Schroeder, H. R., P. O. Vogelhut, R. J. Carrico, R. C. Boguslaski, and R. T. Buckler, *Anal. Chem. 48,* 1933 (1976).
4. Schroeder, H. R., R. C. Boguslaski, R. J. Carrico, and R. T. Buckler, *Methods Enzymol. 57,* 424 (1978).
5. Schroeder, H. R. and F. M. Yeager, *Anal. Chem. 50,* 1114 (1978).
6. Pratt, J. J., M. G. Woldring, and L. Villerius, *J. Immunol. Methods 21,* 179 (1978).
7. Kohen, F., M. Pazzagli, J. B. Kim, H. R. Lindner and R. C. Boguslaski, *FEBS Letts. 104,* 201 (1979).
8. Kohen, F., J. B. Kim, G. Barnard, and H. R. Lindner, *Steroids,* in press (1980).
9. Kohen, F., M. Pazzagli, J. B. Kim, and H. R. Lindner, *Steroids,* in press (1980).
10. Eshhar, Z., F. Kohen, and H. R. Lindner, *Israel J. Med. Sci. 16,* 408 (1980).

# DISSOCIATIVE AND NON-DISSOCIATIVE ELECTRON TRANSFER MECHANISMS IN BIOLUMINESCENCE AND CHEMILUMINESCENCE

E. M. Kosower

Department of Chemistry
Tel-Aviv University
Ramat-Aviv, Tel-Aviv, Israel

Department of Chemistry
State University of New York
Stony Brook, New York 11794

## I. SUMMARY

A new mechanism that involves dissociative electron transfer in the energy transducing step has been proposed for bacterial luciferase catalyzed light emission (6). The proposal involves: 1) dissociation of the 4a-hydroperoxyflavin to a flavin radical and $\cdot O_2^-$, accounting for 570 and 620 nm absorption; 2) $\cdot O_2^-$ addition to the aldehyde carbonyl to form a peroxyl radical; 3) abstraction of H from an enzyme thiol group to form RCH(OOH)OH; 4) thiyl radical abstraction of the H on C in RCH(OOH)OH, a step which can show a $k_H/k_D$ of ca. 4; and 5) dissociative electron-transfer, a highly exothermic step that leads to a protonated flavin excited state, a carboxylic acid and water.

The overall reaction scheme developed for the bioluminescence of firefly luciferin over the years has been interpreted by Schuster as a CIEEL mechanism, which involves in the critical steps (a) an electron-transfer to produce $CO_2$ and the equivalent of an ion-pair and (b) an electron-transfer to form the singlet emitting state (21).

Energy versus reaction coordinate schemes are presented for both mechanisms. In dissociative electron transfer, the reaction which produces the excited state is also the reaction




ISBN 0-12-208820-4

which provides the main driving force, whereas in the CIEEL mechanism, the driving reaction occurs in the step prior to the formation of the excited state.

---

Bacterial luciferase catalyzes the production of blue-green light (Lmax, 490 nm) in a reaction of reduced flavin mononucleotide ($FH_2$), molecular oxygen and a long chain aliphatic aldehyde (RCHO). This occurs via an enzyme:reduced flavin comples (intermediate I) and an enzyme:hydroperoxyflavin complex (intermediate II) (eq. 1) (1-3). Tetradecanal is probably the functional aldehyde in the natural system (4, 5).

$$(1) \quad E + FH_2 = \underset{(I)}{E,FH_2} \xrightarrow{O_2} \underset{(II)}{E,FHOOH} \xrightarrow{RCHO} E,FX^* \longrightarrow h\nu$$

We review here a mechanistic scheme involving dissociative electron transfer in the energy transducing step (6). The scheme includes the recently reported longwavelength absorbing form of intermediate II (7, 8). In addition to the absorption band at 373 nm (2), intermediate II exhibits absorption maxima at 570 and 620 nm. This long wavelength absorption now assigned to a flavin radical (9) arising from a reversible dissociation of the flavin hydroperoxide into the flavin radical and superoxide radical ion (Eq. 2).

$$(2) \quad FHOOH = FH\cdot,\cdot OOH = \cdot FH\cdot,\cdot O_2^- + H^+$$

In the next step of the mechanistic scheme, we propose that the radical pair combines with aldehyde to form a new complex, in which an electrophile (aldehyde) is in proximity to a nucleophile (superoxide ion). Addition to a carbonyl group, especially in the solvent-poor environment at the enzyme active site, could be expected, producing a peroxylhydroxyalkane (Eq. 3), a radical that has sufficient stability to serve as an intermediate in other reactions (10).

$$(3) \quad RCHO + E,FH\cdot,\cdot O_2^- \longrightarrow E,RCHO,FH\cdot,\cdot O_2^- \longrightarrow E,RCH(O^-)OO\cdot,FH\cdot$$

Since it appears likely that the proton should be attached to $N_{10}$ rather than $N_5$ in the emitting state, we postulate that the initially formed peroxyl, flavin radical pair (pair "A") underdoes a conformational rearrangement to an isomeric pair (pair "B") (see the Scheme). Several observations are readily explained in terms of this conformational rearrangement, including the fluorescence behavior of intermediate II.

In the subsequent step, we postulate that an enzyme thiol group donates an H atom to the reactive peroxyl radical, RCH(O⁻)OO•. The importance of the thiol group for the activity of bacterial luciferase has been demonstrated by experiments involving reaction with an alkylating agent such as N-ethylmaleimide (11, 12). The products of the reaction of the peroxyl-hydroxyalkane with the thiol group are a hydroperoxy-hydroxyalkane and a thiyl radical (Eqs. 4 and 4a; flavin not shown).

(4) E,RCH(O⁻)OO• + ESH ⟶ E,RCH(O⁻)OOH + ES•

(4a) E,RCH)OH)OO•+ ES⁻ ⟶ E,RCH)OH)OO⁻ + ES•

We now consider the transfer of the H that is derived from the CHO group of the aldehyde to the thiyl radical. The attack of the thiyl radical on the hydrogen of the CH would be subject to a primary hydrogen isotope effect, and we interpret the observed hydrogen isotope effect on bioluminescence (11) in terms of this hydrogen abstraction reaction (Eq. 5). Although the thiyl radical-hydrogen abstraction reaction is apparently unprecedented among biochemical reactions, it seems to be quite reasonable from the chemical point of view (13).

(5) E,RCH(O⁻)OOH + ES• ⟶ E,RC•(O⁻)OOH + ESH

(flavin not shown)

Finally, we take up the crucial energy transduction step responsible for populating the excited (flavin) state. The carbon-centered aldehyde-derived radical formed by hydrogen abstraction is located near the flavin radical that was formed in Equation 2. We propose that dissociative electron transfer from the highest doubly occupied pi-orbital of the flavin radical to the carbon-centered radical would lead to a protonated excited flavin, a carboxylate anion and a hydroxide ion, as shown in Equation 6. The orbital configuration changes of the flavin in the transformation are indicated below the equation.

(6) E,FH•,RC•(O-)OOH ⟶ E,FH*+,RCOO⁻,OH⁻

$(pi)^2(pi^*)^1$ $(pi)^1(pi^*)^1$

The high electron affinity of the hydroxyl radical (14) and the formation of a second carbon-oxygen bond (15) can be identified as the chief thermodynamic contributions to the exothermicity of the electron transfer reaction. [- $\Delta H$ = 83 kcal/mole, composed of the following contributions: O-O bond cleavage, 51 kcal/mole; second C-O bond from an alcohol-like C-O single bond, 83 kcal/mole (165 kcal/mole - 82 kcal/mole); •OH electron affinity (formation of $OH^-$), 46 kcal/mole; oxidation potential of FH•, 5 kcal/mole].

The final step in the luciferase sequence is the emission of light from the protonated flavin. This species, with the emission peaking at about 490 nm (emission energy, 51 kcal/einstein), was proposed as the bacterial emitter on the basis of its spectroscopic properties (16). The overall mechanistic scheme is summarized in Fig. 1. The approximate energy levels for the most of the species along the reaction pathway are given in Fig. 2.

A non-light-emitting reaction (a "dark reaction") can occur in which reduced flavin reacts with oxygen to form intermediate II, and the latter decomposes to oxidized flavin (FMN) and hydrogen peroxide (17, 18). A simple mechanism for the formation of hydrogen peroxide involves escape of hydrogen peroxide from the ion pair produced by dissociation of the 4a-hydroperoxy adduct, intermediate II, as shown in Equation 7.

$$(7) \quad \text{FHOOH} \longrightarrow \text{FH}^{+}, \text{OOH} \longrightarrow \text{F} + \text{HOOH}$$

Intermediate II exhibits fluorescence (19). Irradiation of intermediate II at its absorption maximum, 373 nm, results in a considerable increase in fluorescence intensity, the final fluorescence spectrum being identical to the bioluminescence emission spectrum. Both the fluorescence and the fluorescence increase have been explained in detail as involving a change from radical pair A to radical pair B (6). That the fluorescence increase on irradiation of intermediate II can be explained in such a straightforward way on the basis of the Scheme lends credence to its validity. It may be supposed that the pair "B" is in equilibrium with the 10a-hydroperoxide adduct of the flavin, and that photodissociation to the ion pair precedes excitation to a fluorescent excited state. Photodissociation may proceed by a single step heterolytic dissociation, or by a two-step process, homolytic dissociation to a radical pair followed by electron transfer (20).

Electron transfer has been recognized as involved in chemiluminescence and bioluminescence. The role of electron transfer in creating the excited state responsible for the bioluminescence from fireflies (i.e., the chemiluminescence of firefly luciferin) is illustrated in Fig. 3 in accordance with the ideas implied and expressed by Schuster (21-23). The energy levels for the various intermediates along the reaction pathway are shown in Fig. 4. Figs. 2 and 4 make clear that the primary difference between the two kinds of electron transfer mechanisms is the timing of the "up-conversion" electron transfer step in relation to the creation of the excited, emitting state. The high energy, emitting state of dissociative electron transfer. In the second mechanism, an electron transfer step is responsible for the "up-conversion" before the emitting state is formed by yet another electron transfer reaction.

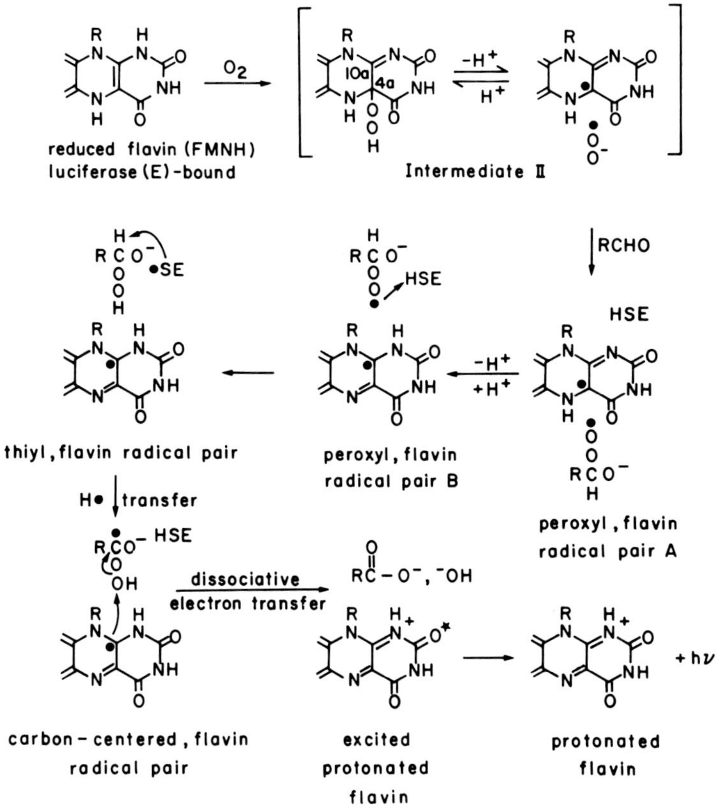

*FIGURE I. The postulated molecular mechanism for the luciferase-mediated bacterial bioluminescence reaction.*

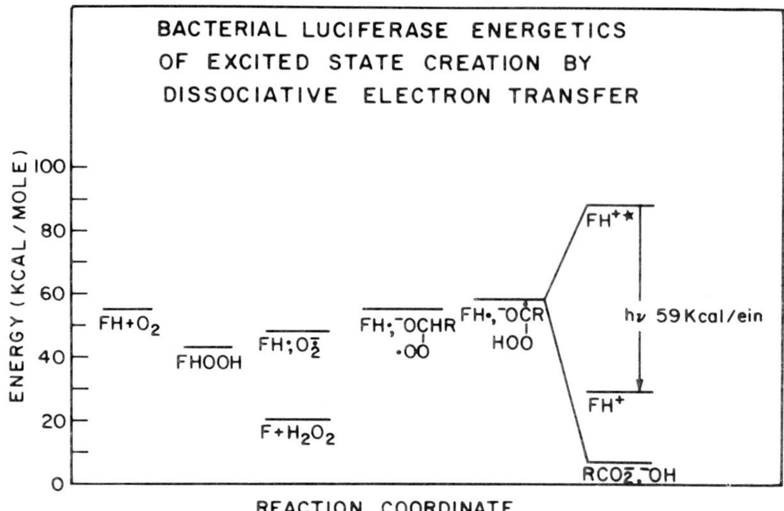


*FIGURE 2. Approximate energy levels for the intermediates along the reaction pathway in the mechanism proposed for the bioluminescence of bacterial luciferase.*

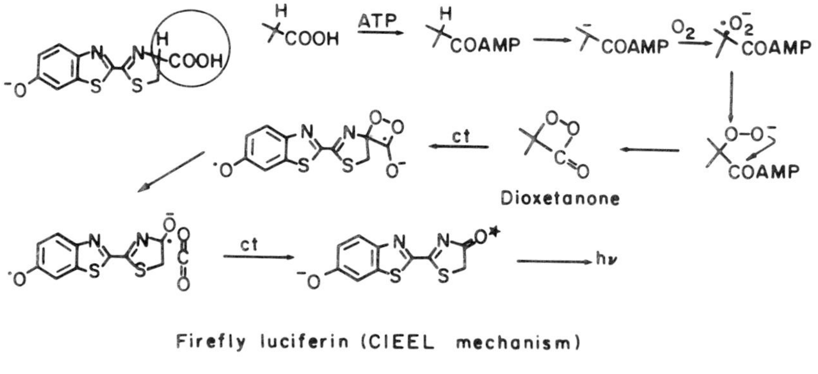

Dioxetanone

Firefly luciferin (CIEEL mechanism)

*FIGURE 3. Mechanism of formation of the excited state responsible for the bioluminescence of fireflies. Many scientists have contributed to the formulation of the chemistry for this pathway.*

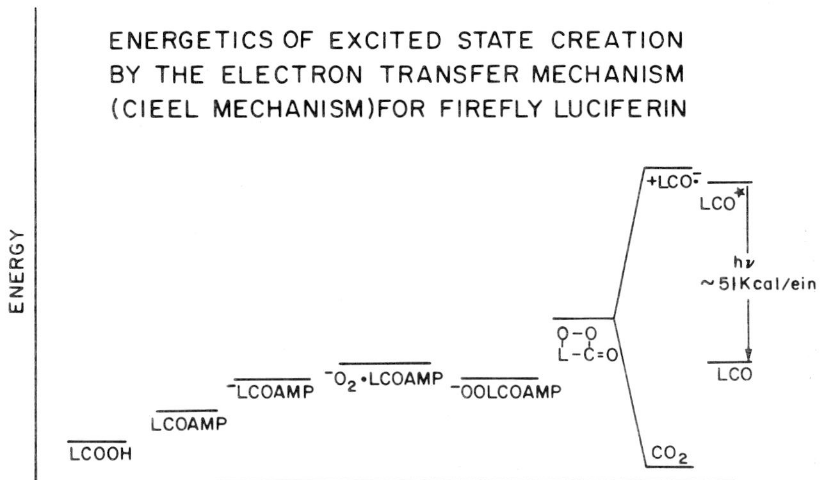


*FIGURE 4. Approximate energy levels for the intermediates along the reaction pathway in the mechanism suggested for the formation of the excited state for firefly luciferin.*

ACKNOWLEDGEMENT

The author expresses his appreciation to Prof. J. Woodland Histings, Biological Laboratories, Harvard University for intensive critical and lengthy discussions covering every detail of the new mechanism.

REFERENCES

1. Hastings, J. W. and Q. H. Gibson, *J. Biol. Chem.* 238, 2357-2554 (1963).
2. Hastings, J. W., C. Balny, C. LePeuch, and P. Douzou, *Proc. Nat'l. Acad. Sci. USA* 70, 3468-72 (1973).
3. Hastings, J. W. and K. H. Nealson, *Ann. Rev. Microbiol.* 31, 549-95 (1978).
4. Ulitzur, S. and J. W. Hastings, *Proc. Nat'l. Acad. Sci. USA* 75, 266-9 (1978).
5. Ulitzur, S. and J. W. Hastings, *Proc. Nat'l. Acad. Sci. USA* 76, 265-267 (1979).
6. Kosower, E. M., *Biochem. Biophys. Res. Commun.* 92, 356-64 (1980).

7. Presswood, R. P. and J. W. Hastings, *Biochem. Biophys. Res. Commun.* 82, 990-996 (1978).
8. Presswood, R. P. and J. W. Hastings, *Photochem. and Photobiol.* 30 (1979).
9. Stankovitch, M. T., C. M. Schopfer, and V. Massey, *J. Biol. Chem.* 253, 4971-9 (1978).
10. Bothe, E., G. Behrens, and D. Schulte-Frohlinde, *Z. Naturforsch* 32b, 886-9 (1977).
11. Nicoli, M. Z., E. A. Meighen, and J. W. Hastings, *J. Biol. Chem.* 249, 2385-92 (1974).
12. Nicoli, M. Z. and J. W. Hastings, *J. Biol. Chem.* 249, 2393-6 (1974).
13. Ohno, A. and S. Oae, *in* "Organic Chemistry of Sulfur", (S. Oae, ed.) pp. 122-6. Plenum Press, New York (1977); Cohen, S. G. and C. H. Wang, *J. Am. Chem. Soc.* 77, 4435 (1955).
14. Hayon, E. and M. Simic, *Accts. Chem. Res.* 7, 114-21 (1974).
15. Berthier, G. and J. Serre, *in* "Chemistry of the Carbonyl Group" (S. Patai, ed.), p. 5. Interscience, New York (1966).
16. Eley, M., J. Lee, J-M. Lhoste, C. Y. Lee, M. J. Cormier, and P. Hemmerich, *Biochem.* 9, 2902-8 (1970).
17. Hastings, J. W., *et al.*, *in* "Bioluminescence in Progress" (F. H. Johnson and Y. Haneda, eds.), pp. 151-86. Princeton Univ. Press (1966).
18. Hastings, J. W. and C. Balny, *J. Biol. Chem.* 250, 7288-7293 (1975).
19. Hastings, J. W. and C. Balny, *Biochem.* 14, 4719-23 (1975).
20. Turro, N. J. *in* "Modern Molecular Photochemistry", pp. 568-9. Benjamin-Cummings Publ. Co., Menolo Park, CA (1978).
21. Schuster, G. B., *Accts. Chem. Res.* 12, 366-73 (1979).
22. Koo, J-Y. and G. B. Schuster, *J. Am. Chem. Soc.* 100, 4496 (1978).
23. Schmidt, S. P. and G. B. Schuster, *J. Am. Chem. Soc.* 102 306-14 (1980).

SITE-SPECIFIC CHEMILUMINESCENT PROBES USED IN THE ANALYSIS OF pH, PEROXIDATION AND FREE RADICALS INTERNALLY IN METABOLICALLY ACTIVE ORGANELLES

Richard D. Lippman

Department of Physical Chemistry
The Royal Institute of Technology
Stockholm, Sweden

and

Department of Medical Cell Biology
Biomedicum, University of Uppsala
Uppsala, Sweden

## I. INTRODUCTION

Since the first international symposium in Brussels in 1978, Merényi and Lind, 1980, (1) in our laboratories have determined with pulse radiolysis the fundamental stoichiometrics of the luminol reaction in their coming article in The Journal of the American Chemical Society. They found that a high reaction occurs when one mole luminol reacts with two moles of superoxide anion radicals ($\cdot O_2^-$) and/or hydroxy radicals ($\cdot$OH) in the initial steps leading to a light emitter (emitters?). These two radicals react with luminol at a rate constant which is several orders of magnitude higher than luminol's reaction with a strong peroxide, while a strong




ISBN 0-12-208820-4

peroxide reacts with luminol several orders of magnitude higher than molecular oxygen. Therefore, the reacting species with the luminol substrate often can be accurately identified by kinetic differentiation.

Following the first mole of $\cdot\overline{O_2}$ or $\cdot OH$ attack, a based stabilized, peroxyhemiketal intermediate is formed as described by Lippman, 1980a, (2). This peroxyhemiketal is of central importance to the understanding of a uniform mechanism in bio- and chemiluminescence since it corresponds to Hastings', 1978 (3) proposed peroxyhemiacetal intermediate of flavin in the bacteria luciferin reaction.

These important developments layed the foundation for new analytical applications of chemiluminescence: Specific chemiluminescent probes can be synthesized to explore various biological phenomena, i.e. free-radical activity, internally in metabolically-active organelles. Furthermore, this new biochemical tool solves some primary experimental difficulties encountered in cell biology: Measurement of the <u>rate</u> of peroxidation reactions and their specific reaction-sites inside organelles *in vitro* and *in vivo*. Figure 1 displays three such organelle possibilities using isoluminol derivatives:

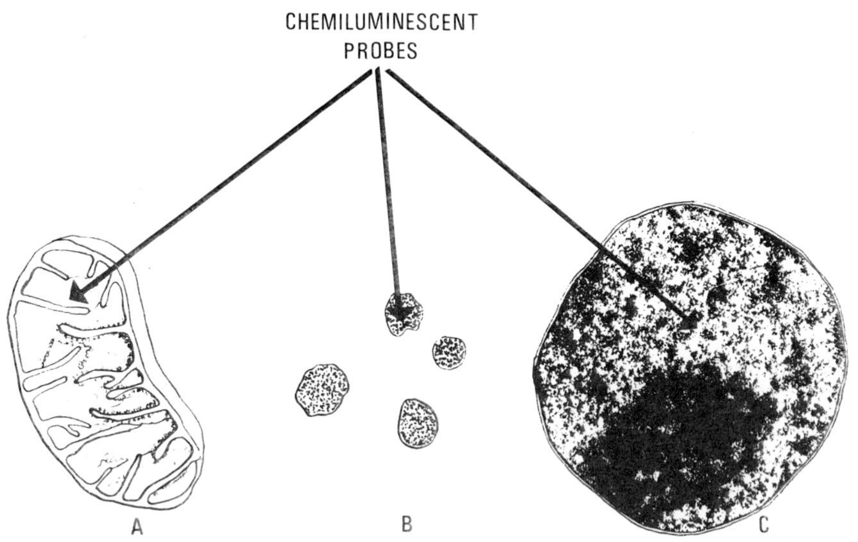


*FIGURE 1. Specially synthesized isoluminol probes used for the internal analysis of metabolically-active organelles. (A) Mitochondrion (B) Lysosomes (C) Cell nucleus (sizes are not drawn proportional to each other).*

Chemiluminescent probe-substrates are sensitive to quantitative determinations of peroxidation and free-radical activities *in vivo* and in metabolically-active organelles, cells, etc. *in vitro*. They are much more site-specific and quantitatively more sensitive than current NMR and pulse radiolytic techniques for radical detection and colorimetric techniques for peroxidation determination. A prominent species of interest in living organisms, the superoxide anion radical, is detected by ESR only in frozen solutions. In general, steady-state concentrations of free radicals are often too low in biological systems to be measured by conventional spectroscopic methods; while peroxidation detection by colorimetry measures only colored reaction products and not the peroxidizing reactions themselves.

## II. INTERNAL PROBING OF MITOCHONDRIA WITH CHEMILUMINESCENCE

In the case of mitochondria (A), one encounters the problem of a double membrane which first must be penetrated by a specific probe. *In vivo,* carnitine is known to esterfy with fatty acids in order to achieve penetration of this double membrane system. This is a facilitated transport process essential for respiratory-chain catabolism.

Lippman, 1980b, (4) synthesized DL-carnitinyl maleate-isoluminol (CML):

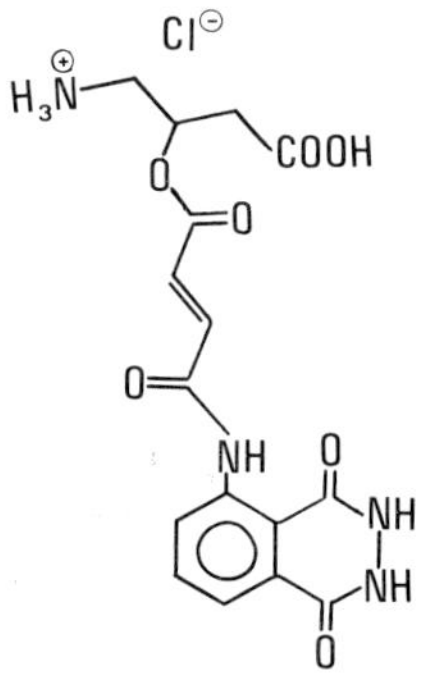

*FIGURE 2. DL-carnitinyl maleate-isoluminol (CML)*

CML was incubated with rat-liver and human-liver mitochondria with insignificant loss of enzyme activities. Superoxide anion radical flux from the respiratory chain was determined at $6 \pm 2\%$ of the total oxygen uptake. This flux is strongly dependent upon intact mitochondrial membranes and shape, histochemical functions, ATP activity and inner matrix pH. Electron microscope examinations showed that CML-impregnated mitochondria were normal.

In another series of gerontological experiments with mitochondria (Fig. 3), a comparison could be made of the effectiveness of antioxidants and other geroprotective mixtures that are known to play effective, free-radical scavenger roles *in vivo* in regard to inhibiting the processes of senescence.

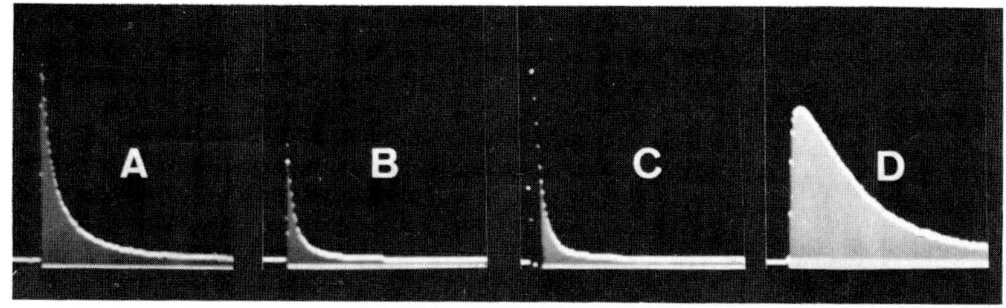


*FIGURE 3. Liver mitochondria from a 70 year old female probed with CML and later incubated with (A) 15 μmole methionine + 15 μmole α-tocopherol, (B) 10 μmole methionine + 10 μmole α-tocoperol + 10 μmole ascorbic acid, (C) 30 μmole copper chloride and (D) control. Rapid exomitochondrial injection of $\cdot O_2^-$ inhibited CML chemiluminescence in the cases of copper and synergistically-scavengering antioxidants.*

This comparison is of significance in gerontology since it is generally accepted that the aging processes can be slowed in man by a conservative estimate of 15-20% given the use of geroprotective thereapy (Comfort, 1978) (5).

Experiments with the CML probe in the mitochondrial inner matrix were extended in a recent paper by Lippman and Agren, 1980 (6). The pH of the inner matrix was determined to be $8.1 \pm 0.2$ as depicted in Figure 4. This direct titration of CML confirmed Mitchell's and Moyle's, 1969 (7) indirect, exomitochondrial, potassium ion determination of a pH of $8.06 \pm 0.08$. Our direct confirmation also added weight to Mitchell's chemiosmotic hypothesis which assumes that oxidative phosphorylation is driven chemically by a pH gradient across the inner mitochondrial membrane.

Again, little loss of enzyme activities was observed, and state 3 and 4 respiration was determined exomitochondrially by ATP leakage and firefly bioluminescence.

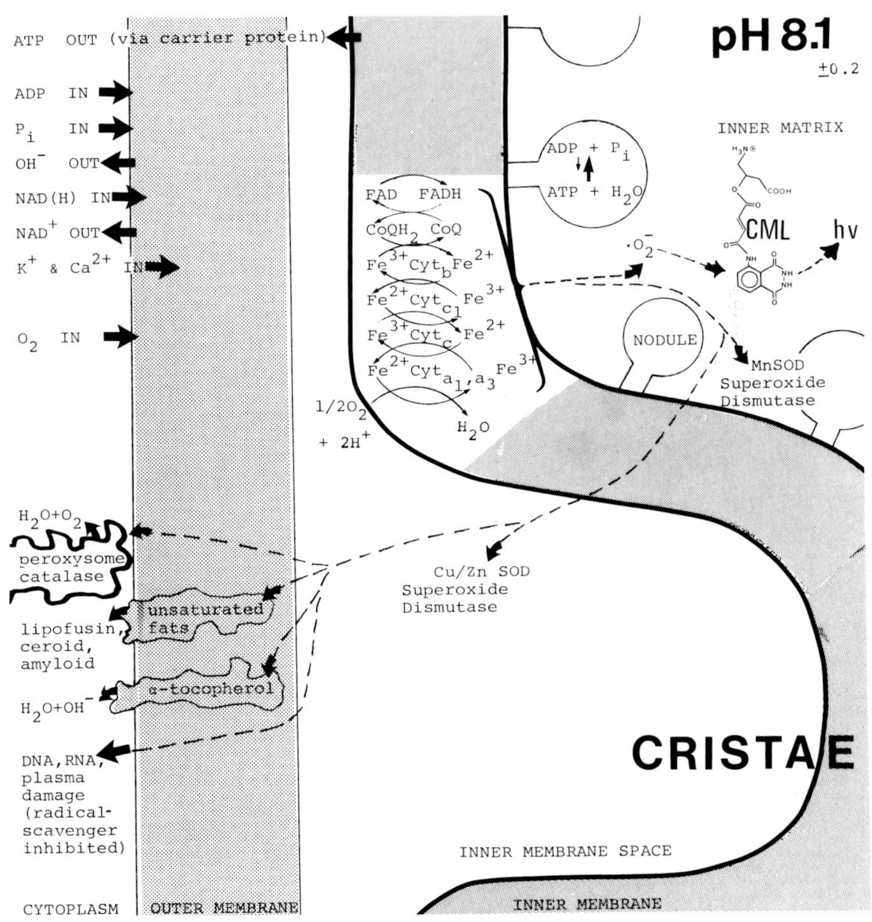


*FIGURE 4. Probing of the mitochondrial inner matrix with CML and direct titration of pH. Superoxide dismutase and other naturally occuring, free-radical scavengers such as α-tocopherol protect mitochondria from respiratory-chain superoxide anion radical flux.*

## III. INTERNAL PROBING OF LYSOSOMES WITH CHEMILUMINESCENCE

In the case of lysosomes (B) (Fig. 1) one encounters an important organelle in cell biology that is responsible for the destruction of unwanted cell debris, i.e. old mitochondria and autolysation of old cells themselves. It is also known that lysosomes easily and quickly engulf and absorb exocellular macromolecules. Therefore, a polysaccharide isoluminol was synthesized which will probe lysosomes of human glial cells and kidney cell extracts. These studies are shedding light upon the *in vivo*, catalytic roles of transition metals and antioxidants involved in lysosomal proteolytic and lipolytic degradations (see Fig. 5).

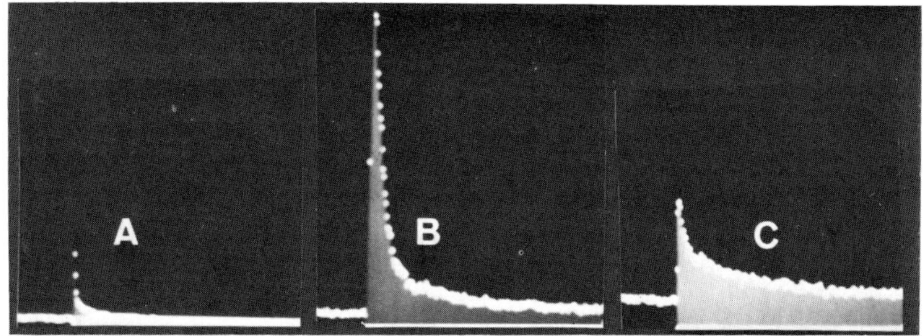


*FIGURE 5. Lysosomes from 787 CG human glial cells (passage 24) were probed with a macromolecular Sepharose-isoluminol substrate and later incubated with 50 μmole (A) α-tocopherol, (B) copper chloride and (C) control. Rapid, exocellular injection of $\cdot O_2^-$ inhibited α-tocopherol protected lysosomes while copper catalysed the luminol reaction.*

Lysosomes are also known to accumulate age pigments, i.e. lipofuscin, hyalin and amyloid. Lipofuscin age pigment occupies about 25% of the intracellular volume of rat neurons at two years of age, while in human neurons this percentage is not attained until about 70 years of age (see Fig. 6).

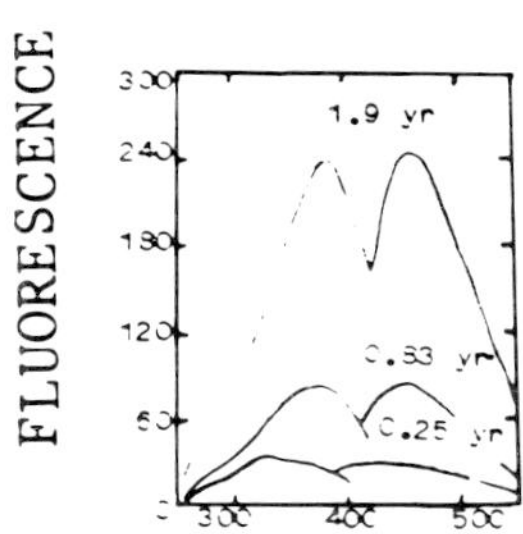


*FIGURE 6. Mouse testes fluorescence spectra at 3 different ages as a result of lipofuscin accumulation.*

These pigmentations are often the peroxidation products of proteins and unsaturated fats abnormally cross-linked and fluorescent. Sepharose BrCN-isoluminol was used to probe glial lysosomes and determine the role of peroxidation in relation to age pigmentation. Preliminary results indicate that a known inhibitor of lipofuscin, namely magnesium orotate, has a greater than fifty percent scavenger effect upon superoxide anion radical intrusion, and peroxidation is catalysed by free copper and iron in metabolically-active glial lysosomes.

## IV. INTERNAL PROBING OF THE CELL NUCLEUS WITH CHEMILUMINESCENCE.

A fruitful area in the use of chemiluminescent probes is in the nucleus of a cell. Here, radiation effects have been well documented, but the infuence of radiation-induced free radicals on pathological responses is largely unknown. For example, a given tolerance level of UV radiation upon DNA can be repaired rapidly by efficient enzyme mechanisms. What is unknown here is the <u>relative</u> sensitivities of various nuclear components given a level of radiation with subsequent free-radical formation. That is to say, what levels of free-radical flux will different RNAs, aromatic acid-containing proteins and the nuclear membrane tolerate as compared to the more reversibly repaired DNA double helix?

These and other fundamental questions bordering on the areas of cancer and gerontological research could perhaps be answered by site-specific chemiluminescent probes. One such probe might be a uracil-isoluminol which would be analogous to fluorouracil. Others are the labelling of other nuclear components withthe author's isoluminolcarbamyl chloride (Fig. 7) whose reaction mechanism in aqueous solution is similar to known acid-chloride labellers.

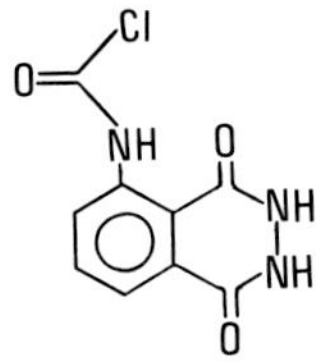

*FIGURE 7. Isoluminol-carbamyl chloride*

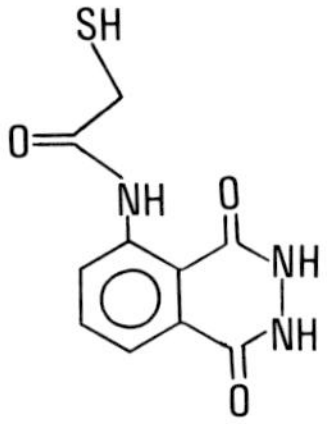

*FIGURE 8. Mercaptoacetyl-isoluminol*

Another possibility is a thiol-isoluminol (Fig. 8 and Ref. 2) since after its reaction with a protein substrate, i.e. an antigen or antibody, the excess reagent is easily removed by a Sepharose 6B column. Still others which may be useful are high molecular weight, aldehyde and thiol specific reagents such as amylose-isoluminol and amylosazonyl semicarbazide-isoluminol (Ref. 2).

Other non-luminol possibilities are bis-phenyl oxalate and acridane phenyl ester derivatives which have been exten-symthesized and used in our laboratories. These esters suffer significant disadvantages in an aqueous biological environment due to difficulties with ester hydrolysis. Frank McCapra proposes that the synthesis of acridane phenyl esters with bulky protective groups near the ester link should inhibit undesirably rapid hydrolysis.

## V INTERNAL PROBING IN OTHER AREAS

Other possible areas of investigation not shown in Fig. 1 are the use of chemiluminescent probes to determine photosensitivities and photoallergies of human skin, especially in relation to melanoma and squamous cell carcinoma. Another is the mechanism of tumor growth accelerations in regard to free radicals: How are mitosis-stimulating and inhibiting substances affected by free radicals and other strong nucleophiles? Lastly and most important is in the study of the long-term oxidations of allysin, lysin, hydroxyallysin and hydroxylysin formed Schiff-bases in elastin and collagen -- irreversible cross-linkings known to cause inelasticity of tissues and hardening of the arteries in aging organisms.

REFERENCES


(1) Merényí, G. and J. Lind, *J. Amer. Chem. Soc.*, in press (1980).

(2) Lippman, R.D., *Anal. Chem. Acta* 116, 181 (1980a).

(3) Hastings, J.W., "1st Int. Symp. Anal. Appl. Biolum. Chemilum.", (E. Schram & P. Stanley, eds.), p. 28-29.

(4) Lippman, R.D., *Experimental Gerontology,* in press(1980b)

(5) Comfort, A. (ed.) "The Biology of Senescence", p. 269. Churchill Livingstone, Edinburgh (1978).

(6) Lippman, R.D. and A. Agren, *Biochem. J.*, submitted for publication, (1980).

(7) Mitchell, P. and J. Moyle, *Eur. J. Biochem.* 7, 480 (1969)

# DIFFERENTIAL EFFECTS OF BACTERIAL AND VIRAL INFECTIONS ON GRANULOCYTE CHEMILUMINESCENCE (CL)

J. P. McCarthy
P. Z. Sobocinski
P. B. Jahrling
D. W. Reichard

U.S. Army Medical Research Institute
of Infectious Diseases
Fort Detrick, Maryland

Previous studies demonstrated enhanced luminol-assisted CL in polymorphonuclear leukocytes (PMN) or rats infected with live, vaccine strain *Francisella tularensis* (LVS) (1). The PMN CL response to *Salmonella typhimurium (St)* infection and endotoxin (LPS) in rats and LVS and a viral infection in guinea pigs (GP) was determined. PMN from rats injected ip with $10^7$ live or heat-killed *St*/100 g b.w. had a 7-fold increase in CL by 24 h *vs* controls. A progressive 9-29-fold increase in PMN CL occurred during the 48 h after ip injection of 250 μg *Escherichia coli* LPS/100 g b.w. CL response to LPS was linearly dose-dependent between 0.01 and 100 μg. GP injected ip with $10^7$ LVS/100 g b.w. had a 34-fold increase in PMN CL at 24 h *vs* controls; however, PMN CL from infected and control GP was lower *vs* rats. In contrast, no enhanced PMN CL occurred in GP on day 3, 8 or 11 after sc injection of Pichinde virus (40,000 PFU) *vs* controls; GP were viremic by day 3 and moribund by day 13. The studies show that *in vivo* stimulation of PMN CL during certain bacterial infections is not species-specific; PMN CL is stimulated *in vivo* by LPS. Finally, PMN CL may be useful in the differentiation of bacterial and viral infections.



ISBN 0-12-208820-4

REFERENCE

1. McCarthy, J. P. and P. Z. Sobocinski, *Fed. Proc. 38*, 1023 (1979).

# ARACHIDONATE - BASED CHEMILUMINESCENCE IN HUMAN GRANULOCYTES AND PLATELETS USING THE MONOLIGHT 301 (DRUG STUDIES)

David Peden
Chris Van Dyke
Cynthia Van Dyke
Knox Van Dyke

Departments of Pharmacology and Toxicology
West Virginia University Medical Center
Morgantown, West Virginia 26506

Vincent Castranova
George Jones

Physiology (ALOSH)
West Virginia University Medical Center
Morgantown, West Virginia 26506

Michael Ringrose

Analytical Luminescence Laboratory Inc.
Westlake Village, California

## I. INTRODUCTION

Various studies have shown that human granulocytes and platelets exhibit chemiluminescence (CL) when metabolizing sodium arachidonate (SA) via cyclo-oxygenase or lipoxygenase pathways (1-5). The metabolites of these pathways are known to have effects upon inflammatory responses (6,7) and are thought to have effects upon arteriosclerosis and thrombosis (8-10).

Van Dyke *et al.* (1,2) have shown that certain non-




ISBN 0-12-208820-4

steroidal anti-inflammatory drugs (NSAIDs) will inhibit CL activity of granulocytes and platelets (platelets isolated via Mills' (3) technique). By using a modification of Van Dyke's techniques in which SA is used to cause CL responses (11) we have done studies on newly developed NSAIDs. Also, using the Monolight 310 photometer, we have been able to make observations concerning the kinetics of the CL reactions of granulocytes and platelets which are not possible when using standard scintillation counters because they are mechanically unsuited for rapid and continuous CL monitoring. The NSAIDs studied were aspirin (ASA), MK-447, 2 phenindione (2 PI), Nordihydioguaiaretic acid (NDG), BW-755C and FPL-55712. Drug concentration was set at 2.8 x $10^{-5}$M.

## II. RESULTS

There are three major observations to be made from these studies: the kinetic plots of the granulocyte and platelet reactions, the similarities between the 2 types of CL reactions and, the inhibition of granulocyte and platelet CL by various NSAIDs. As seen in Figure 1 the granulocyte and platelet CL kinetics are both very rapid and decay in a similar fashion. Tables 1 and 2 provide a statistical profile of the CL inhibitory capacities of NSAIDs with these cell types. One should note the effectiveness of NDG, BW-755C, 2PI and Mk-447 with regards to CL inhibition of both cell types.

*TABLE 1. Granulocyte CL Inhibition by NSAIDS (n=4)*

| *Drug* | *BW-755C* | *NDG* | *2PI* | *Mk-447* | *FPL-55712* |
|---|---|---|---|---|---|
| *% control (mean)* | *0.0* | *0.0* | *0.0* | *6.4* | *85.07* |
| *Standard Error* | - | - | - | *1.12* | *1.181* |
| *Standard Deviation* | - | - | - | *2.36* | *2.36* |
| *T-value* | - | - | - | *83.5* | *12.64* |

*TABLE 2 - Platelet CL Inhibition by Drugs (NSAIDS) (n=3)*

| *Drug* | *BW-755C* | *NDG* | *ASA* |
|---|---|---|---|
| *% control (mean)* | *0.1* | *0.87* | *65.13* |
| *Standard Error* | *0.1* | *0.87* | *2.21* |
| *Standard Deviation* | *0.2* | *1.50* | *3.82* |
| *T-value* | *999.0* | *114.4* | *15.77* |

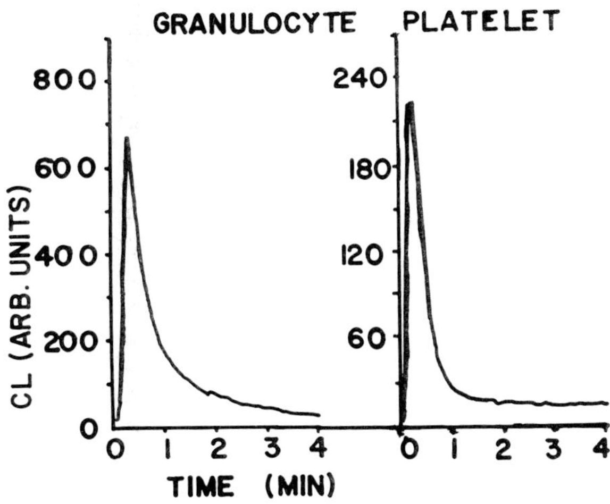


*FIGURE 1. Kinetics of platelet and granulocyte CL reactions.*

## III. DISCUSSION

The two enzyme systems studied, cyclo-oxygenase (CO) and lipoxygenase (LO) metabolize SA to substances which mediate inflammation (9,10). The cyclo-oxygenase pathway metabolizes SA to a prostaglandin form $PGH_2$, which can be metabolized to

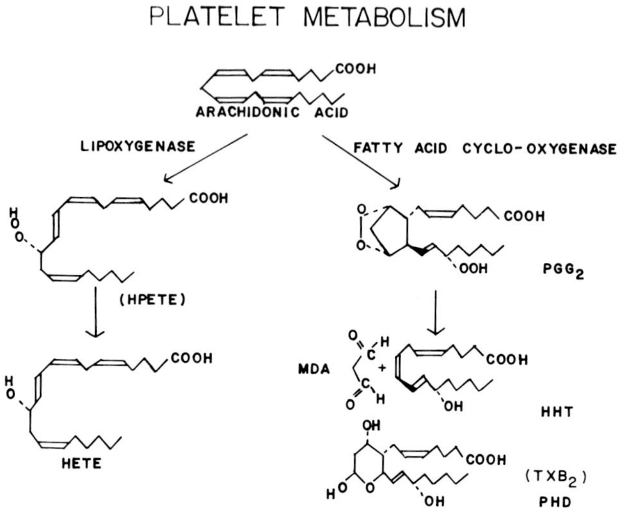

PLATELET METABOLISM

ARACHIDONIC ACID

LIPOXYGENASE

FATTY ACID CYCLO-OXYGENASE

(HPETE)

$PGG_2$

HETE

MDA

HHT

$(TXB_2)$

PHD

FIGURE 2. Metabolism of SA in platelets.

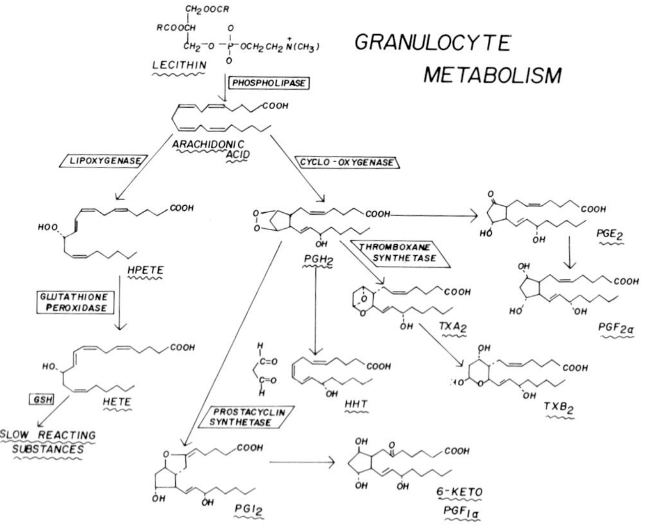


FIGURE 3. Metabolism of SA in granulocytes.

thromboxanes ($TXA_2$ and $TXB_2$) or prostacyclins ($PGI_2$ or $PGX_1$). $TXA_2$ is thought to be involved in platelet (8,10,12,13) aggregation and be a causative factor in thrombosis and arteriosclerosis. $PGI_2$ is believed to have the opposite effect of $TXA_2$ in as it is a vasodilator and a potent anti-aggregating factor (9,13). Lipoxygenase oxidizes SA to form HETE which reacts with glutathione to produce slow-reacting substance A (SRS-A). HETE has also been shown to be leukotactic. See Figures 2 and 3.

CL associated with granulocyte and platelet SA metabolism is thought to be caused by excited state oxygen metabolism (5,11,14) (which will react with luminol to give increased CL) or produced by CO and LO during arachidonate metabolism (4). There is no evidence that in vivo, granulocytes and platelets could store the CO-LO pathway products. When these substances are required, they are very rapidly synthesized after arachidonate is liberated from the phospholipid storage molecules (12). The CL of SA metabolism would be excepted to occur quickly upon introduction of SA. As one can see in Figure 1, this type of CL kinetics occurs in both platelets and granulocytes. This supports the ideas that CL does accompany SA metabolism and that the enzyme systems (for CO and LO activity) in granulocytes and platelets are similar.

Van Dyke et al. (1,2) have shown that CL associated with SA metabolism can be used as a measure of CO and LO activity and NSAID will inhibit the associated CL. Using this modification of the CL NSAIDs screen (1) we have studied the action of various NSAIDs. It was found that aspirin inhibited 40% of granulocyte CL and 35% of platelet CL. In granulocytes it was shown that BW-755C, NDG, and 2PI blocked 100% of control CL levels. Also Mk-447 blocked 93.6% of granulocyte CL. Of those drugs studied in platelets, BW-755C and NDG both inhibited 100% of control level CL (Tables I and II). These results indicate the BW-775C, NDG, 2PI and Mk-447 may be very effective clinical NSAIDs. Also, these results indicate that CL measurements are effective ways to study platelet and granulocyte activity and that variations of this technique may be used to screen drugs, study platelet effects on thrombosis and arteriosclerosis and to develop clinical tests for allergy and phagocytotic disorders such as chronic granulomatous disease (15).

REFERENCES


1. Van Dyke, K., C. Van Dyke, J. Udeinya, C. Brister, and M. Wilson, *Clin. Chem.* 25, 1655 (1979).
2. Van Dyke, K., C. Van Dyke, D. Peden, G. Jones, V. Castranova, E. Bristel, and M. Ringrose, *Microchem. J.* - to be published Dec. 1980.
3. Mills, E. L., J. M. Gerrard, D. Filipovich, J. S. White, and P. G. Quie, *J. of Clin. Invest.* 61, 807-814 (1978).
4. Veldink, G., G. Carssen, S. Slappendel, J. Vliegenthart, and J. Boldingh, *J. Biochem. Biophys. Res. Comm.* 78, 424-428 (1977).
5. Allen, R. C., *Photochem. and Photobiol.* 30, 157 (1979).
6. Flower, R. J., *Handb. Exp. Pharm. (Inflamation)*, Chap. 12, 501 (1978).
7. Zurier, R. B., Saydoff, *Inflamation, Vol. 1, No. 1*, 93-101.
8. Zucker, M. B., *Scientific American* 242, 86-103 (June 1980).
9. Borch, J. W., *Prostaglandins and Therapeutics* 5, *No. 3* (1979).
10. Bierenbaum, M. L., *Prostaglandins and Therapeutics* 5, *No. 3* (1979).
11. Van Dyke, K., M. Trush, M. Wilson, P. Stealey, and P. Miles, *Microchemical J.* 22 (1977).
12. Ferreira, S. H., and J. R. Vane, *Handb. Exp. Pharm. (Anti-Inflamm. Drugs)*, Chap. 31, Vol. 50 II (1978).
13. Willis, A. L., *Handb. Exp. Pharm. (Inflammation)* 50I, Chap. 5 (1978).
14. Nelson, R. D., J. M. Herron, J. R. Schnidtke, R. L. Simmons, *Infection and Immunity* 17, no. 3 (1977).
15. Castranova, V., G. S. Jones, R. N. Philips, D. Peden, and K. Van Dyke, this symposium.

# CHEMILUMINESCENCE DERIVED FROM 4a-(ALKYLPEROXY) FLAVINS

Peter T. Shepherd
Thomas C. Bruice

Department of Chemistry
University of California
Santa Barbara, California

Structure I represents the minimal requirements for chemiluminescence (CL) in the decomposition of 4a-(alkylperoxy) flavins (1). Bioluminescence catalyzed by bacterial luciferase arises from chemical transformations of enzyme-bound I (a) (2).

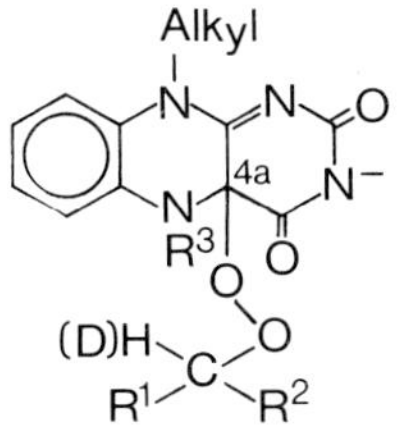

(a) $R^1$ = n-alkyl,
$R^2$ = -OH, $R^3$ = -H

(b) $R^1$ = $C_6H_4CH_3$,
$R^2$ = -H(D), -OH
$R^3$ = $-C_2H_5$

I

Presented here are the results of preliminary experiments which establish that more than one excited species is formed on decomposition of I(b).

The disappearance of I(b) from solution, followed at 365 nm, and the decay in the intensity of light produced, follow the same rate law (1). This is due to the fact that CL arises from minor reactions. Consequently -d[I(b)]/dt is controlled by the major, non-light-emitting pathway (1, 3).

In the absence of added fluorescer, chemiluminescent emission is identical to the fluorescence emission of the parent




ISBN 0-12-208820-4

isoalloxazine ($Fl^*_{ox}$) in all cases studied. $Fl^*_{ox}$ is not produced by energy transfer from another excited species, but is generated directly in a chemi-excitation step from I(b). The $\Phi$ for the reaction is the same whether $R^2$ = -H, or -OH, indicating that the unshared pair of electrons on the peroxy-aldehyde adduct ($R^2$ = -OH) do not contribute kinetically to the formation of $Fl^*_{ox}$. Exchange of -H for -D in structure I(b) leads to an isotope effect upon the quantum yield ($\Phi H/\Phi D$) of 1.7 to 2.0, but does not give a kinetic isotope effect. This proves that a C-H(D) bond is broken in the chemi-excitation process to produce $Fl^*_{ox}$, but that this is a minor pathway for the decomposition of I(b).

The addition of increasing concentrations of $Fl_{ox}$ to the reaction solution brings about an increase in $\Phi$. CL is due to emission from $Fl^*_{ox}$ at all concentrations of $Fl_{ox}$. These results cannot be attributed to energy transfer from chemi-excited $Fl^*_{ox}$ to $Fl_{ox}$, since this would hardly explain the increase in $\Phi$. Thus, another excited species, X*, is produced in the course of the reaction. Moreover, the addition of other fluorescers provides similar results (Fig. 1). In each case the increased CL is due to emission by the added fluorescer. The presence of a fluorescer increases neither the rate constant for the disappearance of I from solution, nor the rate constant for the decay of light emission. Replacement of H- by D- in I(b) results in a decrease in $\Phi_{max}$ by the same amount as in the absence of fluorescer (for fluorescer rubrene $\Phi H/\Phi D$ = 1.7; for fluorescer $Fl_{ox}$ $\Phi H/\Phi D$ = 1.8), but there is no change in the first order rate constants for either the disappearance of I or the diminution in the intensity of light emission. This establishes that the C-H(D) bond is broken on generating X* and that this chemiluminescent process, like that which generates chemi-excited $Fl^*_{ox}$, does not contribute significantly to the disappearance of I(b).

Presently investigations are in progress to determine the nature of X*, a species which must satisfy the two following conditions, i.e., it is non-fluorescent (no CL emission has been observed in the absence of fluorescers, other than that of $Fl^*_{ox}$) and it must be of sufficiently high energy to excite all of the added fluorescers. None of the flavin products of the reaction, isolated by HPLC satisfies these conditions.

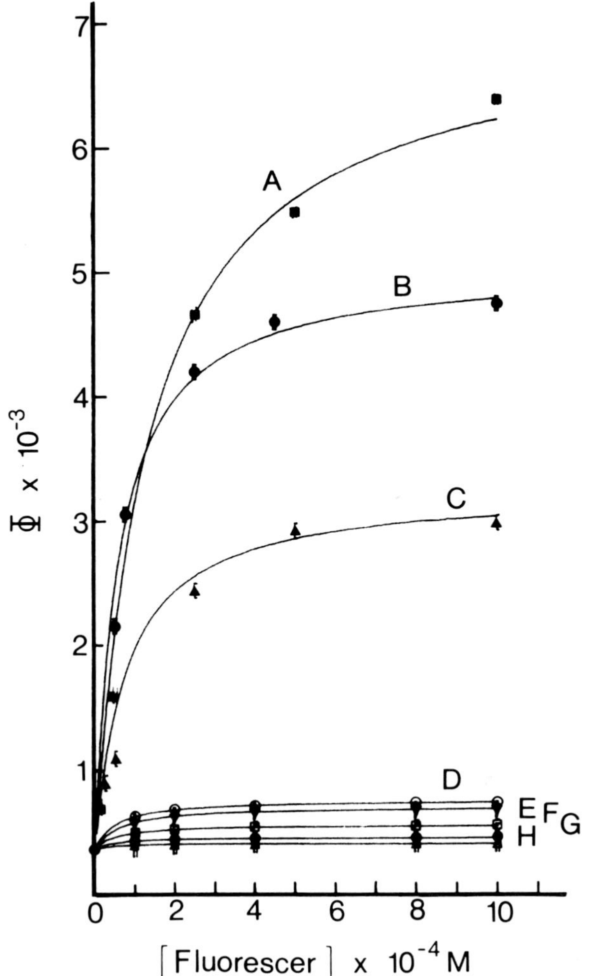


*A = Rhodamine B, B = $Fl_{ox}$, C = 6,7,8-trimethyl lumazine, D = rubrene, E = perylene, F = pyrene, G = coronene, H = 9,10-diphenylanthracene*

*FIGURE I*[a,b,c,d]. *Plot of Φ vs [Fluorescer] for a series of added fluorescers.* [a]*[$N_5$-ethyl flavinium perchlorate] = $10^{-4}$ M,* [b]*Benzyl hydroperoxide = $10^{-2}$ M,* [c]*T = 30°C,* [d]*Solvent = DMF, except for A, where solvent = t-ButOH.*

REFERENCES

1. (a) Kemal, C., and T. C. Bruice, *Proc. Nat'l. Acad. Sci. USA* 73, 995 (1976). (b) Kemal, C., Chan, T. W., and T. C. Bruice , *Ibid*. 74, 405 (1977). (c) Kemal, C., and T. C. Bruice , *J. Am. Chem. Soc.* 99, 7066 (1977).
2. (a) Hastings, J. W. C. Balny, C. Le Peuch, and P. Douzou, *Proc. Nat'l. Acad. Sci. USA* 70, 3468 (1973). (b) Hastings, J. W., and C. Balny, *J. Biol. Chem.* 250, 7288 (1975). (c) Balny, C., and J. W. Hastings, *Biochem.* 14, 4719 (1975). (d) Hastings, J. W., Q. H. Gibson, J. Friedland, and J. Spurich, *in* "Bioluminescence in Progress" (F. H. Johnson and Y. Haneda, eds., ) p. 151. Princeton Univ. Press, New Jersey (1966).
3. Mackiewicz R., D. Sogah, and T. C. Bruice, *J. Am. Chem. Soc.* 101, 5367 (1979).

# B. Bacterial Bioluminescence and Applications

# THE USE OF BENTONITE CLAYS TO INCREASE THE SPEED AND YIELD OF HARVEST OF LUMINOUS BACTERIA BY CONTINUOUS FLOW CENTRIFUGATION[1]

James E. Becvar
Amina Mahomedy
Oscar Dominguez
John Hostak
Bean Burr
A. Bryce Campbell
Frank Loudermilk
Daniel Marquez
Kevin Ayer

Department of Chemistry
The University of Texas at El Paso
El Paso, Texas

In order to establish a program of study into the biochemistry of bacterial bioluminescence at this University, we needed to overcome several hurdles undoubtedly common to the establishment of biochemical research at any institution. To obtain the enzyme necessary for this program of study we needed to grow and harvest large quantities of luminous bacteria. We have been able to culture 60 liter batches of luminous bacteria in a fermentor similar to the original Hastings (1) design at minimal equipment cost by using a reclaimed agitator washing machine. This old-fashioned unit has a removable basin which can be autoclaved. By separately autoclaving the culture medium in small volumes and adding it to the fermentor, a system sterile enough for our purposes is

[1] *This work was supported by grant AH-777 from the Robert A. Welch Foundation, Houston, Texas and by grant RR08012 from the Division of Research Resources, National Institutes of Health.*



ISBN 0-12-208820-4

obtained. In the four fermentations done to date, we have failed to detect any colony-forming contaminants in 0.2 ml samples of medium taken from the fermentor prior to inoculation. Since all inoculations of the luminous species desired are made to the level of about $10^7$ bacteria per ml, this procedure results in essentially culturing only the desired species.

Although good growth was obtained, harvesting the bacteria was initially a problem. Using an old reconditioned De Laval model E-500 cream separator centrifuge, obtained inexpensively from a local dealer in used dairy equipment, we could achieve only about a 20% yield (as measured by reduction in turbidity, $A_{660}$) at a flow rate of one liter per minute. At much slower flow rates, a yield of 50% was approached. Considering the time required, this was still unsatisfactory.

Vintners for years have used a variety of agents for fining (clarifying) wines. One of these agents is bentonite clay. On the possible chance that such a substance might improve the speed and yield of harvesting luminous bacteria by centrifugation, we undertook the series of smaller scale experiments reported here. Only after we were well into these studies did we become aware of the extensive literature (2-7) which pertains to the interaction between bacteria and soil constituents and other solid surfaces. Our studies on the interaction of several species of luminous bacteria with montmorillonite clays demonstrate that these species also adsorb to clays. Moreover, these studies also provide the basis for a practical solution to our harvesting problem. Finally and most importantly, we show that the presence of montmorillonite clays in the cell paste of the harvested bacteria does not inactivate nor reduce the yield of the luciferase from the sedimented bacteria.

The cultures of luminous bacteria used in the test studies reported were grown in 250 ml volumes of NaCl complete medium in one liter flasks shaken on a wrist action shaker. NaCl complete medium contains per liter: 30g NaCl, 7g $Na_2HPO_4 \cdot 7\,H_2O$, 1g $KH_2PO_4$ , 0.5g $(NH_4)_2\,HPO_4$ , 0.1g $MgSO_4$ , 3 ml glycerol, 3g Bactotryptone, 1g yeast extract. In order to investigate as many conditions as possible, small volume (25 ml) samples were harvested using a Damon/IEC Clinical centrifuge at a speed setting (4 or 5) and a duration (300 to 500 seconds) which approximately matched in control experiments the 20% reduction in $A_{660}$ seen in continuous flow centrifugation. Turbidity measurements ($A_{660}$) were made with a 13 mm path test tube in a Bausch and Lomb Spectronic 20.

Stock clay suspensions (25 mg/ml) were prepared by gently stirring the clay in deionized water for 1 to 2 hours prior to use. The clay referred to as bentonite in these studies was a montmorillonite obtained from Dr. Clarence Cooper, La Viña Winery, Chamberino, N.M. 88027. HCl washing of clay was accomplished by twice resuspending a bentonite pellet in 0.1 M HCl followed by four-time resuspension in deionized water. For some studies this acid washed clay was subsequently twice resuspended in 0.1 M ethylenediaminetetraacetate (EDTA), pH 7.0 and rinsed four times in deionized water. NaOH washing was accomplished by two resuspensions in 0.1 M NaOH and four water rinses.

Luciferase activity was measured by the standard assay described elsewhere (8) in *these Proceedings*.

Figure 1 shows the results of several series of experiments in which different amounts of bentonite were added to cultures of several species of luminous bacteria just prior to centrifugation. The far left ordinate gives the $A_{660}$ of the various cultures before centrifugation. The $A_{660}$ after centrifugation of the controls without clay are shown on

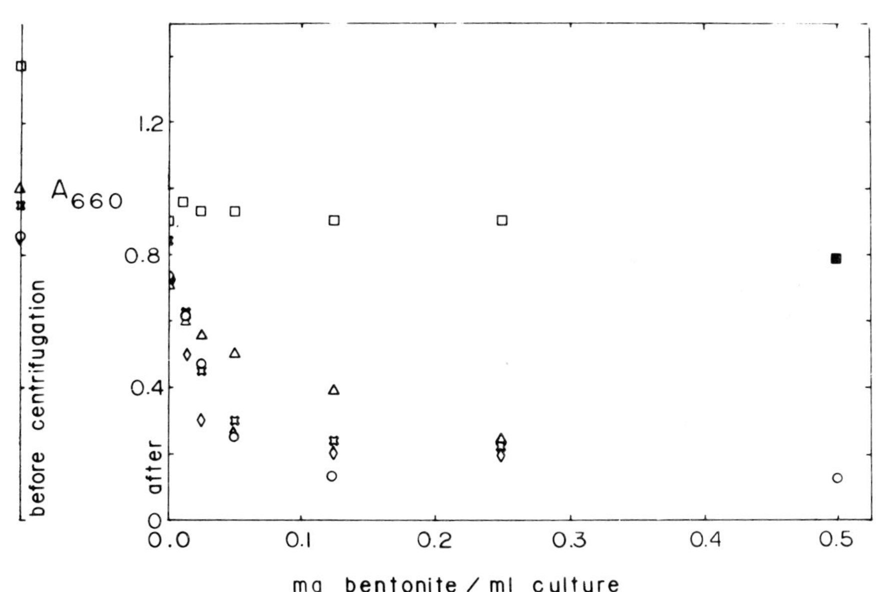


Figure 1. Ability of Bentonite to Flocculate Different Species of Luminous Bacteria

○ P. fischeri
□ B. harveyi
■ B. harveyi, 30 min preincubation
△ P. phosphoreum
¤ P. leiognathi, S-1
◊ P. leiognathi, EGMB

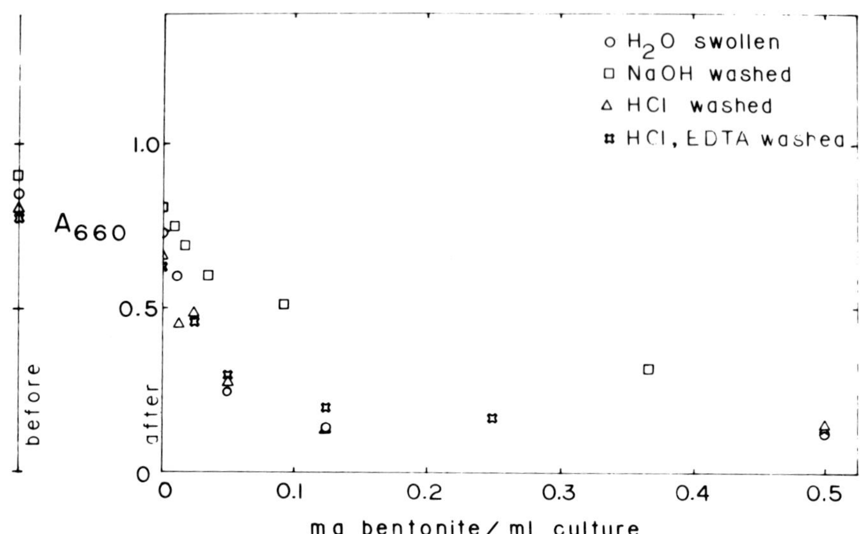


Figure 2. The Effect of Various Pretreatments of Bentonite on the Flocculation of *P. fischeri*

the ordinate at zero mg bentonite per ml culture. A bentonite concentration of 0.1 to 0.25 mg clay per ml of culture is seen to be effective in significantly enhancing the sedimentation of *Photobacterium* species, but not of *Beneckea harveyi*. However, concentrations above 1 mg bentonite per ml of culture does cause effective sedimentation of *B. harveyi* (data not shown).

In an attempt to understand the basis for the effect of clays on *Photobacterium* we tried by pretreatment of the clay in various ways to alter its adsorptive properties. Figure 2 shows the results of several of these studies. None of the pretreatments significantly enhanced the flocculation of *P. fischeri* by clay, but pretreatment by NaOH seemed to decrease the ability of the clay to flocculate the bacteria. It is unclear from these experiments if this was the result of $Na^+$ exchange, but this may have been the case.

Studies with cation exchange materials (carboxymethyl cellulose and carboxymethyl Sephadex) and anion exchange materials (diethylaminoethane cellulose and diethylaminoethane Sephadex) were totally ineffective in enhancing the sedimentation of luminous bacteria in this medium at concentrations of 100 to 1000 times the concentration at which clay begins to show a significant effect in flocculating *P. fischeri*.

TABLE 1. *The Effect of Bentonite on Luciferases in Lysates of Several Species of Luminous Bacteria*

| Species | $A_{660}cm^{-1}$ of culture at harvest | Nature of lysate assayed | Aldehyde used in *in vitro* assays | Activity[a] of control lysate, clay absent, $I_o x 10^{-10}$, q/s | Activity of sample lysate, 0.25 mg clay per ml culture, $I_o x 10^{-10}$, q/s | Activity of sample lysate, 0.50 mg clay per ml culture, $I_o x 10^{-10}$, q/s | Activity of sample lysate, 1.0 mg clay per ml culture, $I_o x 10^{-10}$, q/s |
|---|---|---|---|---|---|---|---|
| *B.h.* | 1.7 | whole | $C_{10}$ | 137 | - | 140 | 153 |
| | | supern. | $C_{10}$ | 176 | - | 136 | 134 |
| *P.f.* | 1.2 | whole | $C_{10}$ | 119 | 140 | - | 106 |
| | | supern. | $C_{10}$ | 102 | 141 | - | 92 |
| *P.l.* | 1.6 | whole | $C_{12}$ | 134 | - | 156 | 143 |
| | | supern. | $C_{12}$ | 147 | - | 169 | 159 |
| *P.p.* | 1.4 | whole | $C_8$ | 1.7 | 1.5 | 1.4 | - |
| | | supern. | $C_8$ | 1.2 | 1.4 | 1.2 | - |

[a] Activities expressed per ml of culture.

Because we are principally concerned with the luciferase within the bacteria, we tested to see if the use of clay in sedimenting the bacteria would in any way interfere with the recovery of luciferase or cause its inactivation. Tests were conducted using control cultures to which no clay was added and using sample cultures to which were added amounts of clay in excess of that needed to produce effective flocculation. Bacteria within controls and samples were completely sedimented by centrifuging at high speed. The pellets were resuspended in cold lysis buffer (1 mM EDTA, pH7.0). Luciferase activities were determined both in whole lysates and in the supernatants of these lysates after clay and cell debris had been removed by centrifugation. The results of these experiments are given in Table I. Clay concentrations at and well above those promoting effective sedimentation of *Photobacterium* species are found to cause no detrimental effect on the luciferase during subsequent isolation steps.

Harvest of 60 liter batches of *P. fischeri* by continuous flow centrifugation was significantly improved by the addition of small amounts of clay to the culture. Up to 80% yield was obtained at 0.25 mg bentonite per ml culture at flow rates of nearly 1 ml per minute.

## REFERENCES


1. Hastings, J. W., personal communication.
2. Lahar, N., *Plant and Soil 17*, 191 (1962).
3. Stotsky, G., and L. T. Rem, *Canad. J. Microbiol. 12*, 547 (1966).
4. Stotsky, G., *Canad. J. Microbiol. 12*, 831 (1966).
5. Daniels, S. L., and L. L. Kempe, *in* "Chemical Engineering in Biology and Medicine" (D. Hershey, ed.), p 391 Plenum Press, New York, (1967).
6. Marshall, K. C., *Biochim Biophys Acta 193*, 472 (1969).
7. Marshall, K. C., *in* "Interfaces in Microbial Ecology", Harvard University Press, Cambridge, Mass., 1976.
8. Becvar, J. E., and L.-H. Wu, *these Proceedings.*

# THE RED ABSORBING FLAVIN SPECIES IN THE REACTION OF BACTERIAL LUCIFERASE WITH $FMNH_2$ AND $O_2$[1]

J.W. Hastings
Robert Presswood

The Biological Laboratories
Harvard University
Cambridge, Massachusetts

Sandro Ghisla
Manfred Kurfürst
Peter Hemmerich

Fachbereich Biologie
Universität Konstanz
Konstanz, West Germany

The reaction of luciferase-bound $FMNH_2$ is known to result in the formation of a long-lived intermediate in the bioluminescent reaction (1). This intermediate was isolated and characterized as the luciferase-peroxyflavin (2), whose structure was later shown to be the flavin 4a-substituted peroxy-adduct (3). In the earlier work this peroxy intermediate had been shown to exhibit a single peak at about 370 nm, a shoulder at about 460 nm, the absorption tailing off around 500 nm, with none above 520 nm.

In more recent publications, however (4,5), it has been reported that the reaction of the luciferase-bound reduced flavin mononucleotide with $O_2$ also results in the appearance

[1] *These studies were supported in part by grants from the U.S. National Science Foundation (PCM 77-19917) and the Deutsche Forschungsgemeinschaft to S.G. J.W.H. was an awardee of the Alexander Von Humboldt Foundation.*



ISBN 0-12-208820-4

of absorption in the 570-610 nm region. Although the spectrum appeared similar to that of the neutral semiquinone (6), no ESR signal was detected in preliminary experiments. Kinetic data suggested that this red absorbing species was related to, possibly in rapid equilibrium with, luciferase intermediates (such as the peroxyflavin) in the pathway leading to light emission.

The present work was undertaken in order to study the formation, properties and possible role in bioluminescence of the material absorbing in the 600 nm region, and to resolve some of the questions raised by the observations described above.

Since no absorbance in the red was noted in the preparations of peroxyflavin earlier isolated and characterized (2, 7,8), we undertook the preparation of the peroxy intermediate

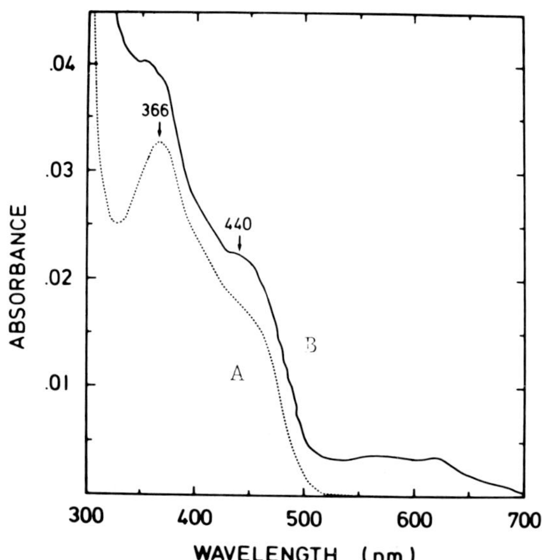


*FIGURE 1. Absorption spectra taken at 0° of the Sephadex G-25 purified luciferase-flavin intermediates formed by reaction of luciferase-$FMNH_2$ with oxygen at 0° (A) when the material reduced with excess dithionite was applied directly to the column and (B) when the material was fully oxidized by exposure to air for 3 minutes prior to application to the column. The first (A) appears to be virtually all in the peroxyflavin form, while the second (B) is a mixture of that with the luciferase neutral flavin semiquinone.*

following the gel filtration procedure of Hastings and Becvar (9). The result was clear (Figure 1, Curve A); peroxyflavin with no red absorbance was formed.

The essential step of the procedure required to obtain a preparation lacking the blue species seemed to be the reduction of the luciferase-flavin complex with excess dithionite and its application to the Sephadex column without prior reoxidation. In this way the luciferase-bound flavin presumably encountered molecular oxygen on the column only after, or in the course of, its separation from small molecules: flavins, dithionite, $H_2O_2$, and other products. As a test of this idea, we made a similar preparation but reoxidized with the air in the test tube at $0^{\circ}C$ three minutes prior to subjecting the material to Sephadex chromatography. Again, the result was clear (Figure 1, Curve B); a similar amount of peroxyflavin was formed but there was, in addition, significant absorption in the 570-610 nm region. The difference spectrum between traces A and B resembles the spectrum for the purified luciferase-neutral semiquinone (see Figure 2); the absence of a trough in trace B at 330 nm is attributed to the significant contribution that the absorption of the semiquinone makes in this region.

The spontaneous decay of the purified luciferase peroxyflavin was followed, as before (8), by the increase in absorption at 440 nm. The half-life at $2^{\circ}$ was about 55 minutes similar to previously reported values under similar conditions. A most significant feature of this experiment was the fact that during the decay there was no development of absorbance in the 570-610 nm region. Thus, under these conditions, no appreciable conversion of the peroxyflavin to the blue species occurred.

As was illustrated in Figure 1, trace B, the oxidation of the reduced flavin-luciferase complex by oxygen resulted in the simultaneous formation of both the peroxy and semiquinone luciferase intermediates. In aerated buffer at $0^{\circ}C$ the subsequent decay of the peroxy compound is more rapid (half-life, 50 min) than that of the semiquinone (half-life, 20 hr; Figure 3 inset). This means that the aged preparations should have only the latter species, which has been found to be so.

Such an aged preparation was subjected to chromatography on a Sephadex G-25 at $2^{\circ}C$. Since the affinity of oxidized FMN for luciferase is low (8,10), the majority of the protein-bound flavin species eluted is the luciferase semiquinone. The spectrum of this Sephadex purified material (Figure 2) is fully characteristic of the neutral flavin semiquinone (6). In the same (aerobic) buffer it decayed isosbestically to FMN, the resulting redox equivalents presumably being taken up by oxygen. Based on the amount of FMN finally formed, and

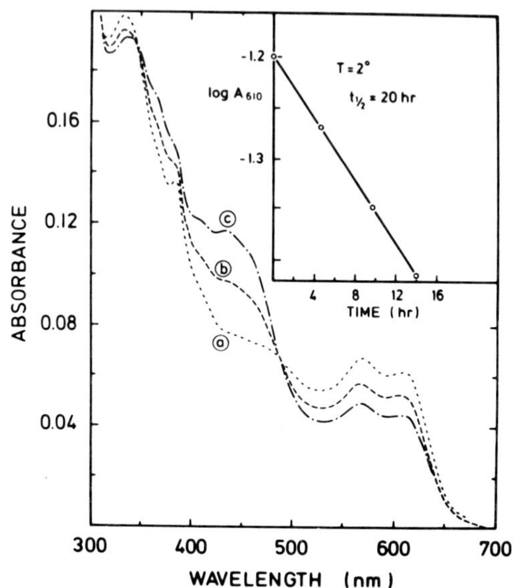


*FIGURE 2. Absorption spectra of the luciferase neutral flavin semiquinone (a) a few minutes after elution from G-25 Sephadex and at two later times (b,c) in the course of its decay to FMN and luciferase. The kinetics of the decay are shown in the inset. To 0.5 mg of luciferase in 0.5 ml was added 0.5 ml of 1 mM $FMNH_2$ (photoreduced) 10 mM EDTA, 0.1 M phosphate buffer, pH 7, under low oxygen tension. After repeated photo reduction and reoxidation, the material was allowed to stand at 0° for 1 hour prior to chromatography on Sephadex G-25.*

assuming that all of the Sephadex purified material was originally in the neutral semiquinone form, the millimolar extinction coefficient for the latter at 610 nm was calculated to be about 4.8.

As shown in the inset of Figure 2, the luciferase semiquinone decayed slowly at 2°C, with a half-lifetime of about 20 hours. Moreover, the decay under these conditions exhibited a high temperature coefficient with an activation energy of about 46 kilocalories. This suggests that the protein may have to change conformation in order to allow either dissociation of the semiquinone or its accessibility to an oxidizing species.

A sample of such a preparation of the blue species was frozen in liquid nitrogen; its ESR signal was measured and found to be characteristic of the neutral flavin radical. A millimolar extinction coefficient of about 4.5 was estimated by reference to a 5-ethyl-riboflavin radical standard.

The luciferase neutral semiquinone radical was also formed by titration with dithionite. Dithionite was added step-wise to a mixture of luciferase (0.65 mM) and FMN (0.15 mM). One equivalent of dithionite was required for the full reduction of the flavin; at half reduction the maximum amount of the blue flavin semiquinone was formed. A millimolar extinction coefficient of 4.2 was estimated.

Based on these results it appeared that the flavin semiquinone would be formed directly by reaction of the reduced and oxidized species. This model was verified by mixing equimolar amounts of FMN and $FMNH_2$ with luciferase at 11° in the absence of oxygen. An appreciable amount of the semiquinone was formed within the first minute; this was followed by a slower increase to an equilibrium value. Under these conditions (absence of oxygen) the semiquinone was stable over a period of 36 hours at temperatures between 0° and 26°, the equilibrium between the semiquinone and the other forms being highly temperature dependent. More than twice as much semiquinone (absorption at 610 nm) was present at 0° as at 26°.

The mechanism of the oxidation of aldehyde by the luciferase peroxyflavin intermediate to produce an excited state is a main question to be addressed; the possible involvement of a blue intermediate had prompted interest, proposals, and speculation (11, 12, 13). The fact that one can obtain preparations of the peroxyflavin lacking appreciable quantities of the radical blue species suggests that the latter is not in fact involved, at least directly, in the bioluminescent reaction.

The luciferase peroxyflavin reacts with long chain aldehydes, such as decanal and dodecanal (14), to give light with a half decay time of only a few minutes at 0° (1,5). Thus the blue species should decay more rapidly in the presence of aldehyde if it is capable of forming the luciferase-peroxyflavin. In fact, in the presence of aldehyde, the blue species was formed equally well at 0° and decayed even more slowly than in the absence of aldehyde. Thus there is no indication to support the suggestion that the neutral flavin semiquinone radical is formed by or is in equilibrium with the luciferase peroxyflavin species as such.

As would be expected, luciferase complexed with the flavin radical is not active for light emission. It is also not capable of reacting with reduced flavin to give light emission. But its decay (Figure 2, inset) is mirrored by an increase in luciferase activity, as assayed by its reaction with $FMNH_2$ and decanal.

REFERENCES

1. Hastings, J.W., and Q.H. Gibson, *J. Biol. Chem. 238,* 2537 (1963).
2. Hastings, J.W., C. Balny, C. Le Peuch, and P. Douzou, *Proc. Nat. Acad. Sci. 70,* 3468 (1973).
3. Ghisla, S., J.W. Hastings, V. Favaudon, and J.M. Lhoste, *Proc. Nat. Sci. 75,* 5860 (1978).
4. Presswood, R.P., and J.W. Hastings, *Biochem. Biophys. Res. Comm. 82,* 990 (1978).
5. Presswood, R.P., and J.W. Hastings, *Photochem. Photobiol. 30,* 93 (1979).
6. Massey, V., and G. Palmer, *Biochemistry 5,* 3181 (1966).
7. Hastings, J.W., and C. Balny, *J. Biol. Chem. 250,* 7288 (1975).
8. Becvar, J.E., S.-C. Tu, and J.W. Hastings, *Biochemistry 17,* 1807 (1978).
9. Hastings, J.W., and J.E. Becvar, *Methods in Enzymology, 57,* 194 (1978).
10. Baldwin, T.O., M.Z. Nicoli, J.E. Becvar, and J.W. Hastings, *J. Biol. Chem. 250,* 2763 (1975).
11. Mager, H.I., and R. Addink, *Tetrahedron Lett. 37,* 3545 (1979).
12. Kosower, E.M., *Biochem. Biophys. Res. Comm. 92,* 356 (1980).
13. Wessiak, A., G.E. Trout, and P. Hemmerich, *Tetrahedron Lett. 21,* 739 (1980).
14. Hastings, J.W., K. Weber, J. Friedland, A. Eberhard, G.W. Mitchell, and A. Gunsalus, *Biochemistry 8,* 4681 (1969).

# BACTERIAL BIOLUMINESCENCE: APPLICATIONS TO ENTOMOLOGY

Edward A. Meighen

Department of Biochemistry
McGill University
Montreal, Quebec

Keith N. Slessor

Department of Chemistry
Simon Fraser University
Burnaby, British Columbia

Gary G. Grant

Forest Pest Management Institute
Canadian Forestry Service
Sault Ste. Marie

## I. INTRODUCTION

### A. *Insect Pheromones*

The study of insect pheromones has undergone a tremendous expansion during the last twenty years. Since the identification in 1959 of 10,12-hexadecadien-1-ol as the sex pheromone of the silkworm moth, *Bombyx mori* (1), the identification of compounds making up the chemical messages of a large number of insect species has been accomplished (2) at a rate that is now approaching exponential proportions.

Almost all insect species utilize highly effective chemical communication systems that depend on the production and release of special compounds known as pheromones. A pheromone, then, is a chemical or group of chemicals that affects




ISBN 0-12-208820-4

the behavior (or some other physiological process) of another insect of the same species. For example, pheromones are used to attract mates (sex pheromones), to mark trails (trail pheromones) or to alert nest mates for defensive purposes (alarm pheromones). Of great practical importance for insect control are the sex pheromones (see below). They have a molecular weight of about 200-350, are moderately volatile and, in the case of many moth and some beetle species, have a long chain aliphatic structure. In recent years, an increasing number of very damaging pest species have been found to release long chain aliphatic aldehydes as part of their sex pheromones (3-5).

Investigations into the synthesis, release and degradation of pheromones in insects and the measurement of levels of airborne pheromone are research areas that have been somewhat neglected because of the lack of an adequate, sensitive and facile technology to handle these problems. Measurement of airborne sex pheromone is of immediate practical importance because they can be used to bait traps that can monitor the level of an insect population or in some cases "trap out" a species, removing it from an area to be protected. Alternatively, sex pheromones can be dispersed into the atmosphere to confuse potential mates, thereby preventing mating and reducing the likelihood of a damaging population in the following year. A rapid, sensitive, and quantitative method for analysing pheromone levels would be invaluable for these problems. The specific and highly sensitive response of bacterial luciferases to aliphatic aldehydes makes the bacterial bioluminescent reaction an ideal tool for the analysis of aldehyde sex pheromones.

### *B. Aldehyde Pheromones*

The structures of the major component of the pheromones of some of the most serious and damaging insect pests in North America are given in Table. I. These insects cause billions of dollars in damage to forestry and farming crops every year. The major component of the pheromone of these insects is a long chain unsaturated aldehyde of fourteen to sixteen carbons in chain length. In addition, the minor components of the pheromones of these insects in many cases are also long chain aldehydes. For example, the pheromone of the eastern spruce budworm is 96% (E)-11-tetradecenal and 4% (Z)-11-tetradecenal (3). Similarly, the pheromone of the Khapra beetle, *Trogoderma granarium,* is a 92:8 mixture of the Z- and E-isomers of 14-methyl-8-hexadecenal (6). Many other insects have also been found to contain long chain aldehydes as their major and/or minor components, including the

*TABLE I. Aldehyde Pheromones*

| Insect | Major Component | Reference |
|---|---|---|
| Spruce Budworm |  [structure] | (3) |
| Corn Earworm |  [structure] | (4) |
| Navel Orangeworm |  [structure] | (5) |
| Tobacco Budworm |  [structure] | (4) |
| Khapra Beetle |  [structure, $CH_3$] | (6) |

European cotton bollworm (7), the orange tortrix moth (8), the cranberry girdler (9), and the silkworm (10), to name only a few. The range of chain length for almost all aldehyde pheromones is between twelve and eighteen carbons, a range that fortuitously is optimal for detection at low levels by bacterial luciferases.

## C. *Detection of Aldehyde Pheromones by Bacterial Luciferases*

Bacterial luciferases respond to very low levels of both unsaturated and saturated aldehydes. Table II compares the bioluminescent response of *B. harveyi* and *P. phosphoreum* luciferase to 100 pmoles of several aldehyde pheromones. The relative bioluminescent activities with the unsaturated aldehyde pheromones are similar to that obtained with tetradecanal, the aldehyde that is believed to be the *in vivo* substrate (11-13) and which gives the maximum bioluminescent response of any saturated aldehyde at low concentration (unpublished data). These results indicate that the bioluminescent reaction catalyzed by bacterial luciferase should be an ideal assay for specifically detecting aldehyde pheromones. Furthermore, the two luciferases respond differently to the various aldehyde pheromones and combined with differences in the rate of luminescent decay provide the basis for qualitatively distinguishing between aldehyde pheromones.

TABLE II. Bioluminescent Response of Bacterial Luciferases to Aldehyde Pheromones[a]

| Aldehyde Pheromone | Luciferase B. harveyi | P. phosphoreum |
|---|---|---|
| Tetradecanal | 100 | 100 |
| E-11-Tetradecenal | 100 | 75 |
| Z-11-Tetradecenal | 80 | 30 |
| Z-11-Hexadecenal | 120 | 70 |
| Z-9-Hexadecenal | 100 | 40 |
| Z-14-Methyl-8-Hexadecenal | 60 | 30 |

[a]Relative light intensities with 100 pmoles of aldehyde in the dithionite assay.

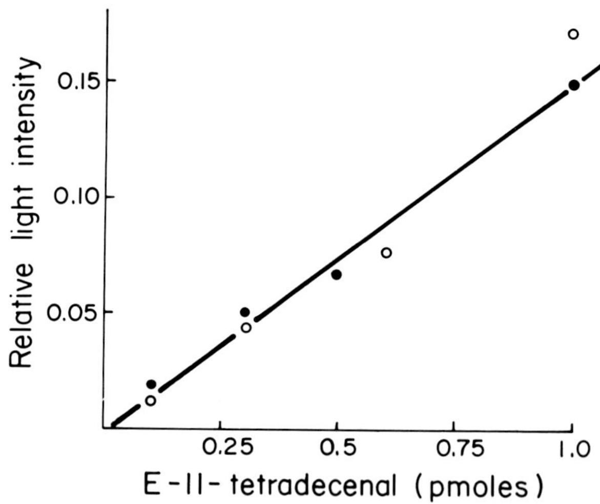


FIGURE 1. Bioluminescent response of B. harveyi (●) or P. phosphoreum (o) luciferase (5 μg) in the dithionite assay to different amounts of the pheromone of the spruce budworm, E-11-tetradecanal. The light intensity was corrected for endogenous luminescence in the absence of added pheromone. The relative responses are dependent on the amount of luciferase in the respective assays.

Figure 1 gives a plot of the bioluminescent responses for two different bacterial luciferases versus the amount of E-11-tetradecenal analyzed. For both luciferases, a linear relationship is obtained between the quantity of pheromone present and the bioluminescent response showing that the bioluminescent assay can easily be used for quantitative analysis. Furthermore, amounts of 11-tetradecenal (the sex pheromone of the spruce budworm) as low as 100 femtomoles can be quantitated using the standard curve, demonstrating that the sensitivity of the assay is extremely high.

### *D. Analysis of Aldehyde Pheromones in Insects*

The sex pheromone of female moths, which attracts the male moth of the same species, is released from a dermal gland located in the intersegmental fold in the posterior abdominal segments. Because these glands often contain about 4 pmoles or more of pheromone on their surface, the bioluminescent assay can be used to detect and measure the pheromone levels in the gland of an individual female moth. Table III gives the average pheromone level in the glands of the female moths of the eastern and western spruce budworm, and the navel orangeworm determined by analyses of individual glands. E-11-Tetradecanal was used as a standard. The results are in good agreement with the levels determined independently by gas liquid chromatography for the eastern spruce budworm (14) and the navel orangeworm (5). We have also recently detected the aldehyde pheromone in the gland of individual female moths of

*TABLE III. Aldehyde Pheromone Levels in Insects*[a]

| *Insect* | *Number Analyzed* | *Average (pmoles/gland)* |
|---|---|---|
| *Eastern Spruce Budworm* | *15* | *10.4* |
| *Western Spruce Budworm* | *10* | *13.5* |
| *Navel Orangeworm* | *14* | *4.0* |

[a]*Heptane extracts of the glands of individual female moths were analyzed by the bioluminescent assay with E-11-tetradecenal as a standard.*

the corn earworm. Analysis of other nonglandular parts of the same insects or alternatively glands of insects that are believed not to contain long chain aldehydes as part of their pheromone, produced very low bioluminescent responses. The sensitivity of the bioluminescent assay will allow us to study the regulation of pheromone levels and the synthesis of pheromones in these pest species.

### *E. Bioluminescent Analysis of Airborne Pheromone*

The analysis of airborne pheromone is required not only to determine the release rates of pheromones from insects but also, and of greater importance, measuring the rates of release and dispersal of pheromone from lures used in traps and from various commercial encapsulated formulations used for mating disruption experiments. This would allow a comparison between pheromone release characteristics and the

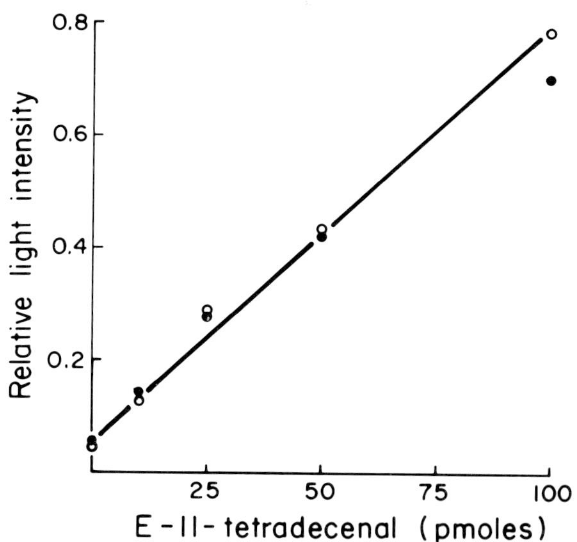


*FIGURE 2. Analysis with B. harveyi (●) or P. phosphoreum (o) luciferase of different amounts of E-11-tetradecenal bound to Porapak Q. The aldehyde pheromone, was extracted with hexane from Porapak Q, the hexane removed by evaporation, and the pheromone dissolved in water before analysis of an aliquot of the extract ( 6%) in the bioluminescent assay.*

behavioral effects obtained on the moths. Application of the bioluminescent assay for this purpose, however, requires a method for quantitatively trapping the pheromone from air that allows subsequent analysis by the bioluminescent assay.

One approach that we are developing is to pass the air containing the aldehyde pheromone through Porapak Q and then extract the absorbed pheromone with an organic solvent (6). Figure 2 shows some initial experiments involving hexane extraction and analysis by the bioluminescent assay of fixed amounts of the aldehyde pheromone, (E)-11-tetradecenal, bound to Porapak Q. The bioluminescent responses with two different luciferases is linearly dependent on the total amount of pheromone bound to the Porapak Q. As little as 10 pmoles was detected in these particular analyses. Recently we have demonstrated that high recoveries of aldehyde pheromone (>50%) can be obtained on absorbing the aldehyde onto Porapak Q from a stream of air and that this approach can be used to detect the release of pheromone from the female moth of the spruce budworm.

## II. SUMMARY AND CONCLUSIONS

The bioluminescent reaction catalyzed by bacterial luciferase provides an extremely sensitive, rapid, and quantitative assay for aldehyde pheromones obtained from some of the most serious insect pests. The assay should be a powerful analytical tool for analyzing the pheromone levels in air as well as in insects.

## REFERENCES


(1) Butenandt, A., R. Beckmann, D. Stamm, and E. Hecker, *Z Naturforsch.* 14b, 283 (1959).

(2) Inscoe, M.N., and M. Beroza, in "Pest Management with Insect Sex Attractants" (M. Beroza, ed.), p. 145. ACS Symposium Series 23, American Chemical Society, Washington, D.C. (1976).

(3) Sanders, C.J. and J. Weatherston, *Can. Entomol.* 108, 1285 (1976).

(4) Klun, J.A., J.R. Plimmer, B.A. Bierl-Leonhardt, A.N. Sparks, and O.L. Chapman, *Science (N.Y.)* 204, 1328 (1979).

(5) Coffelt, J.A., K.W. Vick, P.E. Sonnet, and R.E. Doolittle, *J. Chem. Ecol.* 5, 955 (1979).

(6) Cross, J.H., R.C. Byler, R.F. Cassidy, Jr., R.M. Silverstein, R.E. Greenblatt, W.E. Burkholder, A.R. Levinson, H.Z. Levinson, *J. Chem. Ecol.* *2,* 457 (1976).

(7) Nesbitt, B.F., P.S. Beevor, D.R. Hall, and R. Lester, *J. Insect Physiol. 25, 535 (1979).*

(8) Hill, A.S., R.T. Cardé, H. Kido, and W.L. Roelofs, *J. Chem. Ecol.* *1,* 215 (1975).

(9) McDonough, L.M. and J.A. Kamm, *J. Chem. Ecol.* *5,* 211 (1979).

(10) Kaissling, K.E., G. Kasang, H.J. Bestmann, W. Stransky, and O. Vostrowsky, *Naturwissenschaften* *65,* 382 (1978).

(11) Ulitzur, S. and J.W. Hastings, *Proc. Natl. Acad. Sci. USA* *75,* 266 (1978).

(12) Shimomura, O., F.H. Johnson, and H. Morise, *Proc. Natl. Acad. Sci. USA* *71,* 4666 (1974).

(13) Riendeau, D., and E. Meighen, *J. Biol Chem* *254,* 7488 (1979).

(14) Silk, P.J., S.H. Tan, C.J. Wiesner, R.J. Ross, and G.C. Lonergan, *Environ. Entomol.* (in press).

## ACKNOWLEDGMENTS

Work leading to this publication was funded in part by a USDA Forest Service program entitled "Canadian/United States Spruce Budworm Program and in part by a grant (MT-4314) from the Medical Research Council of Canada.

# THE USE OF THE BACTERIAL LUMINESCENT SYSTEM FOR THE QUANTITATIVE DETERMINATION OF DEHYDROGENASES

H. Watanabe
J. W. Hastings

The Biological Laboratories
Harvard University
Cambridge, Massachusetts

The bacterial bioluminescent system has been employed for analytical measurements in a number of different ways, both *in vitro* and *in vivo*, including the assay of a variety of different enzymes, metabolites, drugs and other substances. A number of contributions in two recent volumes (1,2) can serve as an excellent guide to the literature up to that time.

In the studies presented in this abstract we utilized a novel bacterial luciferase-linked assay system in order to determine the concentrations of several NAD-dependent dehydrogenases. In order to avoid interference, we employed a purified NADPH-specific dehydrogenase (FMN reductase) to provide $FMNH_2$ for the luciferase. This was isolated from the marine luminous bacterium *Beneckea harveyi*, mutant strain MB20 (3), purified as described by Jablonski and De Luca (4). Although this NADPH-FMN reductase had a detectable amount of NADH:NADPH-oxidase activity, it was too little to interfere with the assays at the levels being examined. The reactions employed were as follows:

1. $X_{ox} + NADH + H^+ \xleftrightarrow{DH} X_{red} + NAD^+$

2. $C_{12}\text{-RCOH} + NAD^+ \xleftrightarrow{ADH} C_{12}\text{-RCHO} + NADH$

3. $NADPH + FMN \xleftrightarrow{R'ase} NADP + FMNH_2$




ISBN 0-12-208820-4

4. $FMNH_2 + C_{12}\text{-}RCHO + O_2 \xrightarrow{L'ase} h\nu + C_{12}\text{-}RCOOH + H_2O + FMN$

$X_{ox}$ is the substrate of the NADH specific dehydrogenase (DH) being assayed. Dehydrogenase activity in this reaction (1) results in the production of $NAD^+$, which then stoichiometrically converts long chain alcohol (dodecanol; $C_{12}$-RCOH) to aldehyde (dodecanal; $C_{12}$-RCHO) in reaction (2), catalyzed by yeast alcohol dehydrogenase (ADH). $FMNH_2$ is produced in reaction (3) by the purified NADPH-FMN reductase (R´ase), and the quantity of RCHO produced in reaction (2) is then monitored as light emission in reaction (4), catalyzed by bacterial luciferase (L´ase).

The quantities of luciferase and NADPH-FMN reductase were chosen so that in the test there would be a steady light emission in the coupled assay in the presence of saturating levels of aldehyde. In the experiments reported here we used 8 μgm/ml luciferase and 4 μgm/ml of reductase; alcohol dehydrogenase was 5 μgm/ml in the assays for glutamate dehydrogenase (GlDH) and malate dehydrogenase (MDH), and 50 μgm/ml for lactate dehydrogenase (LDH).

The 12 carbon alcohol (dodecanol) was found to be preferable to the 10 carbon compound (decanol) by virtue of the fact that the light output decayed less rapidly. The fact that such long chain alcohols compete with the long chain aldehydes in binding the luciferase peroxyflavin intermediate (6,7) introduces complications concerning the detailed interpretation of the reaction kinetics. However, the alcohols were evaluated in the complete systems (reactions 1 to 4) so that it was not possible to dissect effects on the last step.

With all of the dehydrogenases studied there was a distinct, sometimes long lag in the development of light emission. This is illustrated in Figure 1 for LDH. This lag was dependent upon the dehydrogenase concentration and ranged from a fraction of a second with 50 μgm/ml LDH to almost an hour with 1 ngm/ml LDH (Table 1). From the subsequent increase in light intensity, attributed to the production of aldehyde, a maximum rate of light intensity increase could be measured (Table 1). This served as a good measure of the dehydrogenase concentration (Figure 2). As shown, this was linear for LDH over a concentration range of 1 ngm/ml to 1 μgm/ml. For GlDH, the linear range was 10 ngm/ml to 5 μgm/ml and for MDH, 0.5 ngm/ml to 100 μgm/ml. In the assay of MDH, the substrate oxaloacetic acid was produced continuously during the assay by using the glutamate-oxaloacetic acid transaminase system.

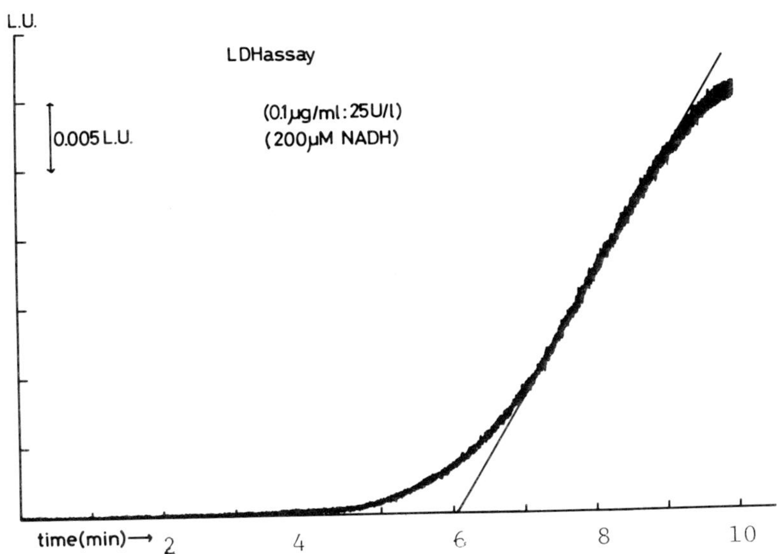


*FIGURE 1. Time course of light output in the assay of lactate dehydrogenase. Beef heart lactate dehydrogenase (LDH) was measured by using the bacterial luciferase-coupled assay system at 25°C. After a lag time of several minutes (see Table I.), an increase in light intensity occurred, whose slope (maximum increase of light intensity) was plotted against dehydrogenase concentration as in Figure 2. 100 ng/ml (25 U/l of LDH were assayed in presence of 200 µM NADH. Assay mixtures are described in legend to Figure 2. Ordinate, light intensity in light units (L.U.); abscissa, time (min.).*

The light emission in the coupled reactions 3 and 4 was high enough (with saturating aldehyde) to measure the different dehydrogenase activities with the soluble enzyme systems. However, a more stable and kinetically longer-lived light emission was obtained using immobilized luciferase and NADPH-FMN reductase (Figure 3). The immobilization of ADH as well gave an equally good, even somewhat better light emission, with regard to intensity and stability.

These two immobilized enzyme preparations were tested in the LDH assay (Figure 4). Even though these provided a more stable and long lived light emission, the assayable range was not very different. On the other hand, the immobilized system might have utility and convenience which would make it superior.

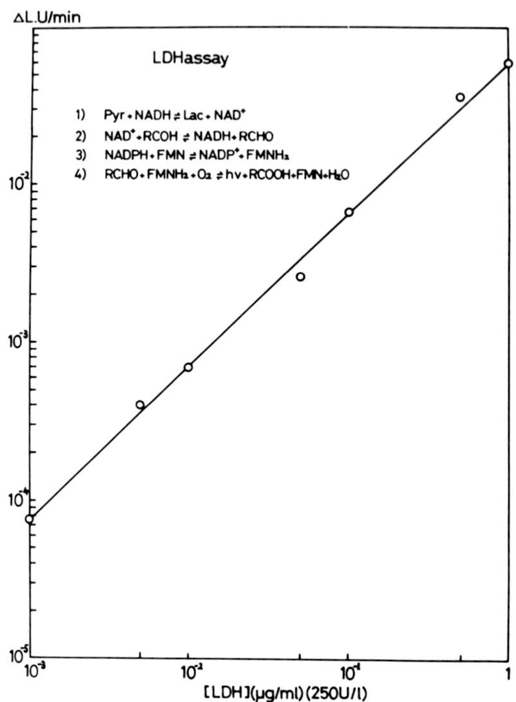


*FIGURE 2. Relation between LDH concentration and light emission. Slopes, as illustrated in Figure 1, were plotted against LDH concentration. Ordinate, maximum light increase rate (L.U./min.); abscissa, LDH concentration. Substrates used in this LDH assay were as follows: 0.01% dodecanol, 2.5 μM FMN, 0.6 mM pyruvate, 0.2 mM NADPH, and 0.2 mM NADPH. Other enzymes used were 50 mg/ml yeast alcohol dehydrogenase (ADH), 4 mg/ml NADPH-FMN reductase and 8 mg/ml luciferase in 50 mM phosphate buffer pH 7.4 containing 0.01% bovine serum albumin (BSA). Reaction was initiated with the addition of 0.1 ml of LDH dilutions in 0.1 M phosphate buffer, pH 7.4, into 0.9 ml of reaction mixtures at 25°C. The yeast alcohol dehydrogenase (300 U/mg), as well as LDH, GlDH, MDH and GOT (from Boehringer-Mannheim), were supplied suspended in 3.2 M $(NH_4)_2SO_4$ solution, and were used without further purification and added without eliminating $(NH_4)_2SO_4$.*

TABLE 1. *Effects of LDH Concentration*

| LDH Conc. | Lag time (min)[a] | Slope[a,b] |
|---|---|---|
| 50 mg/ml | .008 | 140.0 |
| 10 | .04 | 100.0 |
| 5 | .13 | 96.0 |
| 1 | .63 | 59.0 |
| 500 ng/ml | 1.6 | 37.0 |
| 100 | 6.1 | 6.9 |
| 50 | 13.4 | 2.5 |
| 10 | 23.3 | 0.7 |
| 5 | 29.0 | 0.4 |
| 1 | 45.5 | 0.075 |

[a]Refer to Figure 1.
[b]The maximum rate of increase of light intensity, in milli light units/min.

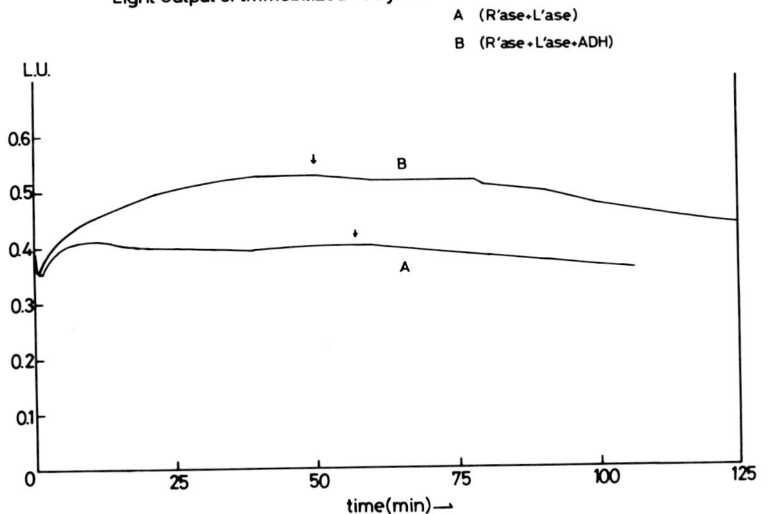


FIGURE 3. *Time course of light output with immobilized enzymes. (A) NADPH-FMN reductase (0.1 mg/ml) and luciferase (1 mg/ml) were immobilized on Sepharose 6B or (B) the same together with ADH (2 mg/ml). These were suspended in 0.1 M phosphate buffer, pH 7, and used to initiate the reaction by adding to a reaction mixture containing 0.01% decanal, 0.2 mM NADPH, 2.5 μM FMN and 75 mM Phosphate buffer, pH 7, containing 0.02% BSA.*

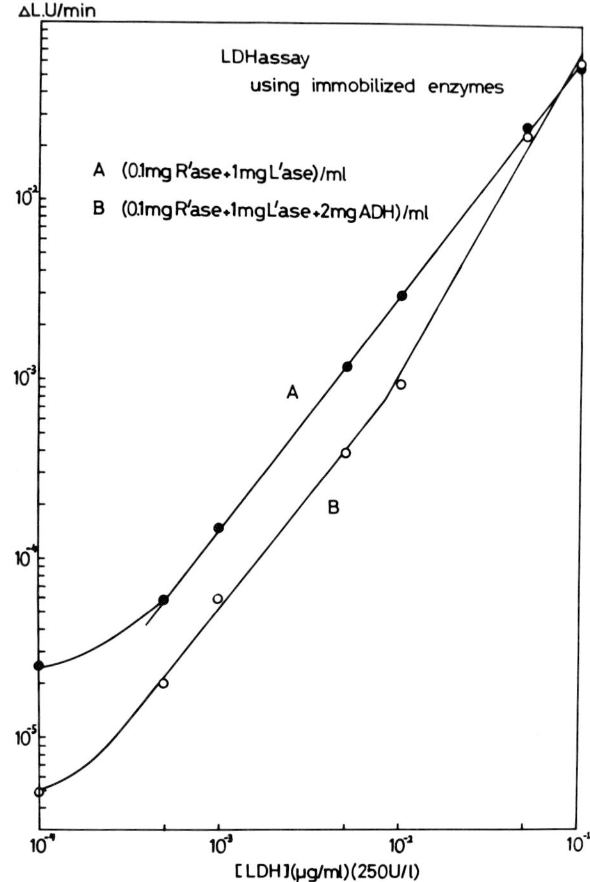


*FIGURE 4. LDH activities were assayed by using enzymes immobilized on Sepharose 6B, the same samples described in Figure 3. (A) NADPH-FMN reductase and luciferase and (B) the same along with yeast ADH. Assay mixtures for LDH were as follows: 50 mM phosphate buffer, pH 7.4, containing 0.01% BSA, 0.1% dodecanol, 2.5 μM FMN, 0.6 mM pyruvate, 0.2 mM NADH, 0.2 mM NADPH, and different concentrations of LDH. When the LDH assay was done using immobilized Sample A, soluble yeast ADH (50 mg/ml) was added. The reactions were initiated by the addition of the immobilized enzyme preparations.*

We believe that these assay procedures will find useful applications in both laboratory and clinical work.

## REFERENCES

1. De Luca, Marlene (ed.), "Methods in Enzymology, 57," 653 pp. Academic Press, New York (1978).
2. Schram, Eric, and Philip Stanley (eds.), "Proceedings of the International Symposium on Analytical Applications of Bioluminescence and Chemiluminescence," 696 pp. State Printing & Publ. Inc., Westlake Village, CA (1979).
3. Jablonski, E., and M. DeLuca, *Biochemistry 17,* 672 (1978).
4. Waters, C. A., and J. W. Hastings, *J. Bact. 131,* 519 (1977).
5. Hastings, J. W., T. O. Baldwin, and M. Z. Nicoli, *Methods in Enzymology 57,* 135 (1978).
6. Hastings, J. W., Q. H. Gibson, J. Friedland, and J. Spudich, *Bioluminescence in Progress 151,* (1966).
7. Tu, S. C., *Biochemistry 18,* 5940 (1979).

# ACTIVITY COUPLING BETWEEN BACTERIAL LUCIFERASE AND FLAVIN ADENINE DINUCLEOTIDE-DEPENDENT SALICYLATE HYDROXYLASE[1]

Shiao-Chun Tu

Department of Biophysical Sciences
University of Houston
Houston, Texas

## I. INTRODUCTION

The bacterial luciferase-catalyzed bioluminescence system is excellent for many analytical applications. Sensitive methods have been devised, with or without coupling with other enzymes, for the determinations of $NAD(P)^+$ and NAD(P)H (1-3), FMN (1.4), proteases (5), oxygen (6), myristic acid and long-chain aldehydes (7), and either substrates (2) or enzymes (8) that lead to the utilization or production of NAD(P)H.

Since bacterial luciferase is highly specific for $FMNH_2$ [2] (9-11), this luminescent system has so far not been successfully applied to the assay of $FAD(H_2)$ or enzymes that require this cofactor. In the one case that FAD was quantitated, the method required a prior hydrolysis of FAD to FMN(4). In the present study we have explored the possibility of coupling bacterial luciferase activity to FAD-dependent enzymes. The successful coupling of luciferase to salicylate hydroxylase is detailed.

[1]*This work was supported by Robert A. Welch Foundation Grant E-738 and National Institute of General Medical Science Grant GM 25953.*

[2]*Abbreviations used: $FMNH_2$ and $FADH_2$, reduced FMN and FAD, respectively; MA and OA, salicylate monooxygenation activity and NAD(P)H oxidation activity, respectively.*




ISBN 0-12-208820-4

## II. MATERIALS AND METHODS

Luciferase (12) and salicylate hydroxylase (13) were purified from cells of *Beneckea harveyi* and *Pseudomonas cepacia*, respectively. Concentrations of salicylate hydroxylase and luciferase were determined by the method of Lowry (14) and based on an absorption coefficient of 1.2 (0.1%, 1 cm) at 280 nm (15), respectively.

One unit of the monooxygenase activity (MA) of salicylate hydroxylase is 1 μmole catechol formation per min and that of the NAD(P)H oxidation activity (OA) is for 1 μmole $NAD(P)^+$ formation per min. Based on known extinction coefficients for $NAD(P)^+$, NAD(P)H, salicylate, and catechol, both MA and OA were measured at 23°C spectrophotometrically at 296 and 340 nm, respectively, in 1 ml 0.05 M phosphate, pH 7, containing salicylate hydroxylase, salicylate, NAD(P)H and, in some cases, flavin. The fractional uncoupling factor is defined as (OA-MA)/OA. In salicylate hydroxylase-luciferase coupled reactions, the light emission was measured using a calibrated photometer (16).

## III. RESULTS

The addition of riboflavin, FMN, or FAD resulted in enhanced activities of NADH oxidation but decreased activities of salicylate monooxygenation for salicylate hydroxylase (Fig. 1). The efficiencies of uncoupling were dependent upon flavin concentrations and were in the order of riboflavin > FMN > FAD. Similar results were also observed with NADPH as a reducing substrate.

A coupled bioluminescence occurs when FMN, decanal, bovine serum albumin and luciferase are added to the salicylate hydroxylase reaction solution (Fig. 2). Substituting NADH with NADPH resulted in changes in the kinetics but not the total light output of the coupled bioluminescence. The omission of bovine serum albumin, which is known to enhance the *in vitro* light emitting activity of *B. harveyi* luciferase (17), caused a 5-fold decrease in the light yield. Approximately 2% of light was detected if salicylate of salicylate hydroxylase was absent, most likely due to a slow non-enzymatic reduction of FMN by NAD(P)H. The omission of any other reacting component resulted in no light emission.

When FMN, salicylate, or NADH was each kept limiting, the total quantum output of the coupled bioluminescence was linearly proportional to the amount of the limiting component

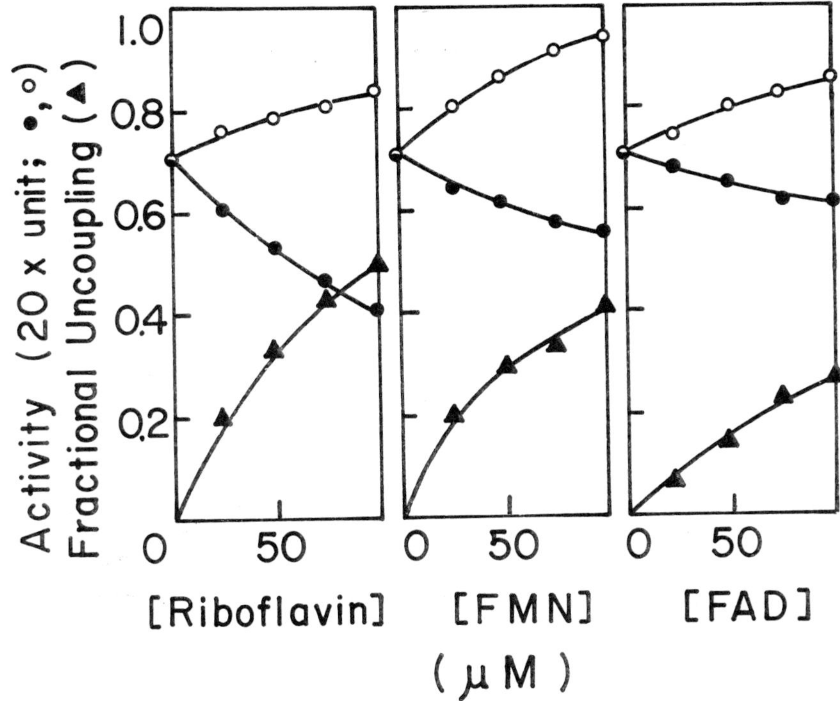


*FIGURE I. Effects of exogenous flavins on salicylate hydroxylase activities. Reactions were carried out at 23°C in 1 ml 0.05 M phosphate, pH 7, containing 3 μg salicylate hydroxylase, 0.25 mM salicylate, 80 μM NADH, and various amounts of flavins as indicated. Symbols are: NADH oxidation activity, (O); salicylate monooxygenation activity, (●); fractional uncoupling (▲).*

(Figure 3 A-C). When salicylate hydroxylase was limiting, the light intensity quickly reached a maximum but decreased very slowly. There was also a significant background emission. After corrections for this background, emission maximal intensities exhibited a good linear relationship with the amounts of hydroxylase used (Figure 3D). The background light can also be significantly reduced by using less NADH in the assay without compromising the sensitivity.

When titrated at various luciferase concentrations, the plot of total quantum output versus luciferase concentration exhibited a typical saturation curve. The maximal light output was determined to be 1.9 x $10^{15}$q, under conditions of limiting NADH (80 μM), by extrapolation to infinite luciferase concentration. Using the spectrophotometric assay,

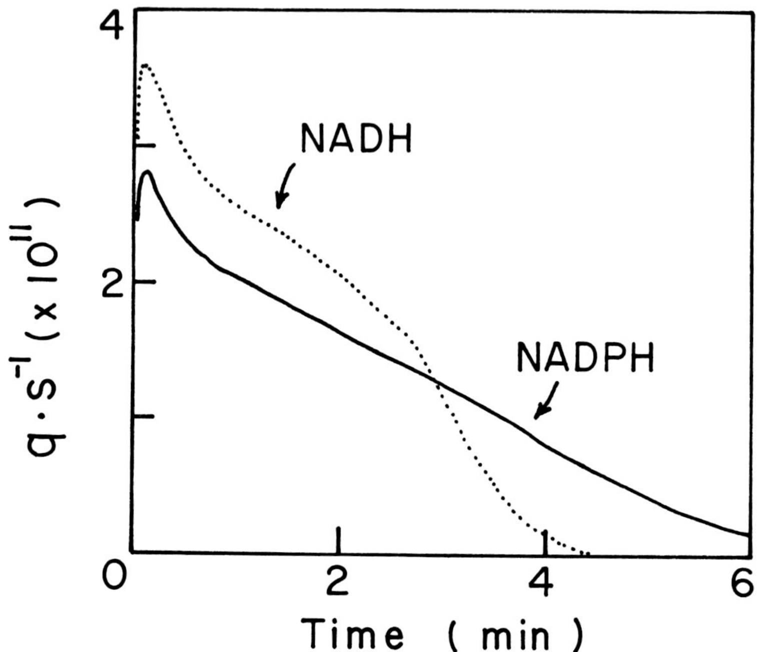


*FIGURE 2. Intensity and kinetics of salicylate hydroxylase-luciferase coupled bioluminescence. Reaction was carried out at 23°C in 1 ml 0.05 M phosphate, pH 7, containing 50 μg luciferase, 2 mg bovine serum albumin, 50 μM decanal, 50 μM FMN, 3 μg salicylate hydroxylase, 0.25 mM salicylate, and 80 μM NADH ( ····· ) or NADPH ( ——— ).*

70% of the total NADH oxidation was found to be channelled to the salicylate hydroxylation. The quantum yield of the coupled bioluminescence at saturating luciferase was thus calculated to be 0.14 based on the amount of NADH oxidized independently of catechol formation.

## IV. DISCUSSION

The reaction mechanism of salicylate hydroxylase involves the reduction of bound FAD by NAD(P)H, the subsequent oxygenation of the bound $FADH_2$, and finally the decarboxylative hydroxylation of salicylate and the regeneration of FAD (18). The addition of increasing amounts of exogenous riboflavin,

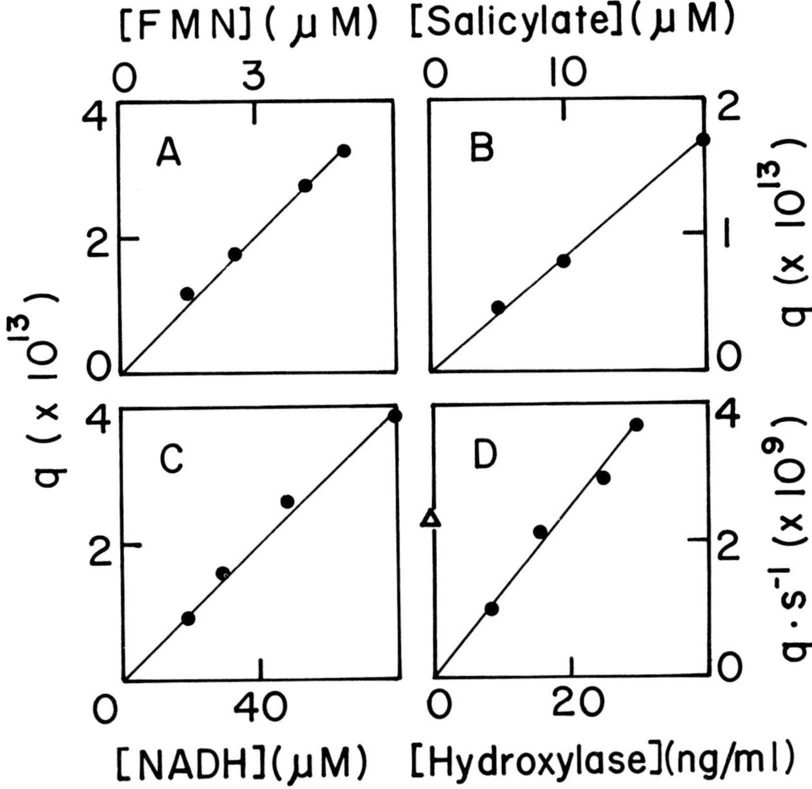


*FIGURE 3. The total light output (q) or the maximal intensity ($q \cdot s^{-1}$) of the coupled bioluminescence reaction in which each of the four components was varied. The reaction conditions were the same as that described in Figure 2. In D, the background light intensity without salicylate hydroxylase (Δ) was subtracted from each measurement.*

FMN, or FAD was found to result in increasing extents of NADH oxidation independent of the salicylate to catechol conversion (Fig. 1). We postulate that the added flavin reacts, in competition with $O_2$, with the bound $FADH_2$ in a reducing equivalent exchange reaction resulting in the regeneration of oxidized holoenzyme, without catechol formation, and free reduced flavin. If FMN is used, the FMNH so formed can then be coupled to bacterial bioluminescence. Since the hydroxylase-bound FAD is only very slowly reduced by NAD(P)H in the absence of salicylate (18), a normal turnover condition is required for an effective activity coupling.

The quantum yield of 0.14 for the coupled bioluminence, based on the quantity of NADH oxidized independently of catechol formation, correlated closely to the known quantum yield of about 0.17 for luciferase based on limiting $FMNH_2$ (19), thus strongly supporting the mechanism postulated. Furthermore, the hydroxylase-bound $FADH_2$ has to be accessible to the added flavin for any reducing equivalent exchange to occur. Based on the observed order of riboflavin > FMN > FAD in their efficiencies of uncoupling (Fig. 1), the salicylate hydroxylase bound $FADH_2$ must be partially exposed during catalysis and its interactions with added flavins may be sensitive to steric and/or charge effects.

The coupled bioluminescence system was shown (Figs. 2, 3) to be useful in quantitating FMN, NAD(P)H, salicylate, and salicylate hydroxylase. In the last case, the coupled bioluminescence assay, under our experimental conditions, was $\geq$ 10-fold more sensitive than the corresponding spectrophotometric assay, and can probably be made much more sensitive under optimal conditions and using a more sensitive photometer.

The technique of coupled bioluminescence demonstrated for salicylate hydroxylase should be, in principle, applicable to other flavoenzymes if the bound reduced flavin cofactors formed during catalysis are accessible to the added FMN. Along this line, D-amino acid oxidase, L-amino acid oxidase, glucose oxidase, and lipoamide dehydrogenase were tested but no significant coupled bioluminescence was observed. On the other hand, anaerobic reductions of added FAD by the FAD-dependent p hydroxybenzoate (20), orcinol (21), and phenol (22) hydroxylases in the presence of NAD(P)H and their respective substrates have been reported. A reducing equivalent exchange has also been recently observed between an FAD-dependent dioxygenase and added flavins (23). It is quite likely that these FAD-dependent enzymes can be similarly coupled to bacterial bioluminescence reaction.

## V. SUMMARY

The NAD(P)H oxidation and substrate monooxygenation activities of the FAD-dependent salicylate hydroxylase can be uncoupled by added flavins, due to a reducing equivalent exchange between the bound $FADH_2$ and the added flavins. When FMN was added, the salicylate hydroxylase activity could be coupled to bacterial bioluminescence reaction. The quantum yield of the coupled bioluminescence, based on the amount of NADH oxidized independently of salicylate monooxygenation, was determined to be 0.14 correlating closely with the known quantum yield of about 0.17 for luciferase based on limiting $FMNH_2$. Potential analytical applications of this coupled bioluminescence system are discussed.

## REFERENCES


1. Stanley, P. E., *Anal. Biochem.* 39, 441 (1971).
2. Brolin, S. E., E. Borglund, L. Tegner, and G. Wettermark, *Anal. Biochem.* 42, 124 (1971).
3. Jablonski, E., M. DeLuca, *Proc Natl. Acad. Sci. U.S.A.* 73, 3848 (1976).
4. Chappelle, E. W., and G. L. Picciolo, *Methods in Enzymology* 18B, 381 (1971).
5. Njus, D. T. O. Baldwin, and J. W. Hastings, *Anal. Biochem.* 61, 280 (1974).
6. Chance, B., and R. Oshino, *Methods in Enzymology* 57, 223 (1978).
7. Ulitzur, S., and J. W. Hastings, *Methods in Enzymology* 57, 189 (1978).
8. Stanley, P. E., *Methods in Enzymology* 57, 181 (1978).
9. Mitchell, G. and J. W. Hastings,*J. Biol. Chem.* 244, 2572 (1969).
10. Meighen, E. A., and R. E. MacKenzie,*Biochemistry* 12, 1482 (1973).
11. Tu, S. -C., J. W. Hastings, and D. B. McCormick, *Fed. Proc.* 36, 722 (1977).
12. Gunsalus-Miguel, A., E. A. Meighen, M. F. Nicoli, K. H. Nealson, and J. W. Hastings, *J. Biol. Chem.* 247, 398 (1972).
13. White-Stevens, R. H., and H. Kamin, *J. Biol. Chem.* 247, 2358 (1972).
14. Lowry, O. H., N. J. Rosebrough, A. L. Farr, and R. J. Randall, *J. Biol. Chem.* 193, 265 (1951).
15. Tu, S. -C., T. O. Baldwin, J. E. Becvar, and J. W. Hastings, *Arch. Biochem. Biophys.* 179, 342 (1977).
16. Mitchell, G., and J. W. Hastings, *Anal. Biochem.* 39, 243 (1971).
17. Cline, T. W., Ph.D. Thesis, Harvard University, Cambridge, MA. (1973).
18. White-Stevens, R. H., H. Kamin, and Q. H. Gibson, *J. Biol. Chem.* 247, 2371 (1972).
19. Becvar, J. E., and J. W. Hastings, *Proc. Natl. Acad. Sci. U.S.A.* 72, 3374 (1975).
20. Hosokawa, K., and R. Y. Stanier, *J. Biol. Chem.* 241, 2453 (1966).

21. Ohta, Y., I. J. Higgins, and D. W. Ribbons, *J. Biol. Chem.* 250, 3814 (1975).
22. Neujahr, H. Y., and K. G. Kjellen, *J. Biol. Chem.* 253, 8835 (1978).
23. Kishore, G., and E. E. Snell, *Biochem. Biophys. Res. Commun.* 87, 518 (1979).

## C. Firefly Luciferase and Applications

# ANALYTICAL ASPECTS OF THE FIREFLY LUCIFERASE REACTION KINETICS[1]

Mushtaq Ahmad
Eric Schram

Institute of Molecular Biology
Vrije Universiteit Brussel
Brussels, Belgium

## I. INTRODUCTION

Depending on the experimental conditions various types of curves can be obtained for the luminescence time-course of firefly luciferase (1-4) and their correct interpretation is therefore important for the analytical applications. In another publication (5) we have proposed a mathematical representation of the luminescence time-course obtained at constant substrate concentration. It is the purpose of the present paper to analyze further some factors affecting this time-course and their importance for the assay of ATP.

## II. MATERIALS AND METHODS

The reagents and experimental conditions were the same as described elsewhere (5).

[1]*This work was supported by a grant of the Belgian Government (Programmatie van het Wetenschapsbeleid).*




ISBN 0-12-208820-4

## III. RESULTS

### *A. Influence of Enzyme Concentration*

The first case to be considered is that in which the substrated are in excess over the enzyme in such a way that their concentration can be considered constant. In the presence of $10^{-6}$M ATP and 2 x $10^{-5}$M luciferin no change in the shape of the curves and the decay rates were observed when raising the enzyme concentration from $10^{-12}$M to $10^{-9}$M. As described elsewhere (5), in the above case the decay can be mathematically interpreted on the basis of a slow release of the end-product from the enzyme rather than on that of an accumulation of this end-product in the solution. (As may be deduced from the data under B, the consumption of ATP is not likely to affect its concentration significantly). The same type of curve is obtained whether ATP or luciferin is used as the limiting substrate. The faster decay rates observed at higher enzyme concentrations do not necessarily imply a faster rate of product inhibition. We could indeed observe that when the ratio of enzyme versus substrate increases and at high concentrations of the other substrate, the consumption of the substrate present in limiting amounts (ATP or luciferin) may become important enough to account for an exponential decrease of the luminescence, without appreciable enzyme inhibition (see

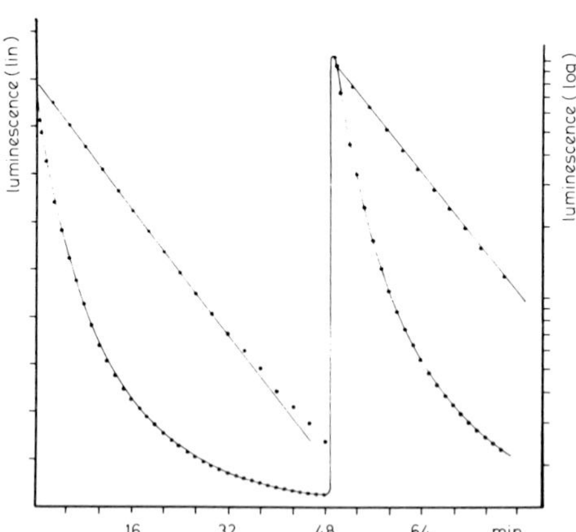


*Fig. 1. Exponential decay of the luminescence due to substrate consumption upon successive additions of ATP ($|E|$ = 4.8 x $10^{-8}$M, $|LH_2|$ = 4.1 x $10^{-4}$M, $|ATP|$ = 5.6 x $10^{-9}$M after a single addition).*

figure 1). At higher luciferase concentrations the luminescence curve may of course also be affected slightly by the corresponding increase of contaminating enzymatic impurities.

It may be anticipated that the luminescence curves observed in practice result from a combination of the several factors described above. According to these observations it does not seem advantageous to increase the enzyme concentration unduly as long as it remains compatible with the sensitivity of the instrument used for the assays. Moreover, higher enzyme concentrations will correspond to a higher inherent background light. When checking over the range $10^{-9}$ - $10^{-12}$M, luciferase showed no significant change in efficiency, in accordance with the results obtained by Denburg and McElroy (6) at much higher concentrations. This indicates that even at very low concentrations the suggested association of two luciferase molecules of molecular weight 50.000 is at least not rate limiting.

### *B. Turn-over of Luciferase*

The long lasting residual luminescence of luciferase preparations has for many years been ascribed to a slow turn-over of the enzyme. The low rate of this turn-over is obviously due, at least in part, to the slow release of the end-product responsible for the inhibition of the enzyme. In order to quantify the turn-over of luciferase, experiments were performed in the presence of $^3$H-ATP and the formed $^3$H-AMP assayed by I.E. chromatography. The following figures are the results of a typical experiment:

| \|E\| | \|$LH_2$\| | \|ATP\| | \|AMP\|[1] | \|AMP\|/\|E\| |
|---|---|---|---|---|
| $6.9\times10^{-9}$ | $1.4\times10^{-4}$ | $2.6\times10^{-6}$ | $1.2\times10^{-7}$ | 11.5 (corrected for blank) |
| $6.9\times10^{-9}$ | - | $2.6\times10^{-6}$ | $0.39\times10^{-7}$ | |

1. *Concentration attained after 35 minutes of incubation.*

### *C. Influence of Pyrophosphate and Triphosphate*

Even in their earliest experiments McElroy and coworkers (7) have already described the inhibiting action of pyrophosphate on the activation of luciferin and its accelerating effect on the release of the end-product. Although both phenomena seem to cancel each other as far as the overall reaction rate is concerned one can take advantage of the fact that the

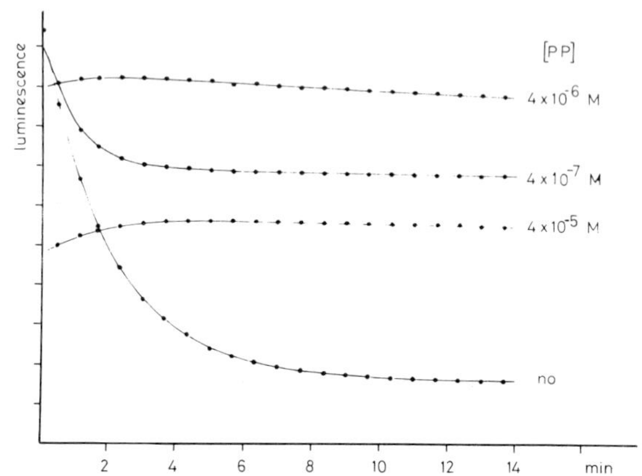


*Fig. 2. Effect of pyrophosphate on the luminescence time-course. ($|E|$ = 4 x $10^{-9}$M, $|LH_2|$ = 3 x $10^{-5}$M, $|ATP|$ = 1.3 x $10^{-6}$M)*

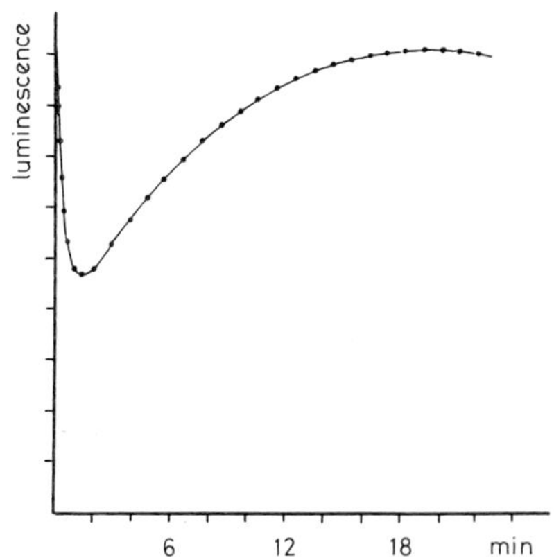


*Fig. 3. Effect of endogenously produced pyrophosphate on luminescence time-course ($|E|$ = 4.84 x $10^{-8}$M, $|LH_2|$ = 4.1 x $10^{-4}$M, $|ATP|$ = 2.87 x $10^{-5}$M).*

release of the end-product occurs at much lower pyrophosphate concentrations than the inhibiting of the activation step (see figure 2).

Addition of 4 x $10^{-6}$M pyrophosphate suppresses the end-product inhibition without appreciable loss of efficiency and appears therefore suitable for analytical purposes. The same effect can be achieved with triphosphate, although at higher concentration (2.5 x $10^{-4}$M), with the additional advantage that it is less easily hydrolyzed by pyrophosphate. As predicted by Cormier and Totter (8), at higher substrate and enzyme concentration and with no added pyrophosphate, sufficient endogenous pyrophosphate is made in order to produce a second rise of the luminescence after the initial peak (see figure 3).

### *D. Effect of Temperature*

The generally accepted value for the optimum temperature of firefly luciferase is about 25°C (9). Using our experimental conditions and detection equipment it was found to be around 19°C with ATP as rate-limiting substrate. Above this temperature the rate of the reaction keeps increasing indicating that the decrease in light yield is due to either a lower quantum efficiency or to a red shift of the emission spectrum resulting in a lower photomultiplier sensitivity. Checking the optimum temperature for individual instruments may therefore appear advisable.

### *E. Measurement of Km*

Because of the rapid decrease of the luminescence at higher ATP concentrations it is not always easy to deduce the initial velocity of the reaction from the luminescence peak. Addition of pyrophosphate may help to circumvent this difficulty (see C) by suppressing the end-product inhibition, but will also affect Km slightly. Better results were obtained by computing the parameters in the formula described in reference 5 and plotting $1/k_1$ (which is proportional to 1/v) against $1/|S|$ (see figure 4). Values of $k_2$ increase at high $|S|$, probably due to endogenous pyrophosphate formation.

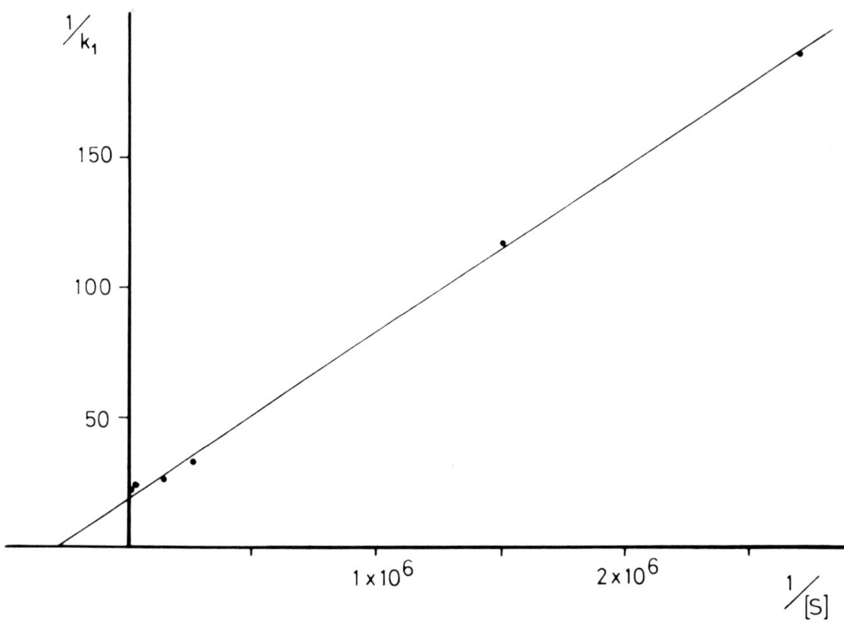


*Fig. 4. Measurement of Km for ATP.* ($|E| = 4 \times 10^{-9}M$, $|LH_2| = 1.17 \times 10^{-4}M$)

## IV. CONCLUSION

Depending on the relative as well as absolute concentrations of enzyme and substrates the decay of the firefly luciferase luminescence can be quantitatively interpreted either by the consumption of the limiting substrate or by the slow release of the end-product from the enzyme. Such considerations define the concentration range of reactants practicable for analytical assays. This range may be extended to higher ATP concentrations by the addition of substances as di- and triphosphate likely to help releasing the end-product without unduly affecting the first steps of the light reaction.

REFERENCES

1. Lundin, A. and A. Thore, *Anal. Biochem. 66,* 47 (1975).
2. Lundin, A., A. Rickardsson, and A. Thore, *Anal. Biochem. 75,* 611 (1976).
3. DeLuca, M., J. Wannlund, and W.D. McElroy, *Anal. Biochem. 95,* 194 (1979).
4. Lemasters, J.J. and C.R. Hackenbrock, *Eur. J. Biochem. 67,* 1 (1976).
5. Schram, E., M. Ahmad, and E. Moreels, these Proceedings.
6. Denburg, J.L. and W.D. McElroy, *Biochemistry 9,* 4619 (1970).
7. McElroy, W.D., J.W. Hastings, J. Coulombre, and V. Sonnenfeld, *Arch. Biochem. Biophys. 46,* 399 (1953).
8. Cormier, M.J. and J.R. Totter, *Photophysiology 4,* 315 (1968).
9. McElroy, W.D. and H.H. Seliger, *in* "Light and Life" (W.D. McElroy and B. Glass, eds.) p. 219. John Hopkins Press, Baltimore (1961).

# BIOLUMINESCENT SCREENING FOR BACTERIURIA

J. A. Lazaroni

Lazaroni Laboratories
Daly City, California

D. Linkley Henry

Process Instrumentation
SAI Technology Company
San Diego, California

## I. BIOLUMINESCENT SCREENING FOR BACTERIURIA BY FIREFLY LUCIFERASE-LUCIFERIN ATP ASSAY

### *A. Principle*

Patients with a progressive urinary tract infection will have a continuously increasing urinary bacterial count and increased ATP concentration in the urine specimen (1). The increase in bacterial count over a fixed time period will indicate possible bacterial infection (2). The bacterial count can be determined by bioluminescent measurement of ATP (3-6).

Urine may contain numerous cellular elements containing ATP. These elements will not increase and can be blanked out in determining increasing bacterial ATP indicating bacterial growth (increased bacterial counts).

### *B. Reagents*

(1) HEPES Buffer, 0.25 M, pH 7.75, "SAIT".
(2) Releasing Reagent, "SAIT".



ISBN 0-12-208820-4

(3) Luciferase enzyme reagent, "SAIT".
Reconstitute with 10 ml HEPES Buffer.
Keep refrigerated.

*C. Standardization Procedure for Bacterial Enumeration by ATP Measurement*

(1) Prepare a suspension of bacteria in HEPES Buffer of approximately $10^6$ org/ml. (3-5 colonies in 2 ml buffer).
(2) Make successive 10 fold dilutions in HEPES Buffer. 1:10; 1:100; 1:1,000; 1:10,000.
(3) Determine corrected luminescent readings on each dilution. Correct for HEPES Buffer blank.
(4) Prepare standard plate counts in duplicate for each dilution.
(5) After 18 hours incubation (35°C), determine numbers of org/ml by comparison of plates vs. dilution.

*D. Calculations*

Calculate K factor for each dilution. In linear range K factor will be equal.

$$K = \frac{\text{luminescent count}}{\text{\# bacteria}}$$

*E. Urine Screen Procedure*

(1) Mix fresh, clean catch urine sample.
(2) Prepare test and blank reaction tubes for each sample.
a. 25 microliters HEPES Buffer to each blank.
b. 50 microliters of releasing reagent to each test.
c. 25 microliters of sample to each test and blank.
(3) Process each pair of tubes (test and blank ) in order.
a. Place tube in photometer.
b. Start reaction with 200 microliters enzyme reagent.
c. Record lumins.
(4) Incubate urine sample 1 hour at 35°C. Repeat tests on incubated sample.

8/16/80

#1

BIOLUMINESCENSE URINE SCREENING
ATP/BACTERIA COUNT IN URINE

Patient #10-9664
Sample quantity: 20μl
Releasing reagent: 50μl
Luciferase-luciferin: 200μl

Microscopic: 4-6 WBC/HPF
Neg. RBC
Neg. Platelets
Neg. Epith. cells

Culture (18 hr. @ 35°C): Positive
>210,000 org/ml

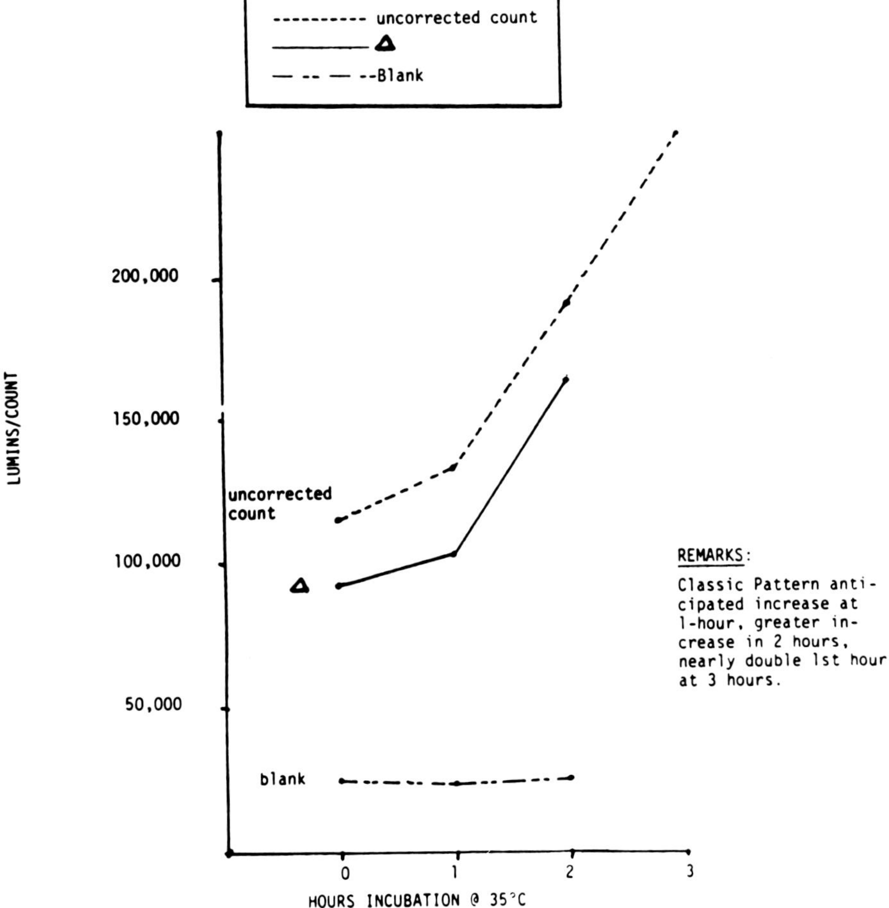

#2

BIOLUMINESCENSE URINE SCREENING

ATP/BACTERIA COUNT IN URINE

8/16/80

Patient #10-9583
Sample quantity: 25μl
Releasing Reagent: 50μl
Luciferin-Luciferase: 200μl

Microscopic: Neg, WBC
Neg. RBC
Neg. Platelets
Neg. Epith cells

Culture (18 hours @ 35°C)
Negative
$10^3$ org/ml

----------- uncorrected count
———————— Δ
— -- — -- Blank

100,000

REMARKS:
Negative urine, classic results. Contamination will frequently die off in urine due to poor quality of nutrient available. pH below 6.0 would also inhibit growth of organisms.

LUMINS/COUNT

50,000

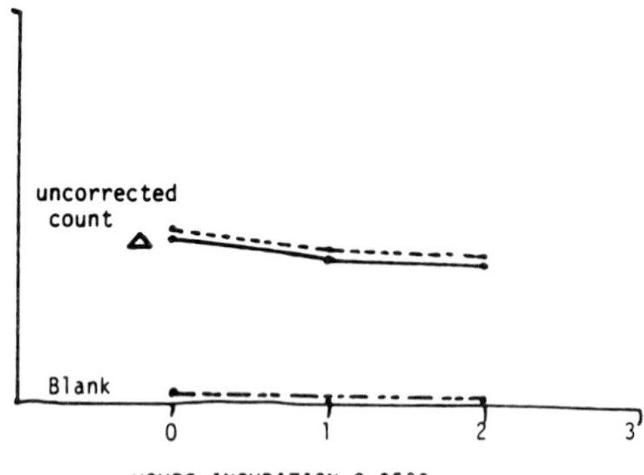


HOURS INCUBATION @ 35°C

8/16/80

#3

BIOLUMINESCENSE URINE SCREENING

ATP/BACTERIA COUNT IN URINE

Patient #10-9614
Sample quantity: 25μl
Releasing Reagent: 50μl
Luciferin-Luciferase: 200μl

Microscopic: 50+ WBC/HPF
Neg. RBC
Few Epith. cells

Culture (18 hr. @ 35°C)
Negative
± 1000 org/ml

| ----------- | uncorrected count |
| ———————— | Δ |
| — -- — -- | Blank |

REMARKS:
Less than 1000 count with large WBC population. High blank count. Low delta proved by negative culture.

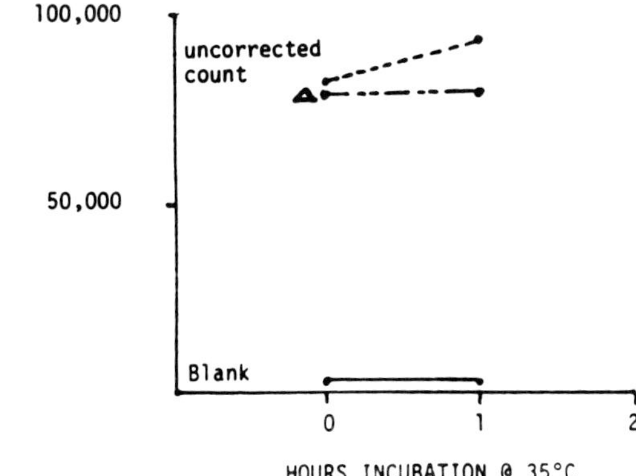

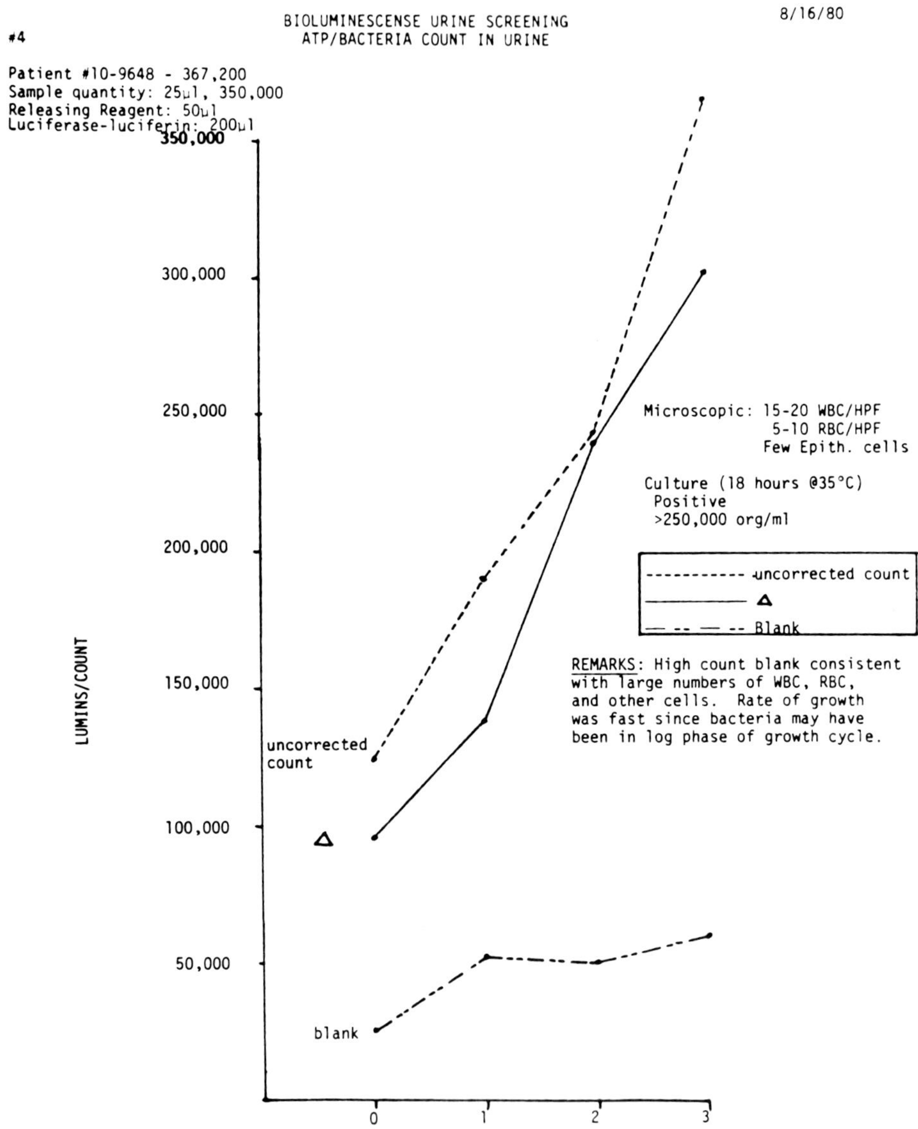

#4
BIOLUMINESCENSE URINE SCREENING
ATP/BACTERIA COUNT IN URINE
8/16/80
Patient #10-9648 - 367,200
Sample quantity: 25µl, 350,000
Releasing Reagent: 50µl
Luciferase-luciferin: 200µl
350,000
300,000
250,000
200,000
150,000
100,000
50,000
LUMINS/COUNT
uncorrected count
Δ
blank
0
1
2
3
HOURS INCUBATION @ 35°C
Microscopic: 15-20 WBC/HPF
5-10 RBC/HPF
Few Epith. cells
Culture (18 hours @35°C)
Positive
>250,000 org/ml
uncorrected count
Δ
Blank
REMARKS: High count blank consistent with large numbers of WBC, RBC, and other cells. Rate of growth was fast since bacteria may have been in log phase of growth cycle.

#5

8/16/80

BIOLUMINESCENSE URINE SCREENING
ATP/BACTERIA COUNT IN URINE

Patient #10-9796
Sample quantity: 20µl
Releasing Reagent: 50µl
Luciferin-Luciferase: 200µl

Microscopic: Neg. WBC
Neg. RBC
Neg. Epith. cells

Culture (18 hours @ 35°C)
Negative
<12,000 org/ml

| Line style | Legend |
| --- | --- |
| ----------- | uncorrected count |
| ———————— | Δ |
| — -- — -- | Blank |

REMARKS: Normal negative urine. This is a classic example of negative urines with no cells seen in microscopic.

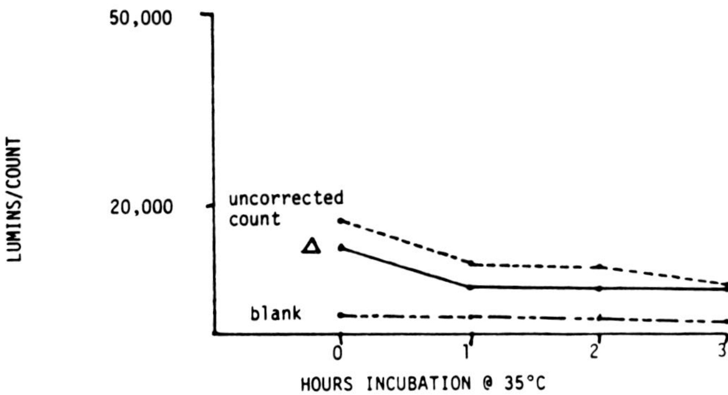

#7 8/16/80

BIOLUMINESCENSE URINE SCREENING

ATP/BACTERIA COUNT IN URINE

Patient #10-10383
Sample quantity: 25μl
Releasing Reagent: 50μl
Luciferin-Luciferase: 200μl

Microscopic: 6-10 WBC/HPF
0-2 RBC/HPF
3-5 Epith cells/HPF

Culture (18 hr. @ 35°C)
Negative
<1000 org/ml

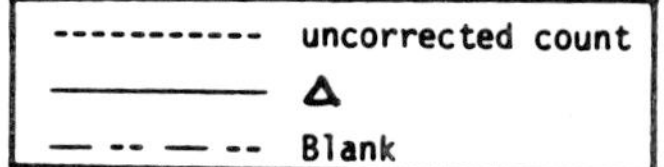


REMARKS: Classic negative. High blank count due to cells. When blank is 50% or more of total count, there is good reason to believe a large quantity of WBCs, RBCs, etc. are present in the urine sample.

The ATP appears to burn off by the end of the first hour.

WBC "seep" ATP over a long period of time 1-2 hours, thus keeping the blank relatively high.

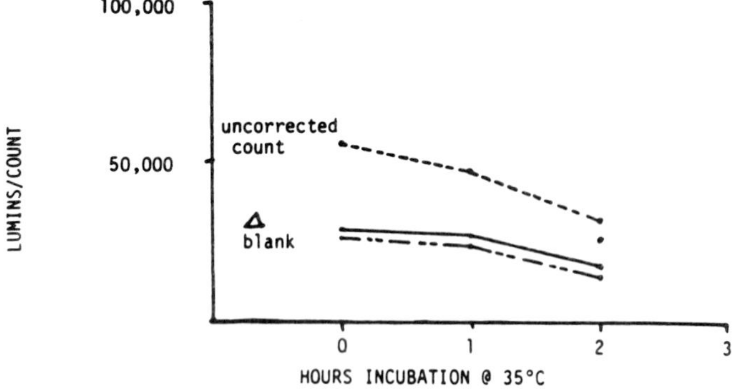

8/16/80

#6

BIOLUMINESCENSE URINE SCREENING
ATP/BACTERIA COUNT IN URINE

Patient #10-9797
Sample quantity: 20μl
Releasing Reagent: 50μl
Luciferin-Luciferase: 200μl

Microscopic; Neg, WBC
Neg. RBC
Few Epith, cells

Culture (18 hours @ 35°C)
Positive
160,000 org/ml

| | |
|---|---|
| ----------- | uncorrected count |
| ———— | Δ |
| — -- — -- | Blank |

REMARKS: Positive culture could be in lag phase of growth, No cells seen in clean catch urine sample (fresh),

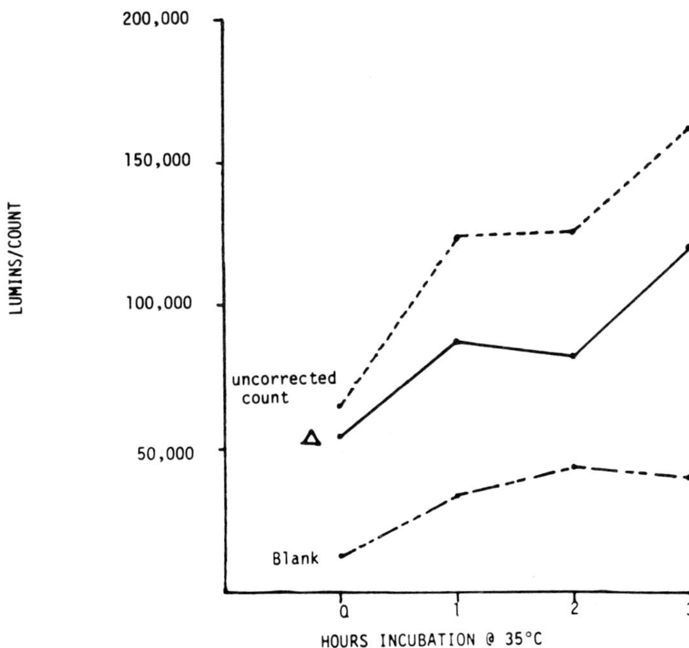

*F. Calculations*

(1) Org/ml = (test lumins-blank lumins) x (K).*
(2) Change in org/ml = fresh sample - incubated sample.

*G. Results*

An increase in org/ml indicates viable growing organisms in sample. Sample showing an increase of 20,000 or more org/ml should be submitted to further evaluation.

See Charts 1-7 for data from actual patient sample runs in a clinical laboratory. The laboratory had full control over method of obtaining a "clean catch" urine to assure minimal contamination of fresh urine specimen. Examples are representative of over 100 urine ATP/bacteria count assays.

*NOTE: ATP Per Bacteria = 1 x $10^{-9}$mcg/cell*

* *See Plate Count Standardization*

*NOTE: Streptococcus Fecalis (Group D) has a large ATP content. The Gram negative rods have comparatively much less.*

*NOTE: Candida Albican does not release ATP Like Bacteria -*

| *FIRST COUNT* | *ONE-HOUR INCUBATION* | *TWO-HOUR INCUBATION* |
|---|---|---|
| *55,600* | *54,500* | *47,000* |
| *BLANK* | *BLANK* | *BLANK* |
| *20,000* | *18,700* | *17,000* |

REFERENCES

1. Kass, E. H., *The Association of American Physicians 69*, (1956).
2. Merritt, A. D., and J. P. Sanford, *Journal of Clinical Laboratory Medicine 52* (1958).
3. Bush, V. N., Grace Lee Picciolo, and Emmett W. Chappelle, *Analytical Applications of Bioluminescence and Chemilumiescence 35*, NASA SP-388 (1975).
4. Kass, E. H., *AMA Archives of Internal Medicine 100* (1957).
5. Barry, A. L., P. B. Smith, and Turk, *Cumitech, American Society for Microbiology.*
6. Thore, A., Ansehn, A. Lundin, and S. Bergman, *Journal or Clinical Microbiology 1* (1975).

# PURIFICATION OF FIREFLY LUCIFERASE BY AMMONIUM SULPHATE PRECIPITATION AND ISOELECTRIC FOCUSING

Arne Lundin
Arne Myhrman

Bioluminescence Centre
LKB-Produkter AB
Bromma, Sweden

Gunilla Linfors

National Defense Research Institute
Umea, Sweden

## I. SUMMARY

A simple two-step procedure for purification of firefly luciferase from crude extracts of firefly tails by ammonium sulphate precipitation and isoelectric focusing is described. Particular emphasis is put on the removal of contaminating enzymes that can cause analytical interference in bioluminescence ATP monitoring according to Lundin et al. (1). The ammonium sulphate precipitate corresponding to 45-55% saturation is used and can be further purified without prior dialysis by isoelectric focusing in a sucrose density gradient. Luciferase precipitates at its isoelectric point and can be collected by centrifugation. The luciferase preparation obtained by this simple purification procedure appears as one band in SDS gel electrophoresis and analytical isoelectric focusing. The median overall yield from the crude extract is 65% and the increase of the specific luciferase activity is 8 fold. Contaminating enzymes are strongly reduced, e.g.




ISBN 0-12-208820-4

adenylate kinase is reduced 5 500 fold. Luciferase purified according to the present paper has been successfully used for ATP monitoring in a variety of applications and is now commercially available (LKB-Wallac, Turku, Finland).

## II. INTRODUCTION

The bioluminescent firefly luciferin luciferase reaction has been used for sensitive assays of ATP since 1947 when McElroy discovered that the reaction required ATP (2). It was shown that the reaction in principle could be used for assay of all metabolites and enzymes participating in ATP converting reactions (3). A number of these applications are described in a recent volume of Methods in Enzymology (4). However, a wide-spread use of assays based on the firefly reaction has been hampered by the lack of purified and standardized luciferase reagents. A particularly troublesome property of previously available reagents has been that the light was emitted as a brief flash. This was due to contaminating enzyme systems in the reagents and to product inhibition of luciferase (5-7).

Recently an improved technique for continuous monitoring of ATP converting reactions by the firefly reaction was described by Lundin et al. (1). Using luciferase purified by gel filtration it was possible, under certain reaction conditions, to obtain a stable light emission proportional to the ATP concentration. Only negligable amounts of ATP are degraded by the luciferase reagent and the addition of ATP results in a stable light emission rather than a flash. This means that the ATP concentration in ATP converting systems can be monitored simply by continuously measuring the light emitted in the firefly reaction.

The ATP monitoring technique described above has been used for a number of sensitive and convenient assays in ATP converting systems as reviewed by Lundin (8). In some applications the levels of adenylate kinase and other ATP converting enzymes contaminating the luciferase preparation obtained by gel filtration was found to be too high and the aim of the present work was to develop a method for preparation of large amounts of highly purified luciferase suitable for assays of enzymes and metabolites by ATP monitoring.

## III. MATERIALS AND METHODS

### A. *Chemicals*

Crude firefly lantern extract (FLE-50), D-luciferin (synthetic), bovine serum albumin, ATP and ADP were obtained from Sigma Chemical Co. GTP was obtained from Böhringer and Ampholine from LKB-Produkter AB (Bromma, Sweden).

### B. *Ammonium Sulphate Precipitation*

A crude lantern extract, 10 vials of FLE-50, corresponding to 500 mg of dried firefly lanterns, was dissolved in 20 ml distilled water. Particulate matter was removed by centrifugation (27,000 g, 4°C, 20 min).

Solid ammonium sulphate was added to the supernatant. Precipitates were collected by centrifugation after stirring for 10 minutes at room temperature. Ammonium sulphate concentrations were expressed in percent of a saturated solution containing 760 g/l.

### C. *Isoelectric Focusing*

The LKB 8100-1 Electrofocusing Column (volume 110 ml) was filled as recommended by the manufacturer. The electrode solutions were 1% $H_2SO_4$ in 46% sucrose and 0.8% KOH in distilled water, respectively. A density gradient was prepared using 50 ml of a solution containing 1.6% Ampholine (pH 5-7) (light solution) and 50 ml of a solution containing 1.6% Ampholine (pH 5-7), 40% sucrose, 1 mM EDTA and 1 mM dithiothreitol (heavy solution). The heavy solution also contained the 45-55% satured ammonium sulphate precipitate. The presence of EDTA and dithiothreitol in this solution counteracts inactivation of luciferase. After focusing (10°C, 60 h, 600 V, < 3 W) the gradient was eluted and collected in 2.5 ml fractions. Fractions with an activity higher than 50% of the peak luciferase activity were pooled.

During focusing most of the luciferase activity precipitates at its isoelectric point and may be collected by centrifugation (27,000 g, 4°C, 20 min). The precipitate was redissolved in 0.1 M tris acetate buffer, pH 7.4. Alternatively the pooled fractions were adjusted to pH 7.4 with 1 M KOH. Both treatments resulted in clear solutions.

### *D. Assay Procedures*

Bioluminescent assays were performed at room temperature using a previously described instrument (5) or the LKB-Wallac Luminometer 1250 (Turku, Finland). All enzyme activities except pyrophosphatase were measured in a solution containing 0.1% bovine serum albumin, 10 mM magnesium acetate, 0.14 mM luciferin and 0.07 M tris acetate buffer, pH 7.75, in a total volume of 1 ml. Nucleotides were dissolved in 0.1 M tris acetate, pH 7.75, containing 2 mM EDTA.

The luciferase activity was determined as the increase of light emission obtained with 20 nM ATP. The activity of adenylate kinase and nucleoside diphosphokinase were measured with the ATP monitoring technique (1). Thus the activities were calculated from the rate of the rise of the bioluminescence obtained after addition of 10 μM ADP and 10 μM ADP + 10 μM GTP, respectively. In fractions with low luciferase activity purified luciferase was added. The nucleoside diphosphokinase activity was corrected for adenylate kinase blank. The total activity of ATP degrading enzymes was measured as the rate of degradation of ATP using 20 μM ATP.

Pyrophosphotase was assayed (after removing interferring Ampholine by gelfiltration), by adding 0.2 ml of the various fractions to 2 ml 0.05 M imidazole buffer (pH 7.3), 0.2 ml 0.05 M pyrophosphate (pH 7.3) and 0.02 ml 0.5 M $MgCl_2$. After incubation (25$^\circ$C, 45 min) liberated phosphate was determined by the method of Fiske and Subbarow as described by Bailey (9).

Protein was determined according to Lowry et al. (10). In order to remove analytical interference from Ampholine the samples were precipitated and washed 4 times in 10% trichloroacetic acid before determination.

### *E. Analytical Isoelectric Focusing and SDS Polyacrylamide Gel Electrophoresis*

Crude lantern extracts and soluble fractions from isoelectric focusing were used without further pretreatment. Particulate fractions from isoelectric focusing were dissolved in distilled water adjusting to pH 7.75 with Ampholine. Ammonium sulphate precipitates were dissolved in minimal amounts of distilled water. Solutions containing large amounts of ammonium sulphate were dialyzed against 0.1 M tris acetate, pH 7.75, containing 1 mM dithiothreitol.

Analytical isoelectric focusing was done on ready-made, thin-layer polyacrylamide gels (LKB Ampholine PAG plate 1804-101, pH 3.5-9.5) according to manufacturer's instructions using the LKB Multiphore 2117 and the LKB Power Supply 2103

(LKB-Produkter AB, Bromma, Sweden). Samples were applied after 30 min pre-focusing and focused for 1 hr at 1200 V. The gels were stained wtih Coomassie Brilliant Blue R-250.

SDS polyacrylamide gel electrophoresis was done in 0.05 M imidazole phosphate buffer, pH 7.0, containing 0.1% SDS using gel rods (4 mm x 90 mm) prepared with 10% acrylamide, 0.3% N,N' - methylenebisacrylamide, 0.15% TEMED and 0.05% ammonium persulphate. Samples were applied after 30 min pre-electrophoresis (1 mA) and the electrophoresis was run for 3 hrs at a constant current (2mA). The gels were stained with Coomassie Brilliant Blue R-250.

## IV. RESULTS

The experiment shown in Fig. 1 was performed to determine a concentration interval of ammonium sulphate for precipitation of firefly luciferase resulting in a high degree of purification and a high yield. Approximately 80% of the luciferase activity precipitated in a concentration interval corresponding to 45-55% saturation. Most of the adenylate kinase and nucleoside diphosphokinase activities precipitated at concentrations above this interval. However, pyrophosphatase and protein precipitated to a fairly high extent within the 45-55% saturation interval.

Fig. 2 shows the isoelectric focusing of the 45-55% fraction obtained by ammonium sulphate precipitation. The luciferase activity focused in a large peak at pH 6.1 and a small peak at pH 5.9 (Fig. 2a). In other experiments the large peak was always found within the interval pH 6.1-6.4. The large peak precipitated during focusing and could be collected by centrifugation.

The isoelectric points of contaminating enzymes were well separated from that of luciferase. Adenylate kinase focused at pH 5.1 and pH > 7 (Fig. 2d). Nucleoside diphosphokinase and pyrophosphatase focused at pH > 7 and pH < 5, respectively (Fig. 2e-f). The plots of optical density and protein concentration (Fig. 2b-c) show several peaks including one corresponding to the luciferase activity.

Luciferase activity and protein concentration were determined in the crude extract (fraction 1), the < 45% ammonium sulphate precipitate (fraction 2), the 45-55% ammonium sulphate precipitate (fraction 3) the > 55% ammonium sulphate precipitate (fraction 4), the total pool from isoelectric focusing (fraction 5) and the soluble and particulate fractions from isoelectric focusing (fractions 6 and 7, resp.).

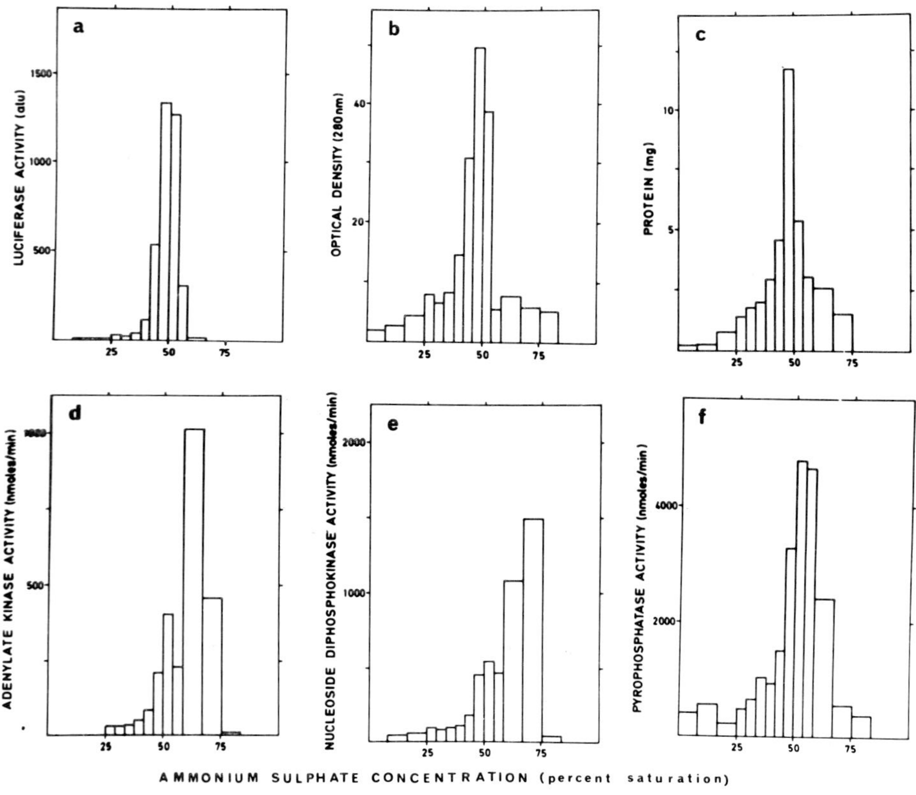


*FIGURE 1. Ammonium sulphate precipitation of crude firefly lantern extract.*

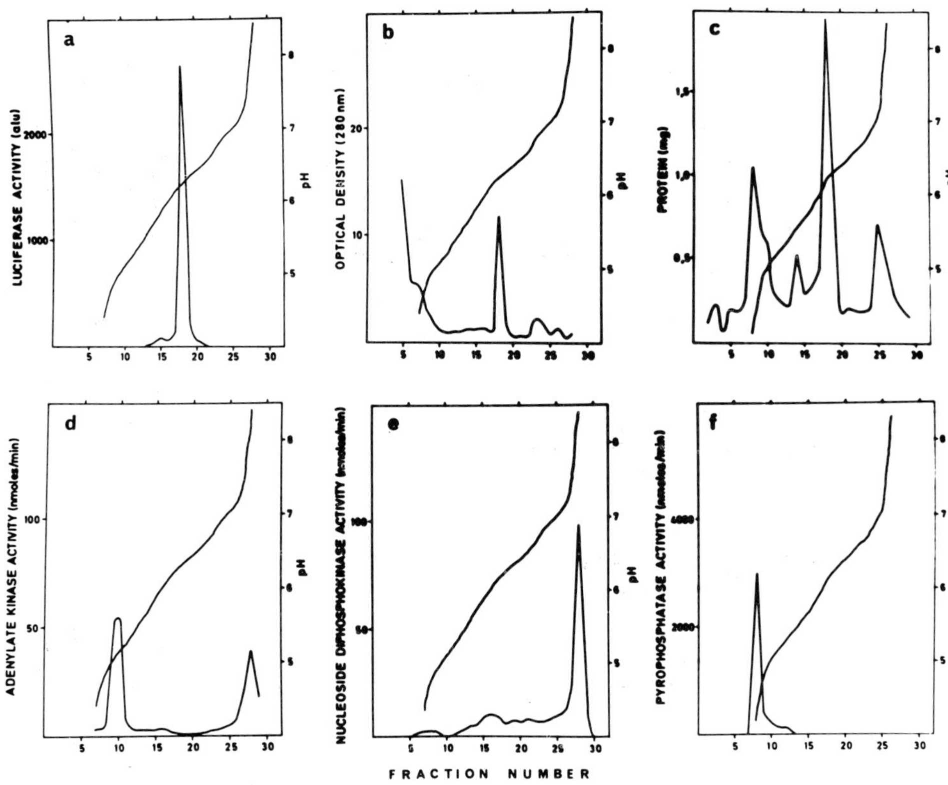


*FIGURE 2. Isoelectric focusing of firefly luciferase purified by ammonium sulphate precipitation. The lines going diagonally across each figure are the pH profiles.*

All the fractions were used to prepare luciferin-luciferase reagents with the same luciferase activity. In these reagents the levels of contaminating enzymes were determined. In Table 1 results on total activities or amounts in the fractions are shown without brackets and results on activities or amounts per ml in the luciferin-luciferase reagents are shown within brackets.

From Table 1 (figures without brackets) one can calculate that the yield of luciferase in the 45-55% ammonium sulphate precipitate (fraction 3) was 80% and that the specific activity increased 3.5 times. The yield in the isoelectric focusing step was only 47% and the specific activity increased 2.4 times. The overall yield was thus only 38% and the increase of the specific activity 8.1 times. However, in 10 routine preparations of luciferase by the described method the median overall yield was 65% (44-81%).

Table 1 (figures within brackets) also shows that the 45-55% ammonium sulphate precipitate (fraction 3) contained much less contaminating enzymes than the crude extract. The levels of these enzymes were further reduced by isoelectric focusing. The particulate fraction of luciferase (fraction 7) collected by centrifugation of the pool obtained by isoelectric focusing was much more pure than the soluble fraction of this pool (fraction 6).

The purity of fractions 1-7 were also studied by analytical isoelectric focusing (Fig. 3A) and by SDS polyacrylamide gel electrophoresis (Fig. 3b). The dramatically increased purity after isoelectric focusing (fraction 5 and 7) as compared to the 45-55% ammonium sulphate precipitate (fraction 3) and the crude extract (fraction 1) is obvious.

## V. DISCUSSION

Methods previously used in the purification of firefly luciferase include ammonium sulphate precipitation (11), chromatography on apatite (11), crystallization against a solution of low ionic strength (11), gel filtration (12-14), ion-exchange chromatography (15) and recently also isoelectric focusing (16) and affinity chromatography (17). The method described in the present paper takes advantage of the following facts:

(1) Luciferase is easily precipitated by ammonium sulphate using a narrow concentration interval.

(2) Luciferase has a narrow peak in isoelectric focusing.

TABLE 1. Purification of luciferase[a]

| Fraction | Luciferase[b] tot. | spec. | Protein | | Adenylate kinase[b] | | Nucleoside diphosphokinase[b] | | ATP degrading enzymes[b] | |
|---|---|---|---|---|---|---|---|---|---|---|
| | alu | alu/mg | mg | (μg/ml) | mU | (μU/ml) | mU | (μU/ml) | mU | (μU/ml) |
| Extraction of lanterns | | | | | | | | | | |
| 1. crude extract | 2578 | 20 | 130 | (5.0) | 11000 | (414) | 19000 | (690) | 5.20 | (0.20) |
| Ammonium sulph. precip. | | | | | | | | | | |
| 2. < 45% satured | 177 | 5 | 38 | (21) | 130 | (73) | 800 | (439) | 0.50 | (0.26) |
| 3. 45-55% satured | 2063 | 69 | 30 | (1.5) | 1900 | (92) | 1900 | (88) | 1.30 | (0.06) |
| 4. > 55% satured | 53 | 2 | 34 | (67) | 60 | (119) | 15000 | (29400) | 0.40 | (0.83) |
| Isoelectric focusing | | | | | | | | | | |
| 5. total pool | 976 | 163 | 6 | (0.6) | 6 | (0.7) | 130 | (14) | 0.40 | (0.04) |
| 6. soluble fraction | 40 | 40 | 1 | (3.1) | 4 | (9.0) | 20 | (46) | 0.04 | (0.09) |
| 7. particulate fraction | 973 | 162 | 6 | (0.6) | 2 | (0.2) | 80 | (8) | 0.20 | (0.02) |

[a]The various fractions were used to prepare reagents with the same luciferase activity. Protein concentrations and activities of contaminating enzymes in these reagents are shown within brackets. Results without brackets refer to total amounts or activities in the fractions.

[b]Luciferase activity was measured using 20 nM ATP and 0.14 mM luciferin and expressed in arbitrary light units (alue). Contaminating enzymes were measured using 10 μM ADP, 10 μM GTP and 20 nM ATP and results are expressed in units (U = μmoles/min).

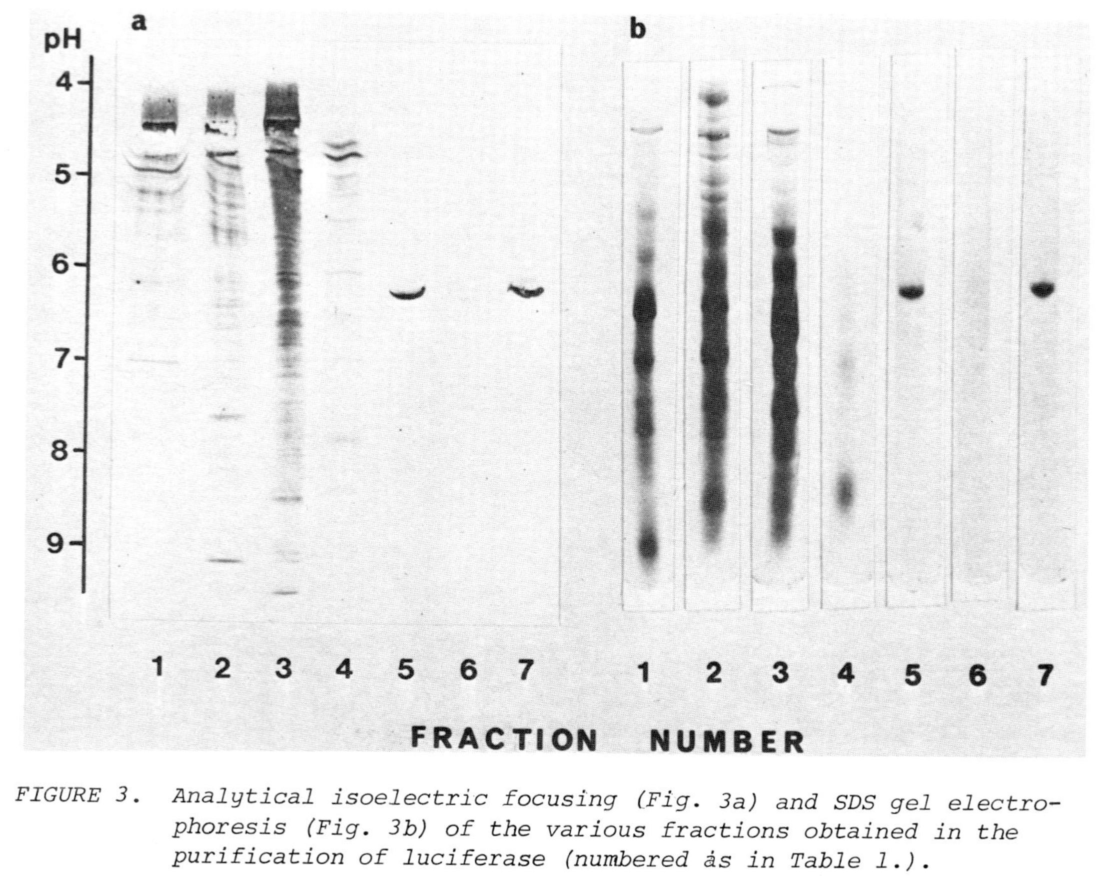


*FIGURE 3. Analytical isoelectric focusing (Fig. 3a) and SDS gel electrophoresis (Fig. 3b) of the various fractions obtained in the purification of luciferase (numbered as in Table 1.).*

(3) Luciferase is an euglobulin and precipitates at its isoelectric point. Thus centrifugation of the peak fractions obtained in the isoelectric focusing discarding the supernatant is a purification step similar to crystallization.

For precipitation of luciferase by ammonium sulphate a concentration interval corresponding to 45-55% saturation is used. This is similar to the 50-60% interval used by Green and McElroy (11). The appropriate ammonium sulphate concentration interval depends on the exact composition of the crude extract and minor variations using different raw materials may be expected.

In the isoelectric focusing of luciferase it was interesting to find a second peak also observed by Denburg and McElroy (18). No explanation as to the nature of this peak is yet available. Results presented by Denburg and McElroy (18) indicate that this peak has nothing to do with the subunit structure of luciferase. The second peak may of course represent another species of firefly and studies are under way to compare luciferase from different species of firefly.

The purification procedure described in the present paper is simple and adequate also for several of milligrams of luciferase. The specific luciferase activity was increased 8 fold as compared to the crude extract. More important from an analytical point of view is that levels of contaminating ATP converting enzymes were strongly reduced, in particular adenylate kinase which was reduced 5500 fold.

In ATP monitoring by firefly luciferase contaminating adenylate kinase is a serious problem since one or more of the substrates and products of this enzyme (ATP, ADP and AMP) are involved in all ATP converting enzyme systems.

The purity of luciferase prepared with the present method (fraction 7) have been compared with the purity of crystalline luciferase prepared according to Green and McElroy (11) and kindly provided by Dr. Marlene DeLuca. Specific luciferase activity and levels of contaminating enzymes were found to be similar.

A reagent containing luciferase purified by a scaled-up version of the described procedure is now commercially available (LKB-Wallac, Turku, Finland). This reagent is routinely used in a large number of laboratories for a variety of applications of ATP monitoring based on the firefly reaction.

ACKNOWLEDGMENTS

The authors want to thank Ms. Eva Forsberg, Mr. Björn Jäderlund, Mr. Hans Fehrnström and Ms. Anne Rickardsson for their assistance inthe experiments reported in this paper, Drs. Herman Haglund and Anders Thore for bringing up the idea to use isoelectric focusing for purification of luciferase and Dr. Marlene DeLuca for providing crystalline firefly luciferase. This work was supported by the Swedish Board for Technical Development and the Swedish Natural Science Research Council.

REFERENCES

(1) Lundin, A., A. Rickardsson, and A. Thore, *Anal. Biochem.* 75, 611 (1976).
(2) McElroy, W.D., *Proc. Nat. Acad. Sci. USA* 33, 342 (1947).
(3) Strehler, B.L. and J.R. Totter, *Arch. Biochem. Biophys.* 40, 28 (1952).
(4) DeLuca, M. (ed.) "Methods in Enzymology," Vol. 57, Academic Press, New York (1978).
(5) Lundin, A. and A. Thore, *Anal. Biochem.* 66, 47 (1975).
(6) DeLuca, M. and W.D. McElroy, *Biochemistry* 13, 921 (1974).
(7) Gates, B.J. and M. DeLuca, *Arch. Biochem. Biophys.* 169, 616 (1975).
(8) Lundin, A. in "Clinical and Biochemical Applications of Luminescence" (L.J. Kricka and F.J.N. Carter, eds.), Marcel Decker, Inc. (in press).
(9) Bailey, K. in "Methoden der Enzymatischen Analyse" (H.V. Bergmeyer, ed.), pp. 644, Verlag Chemie, Weinheim (1962).
(10) Lowry, O.H. N.J. Rosebrough, A.L. Farr, and R.J. Randall, *J. Biol. Chem.* 198, 265 (1951).
(11) Green, A.A. and W.D. McElroy, *Biochem. Biophys. Acta.* 20, 170 (1956).
(12) Nielsen, R. and H. Rassmussen, *Acta Chem. Scand.* 22, 1757 (1968).
(13) Shimomura, O., T. Goto, and F.H. Johnson, *Proc. Nat. Acad. Sci. USA* 74, 2799 (1977).
(14) Momsen, G.,*Anal. Biochem.* 82, 493 (1977).
(15) Bény, M. and M. Dolivo, *FEBS Lett.* 70, 167 (1976).
(16) Lundin, A. in "Methods in Enzymology", (M. DeLuca, ed.) Vol. 57, pp. 56. Academic Press, New York (1978).

(17) Branchini, B.R., T.M. Marschner, and A.M. Montemurro, *Anal. Biochem.* 104, 386 (1980).

(18) Denburg, J.L. and W.D. McElroy, *Biochemistry* 9, 4619 (1970).

# A CONVENIENT AFFINITY CHROMATOGRAPHY-BASED PURIFICATION OF FIREFLY LUCIFERASE

Thomas M. Marschner [1]
Bruce R. Branchini [2]
Angelina M. Montemurro

Department of Chemistry
University of Wisconsin-Parkside
Kenosha, Wisconsin

Recently, there has been increased interest in the analytical applications of bio- and chemiluminescence. The specificity of firefly luciferase for ATP in the conversion of luciferin into the luciferyl adenylate has been used to develop many assays and screening tests in areas such as water pollution monitoring, clinical bacteriology, intracellular metabolism studies, antibiotic-sensitivity testing, etc. The firefly enzyme can be used to directly quantify the amount of ATP in samples, or it can be employed with systems that produce or consume ATP in coupled assays. Since nearly one photon is produces for each molecule of ATP that reacts with the enzyme, ATP measurement offers great sensitivity. Sensitivity is limited, however, by crude preparations of luciferase that contain adenylate kinase and other nucleotide triphosphate transphosphorylases which can produce ATP.

Several partial and complete luciferase purification schemes have been reported. A comparison of the methods cited with respect to yield, reusability of reagents and materials, convenience, product purity, and adaptability to large- or small-scale preparations has appeared (1). Three of the procedures (2-4) yield crystalline enzyme of the highest purity, however, they require at least one lengthy chromatography step or cannot be easily adapted to small-scale preparations.

[1]*Present address: Department of Pharmaceutical Chemistry, University of California, San Francisco, California.*
[2]*Author to whom correspondence should be sent.*



ISBN 0-12-208820-4

We describe here a rapid and convenient affinity chromatography-based procedure for the preparation of crystalline, homogeneous firefly luciferase. One can routinely prepare 3-25 mg of enzyme with the reusable affinity chromatography matrix CH-Sepharose[3] 4b benzylamine.

## I. MATERIALS AND METHODS

The experimental details for the purification of the enzyme have recently appeared in the literature (5).

## II. RESULTS

The data for the preparation of 6.2 mg of crystalline firefly luciferase from 2.2 of freeze-dried *Photinus* lanterns are shown in Table I. A crude enzyme solution was prepared by grinding the lanterns with sand, extracting the resultant powder with acetone, and solubilizing the powder into 25 m*M* Tris buffer, pH 7.9, containing 1 m*M* EDTA. Subsequent dialysis against the same buffer removed luciferin and other low-molecular-weight contaminants. The dialyzed crude enzyme solution (29 ml) was applied to a CH-Sepharose 4b benzylamine column (Chart 1) and chromatographed as shown in Fig. 1. Most of the protein applied to the column was washed off with the equilibrating buffer, 0.02 *M* sodium phosphate, pH 7.8. The initial protein peak, however, contained only 1% of the luciferase activity. A second band of protein emerged when the ionic strength of the buffer was increased to 0.15 *M* sodium phosphate, pH 7.8. This peak contained 85% of the applied luciferase activity, which eluted slightly later than the major protein component (Fig. 1). Fractions 35-44, corresponding to 115-145 ml eluant, were pooled to give a protein sample with specific activity equal to 3.5 units $mg^{-1}$. The combined fractions contained luciferase which has been purified 8.1-fold and represented a 63% recovery of enzyme activity. Subsequent ultrafiltration was nearly quantitative and reduced the volume of the pooled fractions to 3.2 ml with a concomitant increase in the specific activity of the luciferase to 5.3 units $mg^{-1}$.

[3]*Abbreviations used: CH, 6-aminohexanoic acid; BSA, bovine serum albumin; EDC, 1-ethyl-3(3-dimethylaminopropyl)-carbodiimide hydrochloride; SDS, sodium dodecyl sulfate; AH, 1,6-diaminohexane.*

TABLE I. Purification Scheme for Firefly Luciferase

| | Protein[a] (mg) | Volume (ml) | Total activity[b] (units)[c] | Specific activity (units $mg^{-1}$) | Recovery (%) | Purification factor (-fold) |
|---|---|---|---|---|---|---|
| Solubilization[d] and dialysis | 700 | 29 | 300 | 0.43 | 100 | 1.0 |
| CH-Sepharose 4b benzylamine | 55 | 31 | 190 | 3.5 | 63 | 8.1 |
| Ultrafiltration | 34 | 3.2 | 180 | 5.3 | 60 | 12.3 |
| Crystallization | 6.2 | 0.75 | 65 | 10.5 | 22 | 24.4 |

[a] Absorbance at 278 nm is 0.75 for 1 mg $ml^{-1}$ crystallized firefly luciferase (Ref. 7). This value was used for all protein determinations.

[b] Activity units based on photon production in 1 s using the standard firefly luciferin assay described under Materials and Methods.

[c] One unit is equivalent to $10^{13}$ photons $s^{-1}$.

[d] Based on 3.0 g of acetone powder from 2.2 g of freeze-dried Photinus lanterns.

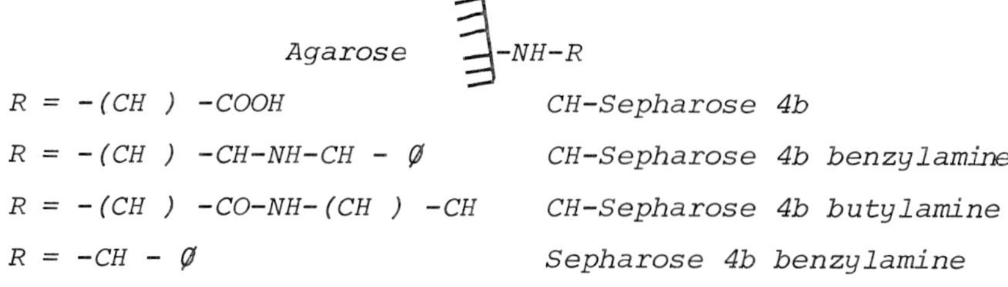


CHART I. Affinity Chromatography Gels

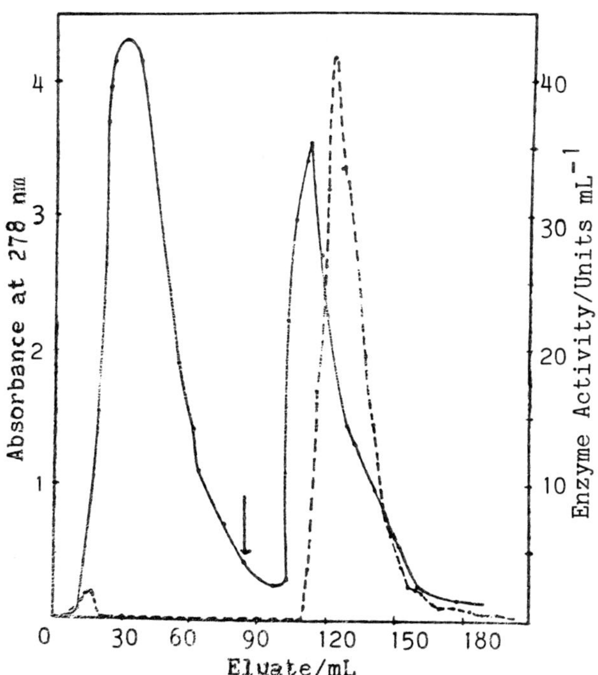


FIGURE 1. CH-Sepharose 4b benzylamine chromatography of dialyzed crude luciferase. The enzyme (from 3.0 g acetone powder) in 25 mM Tris buffer, pH 7.9, containing 1 mM EDTA (29 ml) was applied to a column, 1.5 x 30 cm, containing CH-Sepharose 4b benzylamine (∿20 ml) equilibrated with 0.02 M sodium phosphate buffer, pH 7.8. The column was washed with the equilibrating buffer at a flow rate of 1.2 ml/min and fractions of 3.3 ml were collected. Luciferase was eluted by changing the eluant buffer to 0.15 M sodium phosphate, pH 7.8 (arrow). Absorbance at 278 nm (———) and luciferase activity (----) were measured for each fraction.

Dialysis against 2 m$\underline{M}$ sodium phosphate, 10 m$\underline{M}$ NaCl, and 1 m$\underline{M}$ EDTA at pH 7.3 produced crystalline luciferase in moderate yield. The overall purification was 24.4-fold and the crystalline protein isolated has a specific activity of 10.5 units $mg^{-1}$. In other similar experiments[4] we have obtained enzyme with a specific activity of 85 units $mg^{-1}$. The purity of the crystallized enzyme was verified by SDS-gel electrophoresis which showed only one protein band with a subunit molecular weight of 50,100.

In our studies leading to the development of the CH-Sepharose 4b benzylamine-based purification of luciferase, we performed additional affinity chromatography experiments with the gels shown in Chart 1. CH-Sepharose 4b benzylamine, Sepharose 4b benzylamine, and CH-Sepharose 4b butylamine were derivatized to roughly the same extent as judged by their ability to bind bovine serum albumin (BSA) -- 18, 13, and 15 mg/ml settled gel, respectively. AH-Sepharose 4b did not bind any BSA.

An estimate of the $K_i$ of benzylamine attached to the spacer arm in CH-Sepharose 4b benzylamine was made with $\underline{N}$-benzylacetamide. $\underline{N}$-Benzylacetamide was a competitive reversible inhibitor of luciferase with respect to substrate firefly luciferin with $K_i$ equal to $4.5 \times 10^{-2}\underline{M}$.

Each of the affinity chromatography trials illustrated in Fig. 2 was attempted with dialyzed crude luciferase (6 ml) prepared as described in the CH-Sepharose 4b benzylamine experiments. The columns (4 ml) were equilibrated and washed with 0.02 $\underline{M}$ sodium phosphate, pH 7.8, and the buffer was then changed to 0.15 $\underline{M}$ sodium phosphate, pH 7,8. The data obtained in Fig.2 for luciferase activity recovered are summarized in Table 2. For each affinity column the purification factor and percentage luciferase activity recovered were: Sepharose 4b benzylamine, 5.3-fold and 48%; CH-Sepharose 4b butylamine, 1.1-fold and 11%; and CH-Sepharose 4b, 5.2-fold and 6%.

## III. DISCUSSION

We have described a procedure (Table 1) for preparing 6.2 mg of crystalline firefly luciferase from 2.2 g of firefly lanterns. The key step in the method is affinity chromatography with CH-Sepharose 4b benzylamine. Crystallization proceeds smoothly when the pooled luciferase fractions are concentrated by ultrafiltration and subsequently dialyzed against 2 m$\underline{M}$ sodium phosphate, 10 n$\underline{M}$ NaCl, and 1 n$\underline{M}$ EDTA at pH 7.3. The homogeneity of the crystalline enzyme was confirmed

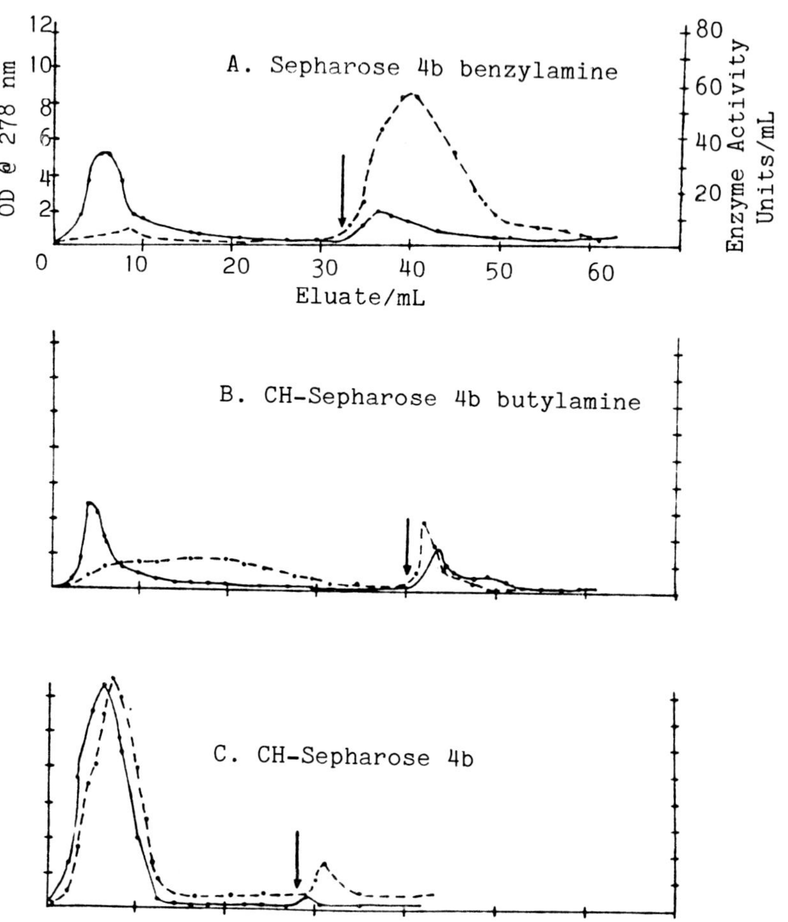


*FIGURE 2. Affinity chromatography of dialyzed crude luciferase. In separate experiments, enzyme (from 0.6 g acetone powder) in 25 mM Tris buffer, pH 7.9 containing 1 mM EDTA (6 ml) was applied to columns, 0.9 x 15 cm, containing ~4 ml of (A) Sepharose 4b benzylmaine; (B) CH-Sepharose 4b butylamine; and (C) CH-Sepharose 4b equilibrated with 0.02 M sodium phosphate buffer, pH 7.8. The columns were washed with the equilibrating buffer at a flow rate of 1.2 ml/min and 1.0-ml fractions were collected. The elution buffer was changed to 0.15 M sodium phosphate, pH 7.8 (arrow). Absorbance at 278 nm (——) and luciferase activity (---) were measured for each fraction.*

*TABLE 2. Summary of Affinity Chromatography Experiments with Firefly Luciferase*

| Affinity column | Percentage luciferase activity recovered | | | Purification factor[a] (fold) |
|---|---|---|---|---|
| | Break-through peak | Peak following eluant change | Total | |
| CH-Sepharose 4b benzylamine[b] | 1.0 | 85 | 86 | 8.1 |
| Sepharose 4b benzylamine[c] | 7 | 48 | 55 | 5.3 |
| CH-Sepharose 4b butylamine[c] | 44 | 11 | 55 | 1.1 |
| CH-Sepharose 4b[c] | 67 | 6 | 78 | 5.2 |

[a] *For peak following eluant change.*
[b] *Data obtained from Fig. 1.*
[c] *Data obtained from Fig. 2.*

by the single protein band observed on SDS-polyacrylamide gel electrophoresis. Our observed subunit molecular weight was 50,100 and was in excellent agreement with reports from high-speed sedimentation-equilibrium ultracentrifugation experiments (6) that luciferase has a molecular weight of 100,000 and consists of two apparently nonidentical subunits of molecular weight 50,000. It is difficult to compare the specific activity of luciferase preparations from other laboratories because there is no generally adopted way to report units of activity. Using the convention of Shimomura et al. (2) in which activity is expressed in photons per second per milligram for the first second of light emission, we expressed the specific activity of our crystalline enzyme preparations as 1.1 to 8.5 x $10^{14}$ photons $s^{-1}$ $mg^{-1}$. The range of specific activity for our protein is in good agreement with the report of 7.8 x $10^{14}$ photons $s^{-1}$ $mg^{-1}$ for three times crystallized and homogeneous protein, one of the most important advantages it offers is overall yield of pure protein from the starting material, freeze-dried *Photinus* lanterns. For example, we obtained (Table 1) a three- to four-fold better overall yield than the widely used protocol of Green and McElroy (1,3) in which 70 mg of crystalline enzyme is generally obtained from 75 to 100 g of firefly tails. It is not possible to compare other purification procedures owing to insufficient literature data. Further positive features of our protocol are realized because it is an affinity chromatogrpahy-based method and it does not require any lengthy chromatographic steps. We can conveniently scale the method down or process more protein by simply washing the column and reusing it. We have used the column described (Fig. 1) five times with no appreciable loss of binding capacity. Further, the CH-Sepharose 4b benzylamine matrix can be easily prepared from commercial reagents that require no prior purification. The use of affinity chromatography has made it possible, therefore, to develop a rapid and convenient method for preparing crystalline firefly luciferase in good yield.

Although we have consistently referred to the CH-Sepharose 4b benzylamine matrix as an affinity chromatography gel, there is strong evidence that it does not behave strictly as a biospecific support. From the elution profile shown in Fig. 1, it is clear that other proteins are retarded and eluted with luciferase during the chromatography step. Additionally, we have found that N-benzylacetamide, a model compound for benzylamine bound to the spacer arm, has very poor affinity, $K_i = 4.5 \times 10^{-2}$ M, for the substrate binding site on the enzyme compared to firefly luciferin, $K_i = 3 \times 10^{-5}$ M (7). However, chromatography with AH-Sepharose 4b benzylamine is slightly more effective than with AH-Sepharose 4b 5,5-dimethylmethoxyluciferin,[4] even though 5,5-dimethylmethoxyluciferin, $K_i = 4 \times 10^{-6}$ M[4], is a far superior ligand to benzylamine.

To gain some understanding of the nature of the chromatography step, we investigated the effectiveness of underivatized CH-Sepharose 4b, CH-Sepharose 4b butylamine, and Sepharose 4b benzylamine in purifying luciferase (Fig. 2 and Table 2). Since nearly all of the luciferase activity recovered in the CH-Sepharose 4b experiment was eluted in the breakthrough peak (Fig. 2), we can safely rule out any significant purification by an underivatized spacer arm which might have contaminated the CH-Sepharose 4b benzylamine. Chromatography with CH-Sepharose 4b did not allow us to evaluate the effectiveness of the hydrophobic nature of the spacer arm because the free carboxyl terminus is charged at pH 7.8 and may act as an ion-exchange site. We considered the possibility that a hydrocarbon arm alone might be sufficient to purify luciferase to the same degree we had observed with CH-Sepharose 4b benzylamine. Substitution of n-butylamine for the benzylamine ligand, however, resulted in a relatively ineffective 1.1-fold purification of luciferase (Table 2). This observation leads us to conclude that we are not observing an important general hydrophobic effect (8) between a saturated hydrocarbon spacer arm and luciferase. In a separate experiment, Sepharose 4b benzylamine-based chromatography produced 5.3 fold purified enzyme in good yield (Table 2). Considering the results of all the experiments illustrated in Fig. 2, we conclude that the aromatic ring moiety is essential for effectively purifying luciferase using our chromatographic conditions. Further, it is advantageous to attach the aromatic group to agarose with a spacer arm (Table 2).

We favor the notion that CH-Sepharose 4b benzylamine functions to bind luciferase more tightly than any other proteins applied to it by a synergistic effect involving hydrophobic interaction between luciferase and the aromatic ring moiety and typical, though perhaps weak, protein-ligand interactions. Since the amino acid composition of firefly luciferase distinguishes it as one of the most hydrophobic enzymes examined (9), it is not surprising that hydrophobic interactions are stronger for luciferase than any of the contaminating proteins. It is not entirely clear why the aromatic ring of benzylamine functions so effectively with our experimental parameters.

We have found the purification scheme described to be reliable and convenient. The method particularly suits our present needs for preparing homogeneous firefly luciferase in sufficient yield for experiments involving ATP analysis and attempts to grow X-ray quality crystals for structure determination.

## REFERENCES


(1) DeLuca, M.A. (ed.) (1978) "Methods in Enzymology," Vol. 57, Academic Press, New York.

(2) Shimomura, O., T. Goto, and F.H. Johnson (1977), *Proc. Nat. Acad. Sci. USA* 74, 2799-2802.

(3) Green, A.A., and W.D. McElroy (1956), *Biochim. Biophys. Acta* 20, 170-176.

(4) Momsen, G. (1977), *Anal. Biochem* 82, 493-502.

(5) Branchini, B.R., T.M. Marschner, and A.M. Montemurro (1980), *Anal. Biochem.* 104, 386-396.

(6) Denburg, J.L. and W.D. McElroy (1970), *Biochemistry* 9, 4619-4624.

(7) Lee, R., and W.D. McElroy (1969), *Biochemistry* 8, 130-134.

(8) Shaltiel, S. (1974) in "Methods in Enzymology" (W.B. Jakoby and M. Wilchek, eds.), Vol. 34, p. 126-140, Academic Press, New York.

(9) Gorus, F., and E. Schram (1979), *Clin. Chem.* 25, 1531-1546.

# BIOLUMINESCENCE MICROASSAY OF CREATINE-KINASE: A COMPARISON WITH A SPECTROPHOTOMETRIC METHOD

Gianni Messeri
Anna L. Caldini
Paola Tozzi

Clinical Chemistry Lab
Arcisp. S. Maria Nuova
Florence, Italy

Mario Pazzagli

Endocrinology Unit
University of Florence
Florence, Italy

## I. INTRODUCTION

Creatine-kinase (EC. 2.7.3.2) is the enzyme which catalyses the transfer of one phosphate group from creatine phosphate (CP) to ADP to form ATP. It is a dimeric molecule and it appears in human tissues under three isoenzymic forms: MM (muscle type), MB (myocardial type) and BB (brain type). The assay of total creatine-kinase (CK) activity in serum has been widely employed for long in the diagnosis of various disorders. Elevated CK activity in serum is associated with lesions of myocardium or skeletal muscles and it is a valuable aid in the diagnosis of acute myocardial infarction and of muscular dystrophy. This test is in great demand in most clinical laboratories and in our hospital (about 5000 beds) 40,000 assays were performed last year.

The most widely used method for the determination of CK measures the increase in adsorption caused by NADPH production by the following reaction system (1):




ISBN 0-12-208820-4

$$\text{ADP} + \text{Creatine-phosphate} \xrightarrow{\text{CK}} \text{Creatine} + \text{ATP}$$

$$\text{ATP} + \text{glucose} \xrightarrow{\text{Hexokinase}} \text{ADP} + \text{glucose-6-P}$$

$$\text{glucose-6-P} + \text{NADP}^+ \xrightarrow{\text{G-6-PDH}^1} \text{6-P-gluconate} + \text{NADPH} + \text{H}^+$$

The main disadvantages of the spectrophotometric assay are the elevated cost of reagents and the required sample volume. As regards muscular dystrophy screening on newborn infants blood, such problems become particularly weighty.

The aim of our work was to assess a cheap method which required reduced blood samples suitable for muscular dystrophy screening. CK activity was determined by continuous monitoring of the ATP producing reaction by firefly-luciferase bioluminescence according to the following scheme:

$$\text{ADP} + \text{Creatine-phosphate} \xrightarrow{\text{CK}} \text{Creatine} + \text{ATP}$$

$$\text{ATP} + \text{luciferin} \xrightarrow{\text{luciferase, Mg}^{++}} \text{AMP-luciferin-luciferase} + \text{PP}$$

$$\text{AMP-luciferin-luciferase} \xrightarrow{O_2} \text{luciferase} + \text{AMP} + CO_2 + \text{oxyluciferin} + \text{photon}$$

Serum adenylate-kinase (EC. 2.4.7.4.3) activity is also able to produce ATP by the following reaction:

$$2\ \text{ADP} \xrightarrow{\text{adenylate-kinase}} \text{ATP} + \text{AMP}$$

In the presence of diadenosine pentaphosphate (DAPP), adenylate-kinase activity is almost totally inhibited and usually it does not exceed 5% of total CK activity. Nevertheless it may be measured and then substracted from total enzymatic activity.

The results we achieved on 54 serum samples by the present method were compared with those from the spectrophotometric assay as regards correlation, precision, reproducibility, linearity, cost and required sample volume.

[1] *G-6-PDH: glucose-6-phosphate dehydrogenase*

## II. MATERIAL AND METHODS

54 serum samples were collected without anticoagulant and stored at +4$^{\circ}$C until assayed.

U.V. assay of total CK activity was performed at 25$^{\circ}$C using the Merckotest (Optimized U.V. test) by a Saitron 903 Spectrophotometer (Saitron s.p.a., Florence, Italy).

Bioluminescence was measured with a Luminometer equipped with a potentiometric recorder (LKB, Bromma, Sweden). Luciferase-luciferin reagent, Lumit PM, was supplied by Lumac Systems, Basel, Switzerland.

A typical bioluminescence reaction graph is reported in Fig. 1. The assay is carried on by preincubating the serum samples (5 μl) with all reagents listed in Tab. 1, except ADP, creatine-phosphate and ATP, for 15 min. After adding ADP, ATP production is started originating from residual adenylate-kinase activity (Fig. 1a). After 30 sec creatine phosphate is added and total ATP producing activity is then recorded (Fig. 1b).

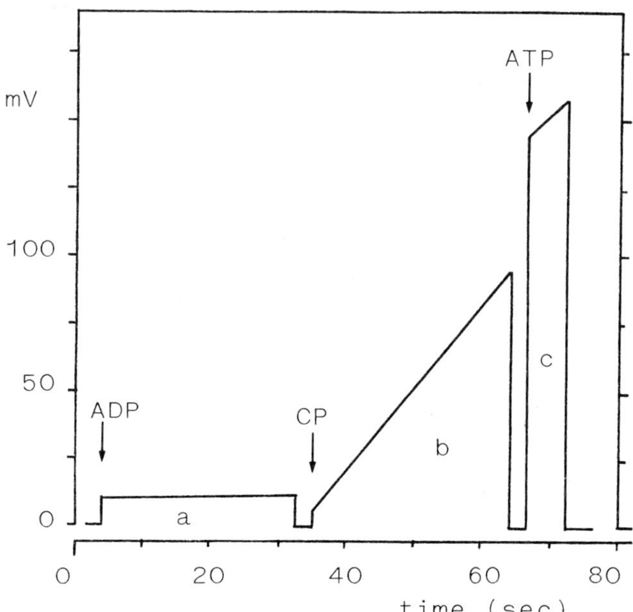


*FIGURE 1. Continuous monitoring of ATP producing reactions by luminescence recording.*

Finally ATP of known concentration is added as internal standard (Fig. 1c).

CK activity is computed according to Lundin (3) from the rate of increase of luminescence after addition of CP. Because of the different reaction condition (2), a control serum of known enzymatic activity was used to correlate results from the two investigated methods.

## III. RESULTS

The sensitivity of the present method allows the accurate measure of enzymatic activities $\geq$ 1mIU/ml. Light emission is linearly related to CK activity up to 500 mIU/ml (Fig. 2).

The correlation between the U.V. and the bioluminescence assay is highly significant (r = 0.981) (Fig. 3).

Precision studies (Tab. II) with the two methods demonstrated that bioluminescent assay is at least equivalent to the spectrophotometric one in both intra and inter assay precision.

*TABLE I. Reagent concentrations in bioluminescence assay (final volume, 0.5 ml).*

| *Reagent* | *Concentration* |
|---|---|
| *Imidazole acetate buffer pH 6.7* | *100 mmol/l* |
| *Mg acetate* | *10 "* |
| *$NaN_3$* | *3 "* |
| *EDTA* | *2 "* |
| *Diadenosine pentaphosphate* | *1 µmol/l* |
| *Dithiothreitol* | *10 mmol/l* |
| *Lumit PM* | *50 µl/0.5 ml* |
| *ADP[a] (Boheringer, Mannheim)* | *10 µmol/l* |
| *Creatine-phosphate* | *10 mmol/l* |
| *ATP* | *50 nmol/l* |

*[a]ADP was purified according to Lundin (3).*

## IV. SUMMARY AND CONCLUSION

The proposed method seems to offer several advantages when compared with the spectrophotometric one. Because of the high sensitivity of the bioluminescence, only five microliters of serum are required and this feature makes the method very suitable for creatine-kinase measurements on newborn infants blood. Bioluminescent assay of creatine-kinase on dried blood spot in screening for Duchenne muscular dystrophy has already been reported (4). Moreover the cost of the bioluminescence test is about one-fourth of the spectrophotometric one and, if considering the large number of assays performed a year, consistent savings become possible.

The diffusion and the development of such technique is, however, strictly connected with the availability of high quality ready-to-use reagents and of automated equipment.

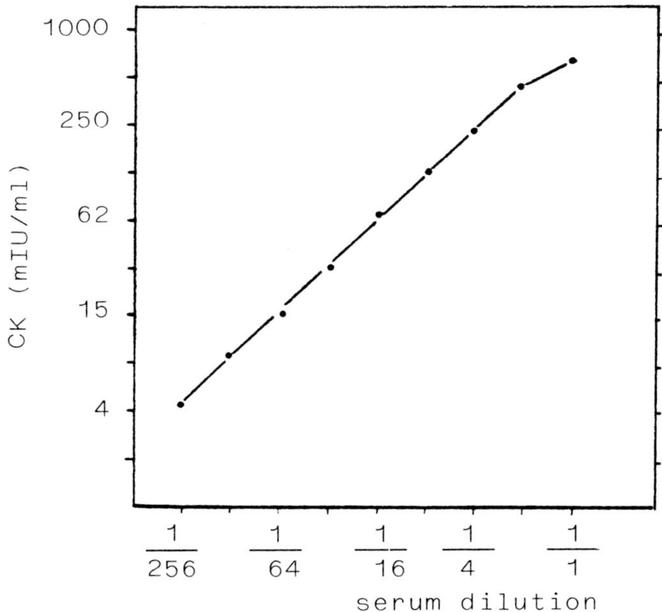


*FIGURE 2. CK activity as determined on serial dilutions of a high activity sample. Dilutions were prepared with a heat-inactivated serum.*

*TABLE II. Intra and inter assay precision for CK activity by spectrophotometric (U.V.) and bioluminescence (BL) assay.*

| | INTER | | | INTRA | | |
|---|---|---|---|---|---|---|
| | mean (mIU/ml) | ± s.d. | CV (%) | mean (mIU/ml) | ± s.d. | CV (%) |
| BL | 50 | 3.7 | 7.5 | 52 | 2.4 | 4.7 |
| | 223 | 12.9 | 5.8 | 229 | 4.6 | 2.0 |
| U.V. | 50 | 5.2 | 10.6 | 56 | 2.0 | 3.6 |
| | 201 | 15.2 | 7.4 | 214 | 5.0 | 2.3 |

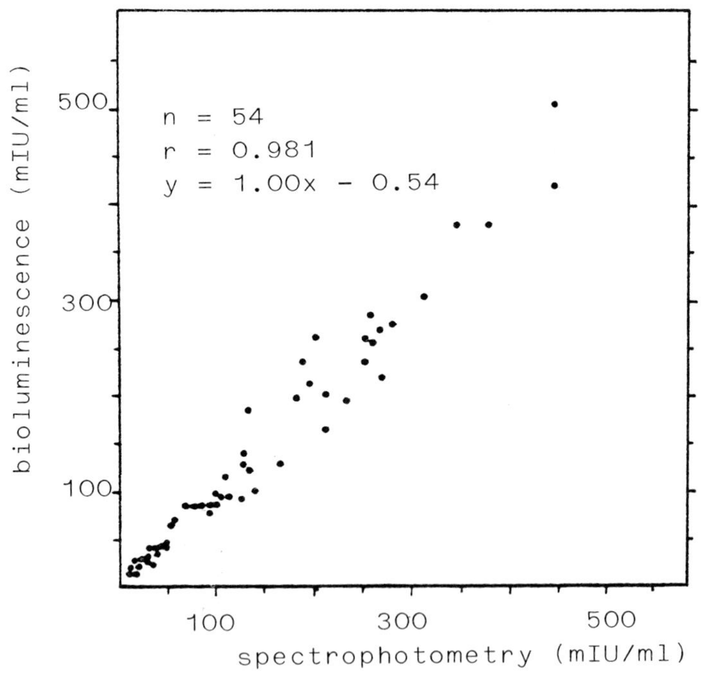


*FIGURE 3. Correlation between bioluminescence and spectrophotometry results.*

REFERENCES

(1) Deutsche Gesellschaft fur Klinische Chemie, *Z. Klin. Chem. u. Klin. Biochem.* 10, 182 (1972).
(2) Lundin, A. in "Methods in Enzymology", 57 (M. DeLuca, ed.) p. 56. Academic Press, New York (1979).
(3) Lundin, A. and I. Styrelius, *Clin. Chim. Acta* 87, 199 (1978).
(4) Dellamonica, Ch., J.M. Robert, J. Cotte, C. Collombell, C. Dorche, *Lancet* 18, 1100 (1978).

# INHIBITORS OF FIREFLY LUCIFERASE IN CLINICAL URINE SPECIMENS

Wright W. Nichols
G.D.W. Curtis
H.H. Johnston

Department of Microbiology
John Radcliffe Hospital
Oxford, U.K.

## I. INTRODUCTION

The measurement of ATP in urine using Firefly luciferin-luciferase has been proposed as a means of estimating bacterial numbers (1) and hence as a means of rapidly detecting significant bacteriuria (2,3,4). Methodologies using Firefly luciferin-luciferase have the potential advantages of speed and space-saving over conventional culture methodologies.

A Firefly luciferin-luciferase test designed to detect significant bacteriuria must ideally satisfy two criteria. First the lower sensitivity limit of the enzyme-photometer detection system must be lower than the bacterial ATP concentrations in specimens showing bacteriuria. Secondly, a bacterial ATP threshold must be defined which divides urine specimens into two populations; those above the threshold having significant bacterial numbers and those below the threshold not having significant bacterial numbers. In practice these ideal criteria are not met and one obtains a mis-match between bioluminescence results and results of conventional culture (2,5). This is usually quantified as "false positivity" and "false negativity."

One source of mis-match between bioluminescence results and culture is that urine specimens variably contain substances which inhibit Firefly luciferase (6) and thereby reducing the apparent ATP concentration. A study of the occurrence,



ISBN 0-12-208820-4

nature and behaviour of Firefly luciferase inhibitors in urine is required in order to define the extent of their interference. This paper describes some results of such a study.

## II. RESULTS AND DISCUSSION

We observed a median inhibition of luciferase of 93% (range 76.0% through 99.5%; n = 20) by urine specimens diluted approximately twofold. This compares to an average inhibition of 60% by tenfold diluted urines reported by Thore et al. (2).

Figure 1 shows how the inhibition of Firefly luciferase varied with the concentration of a typical clinical urine specimen. Luciferase activity decreased exponentially with increasing urine concentration. The experimental points fitted the biexponential equation:

$$\underline{F} = 982 \exp(-0.21P) + 2122 \exp(-0.054\underline{P}) \qquad (1)$$

where:

$\underline{F}$ is the peak light output (arbitrary units),
$\underline{P}$ is the urine concentration (% v/v).

Inhibition by urine is most likely due to the compound inhibition from the combined effects of urinary chloride (7), sulfate and inorganic pyrophosphate (8), urea (Nichols, unpublished) and probably uric acid. The concentration ranges of these substances in normal urine are enough to account for most of the inhibition of Firefly luciferase by clinical urine specimens (9). Furthermore, inhibition by chloride, sulfate or urea was in each case an exponential function of inhibitor concentration (data not shown) as was the case for inhibition by urine (Fig. 1, Equation 1). Since the concentration of each particular inhibitor in a specimen of urine depends on diet and fluid excretion, each urine specimen is expected to inhibit luciferase to a different degree. However, all urine specimens examined in a sample of 20 inhibited luciferase by more than 75% (see above).

What are the consequences of this severe inhibition of luciferase for the estimation of bacterial numbers in urine? If the determination of bacterial ATP concentration is to be accurate then the inhibitors must be removed from the urine by dialysis or gel filtration. Alternatively bacteria may be separated from the soluble constituents of urine by centrifugation or membrane filtration. These procedures would be

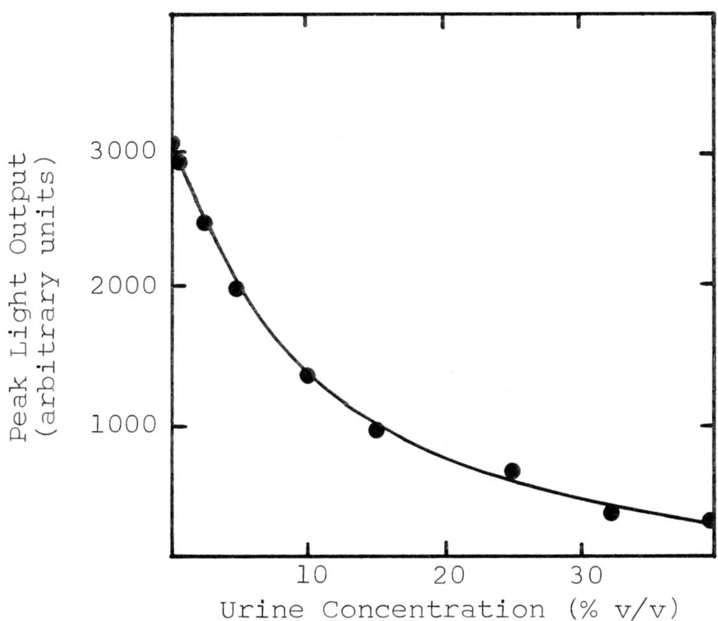


*FIGURE 1. Inhibition of Firefly luciferase by a clinical urine specimen. The specimen was filtered (0.8 μ pore size), heated at 95° for 10 min, cooled and the pH adjusted to 7.75 with Tris base (2.0 M). The reaction cuvette contained 10 ul ATP solution (1 mM); 1 through 80 ul of the treated urine and water to bring the final volume to 100 ul. The cuvette was placed in a Chem Glow photometer (American Instrument Co.) and 100 ul of luciferin-luciferase preparation (Lumit HS special; Lumac B.V., The Netherlands) were injected to start the reaction.*

impractical for a rapid method suitable for routine use in a hospital microbiology laboratory. However, the total removal of inhibitors may not be necessary in practice because the level of inhibition falls sharply with decreasing concentration (Fig. 1). Thus, for example, for the urine shown in Fig. 1, a reduction in concentration of fourfold from 40% to 10% resulted in an increase in peak light output (at constant ATP concentration) of fivefold - from 265 units to 1320 units. That is, one can increase the sensitivity of ATP estimations by dilution of the urine. In estimating a bacterial ATP concentration in a urine specimen, dilution of the

TABLE I. Comparison of two continuous-flow methods of detecting significant bacteriuria in clinical urine specimens

| | Number of specimens | Number of reference positives[a] | False positives[a] (%) | Positivity threshold (nM ATP) | False negatives[a] (%) |
|---|---|---|---|---|---|
| Neat urine | 289 | 63 | 21 | 0.65 | 21 |
| Urine diluted fivefold on sampling | 306 | 59 | 21 | 0.90 | 12 |

[a]The reference test was a semi-quantitative cultural test for which a calibrated microbiological loop was used to spread 0.002 ml amounts of urine on nutrient agar plates (3). The specimen was deemed to be positive if 200 or more colonies were observed following aerobic incubation at 37$^{o}$ for 16 h.

inhibitors necessarily dilutes bacterial ATP. Nevertheless, under such circumstances, the sensitivity increase was still observed (Nichols, unpublished). This is because the increase in light output with decreasing inhibitor concentration was exponential whereas the fall in light output with decreasing ATP concentration was linear (1).

The automated flow system for the detection of significant bacteriuria originally maximized the ratio of urine to luciferase (3,5). The method now incorporates a fivefold dilution step during sampling of the urine to take account of the above observations on luciferase inhibition. This has resulted in improved agreement between the results (Table I). In order to normalize the results obtained with the two flow system arrangements; in each case the flow system's threshold of positivity was determined as the value which yielded 21% false positive results with respect to the results obtained by microbiological culture of the urine specimens (Table I). At a level of 21% false positive results, the automated flow system yielded 21% false negative results without urine dilution and 12% false negative results with urine dilution (Table I.).

## III. SUMMARY AND CONCLUSION

In conclusion, diluting urine inhibitors (simultaneously with bacterial ATP) resulted in an improved "fit" between the continuous-flow, bioluminescence method of detecting significant bacteriuria and results obtained using a standard semi-quantitative cultural technique.

## REFERENCES

(1) Chappele, E.W., and G.V. Levin, *Biochem. Med.* 2, 41 (1968).

(2) Thore, A., A.A. Lundin, and S.J. Bergman, *Clin. Microbiol.* 1, 1 (1975).

(3) Johnston, H.H., C.J. Mitchell, and G.D.W. Curtis, *Lancet (ii)*, 400 (1976).

(4) Alexander, D.N., G.M. Ederer, and J.M. Matsen, *J. Clin. Microbiol.* 3, 42 (1976).

(5) Curtis, G.D.W., and H.H. Johnston in "Proceedings of the International Symposium on Analytical Applications of Bio- and Chemiluminescence" (E. Schram and P. Stanley eds.), p. 488. State Printing and Publishing, Inc.

California (1978).
(6) Conn, R.B., P. Charache, and E.W. Chappele, *Am. J. Clin. Path.* 63, 493 (1975).
(7) Denburg, J.L. and W.D. McElroy, *Arch. Biochem. Biophys.* 141, 668 (1970).
(8) DeLuca, M., J. Wannlund, and W.D. McElroy, *Anal. Biochem.* 95, 194 (1979).
(9) Geigy Scientific Tables, 7th edition (K. Diem and C. Lentner, eds.) p. 661. Geigy Pharmaceuticals, Macclesfield, U.K.

# USE OF A MATHEMATICAL REPRESENTATION FOR THE TIME-COURSE OF THE FIREFLY LUCIFERASE LIGHT REACTION[1]

Eric Schram
Mushtaq Ahmad
Eric Moreels

Institute of Molecular Biology
Vrije Universiteit Brussel
Brussels, Belgium

## I. INTRODUCTION

The proper use of firefly luciferase for analytical purposes is largely dependent on a precise knowledge of the reaction kinetics. Much is known at present about the basic mechanism of the reaction thanks to the pioneering work of McElroy and his group (1). However, much of that work was performed at rather high reactant concentrations where luminescence occurs as a surge of light, followed by a rapid decrease of the luminescence. Because measurement of the luminescence peak height is less likely to be influenced by contaminating enzymes it has also long been the preferred method for the assay of ATP by people using crude luciferase preparations. Thanks to the availability of purified luciferase and of sensitive instruments it has now become customary to work in conditions where no significant decrease of the luminescence is observed, allowing for continuous monitoring and integration of the light intensity over longer periods (2, 3). However, discrepancies still subsist in the literature about the shape and interpretation of the luminescence time course. Several of the parameters involved were there-

[1]*We are thankful to Prof. P. Huybrechts for helpful discussion. This work was supported by a grant of the Belgian Government (Programmatie van het Weterschapsbeleid).*



ISBN 0-12-208820-4

fore reconsidered in our experiments and a mathematical expression derived that accounts for the observed light curves at various enzyme and substrate concentrations, as well as for the slow turn-over of the enzyme.

## II. MATERIALS AND METHODS

### A. *Reagents*

Adenosine triphosphate (cat. no. A6144), D-luciferin (cat. no. L9504) and luciferase type IV (cat. no. L5256) were purchased from Sigma (St. Louis, U.S.A.). Luciferase reagent (LUMIT PM, cat. no. 4103) was obtained from LUMAC (Schaesberg, Holland). Tritium labelled ATP (code no. TRK 336) from the Radiochemical Centre (Amersham, U.K.) was used for the tracer experiments.

The concentration of the ATP and luciferin solutions was determined by spectrophotometry. The Sigma enzyme was standardized by 2,6-TNS titration and its light yield used as a reference for the calibration of other preparations. ATP consumption by contaminating enzymes was checked by adding the tritium labelled nucleotide to luciferase freed of luciferin and analyzing the products formed by ion-exchange radiochromatography. It never amounted to more than a few percent over a period of 30 minutes at the enzyme concentrations used in our experiments.

### B. *Light Measurements*

Luminescence was measured by photon counting using an RCA 8850 photomultiplier tube and ORTEC electronics. Pulses were registered on a 400-channel SA 41 Analyzer (Intertechnique, France).

### C. *Experimental Conditions for ATP Assay*

Assays were performed in TRIS buffer 0.02M, pH 7.4, containing EDTA $10^{-3}$M, $MgCl_2$ $10^{-2}$M, $NaH_2PO_4$ $10^{-3}$M and $NaN_3$ 3.1 x $10^{-3}$M.

The following volumes, added by hand, were used routinely:

200 μl buffer containing 0.15% bovine serumalbumin
20 μl ATP in buffer
20 μl luciferase in buffer containing $4.10^{-4}$M luciferin

A cell-holder with controlled temperature was used for all measurements.

## III. RESULTS

For the present purpose the reaction under consideration can be summarized as follows:

1. $E + LH_2 + ATP + O_2 \rightarrow E\text{-}LO^* + AMP + H_2O + CO_2$

2. $E\text{-}LO^* \rightarrow E\text{-}LO + h\nu$

3. $E\text{-}LO \rightarrow E + LO$

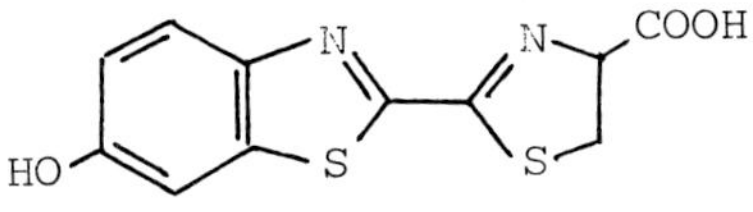

Luciferin ($LH_2$) 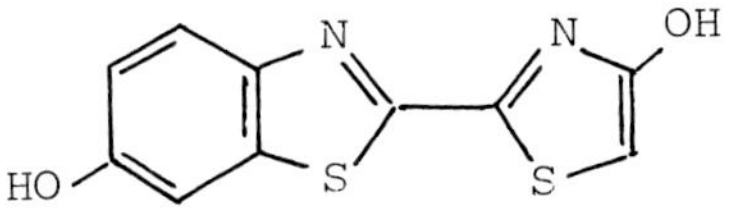 Oxyluciferin (LO)

Reaction 1 involves several successive steps and occurs rather slowly (4). It is followed by the rapid emission of light as shown in reaction 2. For the interpretation of our results it has been assumed that the enzyme is reactivated according to reaction 3 by the slow dissociation of E-LO (at the low concentration of LO present in the medium the reverse reaction may be neglected) and that at any time the luminescence (L) remains proportional to the free enzyme (E) concentration, the concentration of the several substrates remaining constant. After the initial decrease of the luminescence an equilibrium is reached between the formation of the end-product and its release (see figure).

If $k_1$ is the kinetic constant of the luminescent reaction for a given concentration of substrates and $k_2$ the kinetic constant for the release of the end-product we have

$$dE = -k_1\,E\,dt + k_2\,E\text{-}LO\,dt$$

and $E\text{-}LO = E_o - E$; ($E_o$ = initial enzyme concentration)

hence $dE = -k_1\,E\,dt + k_2\,E_o\,dt - k_2\,E\,dt$

$$= -(k_1 + k_2)\,E\,dt + k_2\,E_o\,dt$$

or
$$\frac{dE}{-(k_1 + k_2)\,E + k_2\,E_o} = dt$$

Integration gives:

$$E = \frac{E_o\ (k_2 + k_1 e^{-(k_1 + k_2)t})}{k_1 + k_2} \quad \text{or} \quad L = \frac{L_o(k_2 + k_1 e^{-(k_1 + k_2)t})}{k_1 + k_2}$$

for $t = 0$ $\quad L = L_o$

and for $t = \infty$ $\quad \frac{L_o}{L_e} = 1 + \frac{k_1}{k_2}$ $\quad$ ($L_e$ = steady-state luminescence at equilibrium)

From the graph it can be seen that the experimental figures for the luminescence fit perfectly with the computed curve over more than 25 minutes, supporting our basic assumption. Similar curves were obtained for ATP concentrations ranging from $10^{-5}$M to $10^{-7}$M (below $10^{-7}$M the luminescence decreases but slightly with time). For increasing ATP concentrations the curves become steeper and the plateau is reached more quickly. In that case $L_o$ is also more difficult to observe and its value was therefore obtained by extrapolation.

In our experimental conditions the consumption of substrate was found to be very small and did not contribute to the decay of the luminescence (less than ca. 1% over a period

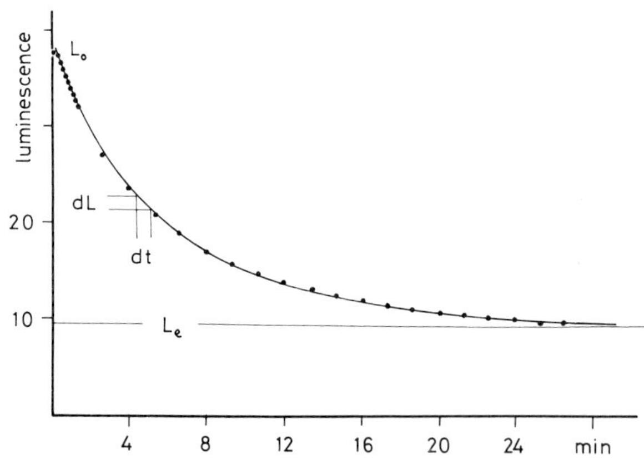


*FIGURE 1. Time course of luciferase light reaction.*

*Final concentrations of reactants were: enzyme 3.85 x* $10^{-9}$*M, ATP 1.16 x* $10^{-6}$*M, luciferin 3.33 x* $10^{-5}$*M.* $k_1 = 2.105 \times 10^{-3}$ $k_2 = 6.571 \times 10^{-4}$*. (Luminescence is expressed in arbitrary units.)*

——— *computed curve* $\qquad$ • • • • • *experimental figures* of 30 minutes). The production of pyrophosphate also remains small and it is not likely to distort the curve by accelerating the regeneration of the enzyme ($k_2$).

From our assumptions it can also be deduced that

$$\lg \frac{L_o - L_e}{L_t - L_e} = k_1 \cdot \frac{L_o}{L_o - L_e} \cdot t$$

When the logarithms in this equation are plotted versus time a straight line is indeed obtained for the experimental values of $L_o$, $L_e$ and $L_t$.

## IV. DISCUSSION

The mathematical expression described in this paper provides an easy way to dissociate and quantify the effects of several factors on the parameters $k_1$ and $k_2$, i.e. on the rate of the light reaction and on the regeneration of the enzyme. Factors that were studied in this way deal with the effect of temperature, pyrophosphate concentrations, etc. It was also possible to dissociate the effect of temperature on the reaction rate and the quantum efficiency. In order to interpret the observed luminescence curves Gates and DeLuca (5) have invoked the presence of two different binding sites on the enzyme. As shown above our own results could be interpreted without resorting to such an assumption. In accordance with early results obtained by McElroy et al. (6) the absence of competitive inhibition by the end-product was confirmed in our experimental conditions by the fact that when fresh enzyme was added to a partially inhibited enzyme preparation, it behaved as if there were no inhibitor present. Absence of competitive inhibition was also shown by Lemasters and Hackenbrock (7,8), using a different approach.

## REFERENCES


1. DeLuca, M., *in* "Adv. Enzymol." (Meister, A., ed.), p. 37, vol. *44*. Wiley, New York (1976).
2. Lundin, A. and A. Thore, *Anal. Biochem. 66*, 47 (1975).
3. Lundin, A., A. Rickardsson, and A. Thore, *Anal. Biochem. 75*, 611 (1976).
4. DeLuca, M. and W. D. McElroy, *Biochemistry 13*, 921 (1974).

5. Gates, B. J. and M. DeLuca, *Arch. Biochem. Biophys. 169*, 616 (1975).
6. McElroy, W. D., J. W. Hastings, J. Coulombre, and V. Sonnenfeld, *Arch. Biochem. Biophys. 46*, 399 (1953).
7. Lemasters, J. J. and C. R. Hackenbrock, *Biochem. Biophys. Res. Commun. 55*, 1262 (1973).
8. Lemasters, J. J. and C. R. Hackenbrock, *Biochemistry 16*, 445 (1977).

# STUDIES ON THE SENSITIVITY OF ATP DETERMINATION USING COMMERCIAL FIREFLY LUCIFERASE[1]

JoAnn J. Webster
Phyllis B. Taylor
Franklin R. Leach

Department of Biochemistry
Oklahoma State University
Stillwater, Oklahoma

## I. INTRODUCTION

There are many variations in the protocol for ATP determination using firefly luciferase. A sample of 109 papers on firefly luciferase was examined for the different buffers used and the following distribution was found (%): arsenate, 64; Tris, 35; glycylglycine, 20; phosphate, 14; and glycine or Mops, 8 each. Since mixtures of buffers were used in several of the assay systems, the frequencies add up to more than 100%. McElroy (1) has commented on the use of arsenate buffered systems; Arsenate is an inhibitor, and if used it lowers the sensitivity of the assay.

Several manufacturers are now producing firefly luciferase reagents and instruments designed for luminescent measurements are being marketed. We have compared the properties of many of the commercial firefly luciferase reagent preparations (2). Because of the availability of new preparations and because of the various assay conditions which have been used, we studied the parameters of the assay with three typical

[1] *Supported in part by EPA Grant R 804613 and Oklahoma Agricultural Experiment Station project 1640. This is manuscript J-3734 from the Oklahoma Agricultural Experiment Station.*




ISBN 0-12-208820-4

luciferase preparations. Sigma FLE-50 was used as a typical lantern extract, DuPont luciferase was used as a typical partially purified luciferase, and Sigma Type IV luciferase was used as an example of crystalline luciferase. The stability of luciferase, luciferin, and ATP was determined under various conditions.

The goal of this project was to develop a sensitive assay procedure using commercially available reagents and instrumentation.

## II. MATERIALS AND METHODS

Light production was measured in a SAI Technology Model 3000 ATP photometer or in a Packard Model 6100 Pico-Lite luminometer. The standard reaction mixture contained (final concentration) 0.025 M Tricine[2] buffer pH 7.8, 5 mM $MgSO_4$, 0.5 mM EDTA, and 0.5 mM DTT.

## III. RESULTS

### A. Reaction System

*1. Buffer.* Webster et al. (3) have compared various buffers in the luciferase reaction system. The order of light production was: Tricine > Hepes = Tris >glycinamide > glycylglycine = Bicine = Hepps = Mops > Tes > phosphate. The buffer used did not influence the pattern of stopped-flow kinetics nor the emission spectrum. The conformation of the enzyme was different in the various buffers as demonstrated by circular dichroism measurements and $K_m$ and $V_{max}$ values. The reaction system contained 0.025 M Tricine at pH 7.8.

*2. Additions - Omissions.* Table I Part A shows the effect of addition of the indicated component to a reaction mixture containg ATP, buffer, and the indicated luciferase-luciferin reagent. The Sigma FLE-50 preparation was

[2]*Abbreviations used: Tricine, 1 N-tris (hydroxymethyl)-methylglycine; dithiothreitol, DTT; bovine serum albumin, BSA; luciferin, $LH_2$.*

TABLE I. *Effect of Additions and Omissions on Luciferase Activity*

| | Activity, Light Units | | |
|---|---|---|---|
| | Enzyme | | |
| Part A<br>Additions | Sigma FLE-50 | DuPont | Sigma Type IV |
| None | 1.1 | 1.3 | 0.2 |
| $+Mg^{2+}$ | 1.6 | 1.7 | 0.1 |
| +EDTA | 0.1 | 0.2 | 0.1 |
| $+LH_2$ | 70.4 | 9.5 | 2.0 |
| +DTT | 1.2 | 1.5 | 0.1 |
| +BSA | 2.4 | 1.7 | 0 |
| Part B<br>Omissions | | | |
| None | 37.9 | 12.0 | 4.7 |
| $-Mg^{2+}$ | 19.9 | 1.1 | 1.7 |
| -EDTA | 32.0 | 11.1 | 4.8 |
| $-LH_2$ | 0.5 | 1.9 | 0 |
| -DTT | 25.0 | 10.5 | 3.7 |
| -BSA | 19.4 | 11.3 | 3.7 |

*100 pg of ATP was used in all reaction mixtures which were 0.025 M in Tricine.*

stimulated by $Mg^{2+}$, luciferin, DTT, and BSA. Light production by the DuPont luciferase-luciferin reagent was stimulated by $Mg^{2+}$ and additional luciferin. Single additions to the Sigma Type IV luciferase gave only minimal increases since most of the required components were absent from the reaction mixture. For Part B the indicated component was omitted from an otherwise complete reaction mixture. $Mg^{2+}$ was required. Bovine serum albumin and luciferin omission from the Sigma preparations decreased light production. While omission of DTT had little effect under the assay conditions, it was added to the reaction mixture because p-chloromercuribenzoate inhibition had established the essentiality of -SH groups.

*B. Measurement System.*

*1. Vial Size.* Changes in the vial size and assay volume used in the SAI Model 3000 photometer change the measured light production. When a 1 ml reaction volume was used, a change from a 20 mm glass scintillation vial to a 10 mm plastic Bio-Vial gave 40% more counts. Reduction of the volume in the Bio-Vial from 0.5 to 0.2 ml gave a 5-fold increase in measured light production. Use of a smaller 6 mm tube failed to increase light production.

*2. Type of Measurement.* DeLuca et al. (4) have suggested that peak height measurement is generally the preferable manner of measuring the reaction. Because integration allows an average of background and a longer period of measurement, this might lead to greater sensitivity. When we determined ATP over the concentration range of 0.2 to 200 ng by peak height determination using voltage measurements, by that regime on the SAI Model 3000 photometer, and by one minute integration, three parallel lines resulted.

*C. Reagent Stability.*

*1. Luciferase.* The DuPont luciferase-luciferin reagent (Table II Part A) was stable in a refrigerator (4°C) for 4 days (prepared in either Tricine or phosphate buffer). When stored longer than 10 days, the preparation was more stable in the phosphate buffer (Part B). When luciferase samples were stored frozen, an average of 74% of the activity was retained with the DuPont preparation and an average of 83% was retained by the Sigma Type IV enzyme. Over a 4 day period there was retention of about 80% activity in both preparations where the same sample was thawed and frozen daily.

*2. ATP.* The essential factor in the longevity of ATP solutions was preparation and maintainance of sterile conditions. Storage in the refrigerator retards some of the destruction as does freezing (Table III). Samples containing 10 μg ATP per ml or greater were stable for at least a year when stored sterile in buffer or frozen.

*3. Luciferin.* When luciferin is stored in the dark, under nitrogen and frozen it is stable for 6 months (Figure 1).

TABLE II. Stability of Luciferase

| Part A | % Activity | |
|---|---|---|
| Hours | Tricine Buffer | Phosphate Buffer |
| 1 | 95 | 86 |
| 5 | 106 | 101 |
| 24 | 102 | 111 |
| 96 | 108 | 106 |
| **Part B** | | |
| Days | | |
| 4 | 109 | 112 |
| 7 | 94 | 115 |
| 14 | 75 | 104 |
| 21 | 57 | 89 |
| 49 | 22 | 65 |

TABLE III. Stability of ATP

| | | % Remaining | | | | |
|---|---|---|---|---|---|---|
| | | Hours | | Weeks | | |
| Solvent | Temp | 8 | 72 | 2 | 24* | 52* |
| Water | $20^{o}$ | 115 | 5 | .001 | 0 | 0 |
| | $4^{o}$ | 101 | 113 | .001 | 0 | 0 |
| | $-15^{o}$ | 93 | 98 | 47 | 70 | 85 |
| Sterile water | $20^{o}$ | 135 | 110 | 86 | - | |
| | $-15^{o}$ | 105 | 69 | 101 | 138 | 93 |
| Sterile tris | $20^{o}$ | 86 | 98 | 96 | - | 114** |

ATP 10 ng per ml except
*diluted to that value from 10 or
**100 μg/ml.

*D. Sensitivity of ATP Determination*

When luciferases are reconstituted and/or supplemented with additional luciferin, there is an increased inherent (with no added ATP) light production. This light production is reduced by aging of the luciferin-luciferase mixture. A freshly reconstituted Firelight preparation produced 9025 counts at zero time decayed to 1086 counts after two hours incubation. Thus, aging is necessary for greatest sensitivity.

The quality of water used is very important. We use tissue culture quality water that has been purified by reverse osmosis, ion exchange resin treatment, glass distillation, Millipore filtration, and autoclaving.

Since fluorescent lights activate impurities in the glass cuvette, which increase background light emission (293 counts/30 sec when protected; 2548 counts/30 sec when exposed to light 1 hr), the measurements are made in a darkened room.

Using either Firelight or Boehringer-Mannheim luciferase in a Packard Pico-Lite photometer, the maximum sensitivity we have achieved reproducibly is 0.1 pg. A typical standard curve is shown in Figure 2.

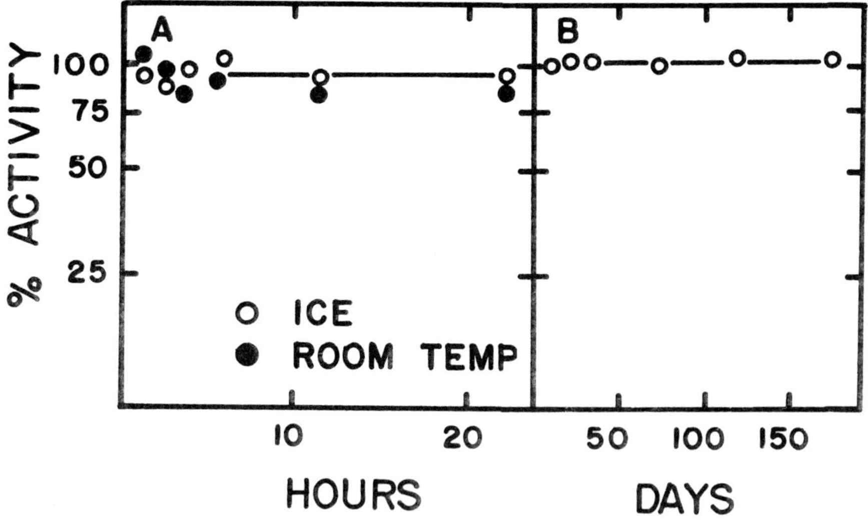


*FIGURE 1. Luciferin stability. Luciferin (5 mg/ml) was dissolved in sterile water, flushed with nitrogen and stored in Biovials wrapped in aluminum foil. For Part B the samples were frozen.*

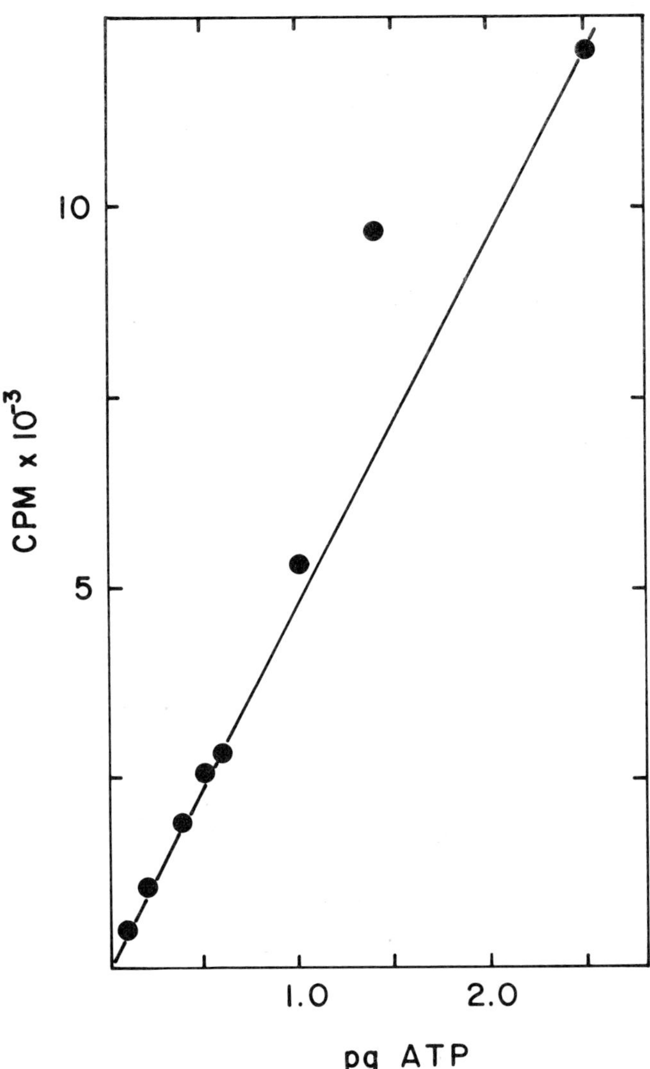


*FIGURE 2. Assay of ATP at low concentrations. Firelight luciferase was reconstituted in Hepes buffer, and aged 3 hr on ice. An equal volume of the enzyme preparation was injected into 100 μliter of sample in Tricine buffer. The Pico-Lite instrument was set for a 1 sec delay and a 30 sec count at 25°C.*

## IV. DISCUSSION AND SUMMARY

In our hands Tricine buffer is superior to any other tested and we believe that this is due to a conformation of luciferase favorable for catalysis (3). $Mg^{2+}$ and luciferin supplementation increase light production. While DTT, EDTA, and BSA do not enhance light production under all circumstances, their addition yields a more reproducible single assay system.

Because of the various geometric relationships between the phototube and the sample container, it is essential to optimize to the proper volume for maximum sensitivity. We find that three parallel lines result when light production is measured by peak height voltage, the peak height mode or integration on a SAI ATP photometer.

The reagents were found relatively stable, but sterility is essential. Using commercial reagents and instrumentation we were able to detect 0.1 pg of ATP with special precautions. With routine conditions the limit of sensitivity was between 1 and 10 pg depending upon the luciferase preparations used.

## REFERENCES


(1) McElroy, W.D. in 2nd Bi-Annual ATP Methodology Symposium (G.A. Borun, ed.), p. 405, SAI Technology Company, San Diego (1977).

(2) Webster, J.J., J.C. Chang, J.L. Howard, and F.R. Leach, *J. Appl. Biochem.* 1, 471 (1980).

(3) Webster, J.J., J.C. Chang, E.R. Manley, H.O.Spivey, and F.R. Leach, *Anal. Biochem.* 105, in press (1980).

(4) DeLuca, M., J. Wannlund, and W.D. McElroy, *Anal. Biochem.* 95, 194 (1979).

# D. General Bioluminescence and Chemiluminescence Including Dinoflagellates, Worms, and Fish

# CHEMILUMINESCENCE OF A 6,7-DIHYDROFLAVIN AND SOME RELATED PTERIDINES

*R. Addink*

Biochemical and Biophysical Laboratory
Delft University of Technology
Delft, The Netherlands

Preparations of 6,7,8-trimethyllumazine produced a flash of light, when treated with hydrogen peroxide in aqueous acidic solution. However, no chemical conversion of the lumazine could then be found. A contamination was responsible for this chemiluminescence (CL). This compound appeared to be 7-hydroxy-6,7-dihydroluminflavin 1 (1). As with hydrogen peroxide in acidic solutions, spontaneous oxidation of 1 in alkaline solutions also led to the 8-oxo compound 2, accompanied by CL (Scheme 1). The mechanism of the removal of the methyl group, under alkaline conditions, was investigated further. As the emission spectrum matched the fluorescence spectrum of 2, it was concluded that 2 was formed in the excited state (2).

The course of the CL is given in fig.2 (curves c1, c3). It was a slow reaction that lasted several days. The time at which the maximum of CL was reached, was dependent on the pH and on the concentration and nature of the buffer used.

Recording UV absorption spectra during the CL reaction at pH 8.7 showed the slow disappearance of the absorption

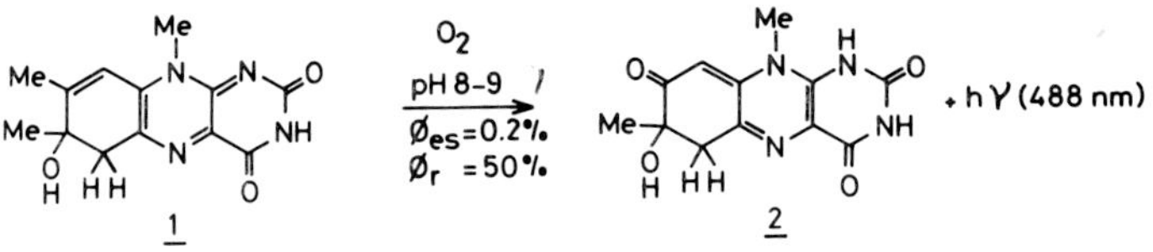

$O_2$, pH 8–9, $\varnothing_{es} = 0.2\%$, $\varnothing_r = 50\%$; 1 → 2 + $h\nu$ (488 nm)

Scheme 1




ISBN 0-12-208820-4

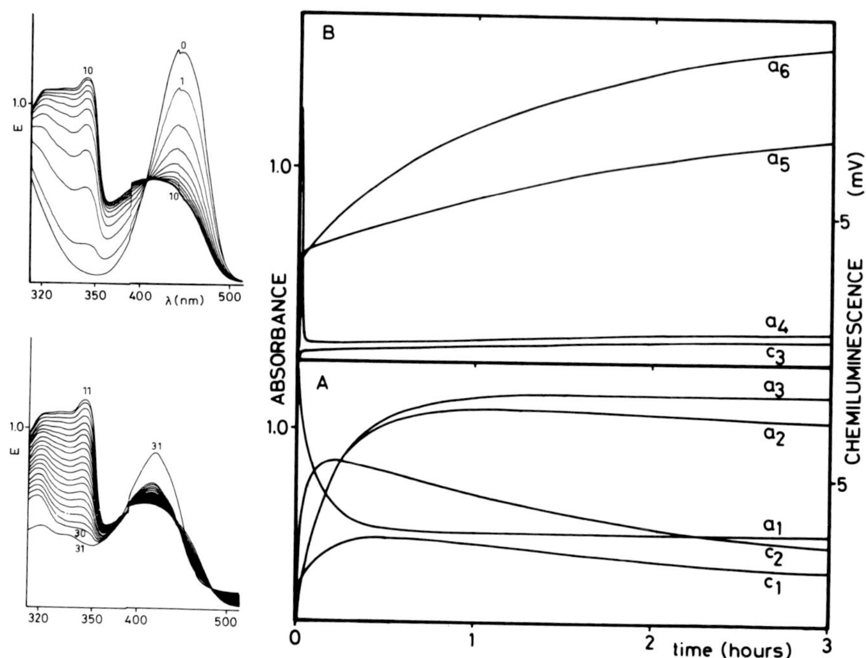


*Fig.1. Repetitive scan of a $6.7 \cdot 10^{-5}$M solution of 1 in water(0), and 0.1M phosphate pH 8.7. A: Spectra were started at 310 nm at 0.5 min(1); every 3.5 min up to 25 min; 32; and 39 min(10) after addition of buffer. (Temperature 25°) B: Continuation, spectra were started at 49 min(11), and every 30 min up to 10.19h (30), and after 24h (31).*

*Fig.2. Course of absorbance (E 445 and E 344) and chemiluminescence at 25° of a $6.7 \cdot 10^{-5}$M solution of 1 in A: 0.1M phosphate pH 8.7. $c_1$, CL, t(max) 29 min; $c_2$, CL after anaerobic accumulation, t(max) 13 min; $a_1$, E 445, aerobic as well as anaerobic; $a_2$, E 344 aerobic, t(max) 1h, (E=1.10); $a_3$, E 344 anaerobic ($E_{max}$=1.17). B: Aerobic, 0.1N KOH; $c_3$, CL, t(max) 3h; $a_4$, E 445; $a_5$, E 344, t(max) 7h, (E=1.30); $a_6$, E 344 in 0.05M NaOMe in MeOH, t(max) 10h, (E=2.0).*

maximum at 445 nm of the starting compound 1 (fig.1A; fig.2A, a1) and the formation of a new absorption maximum at 344 nm (fig.1A; fig.2A, a2). The E 344 reached a maximum value and then decreased again with the concomitant formation of a maximum at 418 nm ascribed to compound 2 (fig.1B). The time at which the intermediate 344 reached its maximum was dependent on the pH and buffer concentration too. It is emphasized that the intermediate 344 is not the direct precursor for the CL reaction, since the CL maximum was reached before E 344 became maximal.

The question arose whether these intermediates would be formed also in the absence of oxygen.

In an anaerobic solution initially the same spectral conversions were observed as in the aerobic experiments. The E 445 decreased in the same manner, but E 344 reached a higher value and did not decrease (fig. 2A, a1, a3). Subsequent addition of oxygen resulted in CL (fig.2A, c2), accompanied by a decrease of E 344. Compared to curve c1, the CL reached a higher value at a shorter time. From this it follows that the two intermediates also were accumulated at pH 8.7 under anaerobic conditions: a precursor for the CL reaction and, in equilibrium with it, the intermediate 344.

Careful acidification of the anaerobic reaction mixture led to the quantitative recovery of the starting compound 1. This indicates that the intermediates are either adducts of 1 with the solvent or tautomers of 1. The influence of the pH and buffer on the formation of the intermediates is in accordance with this suppostion.

In 0.1N KOH a much faster disappearance of E 344 took place (fig.2B, a4) then at pH 8.7. After an initial burst the E 344 was build up slowly (fig.2B, a5). The CL of this solution was very low, again it reached its maximum before E 344 became maximal. This means that in this medium three intermediates were observed separately. In order of formation: an intermediate indicated by the fast disappearance of the visible absorption, the precursor for the CL reaction and the intermediate 344.

When after 205 min the solution was acidified, the spectrum of 1 was quantitatively restored, proving that no appreciable reaction with oxygen had taken place yet.

The influence of solvent appeared from experiments carried out in methanol with methoxide as a base. At low concentrations of methoxide ($10^{-4}$-$5 \cdot 10^{-4}$M), CL and spectral changes were observed, similar to those given in fig.1 for the aqueous solution.

At relatively high methoxide concentrations (0.05M), UV spectral changes occurred (fig.2B, a6), similar to those in 0.1N KOH. However, no CL was observed; and after reaching its maximum value, much higher then in 0.1N KOH, E 344 did not decrease for at least 24 hours. This indicates that 1 was quantitatively converted into the intermediate 344, which was confirmed by NMR. (Decreasing the excess methoxide by addition of acid, again resulted in CL and a concomitant decrease of E 344).

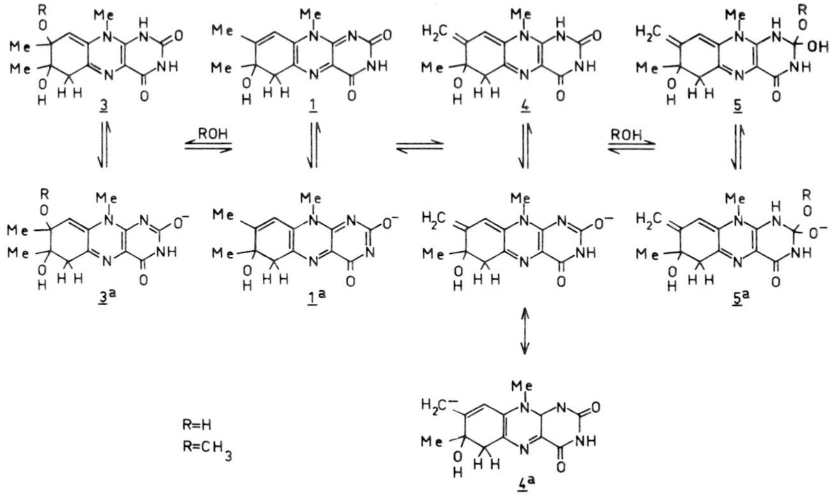

Scheme 2

The molar absorbance of this intermediate in methanol was about 30,000 at 344 nm. Dilution of the methanolic solution and immediately measuring the spectra, gave the molar absorbance of this intermediate at 344 nm in 0.1N KOH and phosphate buffer. It appeared to be the same as in methanol. From this it was calculated that at pH 8.7 under anaerobic conditions (fig.2A, a3) at the highest 60% of the intermediate 344 was formed.

The structure of intermediate 344 was established by NMR to be 5 (Scheme 2). By NMR also the appearance of the adduct 3 was detected. (Details will be published elsewhere (3)). Summarizing, addition of base to 1 shifted the equilibrium between 1 and the adduct 3 and their anionic species to the left, with the concomitant loss of visible absorption. The formation of 5 was revealed by the appearance of the 344 nm maximum. The intermediate $4^a$, formed prior to 5, is the key intermediate in the reaction with oxygen (Scheme 3), as was inferred from the course of the CL and E 344. Compounds 3 and 5 act as storage intermediates, from which the CL reaction is fed slowly.

The formation of the 8-oxo compound 2 in the excited state, as the final product of the CL reaction, is reason to postulate a dioxetane as the immediate precursor in the light giving step. The formation of a second carbonyl compound, formaldehyde, was confirmed.

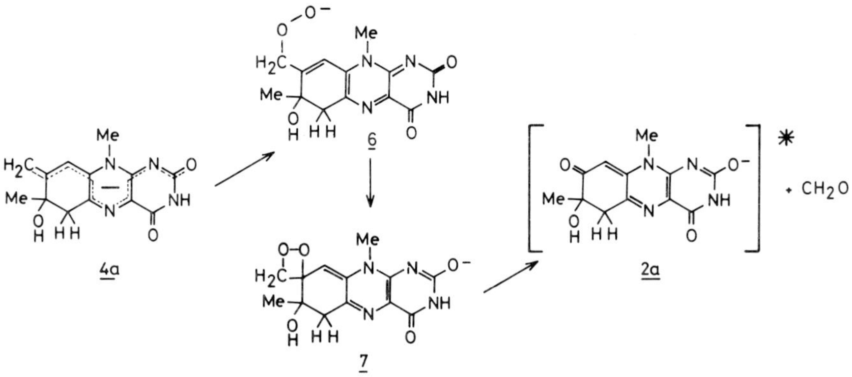

Scheme 3

The proposed route for the formation of the dioxetane 7 is given in Scheme 3. Oxidation of the anion 4a is supposed to give the peroxymethylene anion 6. An intramolecular nucleophilic ringclosure to the dioxetane 7 might then take place, as the 8-position is electrophilic.

AUTOXIDATION OF RELATED PTERIDINES

8-Substituted-6,7-dimethyllumazines and -pterines give the following reactions:
-Hydration at the 7-position (4, 5).
-Formation of a 7-methylene group (6, 7).
-Oxidation by oxygen, giving a 7-oxopteridine (8).

On account of these properties we tested a number of pteridines for CL. In aqueous alkaline solution, autoxidation was not accompanied by CL. In methanol with methoxide no autoxidation occurred. Autoxidation in DMF or DMSO, initiated by addition of a base like $Et_3N$ or t-BuOK, was again accompanied by CL.

The results found in DMF/t-BuOK are summarized in Scheme 4. In the first phase of the reaction, addition of base led to the fast formation of 9 and 12, as judged by UV and NMR spectra of the isolated compounds. In the second phase of the reaction, autoxidation of 9 and 12 finally led to the formation of a 7-oxolumazine 10 or a 7-oxopterine 13 in the excited state. Dioxetanes are proposed as their immediate precursors.

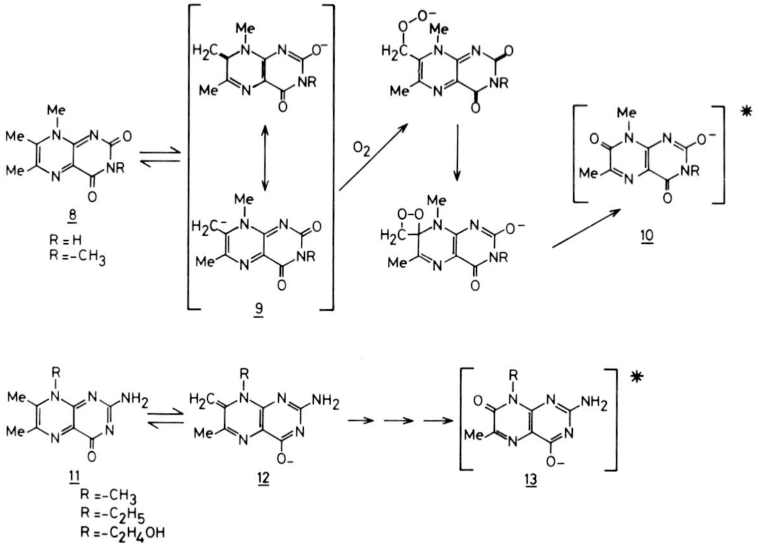

Scheme 4

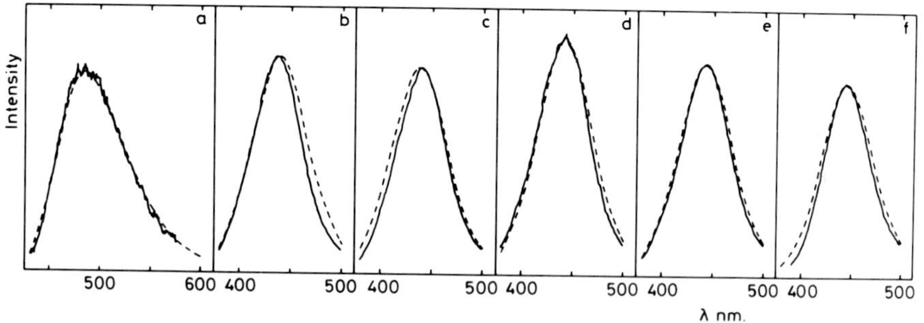


*Fig.3. Solid curves: CL spectra, measured with a bandwidth of 30 nm, of a, 1 in MeOH/NaOMe; in DMF/t-BuOK of b, 8(R=H); c, 8(=Re); d, 11(R=Me); e, 11(R=Et), f, 11 (R=$C_2H_4OH$).*
*Dotted curves: fluorescence spectra of the spent reaction mixtures, identical with those of the authentic 7-oxo compounds.*

The CL spectra matched the fluorescence spectra of the spent reaction mixtures and those of the authentic 7-oxo compounds (fig.3). Also the fluorescence excitation spectra of the spent reaction mixtures and of the authentic 7-oxo compounds were identical. This proves that the 7-oxo compounds were formed in the excited state. The yield of excited states varied between 0.5% and 1%, except for 11($R=C_2H_4OH$), which gave a yield of 0.04%.

The 7-methyl group is essential for the mechanism of these CL reactions. Surprisingly however, addition of t-BuOK to a solution of 3,8-dimethyllumzamine in DMF or DMSO gave CL anyhow. Especially a high concentration of t-BuOK in DMSO gave an efficient CL reaction.

The mechanism of this reaction is not yet clear, but it seems to be connected with the autoxidation of the solvent. In a blank experiment, addition of t-BuOK to DMSO resulted in a fast uptake of oxygen (one equivalent in 15 min). When the lumazine was present a brilliant blue CL was observed. The reaction of the dimsyl anion with oxygen might give peroxy anions, which could attack the electrophilic 7-position of the lumazine, giving a disubstituted peroxide. The formation of dioxetane is difficult to envisage here. Rearrangement via a different mechanism, the CIEEL mechanism for example (9), might then lead to the excited state of 3,8-dimethyl-7-oxolumazine. This was a product in the reaction, it gave a strong blue fluorescence in this medium.

We have also investigated a fully oxidized flavin. Treatment of lumiflavin with t-BuOk in DMF resulted in autoxidation, but no CL was observed. Preliminary investigations showed that lumiflavin-8-carboxylic acid was formed,

## CONCLUSION

The formation of the adduct 3 and the compounds 4 and 5 (Scheme 2) upon treatment of the dihydroflavin 1 with base in anaerobic medium, strongly supports the proposed CL reaction mechanism via a dioxetane as given in Scheme 3.

The general CL autoxidation of the pteridine derivatives proceeding via similar intermediates (9 and 12), further strengthen the validity of this mechanism.

The CL autoxidation of the lumazine lacking the 7-methyl group proceeds via a different mechanism.

## REFERENCES

1. Addink, R., and Berends, W., Tetrahedron 29, 879 (1973).

2. Addink, R., and Berends, W., Tetrahedron 30, 75 (1974).

3. Addink, R., and Berends, W., to be published.

4. Pfleiderer, W., Bunting, J. W., Perrin, D. D., and Nübel, G., Chem. Ber. 99, 3503 (1966).

5. Pfleiderer, W., Bunting, J. W., D. D., and Nübel, G., Chem. Ber. 101, 1072 (1968).

6. Pfleiderer, W., Mengel, R., and Hemmerich, P., Chem. Ber. 104, 2273 (1971).

7. Beach, R. L., and Plaut, G. W. E., J. Org. Chem. 36, 3937 (1971).

8. Rowan, T., and Wood, H. C. S., J. Chem. Soc. C. 452 (1968).

9. Koo, J. Y., and Schuster, G. B., J. Am. Chem. Soc. 99, 6107 (1977).

# *IN VIVO* SPECTROSCOPY OF A BIOLUMINESCENT CELL: *PYROCISTIS LUNULA*

B. Arrio
A. Dupaix
C. Fresneau
B. Lécuyer
P. Volfin

E.R. 118 CNRS, Bât 432
Université de Paris-Sud
Orsay, 91405, France

Until now, the location of the light emitting reaction in dinoflagellates algae like *Noctiluca miliaris, Gonyaulax polyedra* and *Pyrocistis lunula* has been the object of conflicting reports.

The existence of organelles, which would be the light emission centers, has not been unambiguously proved, either by electron microscopy in integral cells or in subcellular fractions obtained by zonal centrifugation. Therefore, scanning and mapping techniques should be of particular interest to obtain some information about location of intracellular components and enzymatic reactions. Photon counting techniques, associated with image intensifiers, are sensitive enough to enable location of light sources as small as 1 um diameter and to provide fluorescence and Raman spectra.




ISBN 0-12-208820-4

# THE COUNTERLIGHTING HYPOTHESIS: *IN SITU* OBSERVATIONS ON ARGYROPELECUS HEMIGYMNUS.

Fernand Baguet

Laboratoire de Physiologie Animale
University of Louvain
Louvain-la-neuve, Belgium

Jacques Piccard

Fondation pour l'étude et protection
de la mer et des lacs
Cully, Switzerland

## I. INTRODUCTION

A possible function of luminescence among marine luminescent fish is based on the preponderance of ventrally-pointing and ventrally-distributed luminous organs. The counterlighting hypothesis suggests that a mesopelagic or a bathypelagic fish might match the dim downwelling residual sunlight from the surface with its ventral bioluminescence (1,2). Being self-illuminated, it merges into the background and the silhouette effect, in ventral view, should be thereby eliminated.

The present work brings the first direct quantitative measures of the parameters controlling the vertical distribution of a luminescent fish, Argyropelecus hemigymnus, and examines to what extent the results are compatible with the counterlighting hypothesis.




ISBN 0-12-208820-4

## II. METHODS

A series of 21 dives was made aboard an autonomous submersible, the mesoscaph FOREL (Jacques Piccard), in the strait of Messina (Sicily) during May 1979.

Light was measured with a photomultiplier PM 270D (International Light) fixed laterally under the plexiglas dome (Fig. 1); the axis of the photomultiplier was parallel to the surface of the water, its window pointing upward.

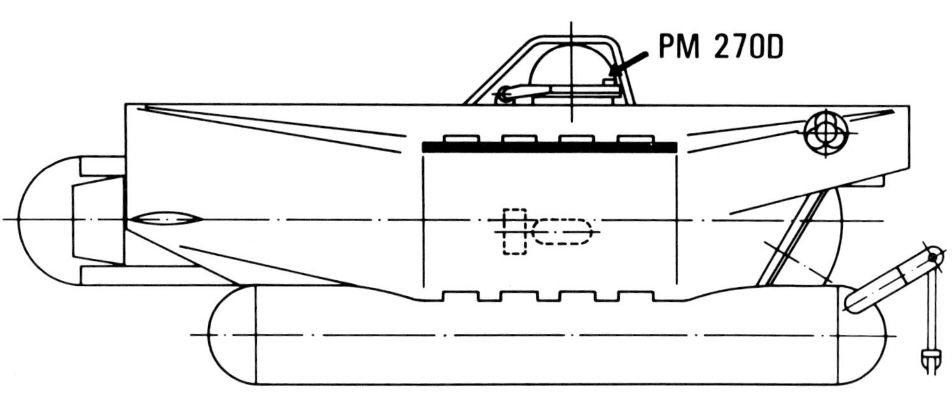


*FIGURE 1.*

Three types of measurements were made aboard the mesoscaph:

1. Transmission of light (430, 470, and 500 nanometers) in the strait from the surface to 500 meters depth, in order to specify the ambient luminescence of the biotope frequented by luminescent fish.

2. Vertical distribution of luminescent fish in the strait at different times of day.
3. Qualitative and quantitative observations of light emission from luminescent fish.

## III. RESULTS

### A. *Light Transmission in the Strait*

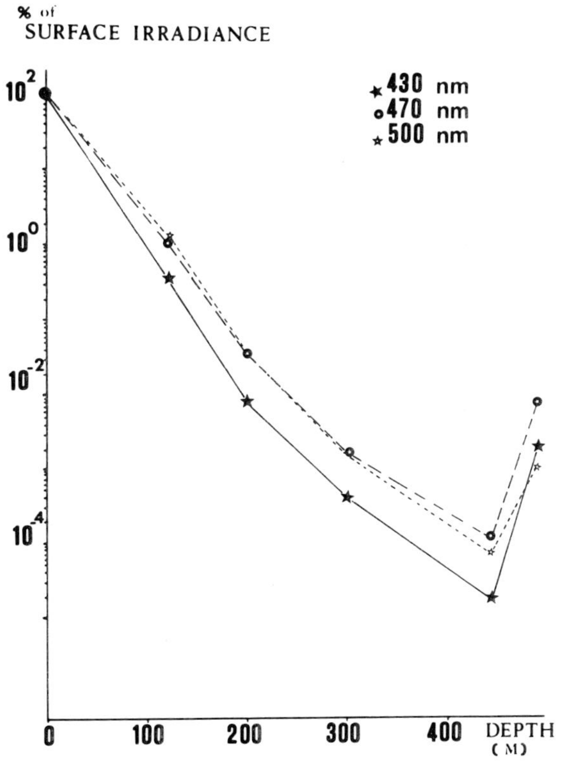


*FIGURE 2. Typical relation between depth and irradiance at three wavelengths expressed as percentage of irradiance values at the surface during a midday dive.*

Figure 2 shows the attenuation of irradiance expressed in % of surface irradiance, measured from the surface at 10.40 h to 490 meters at 13.00 h GMT (38°N 1.7 km, 15°30' E 3.2 km). The magnitude of irradiance decreases with depth to 440 meters; within the next 50 meters irradiance increases rapidly, so that at 490 meters, ambient luminescence was equivalent to the residual sunlight intensity transmitted to 200 meters. In subsequent dives made in neighbouring areas, the layer of light was present at 430 meters at 18.40 h, at 300 meters at 19.15 h and above 120 m after 19.40 h.

*B. Vertical Distribution of Luminescent Fish*

The distribution of Argyropelecus hemigymnus observed during the 72 hours of diving, is limited between 180 and 500 meters; Cyclothone braueri are limited between 330 and 530 meters, while Myctophids were encountered from the surface to 550 meters.

In all our observations, the presence of a fish is correlated with three parameters: the absolute and relative light irradiance at three wavelengths, the depth and the time.

Although there no relation is shown between the absolute intensity of light environment and the presence of Argyropelecus at a given time or at a given depth, we found a striking significant influence of the relative light intensity.

Figure 3 shows that the logarithm of the ambient light expressed as % of the irradiance at surface (470 nm), of the water frequented by Argyropelecus, varies as a function of the time. Regression analysis shows that there is a significant relation between the ambient light level and the time (P 0.001) at which Argyropelecus are encountered:

$$\text{Log \% Irr.} = -10.92 + (0.537 \pm 0.045)\ \text{time}$$

From this general equation we can calculate the relative level of light (470 nm) followed by Argyropelecus at any time of day.

For example, it is calculated that at 13.00 h, Argyropelecus is present in a layer of $1.15 \times 10^{-4}$% of surface irradiance, while at 20.00 h it is in a layer of $6.6 \times 10^{-1}$% of surface irradiance. Knowing the absolute irradiance at the surface at 13.00 and 20.00 h, it is found that the isolume (470 nm) occupied by the fish changes from $1.15 \times 10^{-4}$ μW/cm$^2$ at 13.00 h to $1.3 \times 10^{-6}$ μW/cm$^2$ at 20.00 h.

During this period of time, the isolume followed by Argyropelecus changes by less than two orders of magnitude.

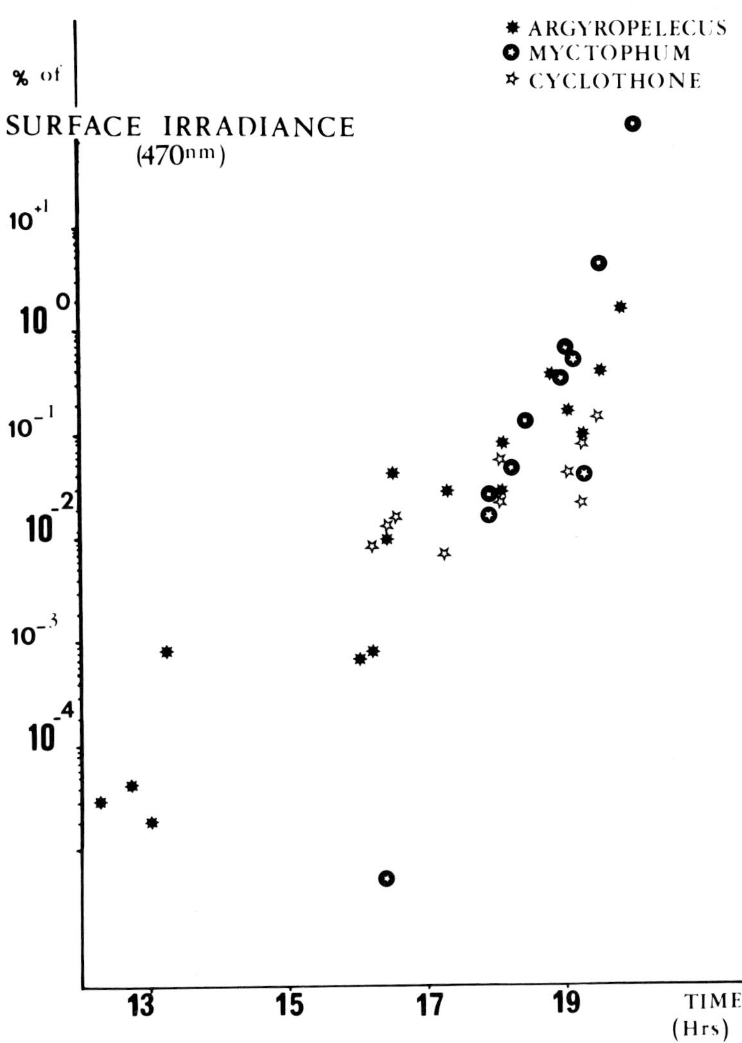


*FIGURE 3. Relationship between the % of surface irradiance (470 nm) and the time at which Argyropelecus ( ✸ ), Myctophids ( ✪ ), and Cyclothone ( ☆ ) are encountered in the strait of Messina.*

For Myctophids and Cyclothone braueri the parameters of the regression equation are significantly different from those calculated for Argyropelecus. During the period extending from 13.00 h to 20.00 h, Myctophids follow an isolume varying from 4.7 x $10^{-11}$ μW/cm$^2$ to 1.5 x $10^{-4}$ μW/cm$^2$. It is suggested that Myctophids do not follow a particular isolume between this period of time.

On the contrary, using the equation elaborated for Cyclothone, it is found that this fish remains in a very close isolume from 13.00 h to 20.00 h: it varies from 3 x $10^{-3}$ μW/cm$^2$ (20.00 h).

### C. *Luminescence of Fish*

Argyropelecus and Myctophids, observed through the plexiglass dome of the Mesoscaph, show a sustained luminescence originating from the ventral photophores of fish and forming a luminous patch. On the other hand, Cyclothone braueri were non-luminous.

We never succeeded to record light produced by Myctophids owing to their rapid swimming. On two occasions we observed and recorded light emitted by Argyropelecus; its distance from the dome was estimated about 20 to 30 cm when the ventral side of the fish went over the window of the multiplier equipped with the 470 nm filter. At that moment a transient signal of 5.4 x $10^{-7}$μW/cm$^2$ was recorded superposed on the background luminescence (9.0 x $10^{-5}$ μW/cm$^2$).

## IV. CONCLUSIONS

The present observations made aboard the Mesoscaph FOREL bring direct evidences that photic environment is, in the strait of Messina, a parameter that affects the vertical distribution of Argyropelecus, Myctophids and Cyclothone. Our visual observations and our light measurements, specify the quantitative relationship between the time or the depth of occurrence of Argyropelecus and the ambiant luminescence.

According to the counterlighting hypothesis, a luminescent fish must exactly compensate the ambiant luminescence by its bioluminescence: in this case we could expect that the presence of such a fish at a given depth is correlated to the absolute value of the ambient light level at this depth. Our results show a significant relation between the presence of Argyropelecus, Myctophids and Cyclothone at a given depth or a given time and the relative but not the absolute level of light.

If we accept that the relationship is specific to luminescent fish, our results do not support the counterlighting hypotehsis just as it is brought forward.

Since Argyropelecus changes its isolume by two order of magnitude between 13.00 and 20.00 h, it is suggested that the animal must modulate its light emission in the same range, and must be able to match the 1.15 x $10^{-4}$ $\mu W/cm^2$ ambient luminescence. According to *in vitro* experiments, living Argyropelecus injected with epinephrine produces a maximal luminescence of 1.7 x $10^{-5}$ $\mu W/cm^2$ at 2.5 cm distance from that photocathode, i.e. 10 times less then the maximal ambient luminescence encountered by the fish during our observation aboard the submersible.

## REFERENCES

(1) Buck, J.B. in "Bioluminescence in action" (P.J. Herring, ed.), p. 419. Academic Press, New York, 1978.

(2) Rauther, M., in "Klassen and Ordnungen des Tierreichs" (H.G. Bronn, ed.) Vol. 6, 1927.

# BIOCHEMICAL STUDIES OF THE BIOLUMINESCENT POLYNOÏD WORMS

A. Dupaix
C. Fresneau
P. Volfin
B. Arrio
B. Lécuyer

E.R. 118 CNRS, Bat. 432
Université de Paris-Sud
Orsay, 91405, France

The molecular mechanism leading to light emission in polynoïd worms has been very little studied up to now. For this purpose, isolation and purification of the fluorescent emitter have been attempted.

Among different resins, best results were obtained when A6 Ultrogel was used to carry out purification. Chromatography of elytra homogenate leads to two main peaks : one of them is excluded in void volume and the other in total volume of the column. The homogenate and the two peaks were characterized by SDS polyacrylamide gel electrophoreses and fluorescence spectra.

From kinetic studies of the luminous reaction in the presence of different triggering agents, we propose a peroxydasic-like mechanism to account for the bioluminescent reaction.



ISBN 0-12-208820-4

# RED FLUORESCENCE OF FISH AND CEPHALOPOD PHOTOPHORES

Peter J. Herring

Institute of Oceanographic Sciences
Wormley, U.K.

## I. INTRODUCTION

The photophores of a number of fishes and squids contain compounds which have a red fluorescence in ultraviolet light. Despite the fact that none of these compounds has yet been unequivocally identified their chemical nature clearly differs considerably from one species to another. They occur in the serial ventral photophores of certain gonostomatid fishes and in the serial and smaller epidermal photophores of some fishes in the suborder Stomiatoidea, as well as the large suborbital photophores of the genera *Pachystomias, Aristostomias,* and *Malacosteus*. Red fluorescence also characterized the photophores of several (probably all) species of the squid genus *Histioteuthis*.

## II. GONOSTOMATID FISHES

Red fluorescence is a feature of the intact photophores of the species of *Ichthyococcus, Polymetme,* and *Valenciennellus*. The aperture of the photophore contains a purple-red chromo-protein, localised in the B cells (1), which is the source of the fluorescence. Pigment from *Valenciennellus* has absorption maxima in aqueous solution at 418, 532, 552, and 600nm with an $E_{418:600}$ of 10.5. The fluorescent prosthetic group is readily extracted in ethyl acetate/acetic acid (3:1 v/v) and after methylation has absorption maxima



ISBN 0-12-208820-4

in neutral chloroform at 410, 555 and 600nm. On silica gel TLC it has an Rf slightly less than that of protoporphyrin IX dimethyl ester.

## III. STOMIATOID FISHES

### A. *Ventral and Epidermal Photophores*

The serial photophores and other small epidermal photophores of all members of the families Astronesthidae, Chauliodontidae, Stomiatidae, Idiacanthidae, Melanostomiatidae and Malacosteidae have a violet, red-fluorescent pigment in their apertures which, like that in the gonostomatids, is associated with the B cell region (1). Many of the regions of 'glandular' luminous tissue in these fishes (2) have a similar associated pigment and fluorescence (e.g. barbel tip, dorsal and lateral patches in *Astronesthes* and the lateral loop on the flanks of *Grammatostomias*).

The pigment is readily extracted as an aqueous chromoprotein solution. That from *Stomias boa* has absorption maxima at 280, 424, 558, and 625nm ($E_{280}$:$_{424}$ = 1.15). The emission spectrum has a single maximum at *ca*. 575nm with excitation peaks at 275, 360 and 405nm. *Chauliodus sloani* material has an almost identical absorption spectrum. Its 421nm peak is reversibly shifted to 428nm on reduction but the 560 peak is not affected. The fluorescence of material from both species is quenched in 5% $H_2SO_4$/MeOH.

### B. *Suborbital Photophores in Malacosteus and Aristostomias*

The suborbital organs of these genera contain substantial amounts of orange-red fluorescent material. It is extracted from *Malacosteus niger* as a dichrois aqueous solution with absorption maxima at 278, 325, (385), 485, 558, and 610nm and an $E_{278:610}$ of 1.77. The fluorescence emission spectrum has peaks at 516, 565, and 635nm and an excitation maximum at ~465nm.

Extracts from *Aristostomias* spp. yield an orange solution with absorption maxima at 422 and 557nm and a shoulder at 460-480nm. Excitation at 360nm gives emission maxima at 508 and 574nm while the 574nm maximimum has an excitation peak at ~475 nm. In neither *Malacosteus* nor *Aristostomias* is there any good evidence for a porphyrin prosthetic group.

## IV. HISTIOTEUTHIS PHOTOPHORES

The purple pigment in the normal epidermal photophores of *Histioteuthis* is not extracted in neutral aqueous solvents but only in acid solution. It has the characteristics of a free porphyrin and after methylation has absorption maxima in neutral chloroform at 408, 507, 542, 596, and 632nm with an $E_{408:507}$ of 8.6. TLC comparisons failed to separate it from authentic protoporphyrin IX dimethyl ester.

## V. CONCLUSIONS

The red fluorescent compounds in different species are neither identical in composition nor probably in function. The observed colour of the light emitted from intact serial and smaller epidermal photophores is blue. The spectral characteristics of the red fluorescent pigments suggest that they may act as narrow bandpass blue filters (3). Such an effect has been experimentally demonstrated in the non-fluorescent photophore pigment of the hatchet fish *Argyropelecus* (4). The value of such a filter in mesopelagic species lies in matching the spectral distribution of emitted light more closely to that of downwelling daylight and hence increasing the effectiveness of ventral counter-illumination. The variety of chemical compounds in different photophore strengthens the assumption that it is their spectral characteristics which are of biological significance. The erratic taxonomic appearance of red fluorescence in gonostomatids is thus merely a reflection of the variety of compounds utilized. The presence of these pigments in the photophores and glandular luminous tissue of the deeper-living stomiatoids is less easy to interpret, for the value of monochromatic light emission is less obvious below the depths of daylight penetration. The pigment is absent from both the simple silhouetting luminous tissue (5) and the large postorbital photophores of most of these fishes. This implies that its role in those photophores where it is present is an accessory one, and that it is not a necessary metabolic component or byproduct of the bioluminescence reaction system.

The three genera with red fluorescent suborbital organs are unusual in that they emit red light from these photophores (6). The postorbital organs in these fishes emit blue light and do not have the red fluorescent material. It therefore seems likely that the red fluorescent material acts as a secondary emitter in the suborbital organs, in a manner analogous to the green fluorescent protein of *Renilla* (7). If this is the case it is to be expected that the in vivo red emission of these fishes will bear some relationship to the fluorescence emission spectra of the extracted chromoproteins.

## REFERENCES

(1) Bassot, J-M., in "Bioluminescence in Progress (F.H. Johnson and Y. Haneda, eds), p. 557. Princeton University press, Princeton (1966).
(2) Herring, P.J., and J.G. Morin in "Bioluminescence in Action" (P.J. Herring, ed.), p. 273. Academic Press, London (1978).
(3) Herring, P.J., *Nature* 267, 788 (1977).
(4) Denton, E.J. and P.J. Herring, *J. Physiol., Lond.* 284, 42P (1978).
(5) O'Day, W.T., *Contr. Sci.* 246, 8pp. (1973).
(6) Denton, E.J., J.B. Gilpin-Brown, and P.G. Wright, *Proc. R. Soc. B.* 182, 145 (1972).
(7) Cormier, M.J. in "Bioluminescence in Action" (P.J. Herring, ed.), p. 75. Academic Press, London (1978).

# OXYGEN-DEPENDENT *STREPTOCOCCUS FAECALIS* CHEMILUMINESCENCE: THE IMPORTANCE OF METABOLISM AND MEDIUM COMPOSITION[1]

Deborah J. Hunter
Robert C. Allen[2]

Department of Clinical Investigation
Brooke Army Medical Center
Fort Sam Houston, Texas

## I. INTRODUCTION

*Streptococcus faecalis*, a member of the lactic acid family of bacteria, lacks a cytochrome system for electron transport and the ability to synthesize protoporphyrin enzymes such as catalase. The microbe can be cultured under anaerobic and aerobic conditions; however, rapid exposure of an anaerobic culture of *S. faecalis*[3] to $O_2$ results in peroxide generation. This activity is catalyzed by a flavoprotein oxidase, NADH: $O_2$ oxidoreductase (1-3). Aerobically cultured *S. faecalis* synthesizes an additional flavoprotein peroxidase, NADH: $H_2O_2$ oxidoreductase, that prevents peroxide accumulation by reduction of peroxide to $H_2O$ (4). *S. faecalis* also synthesizes a manganese containing superoxide dismutase (MnSOD)[4] (5-6). This enzyme scavenges superoxide anion

[1]*The research was supported by Department of Clinical Investigation sponsored project #C-25-78.*

[2]*The opinions or assertions contained herein are the private views of the authors and are not to be construed as reflecting the views of the Department of the Army or the Department of Defense.*

[3]*S. faecalis: Streptococcus faecalis.*

[4]*MnSOD: Manganese superoxide dismutase.*



ISBN 0-12-208820-4

($\cdot O_2^-$)[5] yielding $H_2O_2$ as product (7). The activity of MnSOD increases with increase in culture $O_2$ tension.

Exposure of anaerobic cultures of *S. faecalis* to $O_2$ results in a detectable chemiluminescence, CL[6] (9,10). The present report describes the results of investigations into the role of $O_2$ and its reduction products, $\cdot O_2^-$ and $H_2O_2$, in *S. faecalis* CL.

The effects of culture age (growth phase), medium preparation, and scavenger enzymes were investigated. The observation of CL from autoclaved, anaerobic trypticase soy broth is presented. Isolated *S. faecalis* in phosphate buffered saline require both $O_2$ and a metabolic substrate for CL.

## II. MATERIALS AND METHODS

### *A. Bacterium and Medium*

*S. faecalis* strain number 29212 was obtained from the American Type Culture Collection. Trypticase soy broth (TSB)[7], a soybean-casein digest medium, was purchased from BBL Inc. The TSB was autoclaved at 121°C for 15 min., made anaerobic by gassing with 99.999% $N_2$ for 2 min., and allowed to cool to 23°C before use.

The bacterial density of the culture was determined turbidimetrically using a Coleman Nepho-colorimeter set at 580 nm. Bacteria viability was determined by plate count. Using a given volume and different dilutions of culture, the colonies of *S. faecalis* anaerobically grown on trypticase soy agar were counted, and the count compared to the optical density (O.D.)[8] measurement.

Washed, isolated *S. faecalis* were prepared from TSB cultures in the early exponential phase of growth. The culture was centrifuged (500 x g) for 15 min., the broth decanted, and the *S. faecalis* pellet resuspended in normal saline (8.5 g NaCL/l $H_2O$). The saline wash was repeated and the

[5] $O_2^-$: *Superoxide anion*

[6] *CL: Chemiluminescence*

[7] *TSB: Trypticase soy broth*

[8] *O.D.: Optical density.*

bacteria were resuspended in Dulbecco's PBS[9] (11).

In all the experiments described, one ml of the test sample was transferred to a vial containing one ml of normal saline. The vials were made anaerobic by gassing with 99.999% $N_2$. The total volume per vial was 24 ml.

### B. *Enzymes*

1. Bovine erythrocyte superoxide dismutase ($\cdot O_2^-:\cdot O_2^-$ oxidoreductase) was purchased from Sigma Chemical Co. Its activity was 2.7 units/μg by assay (7).

2. Bovine liver catalase ($H_2O_2:H_2O_2$ oxidoreductase) as a thymol-free powder was purchased from Sigma Chemical Co. Its activity was 11.0 units/μg (12).

3. Myeloperoxidase (donor:$H_2O_2$ oxidoreductase) was purified from human polymorphonuclear leukocytes to an absorbance ratio $A_{430}$nm/$A_{280}$nm of 0.3. Based on extinction coefficient, 40 picomoles were added to the vials where indicated (13).

### C. *Measurement of CL*

A Beckman LS150 scintillation counter was employed for single photon counting. It was equipped with EMI 9829A photomultiplier tubes (bialkali spectral response), and operated in the out-of-coincidence mode with the tritium channel settings. The raw counts were converted to photons by calibrating the counter with a blue photon standard prepared by H. Seliger (14). Counts multiplied by fourteen equal blue photons.

## III. CHEMILUMINESCENCE FROM *S. FAECALIS:* EFFECT OF CULTURE AGE AND ANAEROBIOSIS

The effect of culture age on the CL response to $O_2$ exposure was investigated by establishing a growth curve based on turbidimetric measurement of cell density. Aliquots from the culture were periodically tested for CL response upon $O_2$ exposure over a 20 hour period. A heavy inoculum of an

[9] *PBS: Phosphate buffered saline.*

anaerobic culture was used to initiate growth at time zero. Based on turbidimetric measurement, as depicted in Figure 1, the culture entered into the exponential growth phase within one hour of initiation. Bacterial cell density and bacterial viability counts, measured as colony forming units (CFU)[10], correlated with an $r^2 = 0.99$ over the initial three hours of culture; that is, correlation was maintained up to an O.D. of 0.4. At higher density, total viability decreased with increase in time.

Data for comparing culture density and luminescence response with change in time is presented in Figure 1. The rate of increase in CL correlates with increase in cell

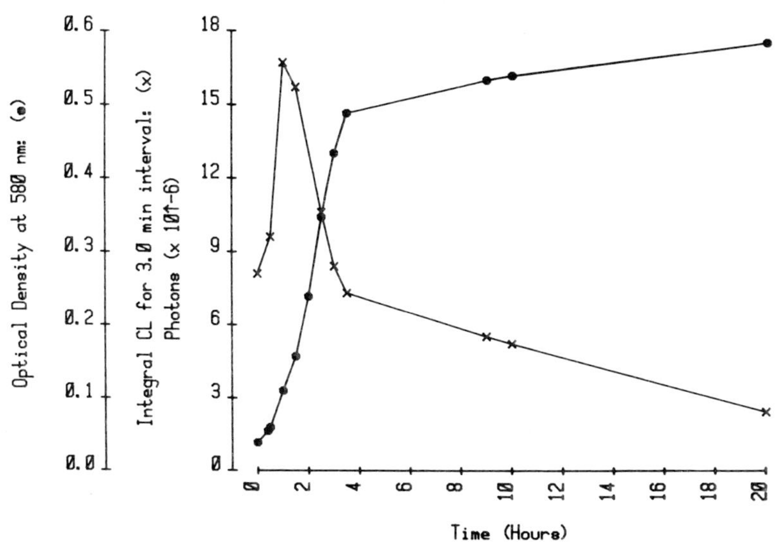


*Figure 1. Effect of culture age on the $O_2$ dependent net CL response from anaerobic cultures of S. faecalis. The net CL for the initial 3 min. period following air exposure and the turbidimetric values for bacterial density are plotted against time. Culture was initiated at time zero by addition of a heavy inoculum of anaerobic S. faecalis to TSB. Viability to plate counts was a linear function of culture density over the range of OD from 0.05 to 0.42. The colony forming units x $10^7$/ml equaled 36 x (OD 580 nm) + 1.6.*

[10] *CFU: Colony forming units*

density for the initial 90 min. of culture. Thus, CL response, cell density, and viability all correlate during the early exponential phase of growth. A maximal CL of approximately 0.2 photons/min./bacterium was obtained at 60 min. of growth. Note the inverse relationship between culture density and CL in late exponential and stationary phases of growth.

The plot of CL from anaerobic and aerobic cultures of *S. faecalis* against culture density is presented in Figure 2. The CL response from the anaerobic culture exceeded that of the aerobic culture by more than one order of magnitude. However under either condition, maximal CL was observed during the early exponential phase of growth.

## IV. *S. FAECALIS* AND TRYPTICASE SOY BROTH CL: THE ROLE OF $O_2$

Anaerobic cultures of *S. faecalis* were grown to a concentration of 1.9 x $10^7$ CFU/ml ($OD_{580}$ 0.095). Under anaerobic conditions, one ml of culture was added to each vial

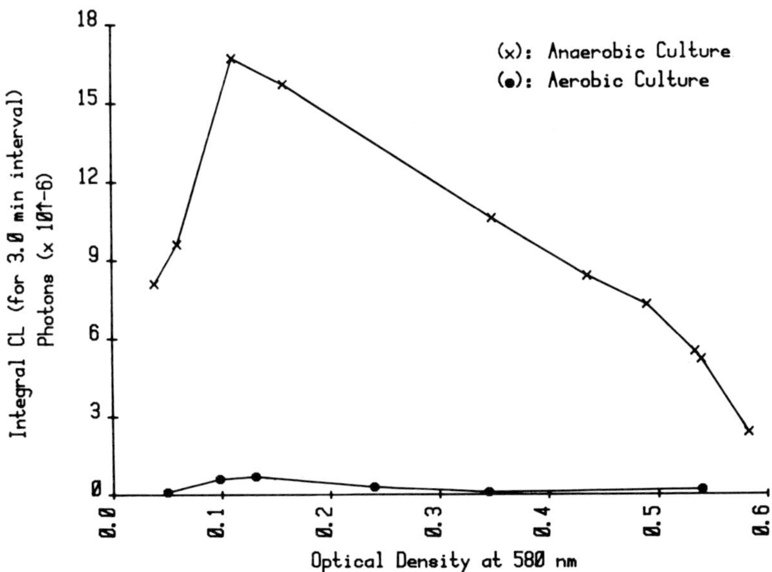


*Figure 2. The plot of $O_2$-dependent net CL responses from anaerobic and aerobic cultures of S. faecalis against culture density. The relationship of turbidimetric reading to viable bacteria is the same as stated in Figure 1.*

containing one ml of saline. Autoclaved, anaerobic TSB (pre *S. faecalis* growth) was prepared in a like manner. Different volumes of 100% $O_2$ were added to vials through a gas-tight rubber stopper and CL was monitored over a 74 min. interval. Based on the vial's gas volume of 22 ml, the range of $O_2$ partial pressures ($PO_2$)[11] tested was 3.4 mm Hg (100 μl 100% $O_2$) to 171 mm Hg (5000 μl /22 ml). The results are presented in Figure 3. The relatively large standard error of the mean illustrates the difficulty in maintaining anaerobic conditions.

CL was also measured following exposure of anaerobic *S. faecalis* culture or TSB to room air (20.9% $O_2$; $PO_2$ 157). The results are presented to the left of and not falling on the x-axis. These values are in agreement with those predicted by $O_2$ titration.

Anaerobically prepared, autoclaved TSB and other media have been reported to generate $\cdot O_2^-$ and peroxide following exposure to $O_2$ (15,16). The observation of CL from anaerobic TSB in the absence of *S. faecalis* is consistent with these reports, and was further investigated.

## V. TRYPTICASE SOY BROTH CHEMILUMINESCENCE

*S. faecalis* were anaerobically cultured in autoclaved TSB until growth reached the early exponential phase. Maintaining anaerobic conditions, an aliquot of the culture was centrifuged, and the *S. faecalis* were washed twice with equal volumes of saline and finally suspended in Dulbecco's PBS without glucose. *S. faecalis* in broth, isolated *S. faecalis* in PBS, and TSB alone (pre and post culture) were tested for CL response following air exposure. The results are presented in Figure 4.

As suggested by the data of the previous section, a relatively large CL was detected from autoclaved, anaerobic TSB either pre or post *S. faecalis* growth. Although not indicated in the graphics, CL was not obtained from unautoclaved, filtered TSB under otherwise identical conditions. This latter observation is in agreement with a previous report (17).

*S. faecalis,* washed free of broth and in the absence of a metabolic substrate, did not yield a detectable CL following $O_2$ exposure (Figure 4 curve b). However, the CL response from *S. faecalis* in TSB was twofold higher than the response from either pre or post TSB alone.

[11] *$P_{O_2}$: $O_2$ partial pressure*

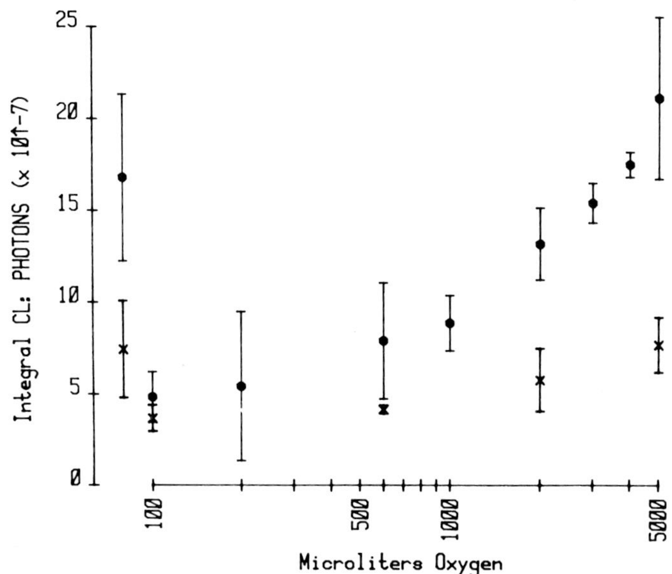


*Figure 3. Integral CL from anaerobic S. faecalis culture and anaerobic TSB plotted against the volume of $O_2$ added per vial. CL is expressed as the integral photons for the 74 min. time interval post-$O_2$ addition. S. faecalis was cultured in TSB to a density of 8 x $10^6$ CFU/ml. Autoclaved, sterile TSB was anaerobically prepared and tested in the same manner. One hundred percent $O_2$ was added to the vials containing one ml of culture or broth, plus one ml of saline. The remaining gas volume of the vial was 22.0 ml; therefore, the range of $O_2$ concentration was from 0.45% to 22.7%/vial. The values to the left of the x-axis are the response to direct air exposure, 20.9% $O_2$.*

## VI. CHEMILUMINESCENCE FROM ISOLATED *S. FAECALIS*: THE REQUIREMENT FOR BOTH $O_2$ AND METABOLIC SUBSTRATE

*S. faecalis* was anaerobically cultured in TSB to early exponential growth phase. The bacteria were then isolated from broth cultures as described in the previous section. The isolated components, *S. faecalis* in PBS and TSB free of bacteria, were exposed to air where indicated, and the integral CL recorded over a 56 min. time interval. The results are depicted by the bar graphs of Figure 5. The samples

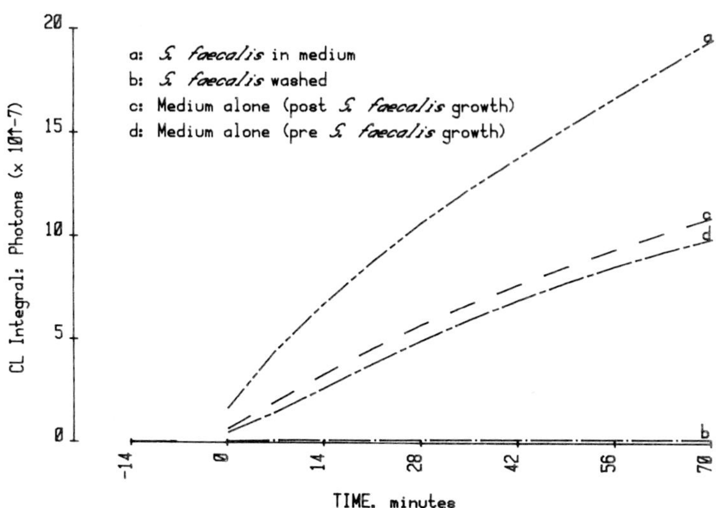


*Figure 4. Plot of the net integral CL from S. faecalis culture in TSB, S. faecalis free of TSB, and bacteria free TSB before and after S. faecalis growth. S. faecalis was anaerobically cultured in TSB to 8 x $10^6$ CFU/ml. S. faecalis was anaerobically isolated from broth, washed with saline, and resuspended in phosphate buffered saline without metabolic substrates to a concentration of 8 x $10^6$ CFU/ml. Pre and post culture TSB was bacteria-free and anaerobic. The specimens were exposed to air at time zero.*

represented by bars a, c, and d received glucose (1.0 mg/ml). The a-bars were not exposed to air.

No CL was detected from *S. faecalis* in the absence of either $O_2$ or glucose. However, CL was detected from isolated *S. faecalis* when both glucose and $O_2$ were present. Although not shown in the figure, glycerol could be substituted for glucose. The addition of glucose to bacteria-free TSB produced the opposite effect; that is, glucose inhibited the $O_2$-dependent CL response.

Myeloperoxidase (MPO)[12] was isolated from human polymorphonuclear leukocytes. This peroxidase is microbicidal if presented with $Cl^-$ as cofactor and $H_2O_2$ as substrate. The oxidative activity of MPO correlates with detectable CL (18,19). In the present experiment, MPO was added where indicated.

[12] *MPO: Myeloperoxidase*

Increase in CL response by the presence of MPO is consistent with the generation of $H_2O_2$ by *S. faecalis* and its presence in isolated TSB.

## VII. CL FROM *S. FAECALIS* CULTURE: INHIBITION BY CATALASE AND SUPEROXIDE DISMUTASE

As previously reported, *S. faecalis* culture CL is inhibited by addition of exogenous catalase and superoxide dismutase (9,10). In the present study, inhibition of CL by either catalase or superoxide dismutase was investigated over a broad range of scavenger enzyme concentrations. One ml aliquots were taken from an anaerobic culture of *S. faecalis* in early exponential phase, and were transferred, under anaerobic conditions, to vials containing either CuSOD[13] or catalase in one ml of saline. CL was recorded for the 65 min. interval following exposure to air. Figure 6 is the plot of integral CL against the log concentration of scavenger enzyme present.

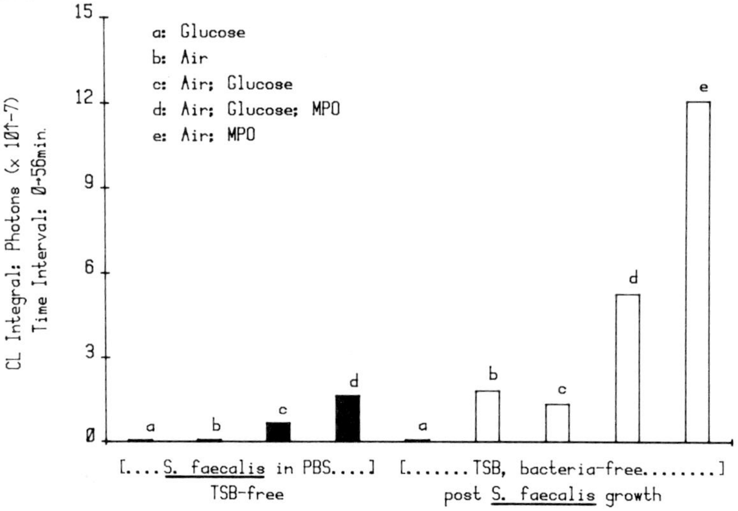

*Figure 5. Bar graph plot of the net integral CL responses from isolated, TSB-free S. faecalis, (3 x $10^7$ CFU/vial) suspended in PBS and post-S. faecalis growth TSB free of air and glucose. Glucose was added to a final concentration of 1 mg/ml.*

[13] *CuSOD: Copper superoxide dismutase*

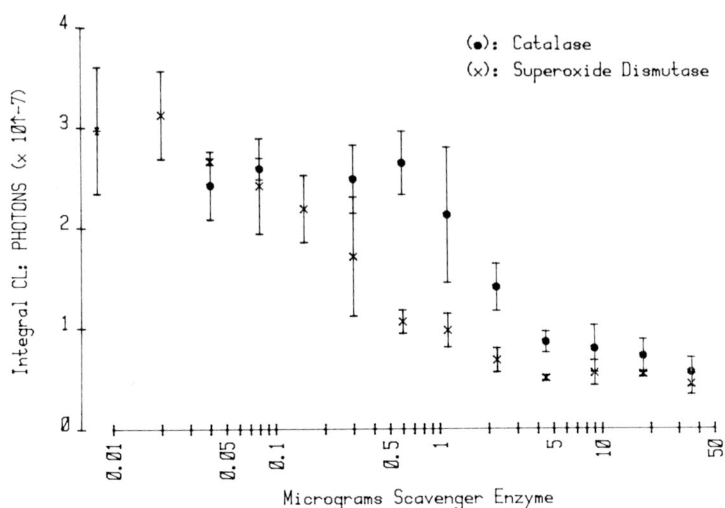


*Figure 6. The CL responses from anaerobic S. faecalis culture plotted against the quantity of scavenger enzyme added. S. faecalis was anaerobically cultured in TSB to an $OD_{580}$ of 0.2, and one ml aliquots were transferred to vials containing the indicated quantity of enzyme in one ml of saline. The net integral CL over the 65 min. post-air exposure period was recorded. Bovine erythrocyte CuSOD had a specific activity of 2.7 units/μg; bovine catalase had a specific activity of 11.0 Sigma units/μg.*

The generation of $H_2O_2$ by *S. faecalis* and its presence in anaerobic, autoclaved TSB was suggested by the MPO studies of the previous section. The inhibition of CL by catalase is further evidence of the presence of $H_2O_2$ and its involvement in CL. Fifty percent inhibition was obtained with a catalase concentration of 1 μg/ml (11 Sigma units/ml).

Exogenous bovine erythrocyte CuSOD exerted a potent inhibition of CL. The percent inhibition was proportional to the log of CuSOD concentration over the range 0.04 to 4.5 μg/ml. Fifty percent inhibition was obtained with a CuSOD concentration of 0.2 μg/ml, 0.5 cytochrome c units/ml (7). The exquisite sensitivity of CL to inhibition by CuSOD is considered as good evidence for the generation of $\cdot O_2^-$ and implies its participation in $O_2$-redox reactions yielding electronically excited products.

## VIII. CONCLUSIONS AND SUMMARY

*Streptococcus faecalis*, a member of the lactic acid family of bacteria, lacks a cytochrome electron transport system and the ability to synthesize protoporphyrin enzymes, such as catalase. This microbe can adapt to growth under both anaerobic and aerobic conditions. Peroxide generation has been reported to follow exposure of anaerobic cultures of *Streptococcus faecalis* to $O_2$. Chemiluminescence is also detected following $O_2$ exposure. The luminescent response to oxygen exposure is dependent upon the degree of prior anaerobiosis and upon the phase of culture growth. Maximum luminescence is observed during the early exponential phase. Exposure of autoclaved trypticase soy broth to $O_2$ yields a chemiluminescence in the absence of *Streptococcus faecalis*. This observation is consistent with reports that $\cdot O_2$ and $H_2O_2$ are produced by such exposure. Chemiluminescence was observed from *Streptococcus faecalis* washed free of broth and resuspended in Dulbecco's phosphate buffered saline; however, unlike the luminescence from autoclaved medium, a metabolic substrate, such as glucose or glycerol, is required for chemiluminescence. Oxygen is required for both bacterial and broth CL. The chemiluminescence from broth cultures of *Streptococcus faecalis* was inhibited by addition of the $H_2O_2$ scavenger enzyme, catalase; and also by the $\cdot O_2^-$-scavenger enzyme, superoxide dismutase. Furthermore, addition of myeloperoxidase to either washed, metabolically active *Streptococcus faecalis*, or to the broth used for growth of *Streptococcus faecalis*, yielded a large luminescent response indicating the generation and accumulation of $H_2O_2$, a substrate required for myeloperoxidase chemiluminescence.

## ACKNOWLEDGMENTS

The authors wish to thank Mrs. D.S. Bratten for assistance in preparing this manuscript.

## REFERENCES

1. Seeley, H. W. and P. J. Vandemark, *J. Bacteriol. 61*, 27 (1951).
2. Seeley, H. W. and C. del Rio Estrada, *J. Bacteriol. 62*, 649 (1951).

3. Dolin, M. I., *in* "The Bacteria", Vol. 2 (eds. I. C. Gunsalus and R. Y. Stanier), p. 425, Academic Press, New York (1961).
4. Dolin, M. I., *J. Biol. Chem. 225,* 557 (1957).
5. McCord, J. M., B. B. Keele, Jr., and I. Fridovich, *Proc. Nat. Acad. Sci. USA 68,* 1024 (1971).
6. Britton, L., D. P. Malinowski, and I. Fridovich, *J. Bacteriol. 132,* 229 (1978).
7. McCord, J. M. and I. Fridovich, *J. Biol. Chem. 244,* 6049 (1969).
8. Gregory, E. M., and I. Fridovich, *J. Bacteriol. 114,* 543 (1973).
9. Allen, R. C., *Photochem. Photobiol. 30,* 157 (1979).
10. Allen, R. C., *in* "The Significance of Superoxide Dismutase", Vol. 1 (eds. J. V. Bannister and H. A. O. Hill) in press, Elsevier, New York (1980).
11. Dulbecco, R., and M. Vogt, *J. Exp. Med. 99,* 167 (1954).
12. Sigma Biochemical Catalog, *Sigma Chemical Company,* 163, (1980).
13. Bakkenist, A. R. J., R. Wever, T. Vulsma, H. Plat, and B. F. Van Gelder, *Biochimica et Biophysica Acta 528,* 45 (1978).
14. Lee, J., and H. H. Seliger, *Photochem. Photobiol. 4,* 1015 (1965).
15. Frölander, F., and J. Carlsson, *J. Clin. Micro. 6,* 117 (1977).
16. Carlsson, J., G. Nyberg, and J. Wrethen, *Appl. Env. Micro. 36,* 223 (1978).
17. Biggley, W. H., R. C. Allen, J. P. Hamman, and H. H. Seliger, *Program 8th Annual Meeting Amer. Soc. Photobiol., Abstract THPM-B2,* 141 (1980).
18. Allen, R. C., *Biochem. Biophys. Res. Commun. 63,* 675 (1975).
19. Allen, R. C., *Biochem. Biophys. Res. Commun. 63,* 684 (1975).

# PRELIMINARY ULTRASTRUCTURAL DESCRIPTION OF COELOMOCYTES OF THE LUMINESCENT OLIGOCHAETE, *Pontodrilus bermudensis* (ANNELIDA)

Barrie G. M. Jamieson

Zoology Department
University of Queensland
Brisbane, Australia

John E. Wampler

Department of Biochemistry
University of Georgia
Athens, Georgia

Michael C. Schultz

Zoology Department
University of Queensland

## I. INTRODUCTION

*Pontodrilus bermudensis* is a marine littoral member of the earthworm family Megascolecidae. On strong electrical, mechanical or chemical stimulation this species exudes a luminous fluid from the mouth and anus. The luminescence is whitish with peak intensity at the wavelength of 550nm. Emission of light by the exuded fluid is further stimulated by addition of hydrogen peroxide and/or *Diplocardia longa* luciferase, response to the enzyme suggesting that *Pontodrilus* luciferin is n-iso-valeryl-3-amino-propanal as in *D. longa* or a close analogue of this (1). The alimentary exudate of *Pontodrilus* has been shown to contain cells identical with coelomocytes despite the absence of dorsal pores through which coelomic fluid is exuded in *Diplocardia*. However, particle-bound or



ISBN 0-12-208820-4

cellular luminescence has been denied for *Pontodrilus* (2; Lynch, personal communication). In *D. longa* luminescence has been directly observed to originate from discrete subcellular loci within large coelomic cells described as free chloragogen cells and in the sessile chloragogue (3). The source of luminescence in *Pontodrilus* requires re-examination and the present study is concerned with ultrastructural description of cells of the exudate and of the coelomic fluid preparatory to videomicroscope and other investigations of this source. Some reference will be made to cytochemical studies which are in progress.

Apart from the study on *Diplocardia* cells, no ultrastructural studies of megascolecid coelomocytes have been published. Constraints of space will not permit detailed comparison of *Pontodrilus* coelomocytes with those of species of other families but a review of the ultrastructural literature is given elsewhere (4).

## II. MATERIAL AND METHODS

Worms collected from Peel Island, Queensland, were maintained for 16 hours in powdered filter paper moistened with filtered sea water to clear the gut. Coelomocytes were obtained either by placing individual worms in ice-cold glutaraldehyde (3%) in 0.1M phosphate buffer, when alimentary exudate was expelled directly into the fixative, or by dorsal incision of the worm in sea water, pipetting off the coelomic fluid and transferring this to glutaraldehyde. Alternatively exudate from electrically stimulated worms was transferred to the fixative. The samples were fixed in the glutaraldehyde for 2 hours, post-fixed in 1% osmium tetroxide in 0.1M buffer for 80 minutes, rinsing in buffer for 30 minutes before and after, and embedded in Spurr's medium after dehydration in a graded ethanol series (30 minutes in each of 20,40,75,90% and absolute). The temperature was maintained at 4°C from glutaraldehyde through 75% alcohol but at room temperature for later alcohols. At each stage samples were centrifuged at 100-300g (100g was found preferable as higher speeds disrupted mucocytes) to allow removal and replacement of the supernatant liquid. Ultrathin sections were cut on an LKB Ultratome IV, stained with uranyl acetate and alkaline lead citrate, mounted on uncoated grids and photographed with an AEI Corinth electron microscope.

## III. RESULTS

### *A. Introduction*

At least nine morphological types are recognizable from their ultrastructure. Of the names applied to these, only two, mucocyte and linocyte, have previously been used. As the cells all resemble lymphocytes in occurring in body fluids outside the vascular system, they are here collectively referred to as lymphocytoids. A subset of these characterized by large numbers of secretory granules will be termed granulocytes, with various distinctive epithets. As in the terminology for human blood cells, granules may occur, though in lesser numbers, in cells not termed granulocytes. Mucocytes and linocytes will be regarded as a distinct subset of the lymphocytoid.

### *B. Ultrastructurally Defined Coelomocyte Types*

*1. Mucocyte (Figs. 3, 11).* Mucocytes are the most abundant of the coelomocyte types; large clusters of them occupy much of the coelom of most segments in paraffin sections. The live cells, in alimentary exudate or released by dissection, rapidly disrupt, shedding their granules, in hyptonic media or on exposure to air if unfixed. They remain intact if expelled into glutaraldehyde or into sea water followed rapidly by fixation. The freed secretory granules become very numerous in untreated exudate and give rise by fusion to strands of moderately electron-dense material which is deduced to be the viscous mucus which characterizes the exudate. When the cells disrupt they also release GER vesicles and mitochondria; the residual free and contained membranes may form configurations resembling portions of the internal "threads" of linocytes (see below). The released granules and other organelles, or partial or whole mucocytes, are commonly phagocytosed by vesicular lymphocytoids in which they are observable in phagocytic vacuoles.

Uncentrifuged or gently centrifuged mucocytes are regular spheres. In resin sections the mean cell diameter is 8.3μm (S.D. 1.7, n 16) and the nuclear diameter 3.2 μm (range 2.8-3.7, n 8), with a cytoplasmic to nuclear ratio in diameters of 3.0 (range 2.5-3.5). Cells in paraffin sections (buffered formalin fixation) are larger, with mean diameter 9.9um (S.D. 1.4, n 68). Live, intact cells, uncompressed on slides, average 13.9μm (S.D. 1.4, n 35). The spherical to ovoid, or somewhat concavo-convex nucleus is invariably strongly eccentric, only a thin layer of cytoplasm intervening between it and the adjacent plasma membrane. Chromatin lines the inner nuclear membrane as small clumps or broad irregular bands and forms scattered patches throughout the nucleus; the outer nuclear membrane, separated by a considerable cisterna, is ribosome-studded.

The mucocyte cytoplasm is packed with approximately equal numbers of large granules of mean diameter 0.6μm (range 0.34-1.02, n 18) and ribosome-studded vesicles (GER) of the same order of size to much smaller. The granules are irregularly or, more commonly, evenly subspherical and are membrane-bound but the thin membrane is observable only in occasional granules and then often for only part of the periphery. The granules vary from weakly to strongly electron-dense, becoming paler as they approach breakdown, and have a finely granular consistency. They are strongly PAS positive, indicating amino- or vic-glycols,and like the cytoplasm, stain strongly for β-glucuronidase. Wright's stain indicates that the cytoplasm is strongly acidophilic. These responses are not shown by other cell-types. The vesicles are negative for the former two tests. Several large, approximately isodiametric mitochondria with mostly short, irregularly disposed cristae are visible in a plane of section. In life and in resin sections several short pseudopodium-like projections are sometimes visible but a phagocytic role is not indicated.

A central Golgi complex has been demonstrated for some cells. The first cisterna, on the concave face, forms a few large interconnected dilatations with moderately electron-dense contents which have a granular structure resembling that of the large, mucous granules in the cytoplasm. These dilatations are here interpreted as prosecretory granules (vesicles) precursory to the cytoplasmic mucus-granules as seen in the mammalian parotid gland (5). A centriole is occasionally seen.

*2. Linocyte (Figs. 4,12).* Linocytes are large, ovoidal cells with a mean longer dimension of 10.9μm (range 8.3-18.2) and cytoplasmic to nuclear ratio of 3.3 (range 1.9-4.9) for a nuclear diameter of 3.4μm (range 2.2-4.9, n 9). The plasma-membrane is lined internally by a thin layer of cytoplasm which in places may be lacking or may be expanded sufficiently to include a rare mitochondrion or one or more small vacuoles. The nucleus, which is strongly eccentric, being almost in contact with the plasma-membrane, is irregularly ovoid or lobed and has dense masses of chromatin forming a wide, periodically constricted peripheral band or peripheral and scattered clumps; a degenerating nucleolus is sometimes recognizable. In cells which seem least degenerate the large space internal to the cytoplasmic sheath is filled by a pleated double membrane which gives the appearance, in section, of a coiled thread wound on itself approximately one hundred times. In fact, however, the loops are linked with each other, mostly laterally, and with the cytoplasmic sheath so that they comprise a complex system of compressed vacuoles which are separated by double membranes which form the so-called thread. In some cells, however, in which the double membrane structure is still retained, the thread forms a single large whorl around a common centre. In other, probably more degenerate cells, the thread appears to consist of loose stacks of many transverse or longitudinal fibres lacking a clear double membrane condition, and may break up or become disorganized. Linocytes are often seen being engulfed or infiltrated by phagocytic lymphocytoids.

3. *Vesicular Lymphocytoid (Fig. 2).* This is a medium-sized cell, of mean diameter 9.1μm (range 7.1-11.2), cytoplasmic to nuclear ratio 2.9 (range 1.9-4.5) and nuclear diameter 3.5μm (range 1.9-5.7, n 10). It is possible that the vesicular lymphocytoid morph may be assumed by more than one of the other coelomocyte types but a typical form is recognizable which has the following characteristics. The nucleus is evenly or irregularly subcircular or somewhat kidney-shaped in outline. It is sometimes embayed by a large vesicle, here termed the paranuclear vesicle, which is a local enlargement of the perinuclear cisterna and like the latter is externally ribosome-studded. Chromatin is distributed evenly in the nucleus as small grains or may show few slight to conspicuous aggregations and a small but distinct granular nucleolus may be visible. Most distinctive are large numbers of moderate-sized to large subcircular to oval GER vesicles (sometimes small and ill-defined) which may or may not be accompanied by the paranuclear vesicle. Forms with small, indistinct GER vesicles have few, often incomplete mitochondria but whether the latter are forming or degenerating is not clear; the fact that these cells sometimes show a well developed Golgi apparatus suggests that they are young vesicular lymphocytoids. Also typical are several large mitochondria per section with round profiles, electron-pale matrices and mostly short, peripheral cristae. Some vacuoles have few or no ribosomes and may show their origin by endocytosis by a narrow, sometimes canalicular connection with the exterior. A few apparent phagosomes with electron-dense diffuse contents, and the occasional multivesicular body, are present.

Some variants, probably referable to this cell type, have large phagosomes containing ingested mucous granules released from mucocytes. A further variant, tentatively to be regarded as a vesicular lymphocytoid, has acquired considerable numbers of, probably intrinsic, electron-dense small oblate granules of the type seen in granulocytes and may show diffuse phagolysosomes with which oblate granules are fusing.

The outline of a typical cell is commonly very irregular, though often without definite pseudopodia, but one or more ragged filopods may be produced. Putative free ribosomes and polyribosomes are moderately numerous scattered throughout the cytoplasm. Larger, moderately abundant granules are probably glycogen.

*4. Macrophagous Lymphocytoid (Fig. 1).* These have mean maximum dimensions (excluding longer pseudopods) of 11.9µm (range 9.7-16.7), a mean nuclear diameter of 3.6µm (range 2.5-5.2) and a cytoplasmic to nuclear ratio of 3.8 (range 2.6-4.3, n 5). The cells are here termed macrophagous lymphocytoids because, in the large size and frequency of ingesta in their phagosomes, they are reminiscent of mammalian macrophages. They appear capable of ingesting intact mucocytes. The term macrophage may be applied for brevity with no implication of homology with vertebrate macrophages. Phagocytosis, although on a smaller scale, is widespread in other lymphocytoids, including granulocytes, but recognition of a distinct type for macrophagous cells is supported by additional morphological features which, in combination, provide a diagnosis for this type. A particularly distinctive feature is the form of the nucleus which has homogeneous, finely granular chromatin and is irregular in outline with abrupt, angular indentations. A nucleolus has been observed. Mitochondria are numerous and large, with pale matrices, and often appear to be in a state of dissolution, sometimes containing myelin figures. In view of the very active phagocytosis, the absence of secretory granules is surprising. This absence does not appear ascribable to utilization of granules which might have been present in a precursory phase as phagosomes containing obviously newly ingested, unchanged cell components are not accompanied by recognizable secretory granules. Some indistinct bodies of medium electron density may, nevertheless,

*Figs. 1 and 2. Coelomocytes of Pontodrilus bermudensis. 1, macrophagous lymphocytoid with ingested mucocyte, X 12 100. 2, vesicular lymphocytoid with paranuclear vesicle, X 12 000.*

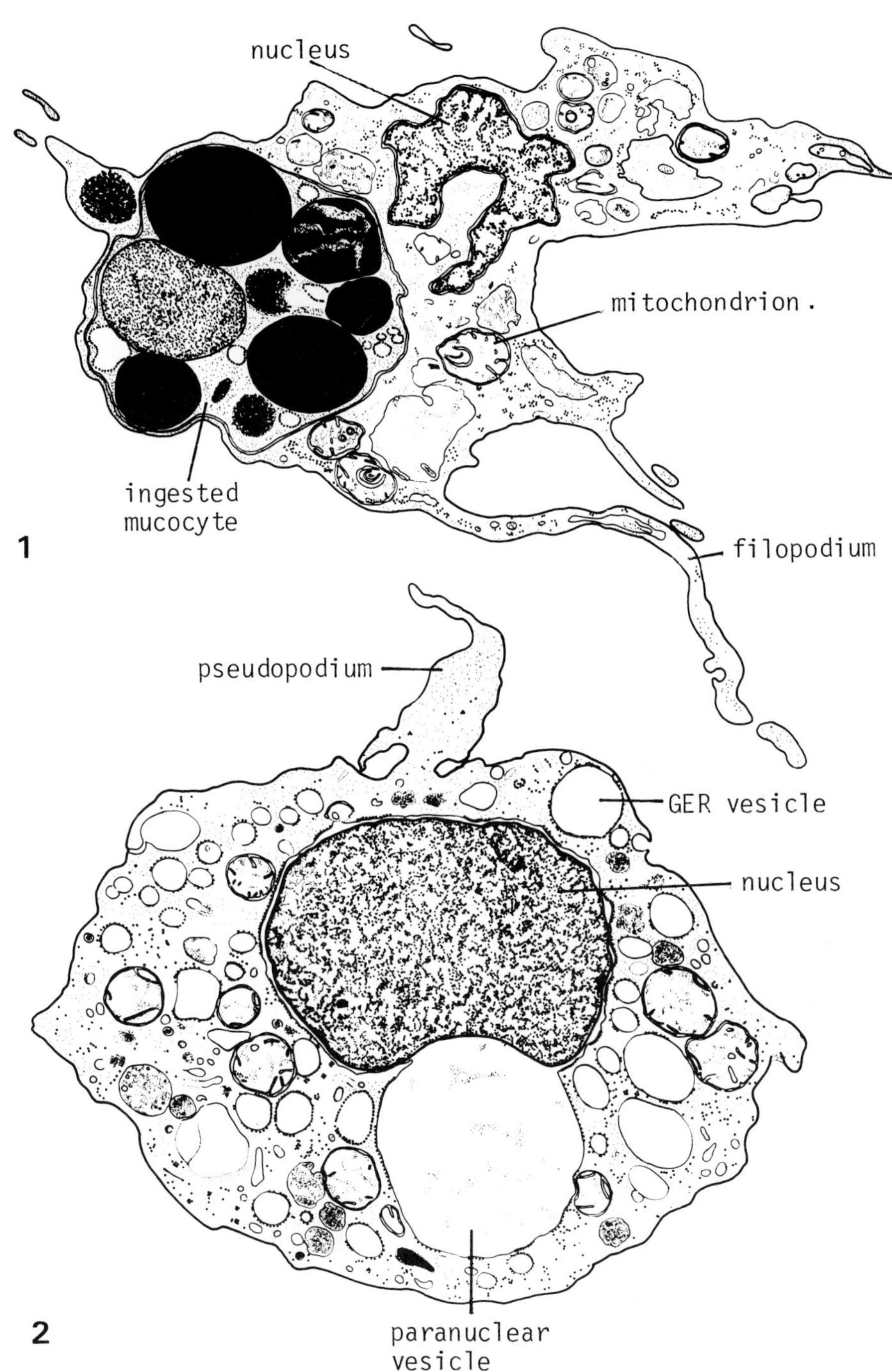
nucleus
mitochondrion.
ingested mucocyte
1
filopodium
pseudopodium
GER vesicle
nucleus
2
paranuclear vesicle

be lysosomes. Long tortuous filopodia are characteristic. Free ribosomes and polyribosomes are numerous throughout the cytoplasm but, unlike the vesicular lymphocytoid, association with the numerous vacuoles or vesicles is very low. One or more active Golgi complexes (with cisternae containing dense contents) from which small Golgi vesicles spread widely in the cytoplasm, and an adjacent centriole, have been seen in some macrophages.

5. *Small Lymphocytoid (Figs. 5,13).* The small lymphocytoid is thus designated because of its morphological resemblance, in its agranular form, to the small lymphocyte of man. The mean diameter is about 7.2μm (range 5.2-8.5) with small cytoplasmic to nuclear ratio, averaging 1.9 (range 1.7-2.6), for a nuclear diameter of 4.0μm (range 2.9-5.9, n 10). The nuclear outline is typically circular, though slightly to considerably indented but may be deeply cleft. Conspicuous large chromatin clumps are peripheral and scattered throughout the nucleus. The outer membrane is ribosome-studded and a large nucleolus may be visible. Very characteristic are the short profiles of narrow GER cisternae, often in single file, concentric with the nucleus. Putative free ribosomes and glycogen granules are scattered throughout the cytoplasm. In what appears to be the simplest (stem?) form mitochondria are absent and dark phagosome-like bodies are few and small though some fusion vacuoles, with nebulous contents, rarely in the form of canaliculi with terminal swellings, are present. Rounded GER vesicles and Golgi complexes are not apparent. Exceptionally a naked paranuclear vesicle is present. The cytoplasm of a typical cell tends to be rounded with a few broad ragged protrusions or broad short filopodia but some cells here classed as, and with the general characteristics of, small lymphocytoids, have numerous long, broad, branching filopods. These cells have acquired moderate numbers of elongate mitochondria with dark matrices and have small numbers of oblate to spherical very electron-dense

*Figs. 3 to 8. Coelomocytes of Pontodrilus bermudensis. 3, mucocyte; note free mucus-granules in the medium, X 7 500. 4, linocyte invested by pseudopodia of a macrophage, X 3 200. 5, small lymphocytoid, X 8 000. 6, medium lymphocytoid, X 6 300. 7, filolobopodial granulocyte; released secretory granules of a mucocyte are coalescing to form mucus at bottom right, X 5 300. 8, lobopodial granulocyte with canaliculi, X 5 800.*

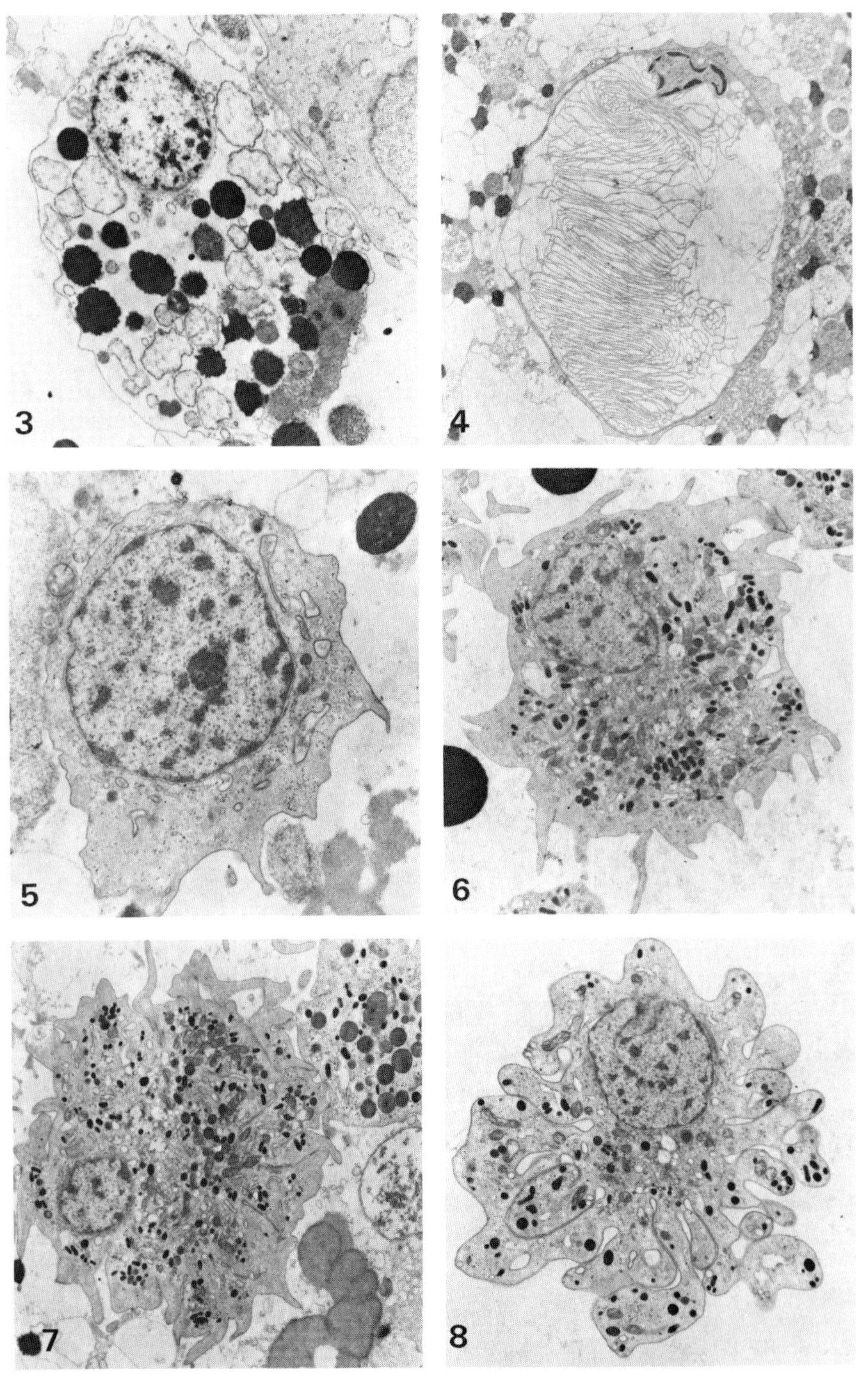
3
4
5
6
7
8

granules. The filopodia and a narrow band of peripheral cytoplasm are largely free of granules or inclusions, other than putative ribosomes and glycogen granules. The secretory granules, including larger spherical ones, do not exceed about 0.3μm. The actively filopodial cells often have large peripheral phagocytic vacuoles (residual bodies) containing minute, bubble like spherules of probably foreign origin. The cytoplasm fringing such vacuoles appears to have lost its bounding membrane in places; dense granules (lysosomes?) enter these vacuoles.

6. *Medium Lymphocytoid (Figs. 6,14,15).* The medium lymphocytoid is so designated because it shows ultrastructural similarities to the small lymphocytoid while being slightly larger. It shows no especially close similarity to the human medium lymphocyte. Origin from the small lymphocytoid is probable. Obvious distinctions from the latter are the slightly greater diameter, mean 8.5μm (range 8.1-9.2), which is due to the slightly increased ratio of the cytoplasmic to nuclear diameter, 2.5 (range 2.0-3.0), for a nuclear diameter of 3.4μm (range 2.8-4.3, n 7), and the large number of granules. Rod-to slightly dumbell-shaped dense granules resemble those of small lymphocytoids in appearance and size and it is not considered that larger, rounded granules reaching 0.4μm in diameter are a different type of inclusion. The nucleus again has patchy peripheral and general chromatin clumps, may exhibit a nucleolus, and has one or more slight to strong indentations, or may be deeply cleft, but tends to be more irregular,

*Figs. 9 to 17 Coelomocytes of Pontodrilus bermudensis. 9, ectoplasmic granulocyte, X 5 100. 10 to 17, details of various types: 10, ectoplasmic granulocyte, X 10 800. 11, mucocyte, X 38 800. 12, linocyte, X 17 800. 13, small lymphocytoid, X 10 000. 14, immature medium lymphocytoid, X 13 400. 15, medium lymphocytoid with tendency to petaloid configuration of pseudopodia, X 23 500. 16, filolobopodial granulocyte, X 10 600. 17, lobopodial granulocyte, X 11 200. c, centriole; ca, canalicule; f, filopod interposed between pseudopodia which are lobopodia with filopodous extension; g, Golgi apparatus; ger, granular endoplasmic reticulum; gerv, granular endoplasmic vesicle; lo, lobopodium; m, mitochondrion; mg, mucus granule; n, nucleus; ps, broad pseudopodium bounded by canaliculi; sg, small, dense granule; t, "thread" of linocyte.*

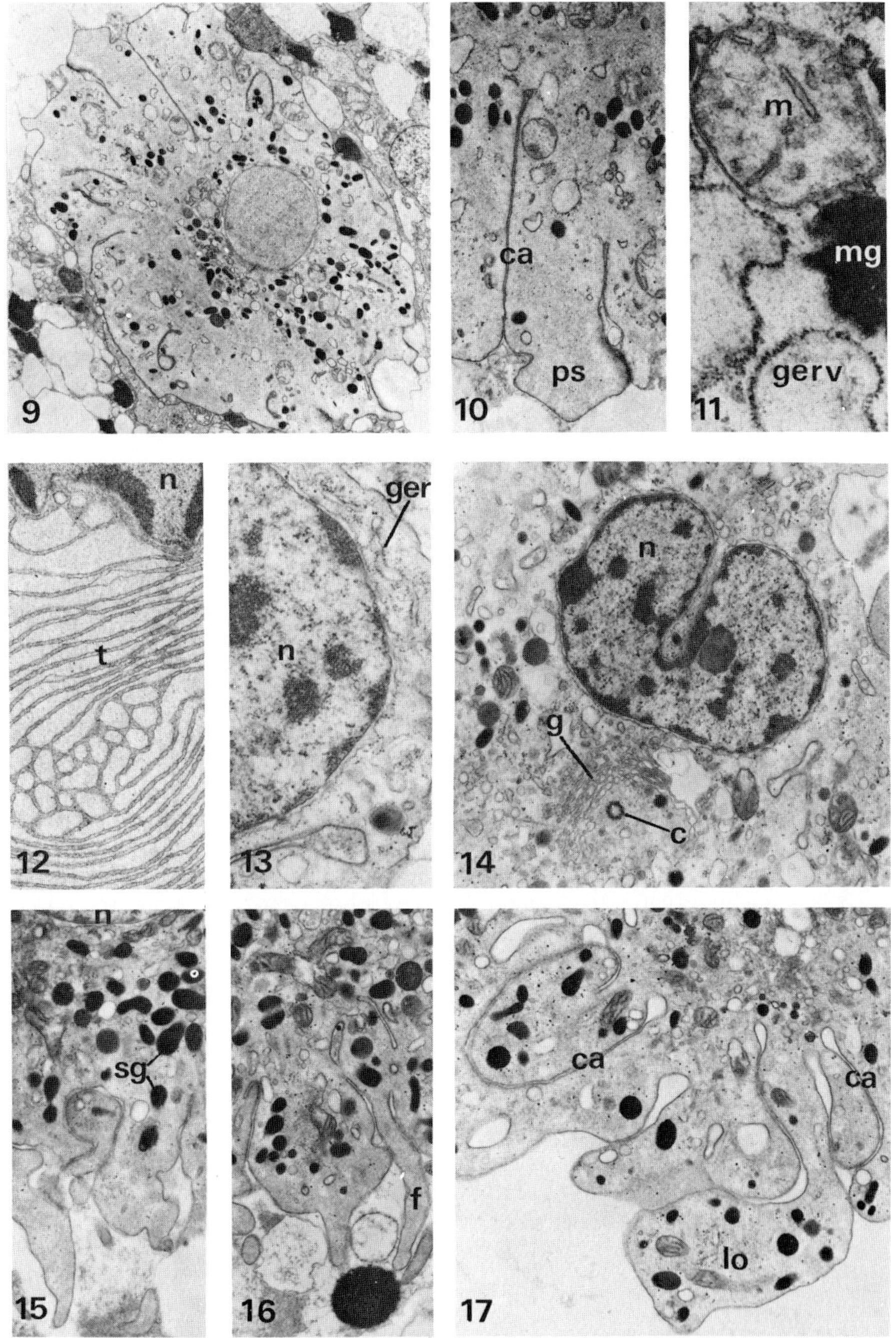
9
10
ca
ps
11
m
mg
gerv
12
n
t
13
ger
n
14
n
g
c
15
n
sg
16
f
17
ca
ca
lo

less circular in outline. Small phagosomes and multivesicular bodies are few. Clear, sperule-containing, residual bodies confluent with the peripheral cytoplasm resemble those of small lymphocytoids. GER cisternae are more dispersed and a few, scattered canaliculi may be present or, in some cells, canaliculi may be frequent between the bases of adjacent filopodia which have some tendency to a petaloid form. Mitochondria are moderately numerous. Putative free ribosomes and glycogen granules are scattered throughout the cytoplasm which again tends to lack inclusions or organelles in a narrow peripheral zone. GER vesicles are few and small.

Some cells with these features but fewer granules show a centriole with encircling Golgi complexes. These probably represent a young stage of the medium lymphocytoid. One cell-type tentatively included with the medium lymphocytoid has numerous large, dense apparent phagolysosomes, few small granules and has microfilament bundles extending into some at least of the filopods which thereby resemble axopods.

7. *Filolobopodial Granulocyte (Figs. 7,16).* The tendency to combine a petaloid configuration of pseudopodia with filopodia which was seen in some medium lymphocytoids is taken further in what is here termed the filolobopodial granulocyte. The obligatorily large number of granules warrants descriptive designation of this lymphocytoid as a granulocyte. It is not unreasonable to consider it a descendant or close collateral of the medium lymphocytoid. It is, however, significantly larger, mean diameter 12.1μm (range 10.0-15.7) an increase achieved by augmentation of the cytoplasmic to nuclear ratio to 3.5 (range 2.7-4.4), with a mean nuclear diameter of 3.6μm (range 2.4-4.6, n 3).

The filolobopodial granulocyte is readily identified by its peculiar pseudopodial structure; broad lobopodia, often bounded by narrow canaliculi which may wind deeply into the cytoplasm, bear or alternate with long filopodia. The appearance of the cell is thus somewhat intermediate between that of the foregoing medium lymphocytoid and the canalicular form of the lobopodial granulocyte described below. Like the former, but unlike the lobopodial cell, the secretory granules and organelles tend to be absent from a narrow peripheral zone. The ultrastructure of the cytoplasm and nucleus is too similar to that described above for the medium lymphocytoid to warrant full description. However, a nucleolus, though possibly present, has yet to be demonstrated. The number of small, rod- to dumbbell-shaped dense granules is greater and large numbers of spherical, moderately dense bodies are present. Whether the large bodies are true granules or are phagolysosomes is not clear. Whereas canaliculi are well developed, comprising a definite lamellar-vacuolar system (*sensu* Droller) (6), GER-cisternae and especially -vesicles are scarce.

8. *Lobopodial Granulocyte (Figs. 8,17).* Granulocytes with small, rod- or dumbbell-shaped and with larger, spherical granules, the numerous pseudopodia of which are mostly large round-ended lobopodia are here termed lobopodial granulocytes. Although common, they are about one sixth as abundant as mucocytes. Two forms are present, one with, the other without canaliculi but this is possibly a transitory distinction. Apart from the lobopodial form, they differ from small or medium lymphocytoids and from filolobopodial granulocytes in that the secretory granules and even the mitochondria, which in all these cells may enter the pseudopodia, are not excluded from the peripheral cytoplasm but frequently approach or abut the inner surface of the plasma-membrane.

Observed diameters vary from as small as the small lymphocytoids (5.7μm) to the largest known for a coelomocyte (18.6μm), mean 10.5μm, with a large cytoplasmic to nuclear ratio, mean 3.2 (range 1.9-4.5), for a mean nuclear diameter of 3.3μm (range 2.4-4.2, n 12). The nucleus is ovoid, variously notched or bi-partite in section. Chromatin clumps line the periphery and are scattered throughout it. A large nucleolus is sometimes visible. Secretory granules are numerous but, of those which are very electron-dense, small narrow oblate granules are relatively less common than in the preceding granulocyte-types whereas larger equally dense granules, circular in cross section and presumably spheroidal, are common. Nevertheless, these dense granules have mean maximum dimensions of only 0.3μm (S.D.0.06, n 55). In addition, however, a considerable proportion of cells have, often abundant, large, subspheroidal moderately electron-dense granules with flecked, granular contents. These range from 0.4-1.2μm, averaging 0.6μm in diameter (S.D.0.2, n 23). As such they are significantly smaller than chloragosomes measured in *Pontodrilus* chloragocytes and identity with these is therefore contraindicated.

Elongate mitochondria with dark matrices are moderately numerous. Golgi complexes, in a circular configuration, with or without associated centriole, are rarely observable. Several large vacuoles, commonly oval to dumbbell-shaped, are characteristic. These lack ribosomes and are here termed fusion vacuoles as, though usually empty, they appear to be the product of fusion of pseudopodia in an endocytotic process. The smaller fusion vacuoles may open by long narrow canaliculi of the lamellar-vacuolar system (one per vacuole) to the exterior. Vacuoles in section commonly contain transected pseudopodia. Canaliculi often form closed circles, enclosing granule-containing cytoplasm. GER cisternae and vesicles, both small, are fairly common. Putative glycogen granules, polyribosomes and free ribosomes are scattered throughout the cytoplasm. Surprisingly, in view of the numbers of lysosome-like dense granules, phagosomes are questionably present and multivesicular bodies are uncommon. Scattered bundles of microfilaments are present and in some variant cells form axial elements of axopod-like, pointed pseudopodia.

9. *Ectoplasmic Granulocyte (Figs. 9,10)*. The ectoplasmic granulocyte is an infrequent cell so-named because a broad peripheral band of cytoplasm, apart from containing scattered putative ribosomes and glycogen granules, may be virtually devoid of granules and other inclusions or there are at least large areas, some of which are peripheral, of such featureless cytoplasm. However, the chief characteristic of this type of granular lymphocytoid is the deep dissection of the cytoplasm by narrow canaliculi which extend inwards from the exterior and end internally in slight dilations. These canaliculi, comprising the lamellar-vacuolar system, separate broad, lobate pseudopodia which project very little beyond the cell periphery and are chiefly recognizable by virtue of their intervening canaliculi.

The cells are large, with mean diameter of 12.9µm (range 10.0-15.7, n 3). While the nuclear diameter remains in the order of 3µm, the cytoplasmic to nuclear ratio is high, at about 4.7. The nucleus is subcircular with ribosome-studded outer membrane and homogenously distributed, finely granular chromatin which lacks appreciable clumping. Its position varies from near central to strongly eccentric. Strongly electron-dense, membrane-bound, circular, oblate, or sometimes dumbbell-shaped granules with mean maximum dimension 0.24µm (S.D.0.06, n 18) are numerous. These granules do not appear to represent different types and therefore no dimorphism is recognized. No Golgi apparatus has been observed. Moderate numbers of mitochondria with circular to bilobed profiles, partial and complete transverse cristae, and matrices resembling the cytoplasm in their weak to moderate density are present. Small, subcircular to irregular GER vesicles, similar ones lacking ribosomes, and V-shaped, circular or tortuous canaliculi in addition to those at the cell margins are frequent. Wisps of microfibrillar material are ubiquitous. Phagosomes have not been seen but multivesicular bodies are present.

## IV. DISCUSSION AND CONCLUSIONS

Of the nine types of coelomocytes in the coelomic fluid or alimentary exudate of *Pontodrilus bermudensis* only two, mucocytes and linocytes, have previously been described for any animal group. Mucocytes are acidophil cells the granules of which are PAS positive (suggesting presence of vic- and/or amino-glycols) and contain β-glucuronidase. Mucocytes have elsewhere been recognized ultrastructurally only in the Enchytraeidae (7 and Richards, personal communication) that described by Richards for *Enchytraeus albidus* being morphologically closest to the mucocyte of *Pontodrilus*. In enchytraeids the granules contain 1,2 glycols and varying amounts of acid mucopolysaccharide. The acidophil cell described ultrastructurally for *Lumbricus terrestris* (Lumbricidae) as the "inclusion-containg cell" (8) also resembles the mucocyte of *Pontodrilus* but differs notably in the small amount of endoplasmic reticulum.

The linocyte was first described in a megascoleid (9) for a cell which had the appearance by light microscopy of being filled by one to several coiled threads. Similar cells were termed nematocytes in *Nematogenia panamaensis* (Ocnerodrilidae) (10) and thread-containing cells in enchytraeids (11). Origin of linocytes from mucocytes after release of secretory granules (7) is here endorsed.

Of the seven *Pontodrilus* cell-types additional to mucocytes and linocytes, none has been recognized by previous workers and none can be identified with certainty with types described ultrastructurally for lumbricids (8,12,13), or enchytraeids (7). Whether homologues do occur or are to be expected in other families is debatable. Similarities with those of other families are apparent but have not been sufficient to justify an identical nomenclature. A brief consideration of genetic relationships of the cell-types within *Pontodrilus* is appropriate for recognition of interfamilial similarities. Origin of linocytes from mucocytes seems established but origin of the mucocytes is not known. The luminescent cell of *Diplocardia longa* is not known. The luminescent cell of *Diplocardia longa* appeared to be a free chloragocyte (3) but ultrastructural investigation of the chloragocytes of *Pontodrilus bermudensis* in connection with the present study excludes such origin of mucocytes in *Pontodrilus*.

Lineages of lymphocytoids, including their subset the granulocyte, can be surmised only tentatively from this purely ultrastructural study. The small lymphocytoid, in its amitochondrial form, appears to be a stem cell which can, however, on further differentiation become phagocytic and probably gives rise to the medium lymphocytoid. The filolobopodial granulocyte may well be a descendant of the medium lymphocytoid. The tendency to exclusion of organelles and granules from a very narrow zone of peripheral cytoplasm in these types and general characteristics such as granule type, and appearance in the filolobopodial form of a well developed lamellar-vacuolar system with a petaloid configuration of pseudopodia suggest that the ectoplasmic granulocyte is derived from or is a close collateral of the filolobopodial cell. The lobopodial granulocyte, with its tendency for organelles and granules to abut the plasma-membrane, is probably not derived from the small lymphocytoid-filolobopoidal lineage, a view supported by the existence of probable precursor cells (not described above) resembling the lobopodial cell but smaller with large cytoplasmic to nuclear ratio and with few or no pseudopodia. The lumbricid cell most closely resembling the lobopodial cell is the type II granulocyte (8) or "transportation amoebocyte" (14) but the type II cell is believed to be derived from the type II and this from the type I lymphocytic coelomocyte whereas the latter two cells approach the small and medium lymphocytoids of *Pontodrilus* in general ultrastructure though not identical with them. The ectoplasmic granulocyte superficially resembles the type I granulocyte of *Lumbricus* (8) which has a wide ectoplasmic band but definite canaliculi are not evident in the latter cell. The macrophagous lymphocytoid is conceivably derived from the vesicular lymphocytoid but the origins of this lineage remain to be elucidated.

Of all the nine cell types, the mucocyte, released granules of which form the viscous mucus of the alimentary exudate, is the most likely source of luminescence. Lobopodial granulocytes are also sufficiently abundant to be considered as possible sources of luminescence and the cooperation of two or more cell types in producing luminescence cannot yet be ruled out. As the luminescence is peroxidative it is curious that no *Pontodrilus* coelomocytes (nor those of lumbricids 8) are peroxidase positive. The possibility that minute inclusions such as the lysosome-like, 0.3μm, small dense granules of *Pontodrilus* granulocytes contain peroxidase but are too small or insufficiently numerous to give a visible positive reaction in smears under the light microscope warrants further investigation.

REFERENCES


1. Wampler, J. E., and B. G. M. Jamieson, *Comp. Biochem. Physiol., 66B,* 43, (1980).
2. Jamieson, B. G. M., and J. E. Wampler, *Aust. J. Zool., 27,* 637 (1979).
3. Rudie, N. G., and J. E. Wampler, *Comp. Biochem. Physiol., 59A,* 1 (1978).
4. Jamieson, B. G. M., *in* "The Ultrastructure of the Ogligochaeta", Academic Press, London, (1981).
5. Leblond, C. P., and G. Bennett, *in* "International Cell Biology" (Brinkley, B. R., and K. R. Porter, eds.), p. 326. Rockefeller Press, New York (1977).
6. Droller, M. J., *Lab. Invest., 29,* 595 (1973).
7. Hess, R. J., *J. Morph., 132,* 335 (1971).
8. Linthicum, D. S., E. A. Stein, D. H. Marks, and E. L. Cooper, *Cell. Tissue Res., 185,* 315 (1977).
9. Benham, W. B., *Quart. J. Micr. Sci., 44,* 565 (1901).
10. Eisen, G., *Proc. Calif. Acad. Sci.,* (3), *2,* 85 (1900).
11. Dehorne, A., *C. R. Acad. Sci. Paris., 179D,* 1079 (1924).
12. Valembois, P., *Bull. Soc. Zool. France, 96,* 59 (1971).
13. Freire, M. A., *Bol. R. Soc. Espan. Hist. Nat. Biol., 74,* 123 (1976).
14. Stang-Voss, C., *Zeit, Zellforsch, Micr. Anat., 117,* 451 (1971).

# ON THE BACTERIAL-ASSOCIATED LIGHT ORGAN IN *CHLOROPHTHALMUS*[1]

Hiroaki Somiya

Faculty of Agriculture
Nagoya University
Nagoya, Japan

## I. INTRODUCTION

*Chlorophthalmus* spp. (Myctophiformes), benthonic or benthopelagic fishes, occur at moderate depth (150-750m) throughout tropical and temperate regions of the world. They are normally under 30cm long (1).

Deep-sea fishes that belong to Myctophiformes have neurally controlled luminescent system, e.g. *Myctophum, Neoscopelus, Lestidium, Coccorella, Benthalbella* and so on. But in the previous paper (2), I reported that *Chlorophthalmus* had an exceptionally small bacterial anal light organ, and presented a brief description on the histological structure of the organ.

So far as I know, literature which deals with the morphological aspect of the symbiotic association between the light organ of the fish (host) and the luminous bacteria is scanty (3). It is the object of this paper to present more detailed information on the light organ of *Chlorophthalmus*, especially the preliminary interpretation on the ultrastructural aspect of the symbiotic association in given.

[1]*This research was supported in part by a grant from Hidaka Foundation for the Promotion of Oceanic Research; I wish to express my thanks to Professor Koji Hidaka in this respect.*




ISBN 0-12-208820-4

## II. MATERIAL AND METHOD

*Chlorophthalmus albatrossis* (ca. 15cm in length) were sampled off Owase (Mie prefecture, Japan). The anal part of the fish was fixed in 2.5% glutaraldehyde and post-fixed in 2% solution of osmium tetroxide. After dehydrating, the pieces of the material were embedded in Quetol 812. Thin sections were made and stained with 2% solution of uranyl acetate and lead (Pb) staining solution. The sections were examined with JEM-100C electron microscope.

## III. OBSERVATION

### *A. Emission Spectrum*

Emission spectrum of the intact anal light organ was measured by Hitach fluorescence spectrophotometer (Fig. 1). Emission peak was observed at 475nm.

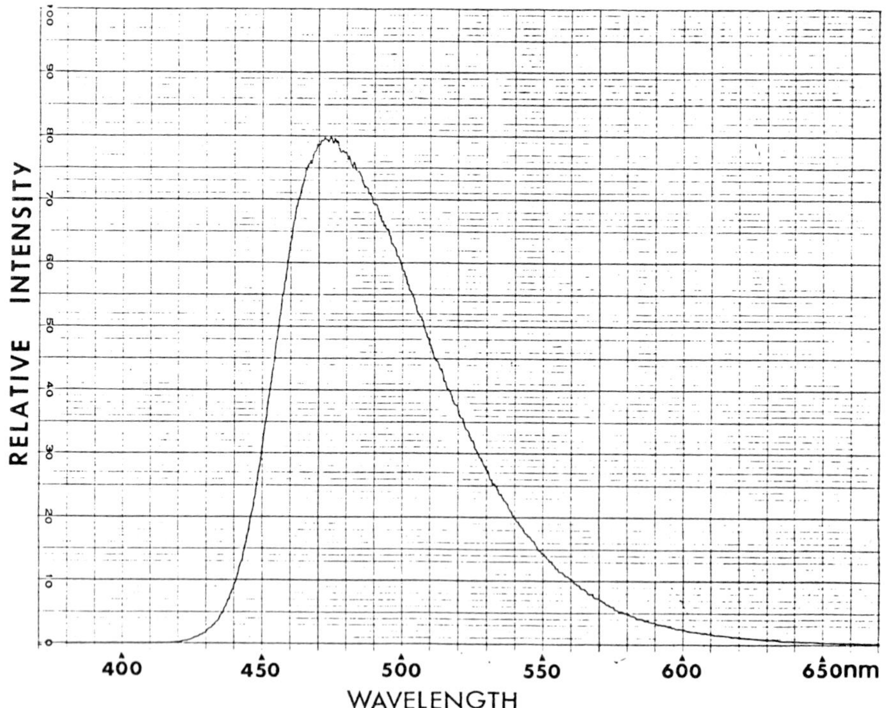


*FIGURE 1. Emission spectrum of the intact light organ is measured peaking at 475nm.*

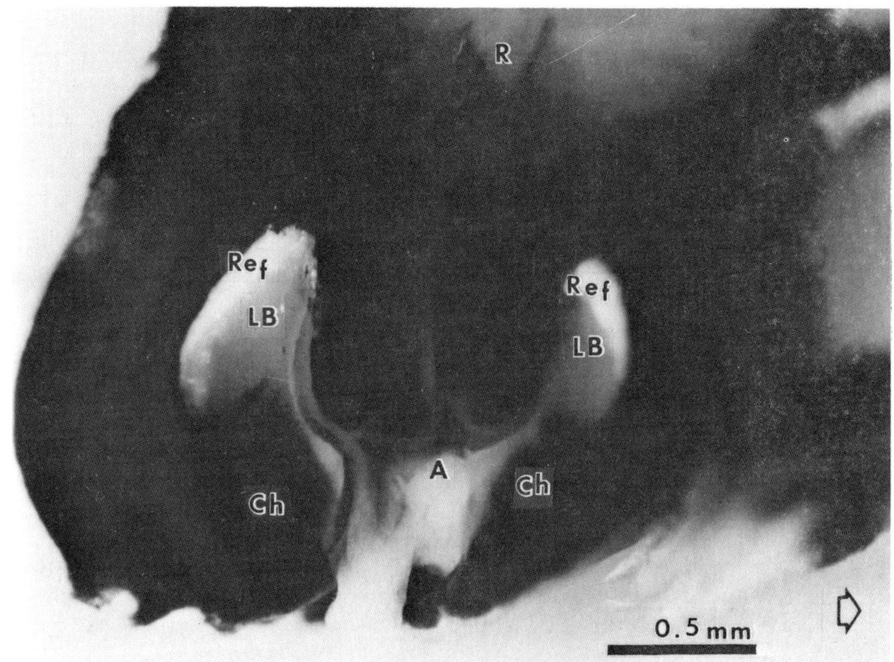


*FIGURE 2. Median section of the light organ showing the position of the reflector (Ref). Note, the caudal part of the organ is larger than the rostral. A:anus, Ch:chromatophores, LB:luminous body, R:rectum, Arrow indicates the rostral.*

### B. *Macro-structure of the Light Organ*

The light organ of the fish is a ring doughnut shape which encircled the anus (2). Here, macro-structure of the organ was examined using a 2.5% glutaraldehyde preserved specimen. The median section of the light organ is shown in Fig. 2. The light organ is composed of three essential elements, luminous body (LB, bacterial colonies and glandular epithelium), reflector (Ref) and chromatophores (Ch). The reflector is found just behind the luminous body. It is composed of many iridophores which contain numerous reflecting platelets. These platelets may be composed of guanine as a major component. Chromatophores may adjust the intensity of the luminescence. Approximate size of the light organ is as follows, diameter: ca.1.6mm and height: ca.1.4mm.

Fig. 2 also indicates that the caudal part of the organ is larger than the rostral.

### C. *Ultrastructure of the Luminous Body*

The luminous body is composed of bacterial colonies and glandular epithelium. In the luminous body, two types of the bacteria were observed. Those are "free" and "associated type" of bacteria. In the main part of the luminous body, the bacteria are mostly associated with "mitochondria and rough ER rich cells" (Fig. 3). From their morphological properties, these cells are possibly derived from the glandular epithelium of the light organ.

In Fig. 3, the bacteria are so tightly associated with the host cell that the bacteria look like as "intracellular." But the bacteria are enclosed by the cell membrane of the cell (Fig. 4). This indicates that the bacteria are extracellular. However, the specific association between the host cell and the bacteria suggests not only physical interaction but also chemical (nutritional) or physiological interaction. In the present observation, parasitic damage due to the presence of the luminous bacteria is not observed in the luminous body. Possibly, the host cell of the light organ may physiologically support the colony of the luminous bacteria.

The luminous bacterium is rod-shaped (approximately 2-4 μm) and Gram-negative. It is easy to culture them in 3% NaCl nutrient agar (pH, 7.2-7.4). Possibly this is psychrotrophic species, *Photobacterium phosphoreum*(4).

## IV. CONCLUSION

1. Light organ of *Chlorophthalmus* is composed of three elements, luminous body (colonies of bacteria and glandular epithelium), reflector, and chromatophores. Emission peak of this organ is at 475nm.

2. In the main part of the luminous body, the luminous bacteria are mostly associated with special cells which contain numerous mitochondria and rough ER. Since the bacteria are enclosed by the cell membrane, it is concluded that the luminous bacteria are extracellular. Bacteria are rod-shaped (approximately 2-4μm) and Gram-negative, possibly a psychrotrophic species, *Photobacterium phosphoreum*.

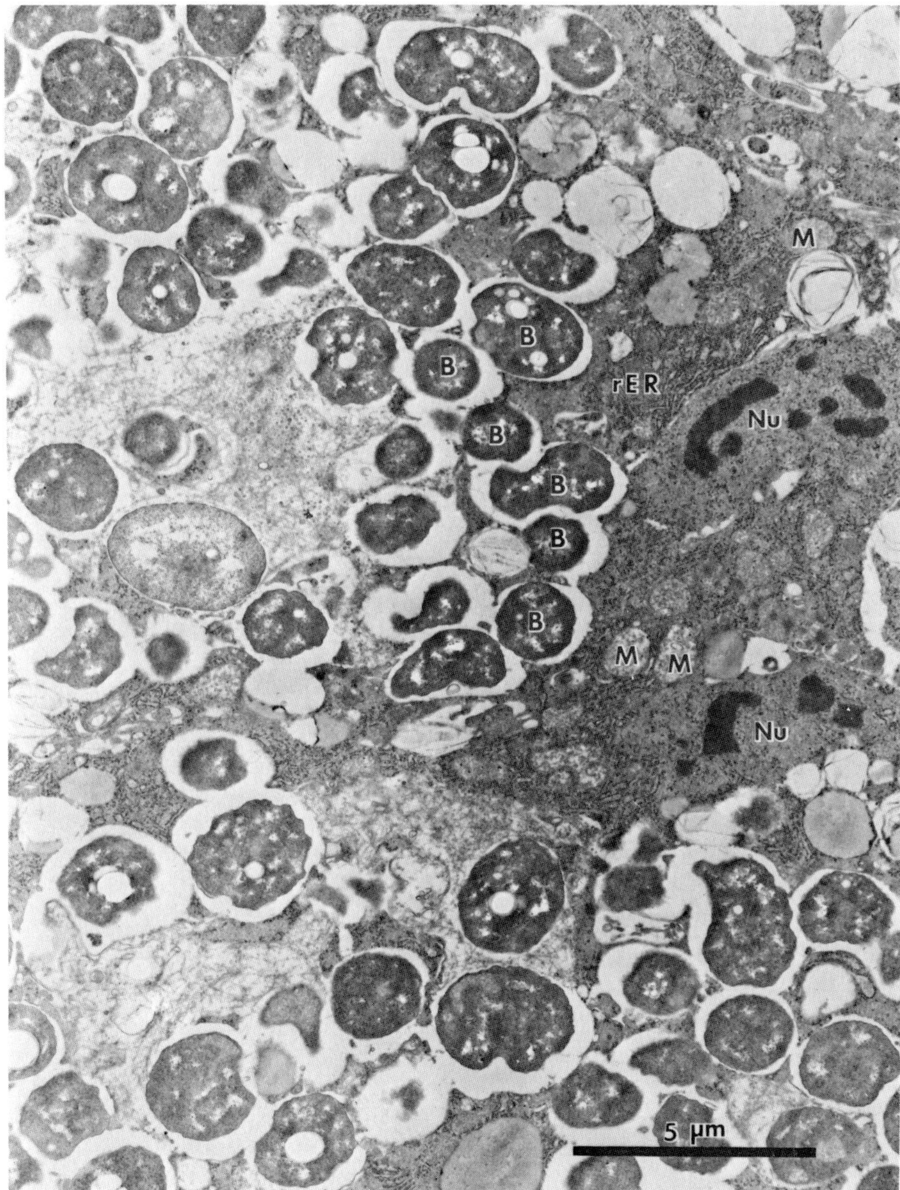


*FIGURE 3. Electronmicrograph of the bacterial colony in the luminous body. The bacteria (B) are mostly associated with 'mitochondria (M) and rough ER (rER) rich cells'. Nu: nucleus of a 'mitochondria and rough ER rich cell'.*

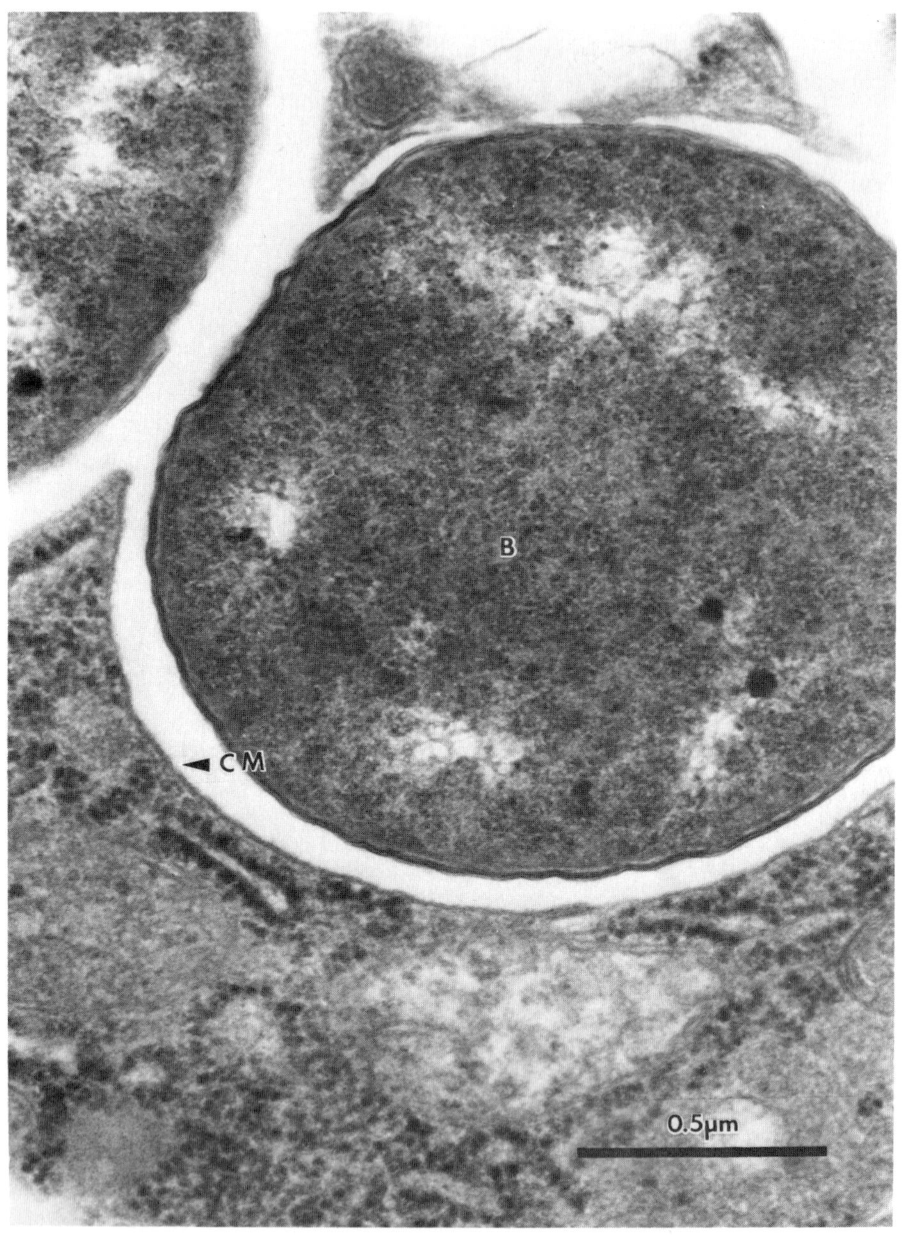


*FIGURE 4. Electronmicrograph of a symbiotic luminous bacterium (B), enclosed by the cell membrane (CM) of a 'mitochondria and rough ER rich cell'.*

## ACKNOWLEDGMENTS

I wish to thank Dr. Y. Haneda, Mrs. K. Koga, Dr. M. Oguri, Professor T. Tamura, Mr. K. Yoshizaki and Dr. N. Wakasugi for their invaluable help in the course of the work.

## REFERENCES

1. Mead, G. M., *Mem. Sears Found. Mar. Res. 1,* pt. 5, 162 (1966).
2. Somiya, H., *Experienta 33,* 906 (1977).
3. Herring, P. J. (ed.), "Bioluminescence in Action." Academic Press, New York (1978).
4. Ruby, E. G. and J. G. Morin, *Deep-Sea Res. 25,* 161 (1978).

# CIRCULAR POLARIZATION OBSERVED IN BIOLUMINESCENCE

H. Wynberg
E.W. Meijer
J.C. Hummelen

Department of Organic Chemistry
University of Groningen
The Netherlands

H.P.J.M. Dekkers
P.H. Schippers

Department of Theoretical Organic Chemistry
University of Leiden
The Netherlands

A.D. Carlson

Department of Neurobiology and Behaviour
University of New York
Stony Brook, New York

The chemistry and biology of firefly luminescence has been studied intensively (1). However, to date no detailed analysis of the polarization of bioluminescence *in vivo* has been performed (2). The present study describes some observations concerning the intensity of circularly polarized bioluminescence. It was found that left and right lanterns of live larvae of the fireflies *Photuris Lucicrescens* and *Photuris Versicolor* emit respectively left and right circularly polarized light. Measurements of the total light emission from both lanterns of the larvae (Fig. 1) gave puzzling and disappointing results. Constant results were obtained when the left and right lanterns were measured



ISBN 0-12-208820-4

separately. The circular polarization of luminescence was measured as the anisotropy factor or g-lum factor $(I_L - I_R)/\frac{1}{2}(I_L + I_R)$. *In vivo* measurements of circular polarization of bioluminescence (CPBL) of 16 firefly larvae lanterns are shown in Fig. 2.

*right lantern* *left lantern*

*FIGURE 1. Sketch of the firefly larva (ventral view), The shaded spots at the bottom represent the lanterns.*

It is concluded that the CPBL of opposite sense must, at least in part, be based on a macroscopic phenomenon rather than on any differences in molecular chirality. The following hypothesis is proposed: The light from a bioluminescent emitter may be (partially) plane polarized due to anisotropy of an absorbing medium or by inhomogeneous formation of excited states due to local molecular organization within a photocyte. This plane polarized light will become elliptically polarized when it passes through a linearly birefringent medium. Oriented biopolymers can serve as such a medium. On a macroscopic scale the larvae, like many other living organisms, have a symmetry plane dividing the lanterns. In this sense the lanterns are enantiomeric, even though their constituent molecules are of the same chirality. It is reasonable to assume that this macroscopic mirror image relationship holds on the level of the membrane structure and orientation of the emitters. This will result in circularly polarized light of opposite sense.

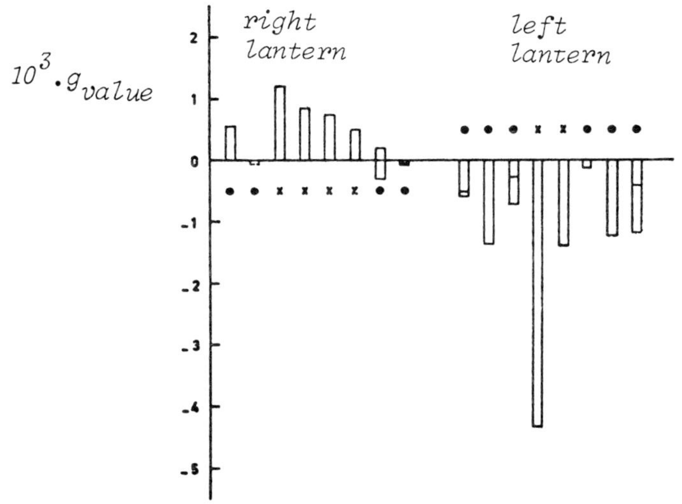


*FIGURE 2. Results of measurements of circular polarization of bioluminescence of firefly larvae lanterns with sufficiently high emission intensity. The height of each bar represents the g- value for individual lanterns of Photuris Lucicrescens (x) or Photuris Versicolor (●). Superimposed bars refer to the effect of orientational changes. The circular polarization is measured at the peak of emission band at a wavelength of 540nm using a spectral bandwidth of 40nm. The rms noise level corresponds with $g \leq 5 \times 10^{-4}$.*

REFERENCES

(1) "Bioluminescence in Action" (ed. P. Herring). Academic Press, Inc.: London (1978).

(2) Wampler, J.E. in "Bioluminescence in Action" (ed. P Herring). Academic Press, Inc.: London (1978), p. 9-48.

# E. Applications of Bioluminescence and Chemiluminescence to Clinical and Biological Problems

ADVANTAGES IN MICROPROCESSING TECHNIQUES IN BIOLUMINESCENCE ASSAYS

*Ambjörn Agren*
*Sven E. Brolin*
*Marianne Lindfors*

Department of Medical Cell Biology
University of Uppsala
Uppsala, Sweden

## INTRODUCTION

The introduction of microprocessing techniques in laboratory routines may be of great value for the further development of bioluminescence assays. Analyses can be completed within a short period of time which should ensure the stability of the light yielding solution. A large amount of data may be collected and it is possible to introduce programmes for detailed use of all the information contained in the individual observations. The assay is facilitated both by semiautomatic feeding of the samples for the measurements and by monitoring the entire analytical sequence. The photokinetic evaluation of single observations must naturally be postponed until each series of analyses has been completed.

## MEASURING AND READ-OUT ARRANGEMENTS

A scheme of the analytical equipment is given in Figure 1.



ISBN 0-12-208820-4

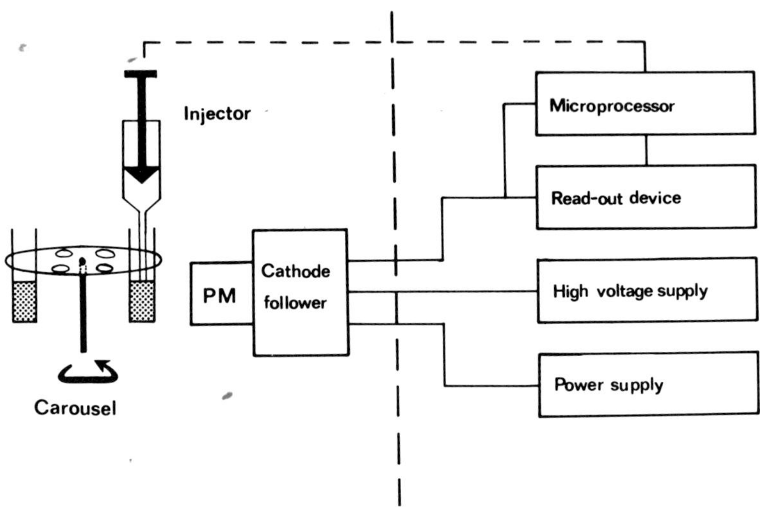


*FIGURE 1. Block diagram of measuring equipment designed for bioluminescence analyses. The samples are placed in a carousel and transported in front of the photomultiplier (PM) at which time the light yielder is injected. The light emission was measured by a photomultiplier connected to the read-out devices.*

Pneumatic mechanics were used to transport samples and also to fill, insert and empty the syringe. A microprocessor, Zilog Z80 A (memory capacity 32 K bytes) was employed for the collection and read out of the analytical data. The computer could also be programmed to monitor the analyses by operating pneumatic valves for both the transportation of the samples and the operation of the syringe.

The kinetic evaluation of single analyses after completion of a whole series of measurements required rapidity and storage capacity which was achieved by combining the microprocessor with a Floppy disk unit (storage capacity 82,000 characters in each diskette).

EVALUATION OF MEASUREMENTS

The convenient analytical performance means that a calibration curve can, if desired, be based on a large number of measurements. The curve can then be determined and automatically plotted as exemplified by determinations of $NAD^+$ concentrations (1), shown in Figure 2.

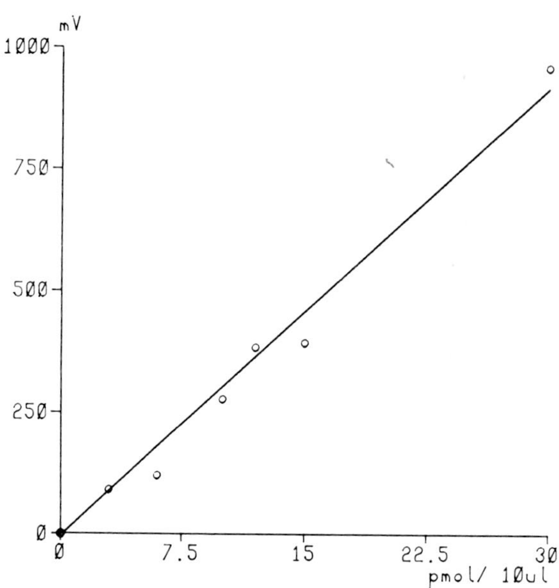


*FIGURE 2. Standard curve relating the amounts of $NAD^+$ to the maximal intensity of the light emission. Data collected and plotted by the microprocessor. Each circle represents the mean of 8 - 15 determinations.*

The automatic plot was achieved using a Houston HI plotter connected to the computer. Each measurement consists of several single recordings from which the maximum value, or integrals of various parts of the light emission curve, can be obtained. Faulty mixing may sometimes result in erroneous measurement values. Failure of this kind affects the build up of the light signal which can be revealed by an automatic plot of its time course. It is particularly important in the development of bioluminescence assays, to check the light formation in this way. An example of an appropriate time course from a determination of $NAD^+$ is shown in Figure 3.

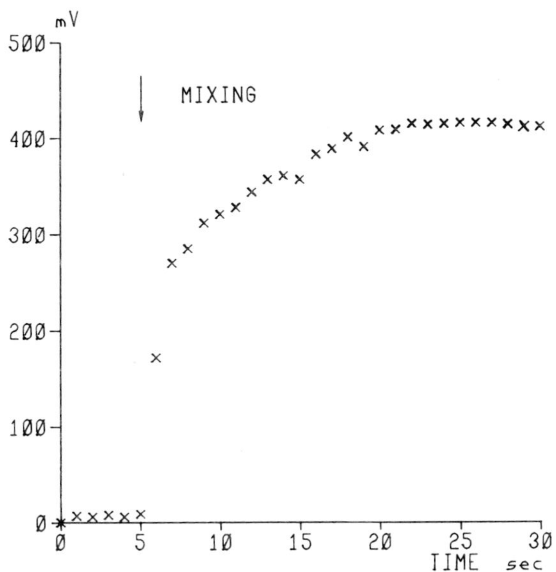


*FIGURE 3. The time course of a single determination of 10 pml $NAD^+$ from the stored standard curve data (cf. Figure 2)*

## APPLICABILITY TEST

The $NAD^+$ concentration was measured in islets of Langerhans isolated from the pancreas of one male mouse. It was a lean mouse of the genetically obese- hyperglycemic strain (gene symbol obob). To evaluate the relationship between nucleotide amounts and sample weights, several islets were pooled to obtain the required amount of tissue. The microprocessor technique was used for drawing the curve shown in Figure 4. The values found in the present study are in good agreement with those obtained using other techniques (2-4).

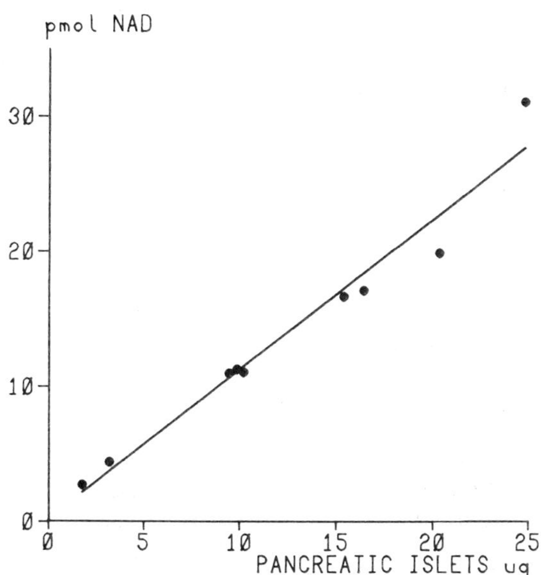


*FIGURE 4. The relationship between $NAD^+$ concentration and the weight of pancreatic islets from a lean mouse. The measurements were collected by the microprocessors, which was programmed for calculating and plotting the curve. Each dot represents a single determination.*

The microprocessor technique was also applied to the estimation of changes in concentrations. The consumption of $NAD^+$ during 3 min of ischemia was measured in small liver-samples drom male lean mice. The samples were homogenized in weak hydrochloric adid in order to extract the oxidized forms of the pyridine nucleotides and to prevent undesired conversion. A simple assessment showed that the concentration of $NAD^+$ decreased drastically from 274. 3 ± 34.3 (n=8) to 64.0 ± 5.7 (n=8) pmol per mg wet weight ($p < 0.001$) during ischemia.

Bioluminescence analyses of both adenine (5,6) and pyridine nucleotides (7) have been greatly simplified by the measuring equipment described in this study. It may be concluded that the use of microprocessors could lead to further development and extended application of bioluminescence analyses.

SUMMARY

The use of microprocessing techniques in bioluminescence assays makes it possible to collect a large amount of analytical data within a short period of time and to evaluate each single observation in detail. In the present investigation, a microprocessor (Zilog Z 80 A) was employed both for semi-automatic measurements and for monitoring the entire analytical sequence. Photokinetic evaluation of single observation must be carried out after completion of a series of analyses. The required storage capacity was obtained by combining the microcomputer with a Floppy disk unit. This permitted convenient and rapid display of the time course of light emission.

The analytical procedure employed has been proved to be of considerable value in measurements of $NAD^+$ concentrations in small samples. The results obtained were in good agreement with those obtained using other micromethods. It may be concluded that the combination of bioluminescence and microprocessor techniques will extend the analytical applications and promote the future development of new assays.

ACKNOWLEDGEMETNS

The generous support of the Swedish Medical Research Council (project number B81-12X-00525), the Swedish Diabetes Association, the Faculty of Medicine at the University of Uppsala and the Alice and Knut Wallenberg Foundation is gratefully acknowledged.

REFERENCES

1. Agren, A., Brolin, S. E., and Hjertén, S., *Biochim. Biophys. Acta 500,* 103 (1977).

2. Brolin, S. E., Borglund, E., Tegnér, L., and Wettermark, G., *Anal. Biochem. 42,* 124 (1971).

3. Matschinsky, F. M., and Ellerman, J. E., *J. Biol. Chem. 243,* 2730 (1968).

4. Malaisse, W. J., Hutton, J. C., Kawazu, S., Herchuelz, A., Valverde, I., and Sener, A., *Diabetologia 16,* 331 (1979).

5. Grill, V., and Agren, A., *Diabetologia, in press* (1980).

6. Wersäll, P., Brolin, S., Petersson, B., Östenson, C.-G., in "Proc. of Int. Symp. of Analytical Applications of Bioluminescence and Chemiluminescence" (M. DeLuca and W. D. McElroy, eds.) Academic Press, New York, (1980).

7. Brolin S. E., Agren, A., and Petersson, B., *Acta Endocrinol (Kbh.), in press* (1980).

# AUTOLUMOGRAPHY: USING BIOLUMINESCFNCE TO "STAIN" FOR ENZYME ACTIVITIES RESOLVED ON POLYACRYLAMIDE GELS[1]

James E. Becvar

Department of Chemistry
The University of Texas at El Paso
El Paso, Texas

## I. INTRODUCTION

Staining methods currently exist for localization of enzyme activities resolved by electrophoresis or isoelectric focusing directly in the resolution gel (1,2). Several methodologies have been developed and used to form colored formazan products from tetrazolium salts at locations in tissues and gels where dehydrogenase activities are found (3-5). The present paper gives a method in which photon emission catalyzed by luciferase in the presence of all necessary substrates exposes film and thereby localizes an enzyme activity within a gel. Examples are presented 1) for the localization of resolved luciferase activities themselves and 2) via a reaction coupled to luciferase for the localization of enzyme activities which produce a substrate used by luciferase. An abstract relating to this method has previously appeared (6).

[1]*This research was supported by grant AH-777 from the Robert A. Welch Foundation, Houston, Texas and by grant RR 08012 of the Division of Research Resources, National Institute of Health.*




ISBN 0-12-208820-4

## II. EXPERIMENTAL

Electrophoresis was conducted at 4°C near neutral pH, under pH conditions which maintained the activity of the enzymes being resolved. For bacterial luciferase and NAD(P)H-FMN oxidoreductase activities, the trailing anion upper reservoir contained 5 mM $tris^+$ $TES^-$, pH 7.0; ($tris^+$ is the cationic form of tris (hydroxymethyl) aminomethane, $TES^-$ is the anionic form of N-tris (hydroxymethyl) methyl-2-aminoethanesulfonic acid). The stacking gel was 4% polyacrylamide and 60 mM $imidazol^+Cl^-$ , pH 6.0, photopolymerized by standard procedures (7,8). The 8% polyacrylamide resolving gel contained 60 mM $tris^+Cl^-$, pH 7.5, polymerized using ammonium persulfate (7,8). For firefly luciferase electrophoresis, the trailing anion solution contained 5 mM $tris^+$ glycylglycine$^-$, pH 8.0, the 4% stacking gel contained 60 mM imidazol $^+Cl^-$, pH 6.5,and the 8% resolving gel contained 60 mM $tris^+Cl^-$, pH 8.0.

Autolumography was accomplished by removing either one or both of the 3¼" x 4" glass lantern plates sandwiching the 1½ mm thick electrophoresis gel, laying the gel or glass plate holding the gel onto a sheet of plastic wrap, in absolute room darkness positioning the gel over a 4" x 5" sheet of Kodak Royal-X pan, Estar thick base 4166 film, and applying approximately 2 ml of the appropriate luminescence initiating solution onto the gel.

For bacterial luciferase the 2 ml of luminescence initiating solution was 0.1 M phosphate buffer, pH 7.0 containing 0.1 mM NADH,5 μM FMN, 0.001% v/v decanal, and 0.5 ml of a partially purified *P. fischeri* NAD(P)H-FMN oxidoreductase activity. The 2 ml of initiating solution for NAD(P)H-FMN oxidoreductase activities was 0.1 M phosphate buffer pH 7.0, containing 0.1 mM NADH, 5 μM FMN, 0.001% decanal, and 10 μg purified *B. harveyi* luciferase. The 2 ml of initiating solution for firefly luciferase was 25 mM $tris^+$ glygly$^-$, pH 8.0, containing 50 μM ATP, 10 mM synthetic firefly luciferin (Sigma), and 5 mM $MgCl_2$. Protein staining with 0.25% Coomassie blue and destaining were done by standard procedures (8).

## III. RESULTS AND DISCUSSION

Figure 1 shows photographs of gels containing electrophoresed partially purified *B. harveyi* luciferase. The samples applied to the 14 successive channels represent the

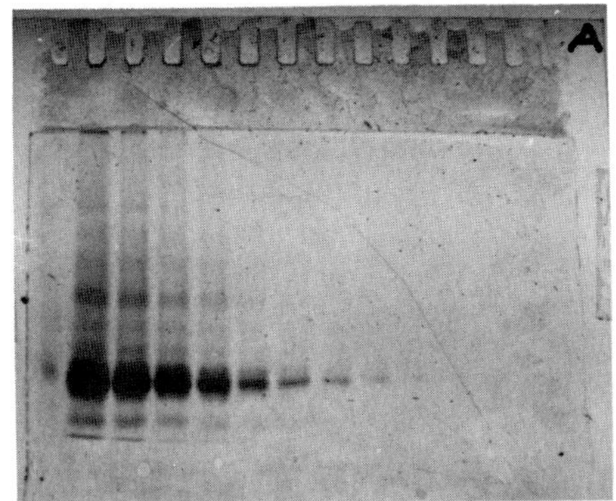


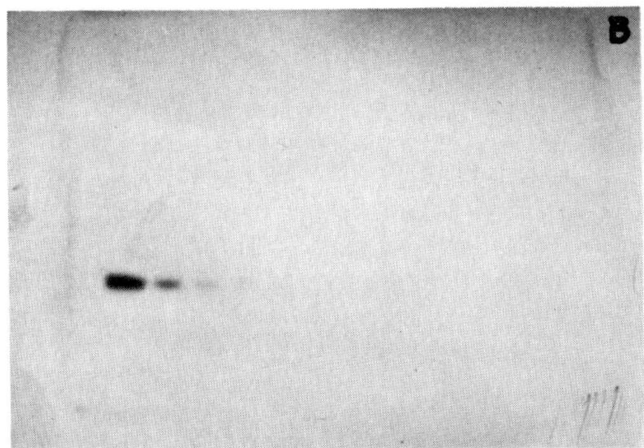


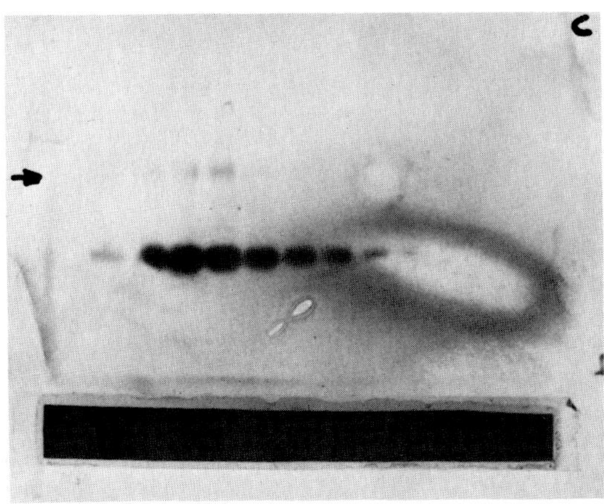


*FIGURE 1. Electropherogram (A) and autolumographs (B and C) of B. harveyi luciferase.*

following μg amounts of luciferase: 0, 20, 10, 5, 2, 1, 0.5, 0.2, 0.1, 0.05, 0.02, 0.01, 0.005, and 0.002. Figure 1A is a Coomassie blue stained gel; luciferase staining can be detected at the level of 0.02 μg sample applied. Figures 1B and 1C are autolumographs of a duplicate gel, for 2 minute and 30 minute exposures respectively. In the 30 minute exposure some of the left side of the gel was not adequately covered with bathing solution. The position of the left hand edge of the gel can be seen outlined in Figure 1C. Using this as a reference, the sensitivity of the method can be determined by counting channels across the gel. In the original autolumograph film sheet, some exposure can be detected at 20 ng of luciferase. In the opinion of the author, far greater sensitivity would be possible if the bathing solution had a higher oxidoreductase activity with the same or lower luciferase background contamination.

A significant observation from these studies and potential future use for the method is the appearance of a minor luminescence at the position of the arrow in Figure 1C. This position corresponds approximately to a stained zone in Figure 1A. The identity of this activity is not known at this time.

Figure 2 shows photographs of a gel on which crude firefly lantern extracts (9) were electrophoresed. Figures 2B and 2C are the 2 minute and 10 minute autolumographs of the same gel subsequently stained with Coomassie blue and shown in Figure 2A. Under the conditions of the experiment, most of the activity in the extract failed to enter the stacking gel and none entered the resolving gel. Electrophoresis at higher pH might be more successful in causing firefly luciferase to enter the resolving gel.

Figure 3 shows the results of experiments in which enzyme activities other than luciferase are localized by coupling to the luciferase reaction. In this electropherogram, NAD(P)H-FMN oxidoreductase activities from *B. harveyi* (left side) and *P. fischeri* (right side) are localized by 10 minute autolumography (Figure 3B) by-means-of *B. harveyi* luciferase and appropriate substrates in the bathing solution. In the gel stained with Coomassie blue (Figure 3A) no sample was applied to channel 12, however in the gel used for the autolumograph, this channel contained a larger amount of sample of the same *P. fischeri* oxidoreductase activity applied to channel 13. Channel 10 contained a different and less active *P. fischeri* oxidoreductase activity too low to be visualized by autolumography.

This method may find general use in the localization of a wide variety of enzymes by coupling to appropriate luciferase reactions. It may be best suited to detecting on gels

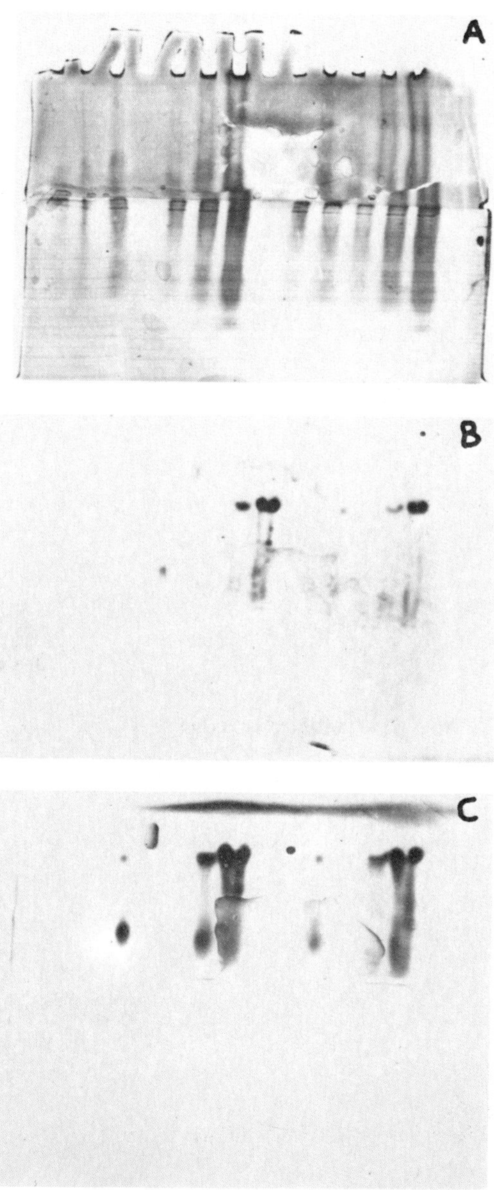


*FIGURE 2. Electropherogram (A) and autolumographs (B and C) of firefly luciferase.*

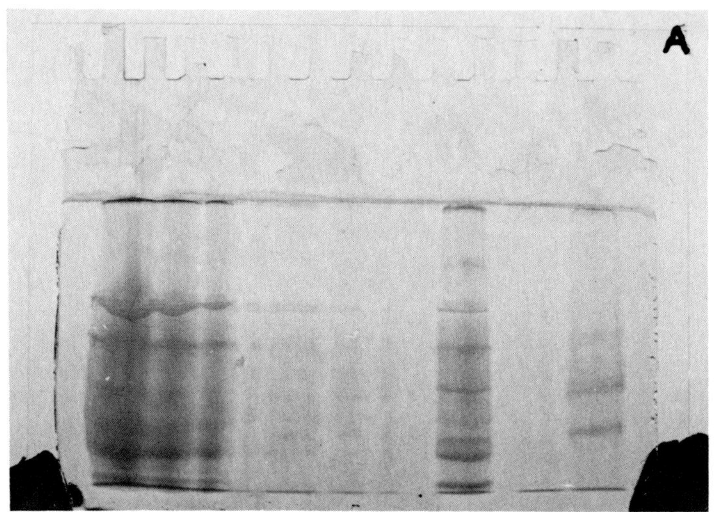


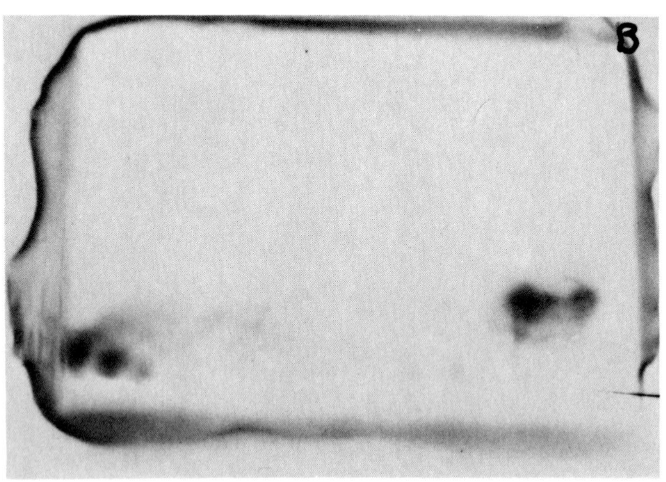


*FIGURE 3. Electropherogram (A) and autolumograph (B) of NADH(P)H-FMN oxidoreductase activities from B. harveyi and P. fischeri.*

the presence of enzyme systems or activities which couple to the luciferase reaction within the luminescent cell.

## REFERENCES

1. Arvidson, S. and T. Wadstrom, *Biochim. Biophys Acta. 310,* 418 (1973).
2. Dahlmann, B. and Kl.D. Jang, *J. Chromatog. 110,* 174 (1975).
3. Nineham, A.W., *Chem. Rev. 55,* 355 (1955).
4. Altman, F.P., *Histochemie 19,* 363 (1969).
5. Tu, S.-C., J.E. Becvar, and J.W. Hastings, *Arch. Bioch. Biophys. 193,* 110 (1979).
6. Becvar, J.E. and D.L. Smithson, *Fed. Proc. 39,* 1603 (1980).
7. Davis, B.J., *Ann. N.Y. Acad. Sci. 121,* 404 (1964).
8. Clark, J.M., Jr., and R.L. Switzer *in* "Experimental Biochemistry", Freeman, San Francisco, 1977.
9. Rasmussen, H.N., *in* "Methods in Enzymology" (M.A. DeLuca, ed.), p. 28, Academic Press, New York, 1978.

# BIOENERGETICS OF INSULIN AND GLUCAGON PRODUCING CELLS EVALUATED IN BIOLUMINESCENCE ASSAYS

*Sven E. Brolin*
*Ambjörn Agren*

Department of Medical Cell Biology
University of Uppsala
Uppsala, Sweden

## INTRODUCTION

In microphysiological research using tissue explants, the potential of bioluminescence techniques should facilitate progress, particularly when important cells are available only in limited amounts. This applies to cells which exert a regulatory function by discharge of granules containing potent signal substances. When we consider the basic problem of how the regulatory function of such small groups of cells, in the nervous and endocrine system, are related to their metabolism we become dependent on the applicability of biochemical micro methods. The relation between secretion and metabolism is a problem of central importance in research on diabetes and it has led to several methodological developments.

Diabetes is one of the most prominent endocrine diseases which affects more than two per cent of the population in the Western world. In concerns a lack of metabolic regulation, with the result that increasing concentrations of glucose fail to elicit release of insulin. Glucose is normally a potent physiological stimulator of the insulin producing B-cells. Thus a greater knowledge of their biochemistry is required if this action is to be satisfactorily understood.

Biochemical analyses of the B-cells are hampered by a limited access to material since they are located in millions of small islets of Langerhans which are distributed throughout the parenchyma of the pancreas. Even by pooling it is difficult to collect more than a few µg of appropriate islet



ISBN 0-12-208820-4

samples.

By application micrometechniques according to Lowry and collaborators (1) the islets of Langerhans have been demonstrated to be enzymatically well equipped for glucose degradation (2-4). Islet glucose utilization has also been studied at various concentrations, using measurements of $C^{14}O_2$ formation from metabolized labelled glucose (5,6). In experiments using Cartesian micro divers it has furthermore been shown that islet oxygen consumption rises strongly with increasing glucose concentrations (7).

## EXPERIMENTAL MODEL

Both B-cells and glucagon-producing $A_2$-cells are located in the islets. An experimental model for comparison of the two cell types has been designed, using streptozotocin for selective destruction of the B-cells, see Figure 1.

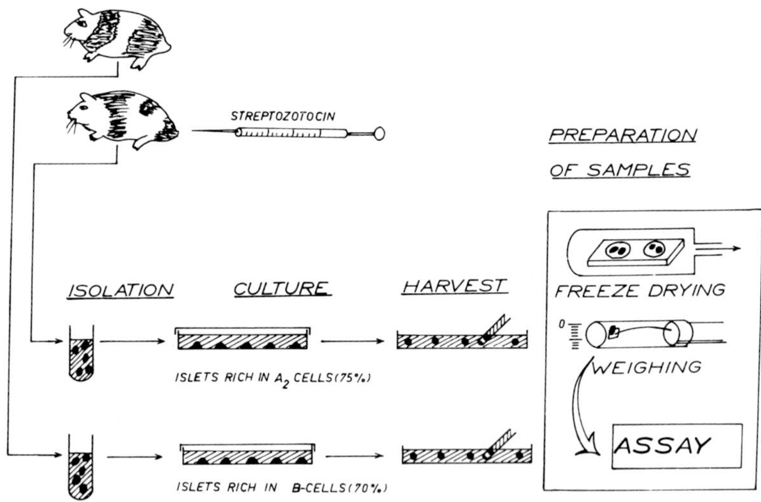


*FIGURE 1. Experimental procedure for preparing normal B-cell rich pancreatic islets for islets enriched in $A_2$-cells. Published with the permission from Diabete & Metabolisme, Paris (8).*

Both the B- and $A_2$cells release hormone granules from a large internal store. This process is stimulated by glucose in the B-cells but inhibited in the $A_2$-cells. Another difference, of deep importance in diabetes, is that the B-cells can be overstimulated by glucose to such an extent that they degenerate, an effect that has not been found in the $A_2$-cells. A raised glucose concentration increased utilization (6) and consumption of oxygen (9),in both cells types. So far, studies of the overall rates of energy fluxes have given basic information on the bioenergetics of the two islet cell types, but further and more detailed studies are required. These investigations could be performed by means of bioluminescence analyses.

## BIOLUMINESCENCE ASSAYS

The first bioluminescence assay of any cell samples weighing less and 1µg concerned determinations of the ATP concentrations in the islets (10). The photokinetic technique developed for theses studies permitted measurements at very low level of detectability. Following further improvements including photon counting single cell assay became possible (11). After the bacterial luciferase system had been shown to be suitable for micro analyses (12,13) the techniques were utilized in biochemical studies of the islets (13,14). For comparisons between insulin and glucagon producing cells bioluminescence has proven to be a convenient and sensitive analytical tool.

### *Adenosine phosphates*

Cells which are dependant on rapid mobilization of energy for movements or generation of signals may possess a short cut for energy provision in which ADP is converted to ATP in the adenylate kinase reaction. The prerequisite for such ATP formation in different islet cells was evaluated by bioluminescence analyses (15). The B-cells displayed a remarkably high activity of adenylate kinase ( EC 2.7.4.3). The activity of the $A_2$-cell was lower than that of the B-cell but still relatively high (Table 1).

*TABLE 1. Adenylate kinase (EC 2.7.4.3) activity of pancreatic islets rich in B- or $A_2$-cells from guinea pigs*

| *Specimens* | *Enzyme activity* [a] | |
|---|---|---|
| | *Freshly isolated* | *Cultured* |
| *B-cell rich islets* | 403 ± 24 (8) | 300 ± 18 (8) |
| *$A_2$-cell rich islets* | 373 ± 32 (7) | 267 ± 16 (6) |

[a]*The results are given as μmol per min and g dry weight ±S.E.M.. The number of animals are given in parentheses.*

Both cell types showed a high concentration of adenosine nucleotides in particular ATP (8). It may be noteworthy that if the $A_2$-cells are supplied with insulin their ATP values are not surpassed by those of the B-cells (16). Although the concentration of ATP is of obvious bioenergetic interest the conversion rate may be more important for the provision of energy. Experiments with hypoxia can be used to test the conversion rates of ATP in comparisons between the B- and $A_2$-cells. Both cell types exhibit rapid changes (8,17), but their total pool of adenosine phosphates remains essentially unchanged.

*Nicotinamide dinucleotides*

Fast conversion of ATP would imply efficient oxidation of NADH in the respiratory pathway resulting in a considerable accumulation of the reduced nucleotide after interruption of the oxygen supply. This has been confirmed in the islets (18). More detailed bioenergetic studies of the islets became possible with the development of a single step bioluminescence assay of NAD+ (19,20), see Figure 2. Thus, it has now become possible to collect enough data to determine the redox state in terms of the [NAD+] / [NADH] ratio. Redox conditions for whole islets have previously been estimated indirectly from the concentrations of substrates determined by enzymatic cycling (21,22).

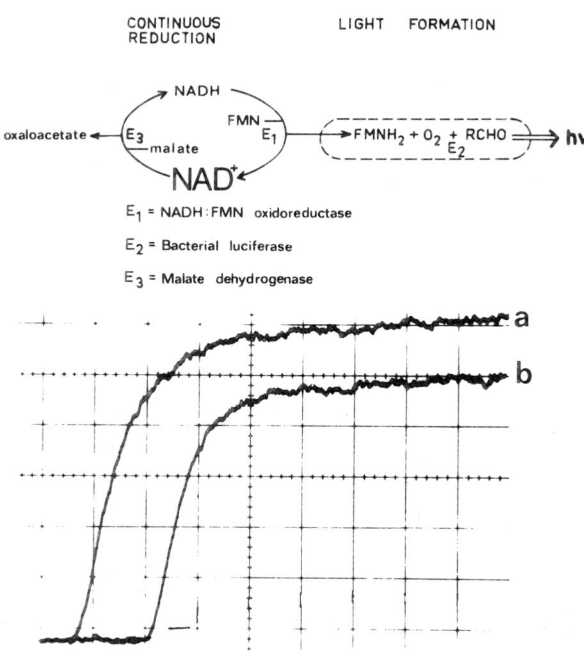


*FIGURE 2. Bioluminescence analyses of NAD+. A continuous reduction of NAD+ results in a continuous formation of NADH and a longlasting light emission is produced (above). Photographs of the oscilloscope screen (below) show the formation of different stable emission levels after addition of 12 (a) and 10 (b) pmol of NAD+.*

Comparisons between the B and $A_2$-cells show some obvious similarities, see Table 2. Energetic differences may nevertheless exist, since the B-cells can be stimulated to such an extent that it leads to exhaustion. Further studies of possible metabolic differences between the B- and $A_2$-cells should include determinations of the rates of change of their redox states at various concentrations of glucose. Such experiments would require the collection and evaluation of a greatly increased amount of analytical data. This can be accomplished by combining the measuring device with a microprocessor equipped with a suitable display system (14).

*TABLE 2. Redox states of pancreatic islets rich in B- or $A_2$-cells from guinea pigs*

| *Specimens* | *NAD+*[a] | *NADH*[a] | *[NAD+] / [NADH]* | |
|---|---|---|---|---|
| *FRESHLY ISOLATED* | | | | |
| *B-cell rich* | 1.1 ± 0.1 | 0.3 ± 0.01 | 3.5 | (n=16) |
| *$A_2$-cell rich* | 0.9 ± 0.1 | 0.3 ± 0.01 | 3.5 | (n=15) |
| *CULTURED FOR 7 DAYS AT 5.5 mM GLUCOSE* | | | | |
| *B-cell rich* | 1.6 ± 0.1 | 0.2 ± 0.01 | 9.8 | (n=16) |
| *$A_2$-cell rich* | 1.3 ± 0.1 | 0.2 ± 0.01 | 7.4 | (n-15) |

[a]*Concentrations of the nucleotides expressed as mmol per kg dry weight ± S.E.M.. The number of animals are given in parentheses.*

SUMMARY

Diabetes is a serious and wide-spred disease, characterized by sequels of disturbed secretory function of the insulin producing cells. In the search for the biochemical deficiency associated with this phenomenon, bioenergetic comparisons with the adjacent but less vulnerable glucagon producing cells may help to reveal the nature of the metabolic disorder. The limited availability of material composed of such cells requires the use of very sensitive micro methods. This has led to the application of bioluminescence techniques and to a further development of their inherent analytical potential.

ACKNOWLEDGEMENTS

Financial support was received from the Swedish Medical Research Council (project number B81-12X-00525) and the Swedish Diabetes Association.

REFERENCES

1. Lowery, O. H., and Passonneau, J. V., in "A flexible system of enzymatic analysis". Academic Press, New York, (1972).

2. Hellman, B., and Täljedal, I.-B., in "Handbook of Physiology, Section 7; Endocrinology I" (D. Steiner and N. Freinkel, eds.), p. 91. American Physiological Society, Washington, (1972).

3. Hellerström, C., and Brolin, S. E., in "Handbook of Experimental Pharmacology vol 32/2" (A. Hasselblatt and F. von Bruckhausen, eds.), p. 57. Springer Verlag, Berlin, (1975).

4. Matschinsky, R. M., in "Handbook of Physiology, Section 7; Endocrinology I" (D. Steiner and N. Freinkel, eds.), p. 199. American Physiological Society, Washington, (1972).

5. Ashcroft, S. J. H., Hedeskov, C. J., and Randle, P. J., *Biochem. J. 118,* 143 (1970).

6. Östenson, C.-G., *Biochen. J. 188,* 201 (1980).

7. Hellerström, C., *Endocrinology 81,* 105 (1967).

8. Östenson, C.-G., Agren, A., Brolin, S. E., and Petersson, B., *Diabete & Metabolisme 6,* 5 (1980).

9 Petersson, B., Hellerström, C., and Gunnarsson, R., *Horm. Metab. Res. 2,* 313 (1970).

10. Wettermark, G., Tegnér, L., Brolin, S. E., and Borglund, E., in "The structure and metabolism of the pancreatic islets" (S. Falkmer, B. Hellman and I.-B. Täljedal, eds.), p. 275. Pergamon Press, Oxford, (1970).

11. Wettermark, G., Stymne, H., Brolin, S. E., and Petersson, B., *Anal. Biochem. 63,* 293 (1975).

12. Stanley, P. E., *Anal. Biochem. 39,* 441 (1971).

13. Brolin, S. E., Borglund, E., Tegner, L., and Wettermark, G., *Anal. Biochem. 42,* 124 (1971).

14. Agren, A., *Dissertation. Acta Univ. Upsalien. 374,* 1 (1980).

15. Brolin, S. E., Wersall, J. P., and Agren, A., in "Proc. Int. Symp. on Analytical Applications of Bioluminescence and Chemiluminescence" (E. Schram and P. Stanley, eds.), p. 458. State Printing & Publishing Inc., Westlake Village, (1979).

16. Östenson, C.-G., *Diabetologia 17,* 325 (1979).

17. Wersäll, P., Brolin, S., Petersson, B., Östenson, C.-G., in "Proc. of Int. Symp. of Analytical Applications of Bioluminescence and Chemiluminescence: (M. DeLuca and W.D. McElroy, eds), Academic Press, New York, (1980).

18. Berne, C., Brolin, S. E., and Agren, A., *Horm. Metab. Res. 5,* 141 (1973).

19. Agren, A., Brolin, S. E., and Hjertén, S., *Biochim. Biophys. Acta 500,* 103 (1977).

20. Hutton, J. C., Sener, A., and Malaisse, W.J., in "Proc. Int. Symp. on Anlytical Applications of Bioluminescence and Chemiluminescence" (E. Schram and P. Stanley, eds.). p. 166. State Printing & Publishing Inc., Westlake

p. 166. State Printing & Publishing Inc., Westlake Village, (1979).

21. Malaisse, W. J., Hutton, J. C., Kawazu, S., Herchuelz, A., Valverde, I., and Sener, A., *Diabetologia 16*, 331 (1979).

22. Ammon, H. P. T., and Verspohl, E. J., *Diabetologia 17*, 41 (1979).

# APPLICATION OF THE PHOTOPROTEIN OBELIN TO THE MEASUREMENT OF FREE $Ca^{2+}$ IN CELLS

*Anthony K. Campbell*
*Maurice B. Hallett*
*Richard A. Daw*
*Malcolm E. T. Ryall*

Department of Medical Biochemistry
Welsh National School of Medicine
Cardiff, U.K.

*Russell C. Hart*

School of Molecular Sciences
University of Sussex
Brighton, U.K.

*Peter J. Herring*

Institute of Oceanographic Sciences
Wormley
Surrey, U.K.

## I. THE IMPORTANCE OF INTRACELLULAR $Ca^{2+}$

During the past 20 years biologists have become increasingly interested in the mechanism by which many electrical and chemical stimuli, acting on the cell surface, can activate processes within the cell. Much evidence, albeit often indirect, exists that the ability of action potentials, hormones and neurotransmitters to stimulate muscle contraction, other forms of cell movement, secretion, cell fertilization and division depends on a rapid increase in cytoplasmic free $Ca^{2+}$ concentration (1,2). Direct



ISBN 0-12-208820-4

evidence for a role of intracellular $Ca^{2+}$ in mediating the effect of these primary stimuli requires the measurement of free $Ca^{2+}$ inside intact cells. Three methods are currently available (1) namely photoproteins, metallochromic indicator dyes and micro-electrodes. However the sensitivity of photoproteins, coupled with their ability to resolve changes in $Ca^{2+}$ distribution in the cell, has resulted in this technique providing most of the novel information about intracellular $Ca^{2+}$ at the present time.

## II. SOURCE OF PHOTOPROTEINS

### A. *Obelin*

Most workers have used the $Ca^{2+}$-activated photoprotein aequorin (3,4,5). However the difficulty of obtaining a regular supply of the jelly fish Aequorea forskalea has led us to search for an alternative source more easily obtained in Europe. The hydroid Obelia geniculata occurs commonly off the coasts of the United Kingdom, growing particularly on various brown seaweeds such as Laminaria spp. It contains a $Ca^{2+}$-activated photoprotein, obelin, which is qualitatively very similar to aequorin, although some quantitative differences have been defined (1,4,5,6). Futhermore sufficient obelin can be prepared from one collection for at least a year's worth of physiological experiments.

### B. *Thalassicolin*

Until recently (7) it was thought that $Ca^{2+}$-activated photoproteins were only to be found in particular luminous organisms from the phyla Cnidaria and Ctenophora. However it has been known for more than a century that, in addition to these animals, extracts from luminous radiolarians (Phylum: Protozoa) can also luminesce in the absence of molecular oxygen. The radiolarian Thalassicola spp. was collected in may thousands in the top 10 - 20 cm of the sea off the west coast of Ireland whilst on the R.R.S. Discovery (see Report on Cruise no. 105). These unicellular organisms emit blue flashes of light when touched and can also be stimulated to luminesce with $K^+$ sea water or ionophore A23187, though these latter stimuli were not always reproducible. Extraction of the organisms in 200 mM tris + 40 mM EDTA, pH 7.0 followed by precipitation of the protein with saturated ammonium sulphate produced a relatively stable preparation of protein which could be stimulated to emit light simply by addition of $Ca^{2+}$. The saturating rate constant at room temperature was approx. 3 $s^{-1}$ and it eluted from DEAE cellulose or G75 sephadex in a manner very similar to obelin (6,8). The light emitted was slightly bluer ($\lambda$ max *ca* 440 nm) than for coelenterate photoproteins. As with the coelenterate photoproteins thalassicolin luminesced strongly in 25 mM $Ca^{2+}$, weakly in 25 mM $Sr^{2+}$ and hardly at all in 25 mM $Ba^{2+}$. No detectable luminescence was seen with 25 mM $Li^{+}$, $Na^{+}$, $K^{+}$ or $Mg^{2+}$. However both $Mg^{2+}$ and $Na^{+}$ inhibited $Ca^{2+}$-activated luminescence. Removal of $O_2$ by vacuum pump, which virtually abolished luminescence from the bacterial or *Cypridina* systems, had no effect on $Ca^{2+}$-activated luminescence from thalassicolin. Futhermore both obelin and thalassicolin could be reactivated using synthetic prosthetic group (9) to produce proteins indistinguishable in properties from the original (Fig. 1). It was concluded that luminous radiolarians contain $Ca^{2+}$-activated photoproteins similar to those found in coelenterates, though in insufficient quantities to be a useful source for physiological experiments.

## III. ENTRAPMENT OF PHOTOPROTEIN INSIDE MEMBRANE VESICLES

The photoprotein obelin has been entrappedwithin various membrane vesicles (Table I) in order to study the effect of various agents on the permeability of membranes to $Ca^{2+}$. Entrapment was quantified by addition of extravesicular $Ca^{2+}$ and then triton X-100 (5). It is important to use scintillation grade triton to minimize luminescence artefacts.

Pigeon erythrocyte 'ghosts' containing obelin have been used to show that adenylate cyclase is inhibited in these cells by more than 50% in the presence of $Ca^{2+}$ = 1 - 10 μM. The 'ghosts' have proved useful in investigating the effects of ionophore A23187 (8), cell antibody + complement (10,11) and fusogenic agents such as Sendai virus or polyethylene glycol (5) or the permeability of biological membranes to $Ca^{2+}$.

*Fig. 1. Reactivation of photoproteins with synthetic prosthetic group. The photoproteins obelin and thalassicolin in 200 mM tris, pH 7.4, were stimulated with 2.5 mM $CaCl_2$ and left for 30 - 60 min. EDTA (5mM) + mercaptoethanol (ca 7 mM) + ca 10 nmole synthetic coelenterate luciferin (8) at 4°C. $Ca^{2+}$-activated luminescence was assayed at defined time intervals. ■ = thalassicolin; □ = obelin*

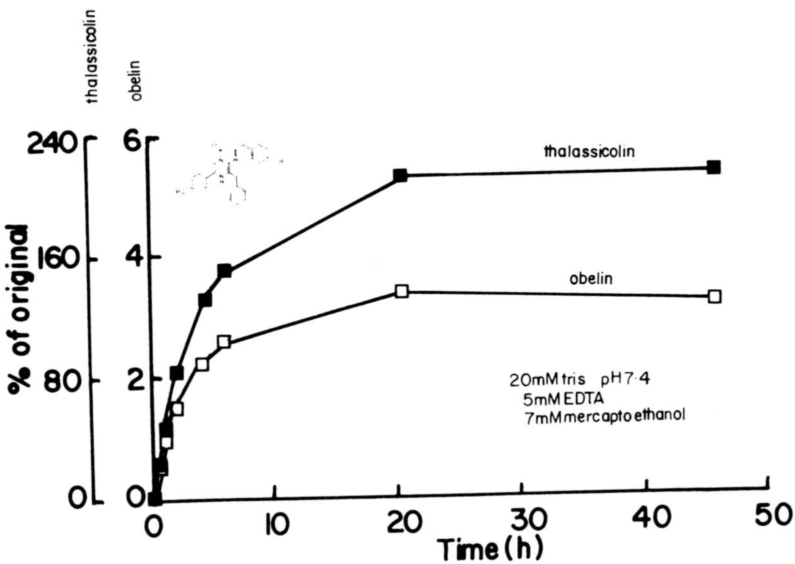


*TABLE I. Membrance vesicles which can entrap obelin*

| Vesicles | Species |
|---|---|
| Erythrocyte 'ghosts' | pigeon, chicken, rat, human |
| Hepatocyte membranes | rat |
| Adipocyte membranes | rat |
| Synaptic membranes | rat |
| Sacroplasmic reticulum | rabbitt |
| Lipsomes | pure phospholipids |

*for references see 1,4,5,8, 12.*

## IV. INCORPORATION OF PHOTOPROTEIN INTO INTACT CELLS

Obelin has been microinjected into several giant cells, for example barnacle muscle and starfish oocytes, where it responsds in a similar manner to aequorin (5). Unfortunately many mammalian cells are too small to microinject. We have investigated the possibility of incorporating photoproteins into intact cells by fusion with membrance vesicles or cell 'ghosts' containing obelin.

### *A. Liposomes*

Liposomes containing obelin are taken up by various cells, including isolated adipocytes (12). Their interaction with cells results in an increase in the permeability of the liposomes to $Ca^{2+}$, as well as the transfer of at least two cell surface antigens into them. At least 50% of the liposomal obelin seems to be taken up by endocytosis and the remainder by adhesion. Using cell antibody + complement as a menas of rapidly increasing intracellular free $Ca^{2+}$ (10) no cytoplasmic obelin could be detected.

### *B. Erythrocyte 'ghosts'*

Sendai virus, in the presence or absence of external $Ca^{2+}$, appeared in induce fusion between erythrocyte 'ghosts' containing Cypridina luciferin and erythrocyte 'ghosts' containing Cypridina luciferase (Fig. 2). Morphologically it was difficult to distinguish between cell fusion and cell adhesion. However light emission from Cypridina luciferin + luciferase enabled cell fusion to be monitored. $Ca^{2+}$ may be necessary to induce fusion at low fusogen concentrations. Using the cell antibody plus complement criterion no obelin transfer could be detected because in the presence of $Ca^{2+}$ a large increase in membrane permeability to $Ca^{2+}$ occurred causing consumption of the obelin, whereas in EGTA the 'ghosts' became very leaky.

## V. CONCLUSIONS

The $Ca^{2+}$-activated photoprotein obelin can be used to quantitatively detect changes in free $Ca^{2+}$ inside cells and membrane vesicles. If these vesicles can be fused with cells then for the first time a method generally applicable to small cells will be available for measuring directly changes in intracellular free $Ca^{2+}$.

*Fig. 2. Fusion of pigeon erythrocyte 'ghosts' by Sendai virus. Two separate populations of sealed pigeon erythrocyte 'ghosts' were prepared (9), one containing a crude Cypridina luciferase extract, the other a solubilized methanol extract of Cypridina (luciferin). Addition of the two populations together, in front of the photomultiplier tube, resulted in a stimulation of luminescence when Sendai virus (△, 75; O, ▲750; ■7500 HAU hema glutinating units) was present. EGTA reduced apparent fusion at low, but not high concentrations of virus. No effect was observable in the absence of virus, nor could the results be explained by release of luciferin + luciferase into the incubation medium. Temp. 37°C.*

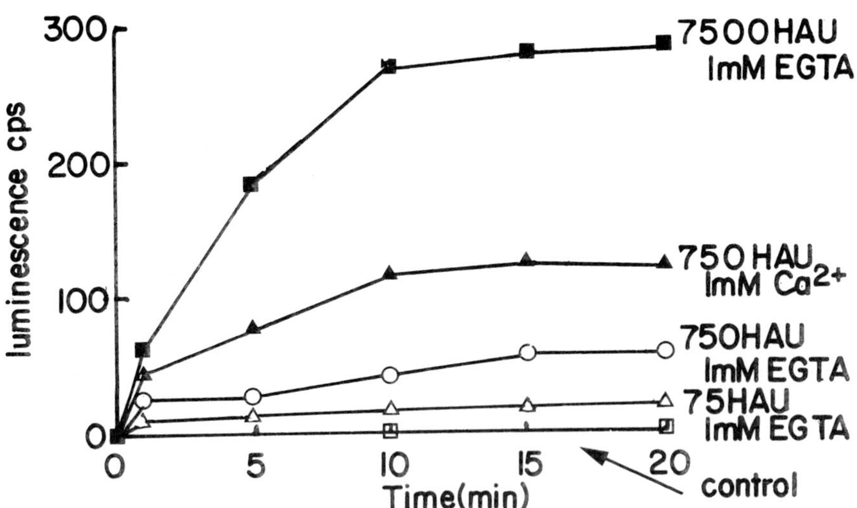

ACKNOWLEDGEMENTS

The Science and Medical Research Councils and the Director and Staff of the Marine Biological Association Laboratory, Plymouth.

REFERENCES

1. Ashely, C. C. and Campbell, A. K. (eds.) "Detection and Measurement of Free $Ca^{2+}$ in Cells". Elsevier-North Holland, Amsterdam (1979).

2. Campbell, A. K., Hallett, M. B., Daw, R. A., Luzio, J. P. and Siddle, K., Biochem. Soc. Trans. 7, 865-869 (1979).

3. Ridgway, E. B. and Ashley, C. C., Biochem. Biophys. Res. Commun. 29, 229-234 (1967).

4. Campbell, A. K. and Simpson, J. S. A. Techniques in Metabolic Research B213, 1-56 (1979).

5. Hallett, M. B. and Campbell, A. K. in "Biomedical Applications of Luminescence" (L. Kricka and T. Carter, eds.) Marcel Dekker, in press.

6. Campbell, A. K., Biochem. J. 143, 411-418 (1974).

7. Herring, P. J., Marine Biology 53, 213-216 (1979).

8. Campbell, A. K. and Dormer, R. L., Biochem. J. 176, 53-66 (1978).

9. Hart, R. C., Matthews, J. C., Hori, K. and Cormier, M. J., Biochemistry 18, 2204-2210 (1979).

10. Campbell, A. K., Daw, R. A. and Luzio, J. P., FEBS Lett. 107, 55-60 (1979).

11. Campbell, A. K., Davies, C. J., Hart, R., McCapra, F. Patel, A., Richardson, A., Ryall, M. E. T., Simpson, J. S. A. and Woodhead, J. S., J. Physiol. proceeding in press.

12. Hallett, M. B. and Campbell, A. K. submitted for publication.

# CHEMILUMINESCENCE OF CYTOTOXIC MACROPHAGES

Martin Ernst
Helmut Lang
Herbert Fischer
Marie-Luise Lohmann-Matthes
Hansjürgen Staudinger

Max-Planck-Institut
für Immunbiologie
Freiburg, FRG

## I. INTRODUCTION

Since the discovery of Allen *et al.* (1) that granulocytes emit chemiluminescence during the phagocytosis-induced respiratory burst, a lot of work has been done to clarify the nature and interrelationship of both phenomena which are not restricted to granulocytes but also occur in monocytes and macrophages (2, 3). By now it seems to be well established that the chemiluminescence phenomenon of phagocytes is tightly connected with the production and release of superoxide, $O_2^-$ and related activated oxygen species like $H_2O_2$, $^1O_2$, and $\cdot OH$, which in the case of granulocytes have been proven to play a major role in the microbicidal activity of these cells (4). Superoxide dependent chemiluminescence has also been measured from granulocytes and macrophages which were incubated with non-particulate stimuli like Concanavalin A, Ca-Ionophor A23187 (3), and cytochalasin E (5) in the presence of luminol (5-amino-2,3-dihydro-1,4-phthalazinedione).

Superoxide and hydrogen perioxide seem to play an important role in the cytotoxic activity of granulocytes (6, 7, 8) and in some cases also on macrophages (7, 8). As both superoxide and hydrogen peroxide in the presence of luminol give rise to chemiluminescence we have studied the effect of cytotoxic activation of macrophages derived from C57Bl mice by means of luminol aided chemiluminescence.




ISBN 0-12-208820-4

## II. MATERIAL AND METHODS

As incubation medium we used modified Eagle's medium without sodium bicarbonate but with 50 mM HEPES (N-2-Hydroxyethylpiperazine-N'-2-ethanosulfonic acid) at pH 7.40 (E-H-medium). Superoxide dismutase (SOD, 2.900 U/mg), catalase (14 000 U/mg) and cytochrome c were obtained from Sigma, Munich. Effector cells: As effector cells we used 5-7 day cultured bone marrow macrophages (9) or peritoneal macrophages rendered activated by i.p. injection of 10 BCG (Bacille Calmette Gúerin) bacteria per mouse and 11 days later i.p. booster injection of $10^4$U/mouse P.P.D. (Purified Protein Derivative of tuberculin). The peritoneal macrophages were harvested 14 days after BCG injection.

Three different cytotoxic systems were used:

1. Bone marrow macrophages were preactivated by one-day preincubation with a highly purified lymphokine, macrophage cytotoxicity factor (MCF) (10), after washing the macrophages, target cells were added.
2. Non-activated bone marrow macrophages were incubated together with antibody coated target cells (ADCC, antibody dependent cellular cytotoxicity).
3. BCG macrophages were activated chemically by addition of 5 ng/ml TPA (12-0 Tetradecanoyl-phorbol-13-acetate, Consol. Midland Corp., Brewster, N.Y.) directly before target cell addition.

As target cells we used P815 cells and P388 cells, the effector to target cell ratio being 10:1. Cytotoxicity assay: Target cells were labelled with $^{51}$Cr, and after coincubation with macrophages, specific release of $^{51}$Cr, was taken as a measure for cytotoxicity.

### A. *Chemiluminescence Measurement*

*a.* Early chemiluminescence measurements were performed using a modified scintillation counter (Packard C2425) in the coincidence off mode containing two selected photomultipliers with high sensitivity in the blue and green region of the light spectrum and being thermostated at 37$^{\circ}$C. $10^6$ macrophages were placed in each vial containing 5 ml E-H-medium. 30 min before chemiluminescence measurement luminol was added (final concentration 40 µg/ml). Samples were measured about every 18 min (depending on the number of samples) each for 0.2 min. Results are expressed as the integrated counts per 15 hours or 6 hours after target cell addition, these counts always containing the relatively high machine background counts.

*b.* Only recently we had the possibility to measure chemiluminescence from cytotoxic macrophages with a new chemiluminescence device (Biolumat LB 95o5, Berthold Co. Wildbad) (11) which allows the simultaneous and continuous measurement of six samples, each of which containing only 2 x $10^5$ macrophages in 0.5 ml E-H-medium.

## *III. RESULTS*

### *A. MCF Activated Bone Marrow Macrophages (Table I)*

*a.* As shown in Table I the background chemiluminescence signal of macrophages after MCF preactivation was consistently higher than the background chemiluminescence signal of non-activated control macrophages.

*b.* The addition of P815 target cells generated a weak but significant long lasting (12-15 hours) increased chemiluminescence signal of MCF activated macrophages. In contrast non-activated macrophages showed only a slightly increased chemiluminescence signal after P815 target cell addition (cf. Table I). Simultaneously performed cytotoxicity assays of parallel samples revealed 15 hours after addition of the target cells a specific cytotoxicity of MCF activated macrophages whereas non-activated macrophages showed no significant specific cytotoxicity (cf. Table I).

TABLE I. Chemiluminescence of Cytotoxic Bone Marrow Macrophages and $^{51}Cr$ Release from P815 Target Cells

| Cells and Conditions | Chemiluminescence integrated | | % $^{51}Cr$ release after | |
|---|---|---|---|---|
| | $10^6$ counts/15 h | $10^6$ counts/6h | 15h | 6h |
| P815 or non-activated macrophages alone | 59.8 ± 7 | 23.9 ± 3 | 27±1 | 15±1 |
| MCF activated macrophages alone | 86.0 ± 8 | 34.4 ± 3 | - | - |
| Non-activated macrophages + P815 | 79.7 ± 6 | | 30±1 | |
| MCF activated macrophages + P815 | 144.3 ± 7 | | 45±1 | |
| Non-activated macrophages + antibody-coated P815 | 301.8 ± 15 | 215.2 ± 6 | | 71±2 |

*TABLE II. Chemiluminescence of BCG Macrophages and $^{51}Cr$ Release from P388 Target Cells*

| Cells and Conditions | Chemiluminescence integrated $10^6$ counts/15h | $\%^{51}Cr$ release after 15h |
|---|---|---|
| P388 alone | 107 ± 10 | 27 ± 2 |
| BCG macrophages alone | 249 ± 25 | -- |
| BCG macrophages + TPA + P388 | 941 ± 30 | 68 ± 2 |
| BCG macrophages + TPA + 0.1 mM cytochrome c + P388 | 267 ± 20 | 36 ± 2 |
| BCG macrophages + TPA + 0.1 mg/ml catalase + P388 | 960 ± 40 | 29.5 ± 2 |
| BCG macrophages + TPA + 0.1 mg/ml SOD + P388 | 821 ± 35 | 66 ± 4 |

### B. *Antibody-Mediated Cellular Cytotoxicity of Bone Marrow Macrophages (ADCC) (Table I)*

Addition of antibody coated target cells to resting macrophages led to a relatively strong chemiluminescence signal in the first three hours after target cell addition. Thereafter the chemiluminescence signal decreased gradually to background levels after a further three hours. Addition of control target cells (not antibody-coated) to macrophages did not change the background chemiluminescence signal significantly.

In parallel with this result the simultaneously performed cytotoxic assay revealed a high specific cytotoxicity within 6 hours after the addition of antibody coated target cells (cf. Table I). The addition of control target cells (not antibody coated) to macrophages did not induce any specific target cell lysis.

### C. *BCG Macrophages (cf. Table II)*

When BCG macrophages were rendered cytotoxic by the addition of TPA a strong increase of the chemiluminescence signal could be observed within the first twenty minutes after the addition of P388 target cells, thereafter the chemiluminescence signal decreased slowly to arrive at background levels after about 12 to 15 hours.

In order to investigate the role of those oxygen compounds which are responsible for the generation of luminol chemiluminescence during macrophage target cell interaction we used superoxide dismutase (SOD), catalase and the superoxide scavenger cytochrome c. Addition of cytochrome c reduced both the TPA-target cell induced chemiluminescence of BCG macrophages as well as the cytotoxic action very strongly, indicating that either superoxide ($O_2^-$) itself or a $O_2^-$ dependent product is responsible for both generation of luminol chemiluminescence and the cytotoxic action.

Addition of SOD, which catalyses the dismutation of superoxide anions to $H_2O_2$ and ground state molecular oxygen, resulted in only a weak reduction of the chemiluminescence signal, and did not alter the cytotoxic action of BCG macrophages. Inversely the addition of catalase which catalyses the decomposition of $H_2O_2$ to molecular oxygen $O_2$ and water, inhibited the cytotoxic action of BCG macrophages fully whereas the chemiluminescence signal was slightly increased. As catalase not only acts as decomposing enzyme on $H_2O_2$, but also enhances the light production of the superoxide luminol reaction (5), one does not expect a strong effect of catalase on the luminescence of a system which contains luminol, superoxide and $H_2O_2$.

Thus the effects of cytochrome c, superoxide dismutase and catalase indicate that the chemiluminescence which can be measured after the addition of TPA and P388 target cells to BCG macrophages is mainly due to the reaction of luminol with superoxide ($O_2^-$), hydrogen peroxide ($H_2O_2$), and perhaps a reaction product of $O_2^-$ and $H_2O_2$, such as hydroxyl radical, $^{\bullet}OH$. The cytotoxic action of BCG macrophages against P388 target cells on the other hand seems to be mainly due to the amount of hydrogen peroxide and perhaps of hydroxyl radical, but not due to the direct action of superoxide anions, since SOD does not inhibit the target cell lysis.

## *IV. CONCLUSION*

Luminol aided chemiluminescence could be measured in three different macrophage cytotoxicity systems, the MCF activation system which leads to long term cytotoxicity (15-wo hours), the antibody dependent macrophage cytotoxicity system, which leads to short term cytotoxicity (6-8 hours), and the chemically induced BCG macrophage cytotoxicity system, which leads to a very strong long term cytotoxicity against P388 target cells. Moreover the chemiluminescence of cytotoxic macrophages seems to reflect their biological function in both kinetics and intensity.

Our results suggest that those oxygen dependent radicals which give rise to chemiluminescence are also involved in the cytotoxic mechanism of macrophages but until now only in the case of BCG macrophages in their cytotoxic action against P388 cells we know that only hydrogen peroxide derived from superoxide anions, both giving rise to chemiluminescence of luminol, participates causally in the cytotoxic action of BCG macrophages.

## REFERENCES

(1) Allen, R.C., R.L. Stjernholm, and R.H. Steele, *Biochem. Biophys. Res. Commun.* 47, 679 (1972).
(2) Nelson, R.D., E.L. Mills, R.L. Simmons, and P.E. Quie, *Infection and Immunity* 14, 129 (1976).
(3) Weidemann, M.J., B.A. Peskar, K. Wrogemann, E.Th. Rietschel, Hj. Staudinger, and H. Fischer, *FEBS Letters* 89, 136 (1978).
(4) Babior, B.M., *N. Engl. J. Med.* 298, 659, 721 (1978).
(5) Ernst, M., unpublished observation
(6) Schmitz, B., H. Mossman, P. Possart, and B.K. Mookerjee,

*Immunobiology* 156, 208 (1979)

(7) Nathan, C.F., L.H. Brukner, S.C. Silverstein, and Z.A. Cohn, *J. Exp. Med.* 149, 84 (1979).

(8) Nathan, C.F., S.C. Silverstein, L.H. Brukner, and Z.A. Cohn, *J. Exp. Med.* 149, 100 (1979).

(9) Meerpohl, H.G., M.L. Lohmann-Matthes, and H. Fischer, *Eur. J. Immunol.* 6, 213 (1976).

(10) Kniep, E.M., B. Kickhöfen, and H. Fischer in "Biochemical Characterization of Lymphokines" p. 149. Academic Press, New York (1980).

(11) Berthold, F.,M. Ernst, H. Fischer, and H. Kubisiak, Six Channel Luminescence Analyzer for Phagocytosis Application. Abstract this conference.

# MEASUREMENTS OF GRANULOCYTE AND PLATELET CHEMILUMINESCENCE IN SMALL SAMPLES OF WHOLE BLOOD

Herbert Fischer[1]
Taizo Kato[2]
Heinrich Wokalek[2]
Martin Ernst[1]
Hartwig Eggert[1]
Ernst Th. Rietschel[1]

Max-Planck-Institut[1]
für Immunbiologie
Freiburg, FRG

Universitätshautklinik[2]
Freiburg, FRG

## I. INTRODUCTION

Measurements of zymosan- or bacteria-induced chemiluminescence of granuloctyes have successfully been used to discover functional and to analyse metabolic defects of these cells in certain diseases (1,2,3). In all studies reported, care was taken to use purified cells and to avoid the complex situation as present in whole blood. This implied that prior to testing the cells had to undergo time-consuming procedures and finally expressed functions which did not mirror their original in vivo capacities. In whole blood or in an environment of local inflammation granulocyte function and metabolism is greatly influenced by the presence of other cells, by mediators generated from platelets, and by antibodies, immune complexes and components of the complement system (4,5).




ISBN 0-12-208820-4

In an attempt to monitor cellular response in an environment resembling the in vivo immune situation an analysis of zymosan-induced chemiluminescence in freshly drawn unfractionated and fractionated blood was performed.

## II. MATERIALS AND METHODS

Blood was drawn from health donors into syringes containing 10 IU heparin per ml blood. For chemiluminescence measurements 0.1 ml heparinized blood were diluted with 0.4 ml Eagle's medium without sodium bicarbonate but containing 50 mM HEPES (N-2-Hydroxyethylpiperazine-N'-2-ethanosulfonic acid) at pH 7.40 (E-H-medium). The diluted blood sample was mixed with luminol (final concentration 40 ug/ml), pipetted into a round bottomed vial, placed into the counting chamber ($37^{\circ}C$) and background chemiluminescence was measured (Biolumat model LB 9500, Berthold, Wildbad, FRG). After 10 minutes zymosan (in general 500 ug in 10 ul E-H-medium) was added. Chemiluminescence was continuously monitored using a Servoger recorder (model 210, BBC Goerz). Counts per minute (cpm) were plotted simultaneously by an interface connected Hewlett Packard calculator (model 97 S).

### *A. Blood Fractionation and Absorption of Plasma*

In order to analyse the role of the various blood components involved in the mediation of zymosan induced chemiluminescence, heparin blood was fractionated. Centrifugation at 250 g for 5 min was used for the separation of sediment from platelet-rich plasma. Platelet-rich plasma was further centrifuged at 1400 g for 10 min to separate platelets from plasma. Plasma in some instances was absorbed by graded amounts (80-1000 ug) of zymosan and after centrifugation (1400 g 10 min) was used as zymosan-absorbed plasma. The different fractions were tested individually and in various combinations for their capacity to generate chemiluminescence following addition of luminol and zymosan.

## III. RESULTS

### *A. Measurements in Blood from Healthy Donor*

As exemplified by an original recording (see Fig. 1) chemiluminescence starts approximately 5 min after addition of zymosan and reaches a maximum after 18-28 min and thereafter declines. A small early peak after 5-8 min was seen only occasionally.

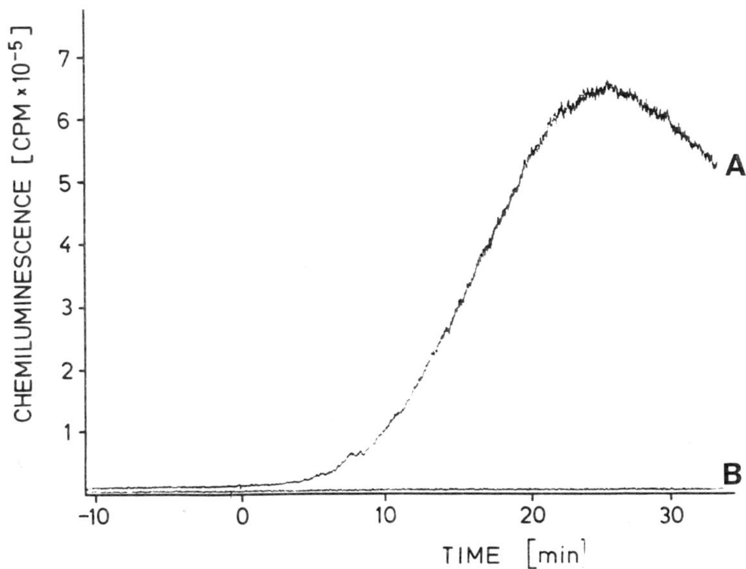


*FIGURE 1. Original recording of chemiluminescence in blood induced by addition of zymosan at time 0. Curve A: plus luminol; Curve B: without luminol*

The chemiluminescence response from blood samples from 15 different donors were comparable in kinetics in intensity. If, however, corrected to cpm/s per $10^3$ granulocytes the standard deviation as calculated from individual samples dramatically decreased, suggesting that granulocytes indeed play a major role in zymosan induced chemiluminescence of fresh blood (see Fig. 2).

## B. *Kinetic Studies with Heparin, Zymosan, and Metabolic Inhibitors*

Blood chemiluminescence was not significantly modified by aspirin, indomethacin (20 μg/ml or sodium cyanide (0.1 mM). However, the blood chemiluminescence response was dependent on the heparin dose; higher amounts led to its suppression. It was furthermore dependent on the zymosan dose, larger doses leading to enhanced chemiluminescence and an earlier appearance of the chemiluminescence peak.

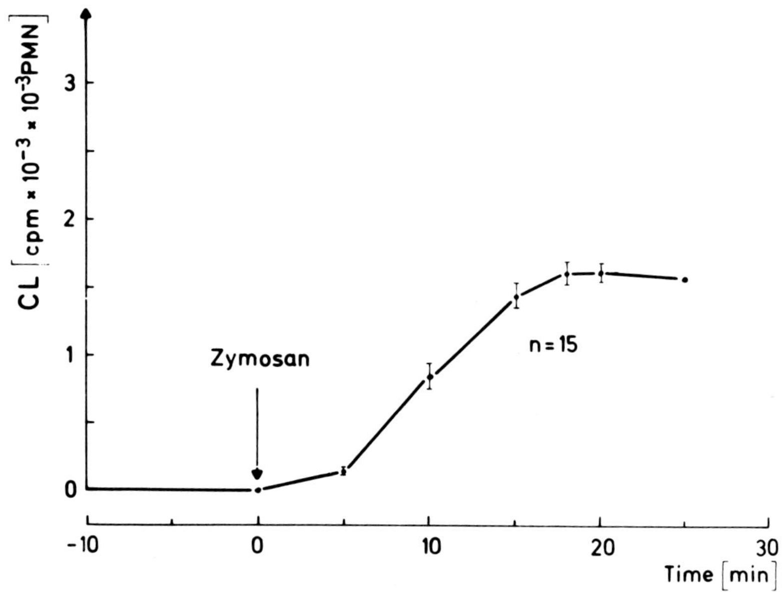


*FIGURE 2. Zymosan-induced chemiluminescence in fresh blood of 15 human volunteers; actual counts corrected to counts per $10^3$ granulocytes.*

### C. *Analysis of Isolated Blood Components*

Analysis of isolated blood components showed that the early minor peak (5-8 min) of zymosan-induced chemiluminescence was dependent on the presence of thrombocytes, while the major peak (18-28 min) was due to granulocytes. Granulocyte dependent chemiluminescence accounted for approximately 90% of the total chemiluminescence observed with fresh blood samples.

### D. *Effect of Opsonizing Plasma Factors*

In order to study the role of plasma factors in the induction of granulocyte and platelet chemiluminescence in whole blood, plasma was absorbed by increasing amounts of zymosan. When completely absorbed plasma was added to freshly prepared sediment the chemiluminescence response to zymosan was decreased to 15% of the response seen with non-absorbed plasma. These results show that opsonizing factors greatly contribute to the chemiluminescence response in whole blood.

### E. *Granulocyte Platelet Interaction in Standing Blood*

All experiments thus far described have been performed with blood samples which were either immediately analyzed following venipuncture or kept on ice prior to use. If in contrast, blood was incubated at 37$^{o}$C for 1-3 hours, an enhanced chemiluminescence response was observed. This phenomenon is due to the increasing aggregability of platelets (6,7) which leads to granulocyte platelet interaction. In presence of 20 ug/ml of aspiring, which irreversibly blocks platelet cyclooxygenase and also platelet chemiluminescence (8) enhancement of the chemiluminescence response in standing blood is substantially reduced.

## IV. CONCLUSIONS

Our experiments have shown, that measurements of zymosan-induced chemiluminescence can readily be performed in freshly drawn non-fractionated blood when luminol is present. In absence of luminol all photons emitted are quenched or absorbed by erythrocytes.

The curves obtained from a single individual at various times are fairly reproducible provided the time intervals following venipuncture being kept constant or, alternatively, blood samples are kept on ice.

Differences in chemiluminescence intensities, which are obvious when different individuals are compared, are mainly due to the different numbers of granulocytes. So, the bulk of chemiluminescence evoked in blood by addition of zymosan is due to the stimulation of granulocytes. Platelets, which are also stimulated by zymosan and emit photons when luminol is present, contribute to the overall chemiluminescence with a share of less than 10%.

The intensity of granulocyte and platelet chemiluminescence in blood is greatly dependent on the presence of factors in plasma, which can be removed stepwise by absorption with increasing doses of zymosan. Here we touch upon one of the most important features of chemiluminescence measurements in whole blood: They open the possibility to measure with ease the dynamics of opsonizing factors as present or absent in a given patient.

Another important feature which might also have possibilities for future clinical application is given by the fact that variations of platelet aggregability can be measured by a simple two step procedure.

## REFERENCES


(1) Rosen, H., S.J. Klebanoff, *J. Clin. Invest.* 58, 50 (1976).

(2) Harvath, L, F.R., *New Engl. J. Med.* 300, 1130 (1979).

(3) Barbour, A.G., C.D. Allred, C.O. Solberg, H.R. Hill, *J. Inf. Dis.* 141, 14 (1980).

(4) Wright, A.E., S.R. Doughlas, *Proc. Royal Soc.* 72, 357 (1903).

(5) Goldstein, I.M., D. Ross, H.B. Kaplan, G. Weismann, *J. Clin Invest.* 56, 1155 (1975).

(6) Warlow, Ch., A. Corina, D. Ogston, A.S. Douglas, *Thrombos. Diathes Haemorrh (Stuttg.)* 31, 133 (1974).

(7) Glasg, M. E. Hamilton, *The Lancet* 542 (1967).

(8) Mills, E.L., J.M. Gerrard, D. Filipovich, J.D. White, P.G. Quie, *J. Clin. Invest.* 61, 807 (1978).

# FLOW INJECTION ANALYSIS WITH CHEMILUMINESCENCE DETECTION; RECENT ADVANCES AND CLINICAL APPLICATIONS[1]

Mary Lynn Grayeski
Jerry Mullin
W. Rudolf Seitz

Department of Chemistry
University of New Hampshire
Durham, New Hampshire

Elizabeth Zygowicz

United Technologies
East Hartford, Connecticut

Flow injection methods allow simple, versatile, rapid low-cost automated analysis (1). In flow injection methods solutions mix by natural hydronamic processes as they flow through small coils of tubing. By appropriate choice of flow rate and tubing diameter it is possible to achieve rapid mixing without extensive sample spreading. Because flow injection methods provide for rapid reproducible mixing, they are well suited for analytical chemiluminescence (CL) measurements. In the initial report of flow injection analysis with CL detection, hydrogen peroxide samples were injected into a flowing stream of a single copper-luminol-base reagent (1). Measurements with this system could be made rapidly with good precision. However, we have subsequently encountered serious interferences when attempting to couple this chemistry to analyses for glucose and uric acid in serum using glucose oxidase and uricase to catalyze peroxide formation. Glucose

[1] *Partial Financial Support for this research was provided by NSF Grant (HE-7825192).*



ISBN 0-12-208820-4

interferes with uric acid analysis and uric acid interferes with glucose analysis. In both cases, the measured value is less than the true value. We suspect that glucose and uric acid are interacting with the Cu(II) catalyst.

To find a suitable system for interference-free peroxide analysis, we have adapted the ferricyanide-hydrogen peroxide-luminol system (2,3) to flow injection. This requires the so-called "merging zone" principle. The system is diagrammed in figure 1.

This system allows rapid measurement of peroxide. In other studies, it has been shown that this chemistry can be coupled to interference free measurements of glucose in serum (2,3). There is a large background signal due to the reaction of ferricyanide-oxygen-luminol, however, it is sufficiently constant to be readily accounted for. Using this system the detection Lim. for peroxide is less than $10^{-7}$M, and four samples can be measured per minute.

We have also been interested in ATP measurements using flow injection. The problem for ATP analysis is to minimize reagent consumption. Two configurations have been employed, illustrated in figures 2 and 3.

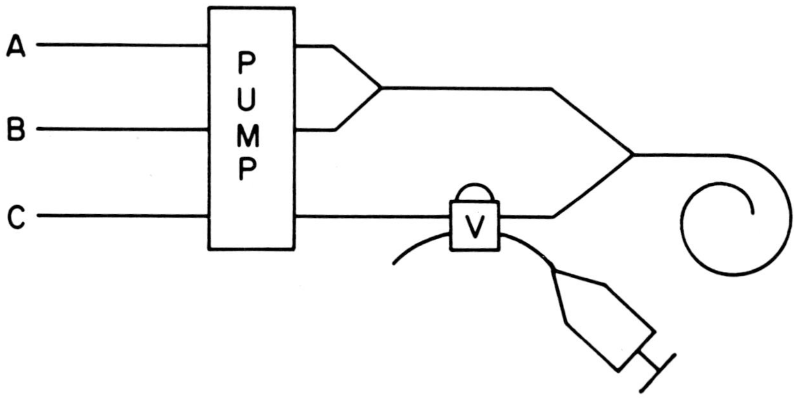


*Fig. 1. Diagram of system for flow injection analysis of hydrogen peroxide. Solutions are as follows: A-ferricyanide, B-luminol in boric acid-KOH buffer, and C-background buffer. Sample is injected using valve V and mixes with the ferricyanide-luminol reagent in the coil, which is positioned in front of a photomultiplier detector.*

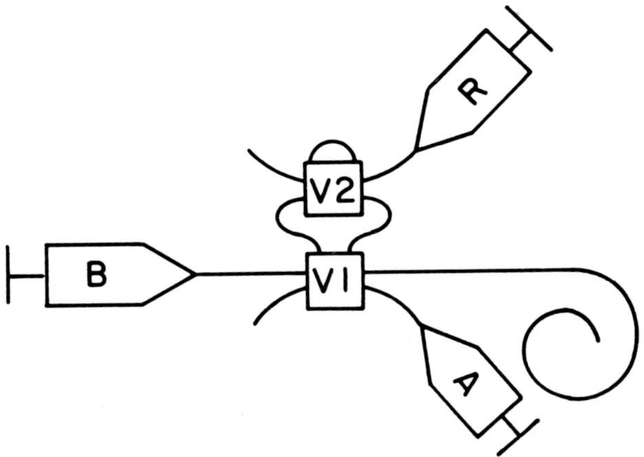


*Fig. 2. First configuration for flow injection analysis of ATP. Background solution is pumped through flow system. $V_2$ is used to inject a slug of luciferase reagent, R, inside a larger slug of analyte, A, injected through V1. They mix in coil positioned in front of detector.*

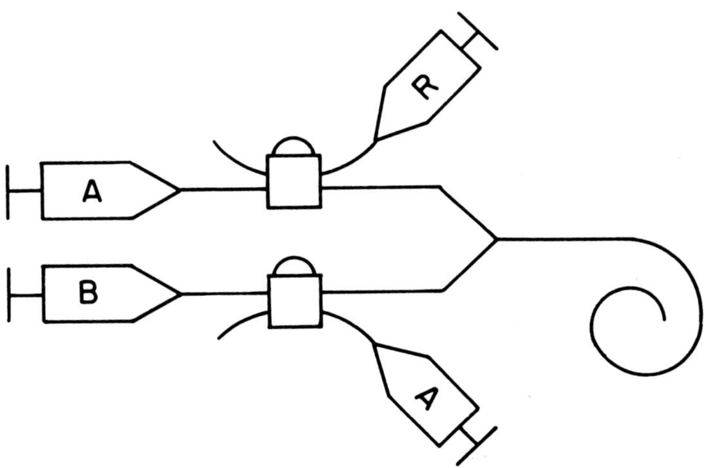


*Fig. 3. Second configuration for flow injection analysis of ATP. This involves simultaneously injecting slugs of R and A into background flow streams, A and B.*

Both systems provide for rapid analysis of ATP. The second system consumes less sample.

## REFERENCES

1. Rule, G. and W.R. Seitz, *Clin. Chem., 25,* 1635 (1979).
2. Auses, J.P., S.L. Cook, and J.T. Maloy, *Anal. Chem., 47,* 244 (1975).
3. Bostick, D.T. and D.M. Hercules, *Anal. Chem., 47,* 447 (1975).

# ULTRASENSITIVE ASSAY OF NAD/NADH WITH FIREFLY LUCIFERASE

Lars-Åke Idahl

Department of Histology and Cell Biology
University of Umeå
Umeå, Sweden

## I. INTRODUCTION

Numerous biological reactions involve NAD or NADP as a cofactor. An important aspect of this property of pyridine nucleotides is that they can be used to measure a large number of metabolites or enzymes.

Sometimes very high sensitivity is necessary in order to detect the amount of nucleotide which has been oxidized or reduced in an enzymatic assay. The demand for sensitivity may be met by chemical amplification in an enzymatic cycling system. Here the pyridine nucleotide is alternatively oxidized by one enzyme, then reduced back again by another in a cycling fashion. After a sufficient number of cycles, one of the products which has accumulated is measured in a final indicator reaction. Thus there is a substantial gain in sensitivity.

A cycling method of NAD was described in 1961 (1) which used lactic dehydrogenase (EC 1.1.1.27) and glutamic dehydrogenase (EC 1.4.1.3, GDH). This method provided cycling rates up to 8000 per hour. However, this method has several disadvantages and it was later modified by substituting glyceraldehyde-3-P dehydrogenase (EC 1.2.1.12, GAPDH) for lactic dehydrogenase (2). A distinct improvement was obtained with a cycling method which uses malic and alcohol dehydrogenases (EC 1.1.1.37, and EC 1.1.1.1) and provides cycling rates greater than 30000 per hour (3).




ISBN 0-12-208820-4

These cycling systems use fluorometric detection of NAD or NADH produced in the final indicator reaction. Although native fluorescence of pyridine nucleotides is very sensitive detection method with an analytical limit of $10^{-6}$ mol/l (3), it is superseded by the exceedingly sensitive detection of ATP by bioluminescence which has an analytical limit better than $10^{-10}$ mol/l (4).

The present abstract describes an assay for NAD which uses ATP-induced bioluminescence as the final indicator reaction. ATP is produced from ADP equivalent to the amount of NAD present in the reaction mixture by the enzymes glyceraldehyde-3-P dehydrogenase (EC 1.2.1.12, GAPDH) and 3-phosphoglycerate kinase (EC 2.7.2.3, PGK) in the initial reagent:

$$NAD^+ + \text{glyceraldehyde-3-P} + P_i \longrightarrow \text{glycerate-1,3-}P_2 + NADH + H^+$$

$$\text{glycerate-1,3-}P_2 + ADP \longrightarrow \text{glycerate-3-P} + ATP$$

Furthermore, the reagent may be converted to a cycling system for NAD by adding an oxidizing step catalyzed by glutamic dehydrogenase (EC 1.4.1.3, GDH):

$$NADH + \alpha\text{-oxoglutarate} + NH_4^+ \longrightarrow \text{glutamate} + NAD^+$$

ATP is accumulated during the NAD-cycling, hence increasing the sensitivity.

The present method should have several advantages compared to those previously described due to the higher sensitivity of the indicator reaction, since only a moderate number of cycles, if any, are required to achieve comparable sensitivity.

## II. MATERIALS

Firefly luciferase was prepared from desiccated firefly lanterns (FFT, Sigma Chemical Co.) as earlier described (4). D(-)-luciferin was from Sigma Chemical Co. Other enzymes and biochemicals were obtained from Boehringer, Mannheim. Chemicals were of analytical grade.

ADP was purified with ion exchange chromatography to remove traces of ATP (5).

## III. METHODS

The initial reagent for assay of NAD contains per ml: 6 μmol $K_2HPO_4$, 3 μmol $MgSo_4$, 1 μmol EDTA, 0.2 μmol glyceraldehyde-3-P, 0.25 μmol ADP, 3 μg GAPDH, 1.5 μg PGK and 1 mg electrophoretically pure crystalline bovine serum albumin. The reagent is buffered with 100 mmol/l N-2-hydroxyethylpiperazine-N-2-ethanesulphonic acid (HEPES) adjusted to pH 7.5 with KOH.

Two μl each of sample and reagent are mixed in a rack of teflon oil-wells (6) and incubated for 30 min at room temperature. After the incubation period, 20 μl 25 mmol/l HEPES buffer, pH 7.5 is added and the teflon rack is heated at 100°C for 10 min. The rack is cooled and a 10 μl portion is transferred to a test tube for determination of the ATP content (4).

A cycling system for NAD/NADH is obtained by supplementing the initial reagent with 6 μmol α-oxoglytarate, 10 μmol ammonium acetate and 15 μg GDH per ml. The content of GAPDH is increased to 50 μg per ml. The cycling reagent is mixed with acid-washed charcoal (1 mg per ml) and incubated for 30 min at room temperature after which the charcoal is removed by filtration. Note that ADP is added after the charcoal treatment. This treatment reduces the blank by removing NAD and ATP from the enzymes.

The assay procedure for assay of NAD using the cycling system is identical with that described above except that the initial reaction, the cycling step, is allowed to proceed for 60 min at 38°C.

## IV. RESULTS AND DISCUSSION

Figure 1 shows the result of a determination of NAD standards with the non-cycling procedure. The NAD values have been corrected for the loss due to the fact that only a portion of the initial reagent is transferred to the ATP-assay. The overall blank corresponds to 0.5 pmol NAD or less than $10^{-8}$mol/l in the indicator reaction. The recovery is about 90%. GAPDH is specific for NAD. Note that only the oxidized form of the pyridine nucleotide is determined in the non-cycling procedure. NADH has to be oxidized prior to determination or assayed by the cycling procedure.

The presence of ATP in the sample will result in erroneous values. ATP may, however, be removed by apyrase treatment (4) without destruction of NAD.

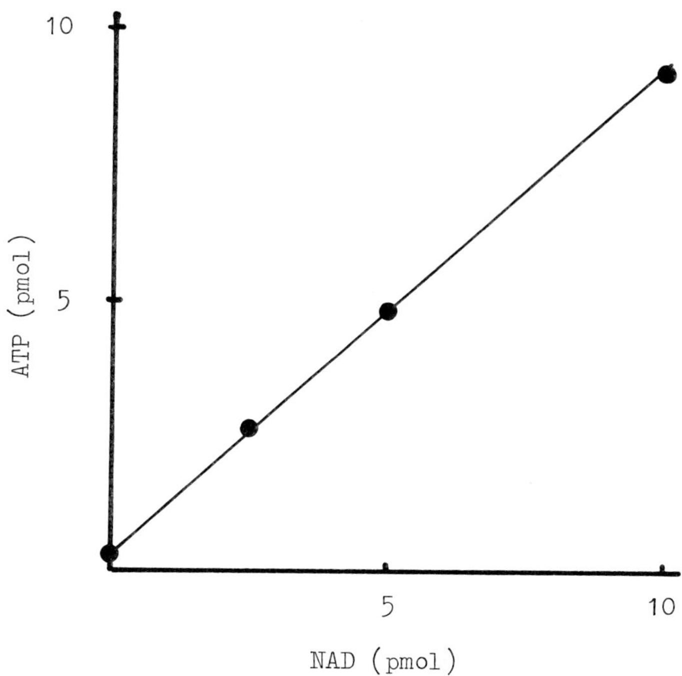


*FIGURE 1. Correlation between NAD and production of ATP in the non-cycling assay.*

The cycling reagent uses GAPDH and GDH for the cyclic oxidation and reduction of NAD/NADH. These enzymes have been shown to constitute a satisfactory enzyme couple for cycling (2). The assay is specific for NAD/NADH due to GAPDH. The amounts of enzymes used in the present method results in a mdoerate cycling rate. Note that the assay measures the sum of NAD + NADH although treatment with alkali or acid prior to determination will selectively destroy one or the other of the pyridine forms (1).

Due to selective chemical amplification of the NAD/NADH content of the sample it is usually not necessary to remove ATP from the sample.

An example of NADH standards assayed with the present cycling method is shown in Fig. 2. The values are corrected for loss in the procedure (see above). The cycling rate was about 800 cycles per hour. The overall blank corresponded to 0.5 fmol ($10^{-15}$ mol) which indicates that the present cycling method is at least as sensitive as previously described methods. It has furthermore the advantage of being less time-consuming allowing a larger number of samples in one run.

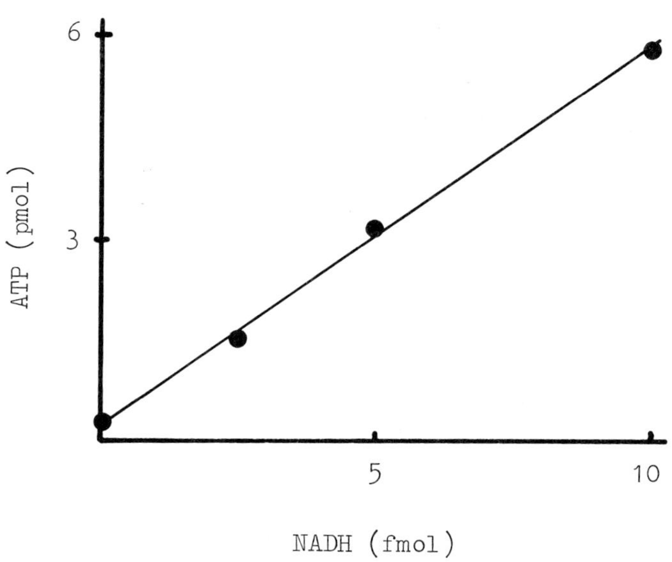


*FIGURE 2. Correlation between NAD and production of ATP in the cycling method.*

## V. ACKNOWLEDGEMENTS

This work was supported by the Swedish Medical Research Council (12X-4963) and Magnus Bergvall Foundation.

## REFERENCES

(1) Lowry, O.H., J.V. Passonneau, D.W. Schulz, and M.K. Rock, *J. Biol. Chem.* 236, 2746 (1961).

(2) Matschinsky, F.M. in "Methods in Enzymology" (D.B. McCormick and L.D. Wright, eds.) Vol. 18, Part B, p. 3. Academic Press, New York (1971).

(3) Kato, T., S.J. Berger, J.A. Carter, and O.H. Lowry, *Anal. Biochem.* 53, 86 (1973).

(4) Idahl, L-A., in "Proceedings of the International Symposium on Analytical Applications of Bioluminescence and Chemiluminescence" (E. Schram, and P. Stanley, eds) p. 401. State Printing and Publishing, Inc., Westlake Village (1979).

(5) Munch-Petersen, A., and J. Neuhard, *Biochem. Biophys. Acta* 80, 542 (1964).

(6) Matschinsky, F.M. in "Recent Advances in Quantitative Histo- and Cytochemistry" (U.C. Dubach and U. Schmidt, eds.) p. 143. Hans Huber Publishers, Bern (1971).

# A CLINICAL USE OF SOLID PHASE SEPHAROSE-ISOLUMINOL FOR SENSITIVE THIOL ASSAY OF SERUM CHOLINESTERASE

Richard D. Lippman

Department of Physical Chemistry
The Royal Institute of Technology
Stockholm, Sweden

and

Department of Medical Cell Biology
Biomedicum, University of Uppsala
Uppsala, Sweden

## I. INTRODUCTION

Effective exchange reactions are known to occur between thiols and disulfides in basic solution according to:

$$R^1S^- + R^2S\text{-}SR^3 \longleftrightarrow R^1S\text{-}SR^2 + R^3S^-$$

If $R^1S^-$ is the thiol or other strong nucleophile to be determined, and the disulfide $R^2S\text{-}SR^3$ is a chemiluminescent thiol covalently bound to a solid thiol polysaccharide, e.g. Sepharose 6B, a thiol-disulfide interchange will occur in weakly alkaline solutions; thus releasing into solution the chemiluminescent thiol in proportion to the amount of thiol bound. This chemiluminescent thiol solution is then easily and quickly filtered from the insoluble Sepharose matrix and the washings can be measured by conventional detection methods.




ISBN 0-12-208820-4

This system was created by Lippman, 1980a (1) and the chemiluminescent thiol used was mercaptoacetylisoluminol. In this short communication, a clinical application of this system is the determination of serum cholinesterase or CHS (EC 3.1.1.8; acylcholine acyl-hydrolase). This is diagrammed in Figure 1. CHS is of clinical significance since it is an indicator of possible insecticide poisoning and atypical forms of the enzyme. Decreases in activity of up to 70% are known to occur in muscular dystrophy, acute hepatitis, chronic renal disease, pulmonary embolisms, pregnancy, acute infections and carcinoma with metastases in the liver. Increases in activity are known in nephrotic syndrome, obese diabetics, hydrocephalus, Guillain-Barre disease, multiple sclerosis, meningitis, brain tumors, hemochomatosis and throtoxicosis.

CHS activity in normal people varies widely between 4 and 12 U/ml serum at 37°C. Generally, this is determined colorimetrically using an acetylthiocholine ester as a substrate. The enzymatic hydrolysis of this thioester yields a free thiol, the thiocholine ion, which in turn is coupled to a

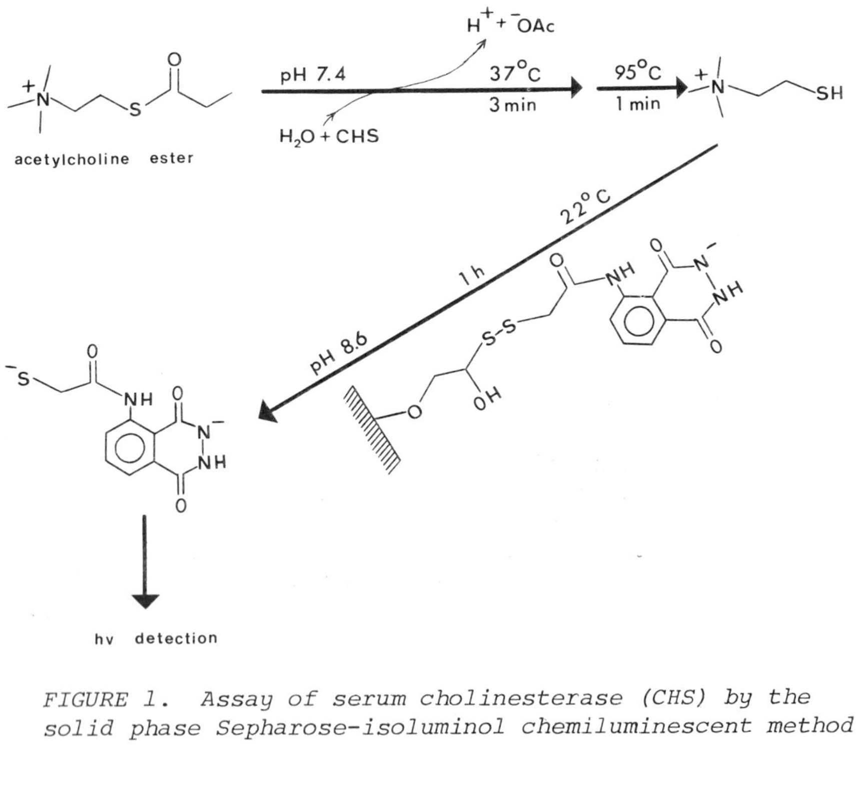

acetylcholine ester

pH 7.4 $H^+ + {}^-OAc$ 37°C 3 min 95°C 1 min

$H_2O$ + CHS

22° C 1 h pH 8.6

hv detection

*FIGURE 1. Assay of serum cholinesterase (CHS) by the solid phase Sepharose-isoluminol chemiluminescent method.*

thiol-disulfide interchange reaction using Ellman's reagent (DTNB = 5,5'-dithiobis(2-nitrobenzoic acid) to yield finally the chromogen, 5-mercapto-2-nitrobenzoic acid which has an absorbance maximum at 410 nm.

## II. METHODS

All reagents were purchased from Sigma Chemical Co., St. Louis, MO unless otherwise stated. 400 μl of substrate (10 mmole acetylthiocholine iodide in 100 ml 0.01 M Tris-HCl buffer, pH 7.4, plus 1 μmole EDTA) was pipeted to each sample tube and double blanks. All tubes were equilibrated at 37°C for 10 min. Serum specimens were diluted 100-fold with the same buffer and 1 ml aliquots were added to each tube. The reaction was allowed to proceed exactly 3 min at 37°C. All tubes were quickly thermostated in boiling water for 1 min in order to stop the reaction. Upon cooling to room temperature, 135 mg of water-pump dryed Sepharose 6B-(mercapto-acetyl)-isoluminol (ref. 1) suspended in 200 μl of 0.1M boric acid-sodium hydroxide buffer, pH 8.6, plus 1 μmole EDTA was added to each tube. The tubes were turned continuously on a wheel at ambient temperature for exactly 1 h, the Sepharose was filtered off, and the wash waters were divided into 200 μl aliquots for chemiluminescent assay.

## III. INSTRUMENTATION

Chemiluminescent measurements were determined using photo-counting instrumentation according to Wettermark et al., 1975, (2) and using reagents such as microperoxidase according to Lippman, 1980b, (3). Similar results were obtained with conventional clinical instruments using photodiode detectors.

Automated sampling was achieved by connecting two Hamilton[1] Microlab injection-syringe pistols together with a flat-band lead and eight-pole Cannon contacts. The external +5V digital pulse from the Hamilton's minicalculator was sufficient to simultaneously drive both injection-syringe pistols the same distance as when only one pistol is engaged. Syringes could be selected from 50 μl to 5 ml capacities. They were connected to the two threaded inlets of an Aminco-Morrow stopped-flow cell (150 μl sample volume with a possible dead time down to 3 msec). One side of the cell's light-path

[1] *Hamilton Bonaduz AG, Bonaduz, Switzerland*

inlet was covered by a tiny mirror while the light-path outlet was positioned in front of a photomultiplier (ref.2). Thus, dozens of measurements could be made easily for each sample when ca 75 µl solution (A) sample in the first syringe) was rapidly mixed in two 90° bends in the stopped-flow cell with ca 75 µl solution (B) (the oxidant). Discharge of spent reagents was also achieved automatically when new reagents were injected into the cell.

Another advantage of this system which was not needed in these experiments is that the mirror from the cell's light-path inlet can be replaced with an optical fiber which in turn is linked to a double grating monochromator and a fixed halogen light source. This would allow continuous or stopped-flow spectroscopic measurements.

## IV. RESULTS AND CONCLUSIONS

A calibration curve is depicted in Figure 2 together with some values determined from normal serum.

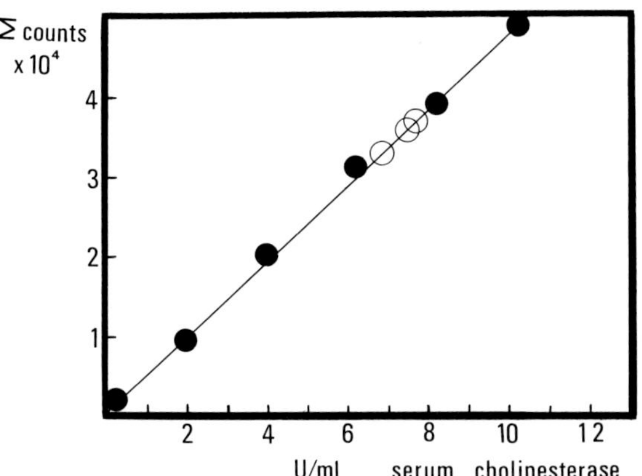


*FIGURE 2. Standards used to determine the above calibration curve for cholinesterase. Cholinesterase values obtained from normal serum.*

In regard to the chemiluminescent reactions, no significant quenching effects were observed from the 100-fold diluted serum nor from other reagents used. A limit of detection of sub-nanomole quantities was sufficient for a reliable CHG assay. Presumably, a fine tuning of this method would produce a near-picomole limit of detection which may be a valued goal in detection of trace CHS activities in various tissues, e.g. spleen and nerve cell endings. These last two sources typify another type of CHS, namely EC 3.1.1.7 acetylcholine acetyl-hydrolase, which could be more thoroughly explored and examined using this sensitive technique.

## REFERENCES

(1) Lippman, R.D., *Anal. Chim. Acta.* 116, 181 (1980).

(2) Wettermark, G., H. Stymne, S.E. Brolin, and B. Petterson, *Anal. Biochem.* 63, 293 (1975).

(3) Lippman, R.D., *Experimental Gerontology,* in press (1980).

# CHEMILUMINESCENCE EXPERIMENTS ON A MODEL OF THE OLIGODYNAMIC ACTIVITY OF SILVER

Jerzy D. Meduski
Jerzy W. Meduski
James De La Rosa
Alexander Goetz*

Nutritional Research Laboratory
University of Southern California
School of Medicine
Los Angeles, California

## I. INTRODUCTION

The term "oligodynamic activity" is used to describe the germicidal property of aqueous solutions containing extremely low quantities of silver without obviously present anions. This usage excludes both ionizable salts of silver of known antiseptic power (e.g. nitrate, lactate or picrate) and colloidal silver preparations used as bacteriostatic drugs which contain high concentrations of nonionized silver. Protargin, for instance, contains from 19 to 23% silver.

The first modern observation of the oligodynamic activity of silver was made by Raulin (1), who noticed that microorganisms cannot develop in silver containers. Naegeli confirmed this (2), named the phenomenon the oligodynamic activity of silver, and found that copper exhibits similar properties. It has been thought that metallic silver dissolves in water and produces the oligodynamic effect. Krepelka and Toul found that distilled water contains, after equilibration with silver surfaces, 0.037 mg of silver per 1 liter, (3). The mechanism of the oligodynamic effect remains obscure (4). The experiments presented here are an attempt to elucidate this mechanism using chemiluminescent reactions.

* Deceased



ISBN 0-12-208820-4

## II. EXPERIMENTAL

There are several methods of producing the oligodynamic activity of silver. Heating metallic silver in air, in oxygen, or submerging it in diluted solutions of hydrogen peroxide lead to silver activation. Interestingly, one of the methods of inactivation is heating silver in nitrogen.

We decided to produce the oligodynamic effect by impregnating a neutral substance which exposes a large surface (e.g. silica gel, or lampblack) with small quantities of silver according to the procedure developed by Goetz (6). The used preparations contain: 45.4 nmoles of Ag per 10 mg of silicon-based Goetz Oxygen Catalyst, GOC(Si), and 2.69 micromoles of Ag per 10 mg of carbon-based catalyst, GOC(C). As a silicon base we used $SiO_2$ in form of Syloid #308, (Grace Chem. Co.); as a carbon base: Lampblack B-5, (Monsanto Co.).

The presence of the oligodynamic effect in our preparations has been demonstrated using growth inhibition in cultures of E. coli, Isochrysis galbana, and Scrippsiella trochoidea.

The chemiluminescence has been measured using single photoelectron counting with RCA C31034A photomultiplier tube and a set of cut-off filters (Wratten type, Eastman Kodak Co.) for the emission spectrum determination.

## III. RESULTS AND DISCUSSION

Our experimental approach evolved from our hypothesis that the biologically active species produced by the oligodynamic activity of silver is $H_2O_2$. To check this we used a reaction:

$$H_2O_2 + NaOCl \rightarrow O_2^* + H_2O + Cl^- \qquad (\#1)$$

studied by Khan and Kasha (7 & 8). It generates the metastable molecular oxygen singlet states: $^1\Sigma g^+$ and $^1\Delta g$ represented by $O_2$* in the equation #1. The transitions $^1\Sigma g^+ \rightarrow {}^3\Sigma g^-$ and $2(^1\Delta g) \rightarrow 2(^3\Sigma g^-)$ generate chemiluminescence.

The experiments supported our hypothesis. GOC(C) and GOC(Si) do generate chemiluminescence when reacted with NaOCl in place of $H_2O_2$(Table 1).

The emission spectrum shows maxima corresponding to $2(^1\Delta g)$ and $^1\Sigma g^-$ species of the singlet molecular oxygen (Figure 1).

To further support our findings, we attempted to observe a transfer mechanism involving singlet oxygen and a fluorescer triplet (Tables 2 & 3) observed in the $H_2O_2$ + NaOCl system.

Fluorescein potentiates the light emission response of systems: GOC(C) + NaOCl and GOC(Si) + NaOCl.

An experiment summarized in Table 4 provided additional support of our hypothesis. The replacement of water with $D_2O$ would increase the light emission due to the relative inability of $D_2O$ to absorb vibrational energy from singlet oxygen

*TABLE 1. Emission Responses of GOC(C), GOC(Si), Charcoal, Silica Gel, and $AgO_2$ with NaOCl*[a]

| *Exp. #* | *Reactants* | *Peak Emission (cts/sec)* |
|---|---|---|
| 1 | GOC(C) + NaOCl | 60 |
| 2 | Charcoal + NaOCl | 0 |
| 3 | GOC(Si) + NaOCl | 450 |
| 4 | Silica Gel + NaOCl | 0 |
| 5 | $AgO_2$ + NaOCl | 0 |

[a] *50 mg GOC(C), 50 mg charcoal, 20 mg silica gel, 20 mg $Ag_2O$, 1 ml 0.672 M NaOCl (exp. #1 & #2), 1 ml 6.72 mM NaOCl (exp. #3, 4, & 5) have been used.*

*TABLE 2. Emission Response of GOC(C) with NaOCl and Fluorescein*[a]

| *Exp. #* | *Reactants* | *Peak Emission (cts/sec)* |
|---|---|---|
| 1 | GOC(C) + NaOCl | 40 |
| 2 | GOC(C) + Fluorescein + NaOCl | $3.5 \times 10^4$ |
| 3 | $H_2O_2$ + Fluorescein + NaOCl | $1.1 \times 10^5$ |

[a] *5 mg GOC(C); 5 ml 0.336 M NaOCl; 1 ml 0.9 mM Fluorescein; 0.02 ml 0.1 M $H_2O_2$.*

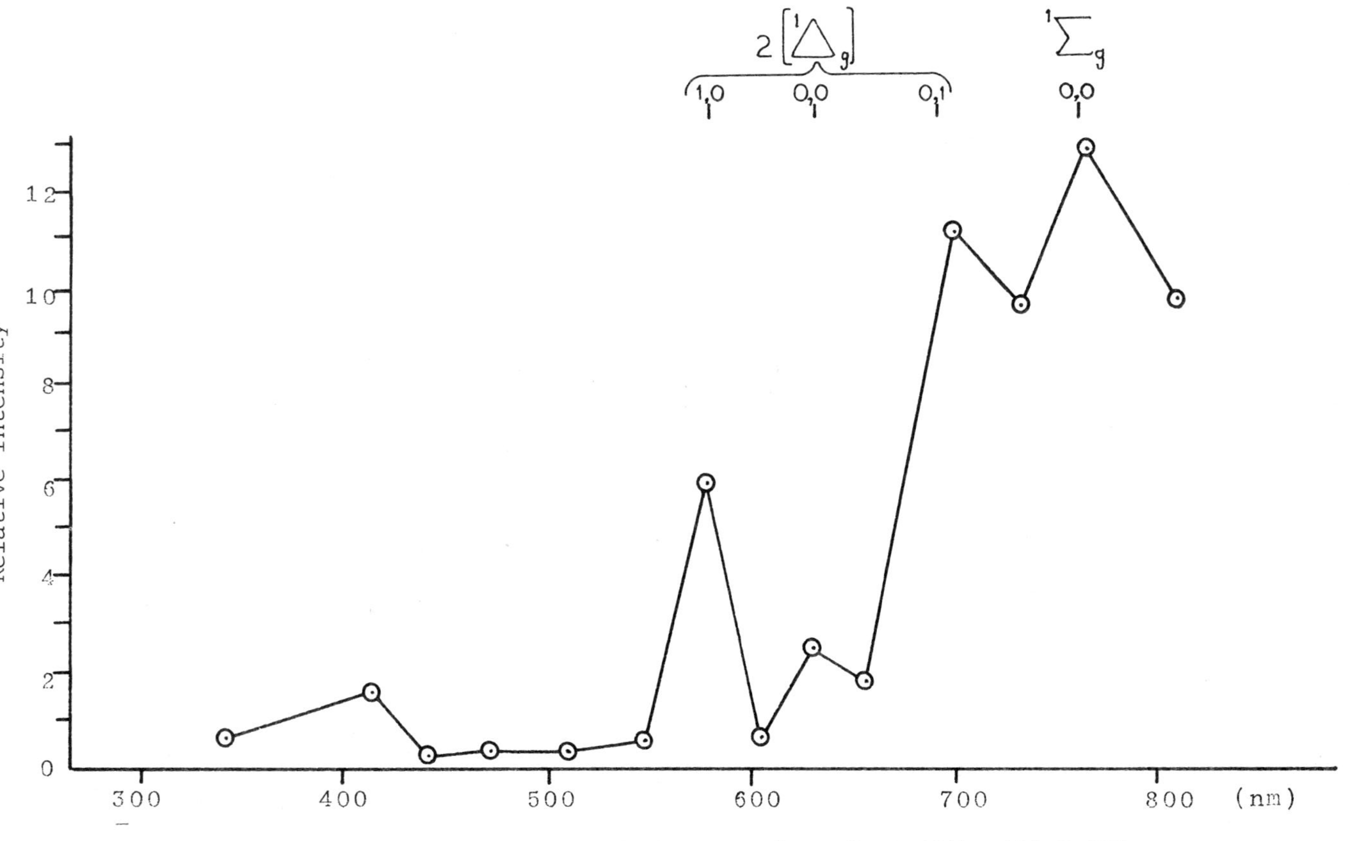


*Fig. 1. Emission spectrum of the reaction of GOC(Si) with NaOCl*

*TABLE 3. Emission Response of GOC(Si) with NaOCl and Fluorescein*[a]

| *Exp. #* | *Reactants* | *Peak Emission (cts/sec)* |
|---|---|---|
| *1* | *GOC(Si) + NaOCl* | *451* |
| *2* | *GOC(Si) + Fluorescein + NaOCl* | $1.9 \times 10^4$ |
| *3* | $H_2O_2$ *+ Fluorescein + NaOCl* | $7.1 \times 10^3$ |

[a] *20 mg GOC(Si); 0.02 ml 0.1 M* $H_2O_2$*; 0.1 ml 3 mM Fluorescein; 1 ml 6.72 mM NaOCl.*

*TABLE 4. Enhancement and Quenching of Light Emission by GOC(Si) - NaOCl System*[a]

| *Exp. #* | *Reactants* | *Emission (integrated over 400 (seconds)* |
|---|---|---|
| *1* | *GOC(Si) + NaOCl (in* $H_2O$*)* | *100 %* |
| *2* | *GOC(Si) + NaOCl (in* $D_2O$*)* | *115 %* |
| *3* | *GOC(Si) + DABCO + NaOCl (in* $H_2O$*)* | *44 %* |

[a] *80 mg GOC(Si); 1 ml 6.72 mM NaOCl; 99%* $D_2O$ *and 1%* $H_2O$ *was a solvent in exp. #2; 0.1 ml 0.2 M aqueous solution of 1,4-diazobicyclo-(2,2,2)-octane (DABCO) was used in exp. #3.*

compared with that of water. This would occur only in the presence of singlet oxygen. Our results supported this premise: we have obtained the $D_2O$ effect.

Among the quenchers of singlet oxygen we have tried 1,4-diazobicyclo-(2,2,2) octane (DABCO) and we have observed a quenching effect in our system (Table 4).

The experiments summarized above indicate that the origin of the oligodynamic action of silver is the introduction on silver surfaces of the Ag/Ag+ couple:

$$Ag = Ag^+ + \bar{e} \qquad (E^\circ = -0.7995\ V)$$

a reaction capable of generating free radical species which would produce $H_2O_2$, possibly as a result of a reaction sequence:

$$\bar{e} + H_2O \rightarrow (H_2O)^- \rightarrow H\cdot + (:OH)^-$$

$$H\cdot + O_2 \rightarrow H\text{-}O\text{-}O\cdot$$

$$H\text{-}O\text{-}O\cdot + H\cdot \rightarrow H_2O_2$$

Hydrogen peroxide would then participate in the reaction with NaOCl (equation #1) carried out by us to prove our hypothesis.

## IV. REFERENCES


1. Raulin, J., *Ann. de Sci. Nat.*, 11,220 (1869).
2. Naegeli, C., *Neue Denkschriften d. allg. Naturforsch. Ges.*, 33,1 (1893).
3. Krepelka, H. and F. Toul, *Coll. Trav. Chim. Tchécoslov.*, 1, 155 (1929).
4. Harvey, S. C., *in* "The Pharmacological Basis of Therapeutics" (L. S. Goodman and A. Gilman, eds.) 999, 5th ed., The MacMillan Co., London, (1975).
5. Goetz, A., R. L. Tracy, F. S. Harris, *in* "Silver in Industry" (L. Addicks, ed.), 405, Reinhold Publ. Co., New York, (1940).
6. Goetz, A., *in* "Papers of Alexander Goetz, 1897-1970", California Institute of Technology, Archives, 1-32, Pasadena, California, 91125.
7. Khan, A. U. and M. Kasha, *Nature*, 204, 241, (1964).
8. Kasha, M. and A. U. Khan, *in* "International Conference on Singlet Oxygen and its Role in Environmental Sciences", *Annals of the N.Y. Acad. Sci.*, 171, 7 (1970).

# CHEMILUMINESCENT IMMUNOASSAY FOR THE DETECTION OF VIRUS/ANTIBODY AGGREGATES

Robert J. Miller, Jr.
Douglas W. Reichard

U. S. Army Medical Research Institute of Infectious Diseases
Fort Detrick, Maryland

## I. INTRODUCTION

Radioimmunoassay (RIA) (1), the enzyme-linked immunosorbent assay (ELISA) (2-5), and chemiluminescent assays (6-8) have been used for the identification and detection of antibodies, antigens, and microorganisms. We previously reported that a chemiluminescent immunoassay (CLIA) using antibody covalently bound to polyacrylamide beads was useful in the identification of low concentrations of virus (9, 10).

We are reporting the use of the luminol-$H_2O_2$ chemiluminescent assay for the analysis of gradient fractions containing free and bound (virus-complexed) horseradish peroxidase (HRP)-labeled antibody as an indirect means of detection of live attenuated Venezuelan equine encephalomyelitis vaccine strain TC-83 (VEE).

## II. MATERIALS AND METHODS

### *A. Antibody and HRP-Labeled Antibody Preparation*

Monkey IgG anti-VEE antibody was purified from hyperimmune antiserum by $(NH_4)_2SO_4$ precipitation and protein A column chromatography (10). Enzyme conjugate was prepared according to the procedure of Nakane and Kawaoi (13) and purified as previously reported (10). Antibody-HRP conjugate (0.4 mg/ml) was stored at 4°C in 1% BSA-Hanks' balanced salt solution,



ISBN 0-12-208820-4

pH 7.6 (BSA-H).

### *B. Radioiodination of Antibody*

The purified IgG fraction of monkey anti-VEE was iodinated using Na[$^{125}$I] (specific activity: 1.3 μCi/μg antibody) following the lactoperoxidase procedure described in BioRad Laboratories Product Information 1060.

### *C. Virus Preparation*

VEE virus, passaged once in BHK-21 cells, was labeled with tritium by the addition of labeled L-amino acids to medium 199 containing Earle's salts and 1:40 concentration of amino acids (1). Cell culture supernatants were havested, concentrated, and purified by rate zonal centrifugation (14). Infectious virus was assayed by counting plaque-forming units (PFU) in primary duck embryo cell culture monolayers. Dilutions of virus were made in BSA-H.

### *D. Gradient Assay Conditions*

Virus/antibody mixtures were incubated for 1 h at 37°C with mild agitation. Following incubation, 200-μl samples of virus/antibody mixture were applied to linear sucrose gradients (10-30% sucrose, w/w) in TBES buffer (0.089 M tris, 0.089 M boric acid, 0.0022 M EDTA, 0.15 M NaCl, pH 8.4) formed in 17-ml tubes over a 1.0-ml cushion of 43% w/w sucrose. Gradients were centrifuged in a SW 27 rotor at 27,000 rpm at 4°C for 2 h. Fractions (1.0 ml) were collected from the gradient mechanically and assayed.

### *E. Radioimmunoassay*

Tritiated VEE specimens were analyzed by counting 100-μl samples of gradient fractions in 6 ml of Scintolute and Scintosol (4:1) on a Searle Mark III scintillation counter.

Analysis of 100-μl samples from gradient fractions containing virus/[$^{125}$I]-labeled antibody were counted on a Searle model 1185 gamma counter.

### *F. Chemiluminescent Assay*

Samples (100 μl) from gradient fractions containing virus/HRP-labeled antibody complexes were pipetted into 6 x 50-mm borosilicate glass reaction cuvettes and 100 μl of 1.5 mM luminol (in 0.1 N NaOH and o.1% EDTA) were added. Following 5-10 min incubation, the cuvettes were placed in the reaction chamber of a Chem-Glow photomultiplier microphotometer (Aminco, Rockville, Md) and 100 μl of 3% $H_2O_2$ injected. Peak flash-height response was measured on a potentiometric recorder (50 mV setting).

## III. RESULTS AND DISCUSSION

Figure 1A represents the sedimentation profile of tritiated VEE in a 10-30% sucrose gradient. Figure 1B is the radioactive sedimentation profile of VEE incubated with [$^{125}$I]-labeled monkey IgG anti-VEE. Figure 1C is the profile of VEE incubated with HRP-labeled monkey IgG anti-VEE as measured by CLIA, with a detection limit of 0.5 x $10^3$ PFU/ml.

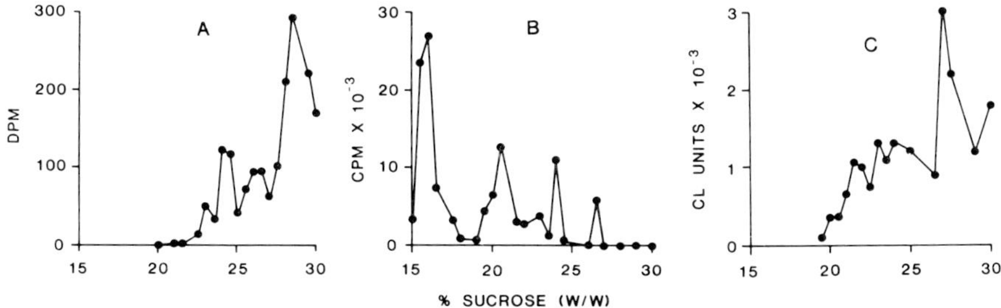


*Figure 1. Sedimentation profiles of VEE virus or virus/antibody aggregates in sucrose density gradients.*
*(A) Virus only (50 μl of $10^8$ PFU/ml of tritiated VEE).*
*(B) Virus plus radiolabeled antibody (100 μl of $10^6$ PFU/ml of VEE + 1.0 ml radioiodinated antibody (3.6 μCi/ml).*
*(C) Virus plus HRP-labeled antibody (100 μl of $10^4$ PFU/ml of VEE + 1.0 ml HRP-labeled antibody (40 μg/ml).*
*Data points for 1B and 1C are the result of subtracting control labeled antibody values from experimental observations.*

Although RIA and CLIA have the same relative sensitivity (unpublished data), CLIA offers several advantages. The CLIA-luminol system does not require the use of radioactive materials, as does RIA. The luminol-based assay also does not require an enzymatically active label as does ELISA. The chemiluminescent response is dependent on heme concentration (15-17) present in the conjugate.

The combination of virus with antibody resulted in the formation of aggregates, as demonstrated by electron microscopy (unpublished data). The shoulders and various peaks in the 20-30% sucrose range represent the normal distribution of aggregates (i.e., pairs and triplets) commonly seen in VEE virus populations purified by the procedures described.

The use of a chemiluminescent assay for analysis of virus/antibody complexes separated by sucrose density gradient fractionation has been shown to be a sensitive, qualitative measurement for the presence of virus and is thus comparable to the RIA analysis. The applicability of such a procedure to clinical specimen analysis is apparent. Presently, viral identification depends upon morphological description and cell culturing coupled with immunological techniques. This process usually requires days for identification. The procedure described here requires less than 8 h for detection of virus.

The procedure will be examined further to determine its specificity for the identification of closely related viruses.

## REFERENCES


1. Jahrling, P. B., R. A. Hesse, J. F. Metzger, *J. Clin. Microbiol.* 8, 54 (1978).
2. Voller, A., *Diagnostic Horizons* 2, 1 (1978).
3. Drow, D. L., D. G. Maki, D. D. Manning, *J. Clin. Microbiol.* 10, 442 (1979).
4. Ruitenberg, E. J., P. A. Steerenberg, B. J. M. Brosi, and J. Buys, *Bull. W.H.O.* 51, 108 (1974).
5. Miranda, Q. R., G. D. Bailey, A. S. Fraser, and H. J. Tenosa, *J. Infect. Dis.*, 136, S304 (1977).
6. Halmann, M., B. Velan, and T. Sery, *Appl. Environ. Microbiol.* 34, 473 (1977).
7. Velan, B., H. Schupper, T. Sery, and M. Halmann, *in* "International Symposium on Analytical Applications of Bioluminescence and Chemiluminescence Proceedings," p. 431. State Printing & Publishing Co., Westlake Village, California (1979).
8. Reichard, D. W., R. J. Miller Jr., *Fed. Proc.* 38, *1013* (1979).
9. Miller, R. J., Jr., and D. W. Reichard, *Fed. Proc.* 39, 919 (1980).
10. Reichard, D. W., and R. J. Miller, *Army Science Conf. Proc.* (1980) in press.
11. Brakke, M. K., *Adv. Virus Res.* 7, 193 (1960).
12. Schumaker, V. N., *Adv. Biol. Med. Phys.* 11, 245 (1967).
13. Nakane, P. K., and A. Kawaoi, *J. Histochem. Cytochem.* 22, 1084 91974).

14. Jahrling, P. B., and L. Gorelkin, *J. Infect. Dis.* 132, 667 (1975).
15. Neufeld, H. A., C. J. Conklin, and R. D. Towner, *Anal. Biochem.* 12, 303 (1965).
16. Ewetz, L., and A. Thore, *Anal. Biochem.* 71, 564 (1976).
17. Neufeld, H. A., L. T. Carleton, and S. Witz, *Army Science Conf. Proc.* 2, 157 (1966).

# AN IMMUNOASSAY FOR PLASMA CORTISOL BASED ON CHEMILUMINESCENCE

M. Pazzagli

Endocrinilogy Unit
University of Florence
Italy

F. Kohen
J. B. Kim
H. R. Lindner

Department of Hormone Research
The Weizmann Institute of Science
Rehovot, Israel

## I. INTRODUCTION

Recent studies from our laboratory (1-3) have shown that chemiluminescence immunoassay can be a feasible alternative to radioimmunoassay (RIA) of steroids. The ligand-chemiluminescent marker conjugates are relatively easy to prepare, are stable and can be measured at pM levels (4-5). Moreover, assays of this type do not necessitate a phase separation step (1-5). We shall describe here an assay procedure for plasma cortisol which utilizes chemiluminescence as the end point.

For this purpose the chemiluminescent marker aminopentyl ethyl isoluminol (APEI) is covalently attached through its free amino group to a carboxy derivative of cortisol (Fig. 1). The resulting cortisol-chemiluminescent marker conjugate emits light upon oxidation with microperoxidase and $H_2O_2$. When the cortisol-APEI conjugate is bound to anti-cortisol IgG, the peak light intensity of the conjugate is decreased, but the total light yield is slightly increased, due to an increase in light production during the decay part (DP) of the reaction.




ISBN 0-12-208820-4

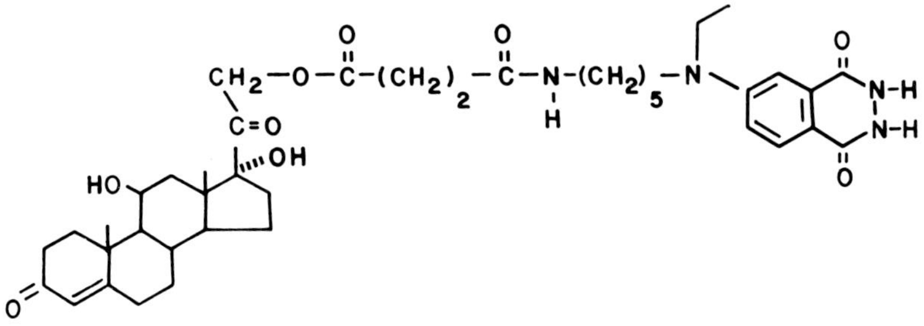

CORTISOL- APEI CONJUGATE

*FIGURE 1. Proposed structure for cortisol-aminopentyl ethyl isoluminol (APEI) conjugate.*

The binding of cortisol-APEI and consequent shift of the light emission curve to the right is prevented by the addition of free cortisol in a competitive manner. Results of an immunoassay for plasma cortisol based on the properties of this system are reported here.

## II. MATERIALS AND METHODS

11β,17,21-Trihydroxy-4-pregnene-3,20-dione-21-hemisuccinate was attached covalently to 6[N-(5-aminopentyl)-N-ethyl]-amino-2,3-dihydrophthalazine-1,4-dione (aminopentyl ethyl isoluminol, APEI) to yield cortisol-APEI conjugate (3). Antiserum to cortisol was raised in rabbits, using a cortisol-21-hemisuccinate thyroglobulin conjugate as the immunogen. Anticortisol IgG fraction was prepared by ammonium sulfate precipitation, followed by DEAE-52 cellulose column chromatography and dialysis against phosphate-buffered saline (6). All other reagents and assay procedures were as previously described (1-3), unless specified otherwise.

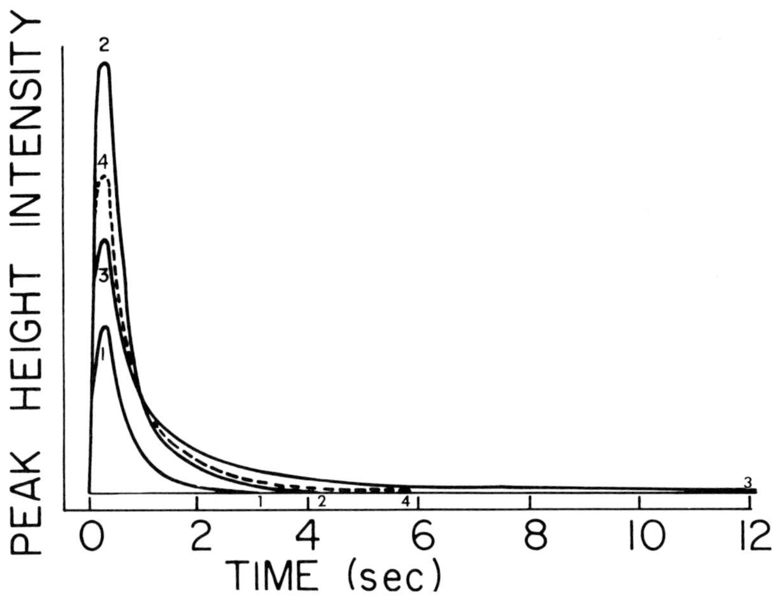


*FIGURE 2. Oscilloscope tracings of the light signal obtained upon oxidation of cortisol-APEI conjugate in the presence of specific antibody with or without addition of cortisol.*

*Blank: (1) enzyme solution (microperoxidase) only was mixed with $H_2O_2$; (2) signal due to the oxidation of 25 pg of cortisol-APEI conjugate; (3) the conjugate (25 pg) was oxidized in the presence of anti-cortisol IgG (0.23 pmol/tube); (4) the conjugate (25 pg) was oxidized in the presence of anti-cortisol IgG (0.23 pmol/tube) and cortisol (625 pg). Oscilloscope settings: speed 1 sec/1.22 cm; sensitivity 3 x 100 mV/1.22 cm.*

Light emission was measured with a Luminometer Model 2080 (Lumac Systems, Basel).

## III. RESULTS

### A. *Chemiluminescence immunoassay for cortisol*

Cortisol-APEI conjugate produced light upon oxidation by a $H_2O_2$-microperoxidase system at pH 8.6. Decay part light emission measured 2 sec after addition of oxidant increased

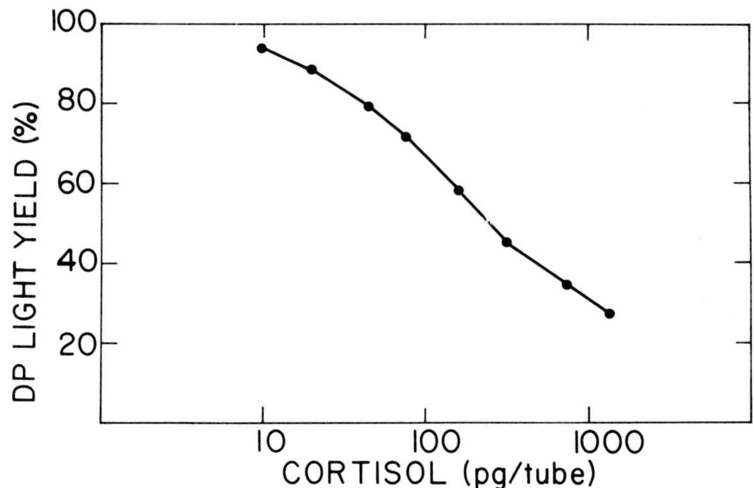


*FIGURE 3. Representative dose-response curve for cortisol measured by chemiluminescence immunoassay.*

*Varying amounts of cortisol were incubated with 0.23 pmol of anti-cortisol IgG in a total volume of 200 µl of assay buffer for 20 min at room temperature and 20 min at 4°C. Cortisol-APEI conjugate (25 pg in 100 µl of assay buffer) was then added, and the incubation was continued for another 90 min at 4°C before adding the enzyme solution (100 µl) and oxidant. The difference in decay part light yield between the maximal enhancement and basal control tubes was taken as 100% and is plotted here against log-dose of cortisol.*

linearly with cortisol-APEI concentration up to 160 pg. The lower limit of detection was 4 pg/tube. When cortisol-APEI conjugate was bound to anti-cortisol IgG, the specific antibody reduced the peak height of the light response, but enhanced the light emission during the decay part (DP) of the reaction, with only a small increase in total light yield (Fig. 2, curve 3 vs. curve 2). Addition of free cortisol (Fig. 2, curve 4) reduced this effect. Using cortisol-APEI conjugate in this system, an assay curve was obtained with a linear range of 20-1000 pg steroid/assay tube (Fig. 3).

*B. Reliability of the cortisol assay.*

The *specificity* of the immunoassay using anti-cortisol IgG and chemiluminescence was similar to that observed when using the same antiserum in RIA: light emission was unaffected by addition of progesterone, estradiol, testosterone, dexamethasone, spironolactone or bethamethasone; minor cross reaction was observed in the chemiluminescence immunoassay with 11-desoxycortisol (1%), corticosterone (3%), desoxycorticosterone (0.1%), 17α-OH-progesterone (14%) and prednisolone (1%) (3). These compounds gave a similar degree of cross-reaction in the RIA.

The *precision* of the assay was evaluated by performing replicate determinations using 0.1 ml samples on one pool of normal male plasma on the same day or repeated determinations on separate occasions. Methylene chloride extracts from 0.1 ml samples were used. The within-assay coefficient of variations was 8.3% with a mean cortisol content (μg/100 ml plasma) of 7.5 ± 1.55 S.D. (n = 11). Addition of methylene chloride extracts from 0.1 ml plasma to the enzyme solution and oxidant in the absence of cortisol-APEI conjugate generated no light signal. Likewise, methylene chloride extracts of distilled water samples gave no significant light emission in the assay (<4 pg/tube).

*C. Comparison of plasma cortisol levels as determined by chemiluminescence immunoassay or RIA.*

Plasma samples (n = 20) from normal individuals, from patients with Cushing's Syndrome or with Addison's Disease and from patients under ACTH treatment or dexamethasone suppression were extracted with methylene chloride, and the extracts were assayed for cortisol by radioimmunoassay and by chemiluminescence immunoassay, using the same anti-serum. The results of the two methods agreed well: $r = 0.98$; $y = 0.96\ x - 0.28$, where y represents values determined by chemiluminescence immunoassay.

## IV DISCUSSION

This paper describes the development of an immunoassay procedure for plasma cortisol based on chemiluminescence. We have recently described a chemiluminescence assay procedure for the determination of progesterone (1,7) and estriol-16α-glucuronide (2,8). In the latter two systems, the homologous antibody caused an increase in both the peak height of light emission and the total light yield of the steroid-isoluminol conjugate. The system described in the present paper behaves differently in that the peak height of the light response is not augmented, but on the contrary, reduced and total light yield is only slightly increased; however, the antibody caused a delay in light emission, resulting in a shift of the response curve to the right. This antibody effect forms the basis of the assay described here. Its theoretical basis remains to be elucidated. Thus, we do not know whether the discrepant behavior is a peculiar property of the particular steroid conjugate used as the labeled ligand or is due to the type of antibody used.

The cortisol assay described here can be accomplished within 2 h. The sensitivity achieved (20 pg/tube) approaches that obtained by radioimmunoassay, and is satisfactory for determination of cortisol in human plasma. However, it should be pointed out that these are pilot experiments and the new methodology still requires validation using a wider range of normal and pathological biological material.

## ACKNOWLEDGMENTS

This work was supported by grants to H.R.L. from the World Health Organization, the Ford Foundation and the Rockefeller Foundation. M.P. was a Visiting Fellow from the Endocrinology Unit, University of Florence, at the Weizmann Institute. H.R.L. is the Adlai E. Stevenson III Professor of Endocrinology and Reproductive Biology at the Weizmann Institute of Science and a Scholar-in-Residence at the Fogarty International Center of the National Institutes of Health.

## REFERENCES

1. Kohen, F., M. Pazzagli, J. B. Kim, H. R. Lindner, and R. C. Boguslaski, *FEBS Letts. 104,* 201 (1979).
2. Kohen, F., J. B. Kim, G. Barnard, and H. R. Lindner, *Steroids* (in press) (1980).
3. Kohen, F, M. Pazzagli, J. B. Kim, and H. R. Lindner, *Steroids* (in press) (1980).
4. Schroeder, H. R., R. C. Boguslaski, R. J. Carrico, and R. T. Buckler, *Methods Enzymol. 57,* 424 (1978).
5. Schroeder, H. R. and F. M. Yaeger, *Anal. Chem. 50,* 1114 (1978).

6. Campbell, D. H., J. S. Garvey, N. E. Cremer, and D. H. Sussdorf, *in* "Methods in Immunology," 2nd ed., pp. 189-198, Benjamin, New York (1970).
7. Kohen, F., J. B. Kim, and H. R. Lindner, *in* "Proceedings of Second International Symposium on Luminescence," this volume, Academic Press, New York, in press.
8. Kohen, F., J B. Kim , G. Barnard, and H. R. Lindner, *in* "Proceedings of Second International Symposium on Luminescence," this volume, Academic Press, New York, in press.

DETERMINATION OF HEMOGLOBIN AND OTHER HEMOPROTEINS IN SERUM BY LUMINOL CHEMILUMINESCENCE

Thomas Olsson
Kurt Bergström
Anders Thore

Department of Clinical Chemistry
Huddinge University Hospital
Huddinge, Sweden

## I. INTRODUCTION

There is a pronounced need for sensitive and simple determinations of hemoproteins in serum e.g. hemoglobin in order to detect hemolytic reactions caused by diseases or mechanical treatment of blood. Hitherto, spectrophotometric detection principles have been used, based on peroxidase activity of protohemin contained in the hemoglobin molecule. Several chromogenic substances have been used as substrates e.g. benzidine (1) or tetramethyl benzidine (2,3,4). However, benzidine is not suitable due to potential cancer hazards (2). The attained sensitivity has been 1-5 mg hemoglobin per liter serum (3,4) using tetramethyl benzidine.

The chemiluminescent luminol reaction has been described as a sensitive tool for hemoprotein determination (5,6). The sensitivity in model systems has been around $5 \times 10^{-12}$ M of hemin i.e. about $1 \times 10^{-6}$ g of hemoglobin per liter. These studies have been performed in the absence of interfering substances, contrary to conditions expected in biological samples e.g. serum specimens.

The present study was undertaken in order to evaluate the possible application of the chemiluminescent luminol reaction to determine hemoproteins in serum, with special reference to hemoglobin.



ISBN 0-12-208820-4

## II. MATERIALS AND METHODS

### *A. Reagents*

Luminol (3-aminophthalhydrazide) was obtained from Merck and $3,3^1,5,5^1$-tetramethyl benzidine (TMB) was from Aldrich. Myoglobin (sperm whale), catalase (bovine liver, E.C. No. 1. 11.1.6) and protohemin (bovine) were from Sigma. Hemoglobin (Hb) was prepared from washed human erythrocytes as described by Hanks et al. (1). Hemoglobin was determined by the cyan-methemoglobin procedure (7) and standard solutions (100 mg Hb/l) were made up in serum or deionized water. All other reagents were of analytical purity. All solutions were made up in deionized water (conductivity $<1$ µS). Serum samples were randomly collected at the department of clinical chemistry.

### *B. Methods*

The luminol assay of hemoproteins was carried out in a luminometer (LKB Produkter AB, Stockholm, Sweden) essentially as described previously (5,8). Sample (10 µl) was added to 1 ml of NaOH (0.1 M), 0.2 ml of alkaline luminol reagent (1.2 mM) was added and the reaction was initiated by addition of 0.2 ml of perborate (50 mM).

The spectrophotometric determination of hemoglobin with TMB was performed according to Standefer and Vanderjagt (3) with minor modifications. Sample (10 µl) was added to 0.5 ml of TMB (5 mg/ml in 90% acetic acid) and the peroxidase activity was initiated by addition of 0.5 ml of $H_2O_2$ (1%). After 25 min (25$^\circ$C) the reaction was stopped by addition of 5 ml of acetic acid (90%). The colour was measured spectrophotometrically at 375 nm.

Protohemin was extracted from serum by acidified acetone as described previously (8,9). Serum (0.5 ml) was treated twice with acetone containing 1% of HCl (4.5 ml) at 0$^\circ$C, the first time for 6 h and the second time for 2 h. Precipitated proteins were removed by centrifugation and protohemin was taken into diethyl ether after twofold dilution of the acetone phase with 1% HCl. Diethyl ether was evaporated to dryness and residual material was dissolved in 0.05 M NaOH and immediately assayed by luminol or TMB reactions.

## III. RESULTS

### A. *Sensitivity and Dynamic Range of Luminol and TMB Reactions*

In the first experiment the sensitivity and dynamic range of the luminol and tetramethyl benzidine (TMB) reactions were compared in a model system (Fig. 1) based on dilutions of hemoglobin in deionized water. Luminol was by far the most sensitive method. The limit of sensitivity was about 3 x $10^{-4}$ mg/l compared to 2 mg/l with the TMB reaction.

The upper limit of the dynamic range was at least 4 x $10^{-1}$ with the luminol reaction and about 2 x $10^{2}$ mg/l with the TMB reaction. However, it is of course possible to dilute samples with high content of hemoglobin, as in clinical serum specimens (>1 mg/l), to fall within the range of the luminol reaction.

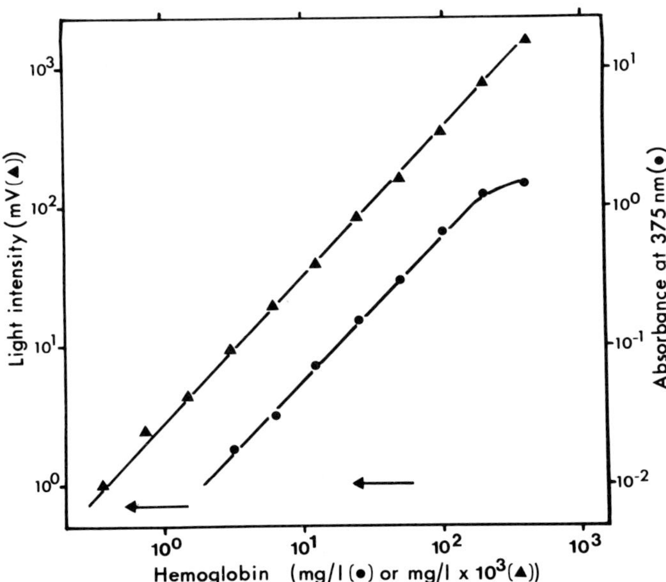


*FIGURE 1. Determination of hemoglobin by the luminol (▲) or the TMB reaction (●). Hemoglobin was diluted in deionized water (400-0.2 mg/l). Samples were further diluted $10^3$ before addition to NaOH and determination with luminol. The limits of sensitivity were set by background reactions. The limiting level was defined as the upper limit of a 95% confidence interval for the average plus 10% and has been indicated by arrows for the two reactions, respectively.*

### B. Interference of Serum with the Luminol Reaction

In the following experiment some serum specimens were diluted in deionized water and assayed for material reactive in the luminol reaction. Very high light intensities were obtained, above the operative range of the luminometer, with undiluted or slightly diluted samples. Dilutions of $10^3$ or more could usually be measured with the used experimental set-up. However, upon dilution light intensity was decreased considerably less than proportional.

This effect was due to inhibition caused by serum components, probably proteins, as can be seen in Fig. 2. In the presented experiment a constant addition of hemoglobin was assayed together with different amounts of serum or serum components (constant addition technique). Light intensity from added serum was corrected for and the inhibition was calculated by comparison with light intensity from hemoglobin in the absence of serum or serum components.

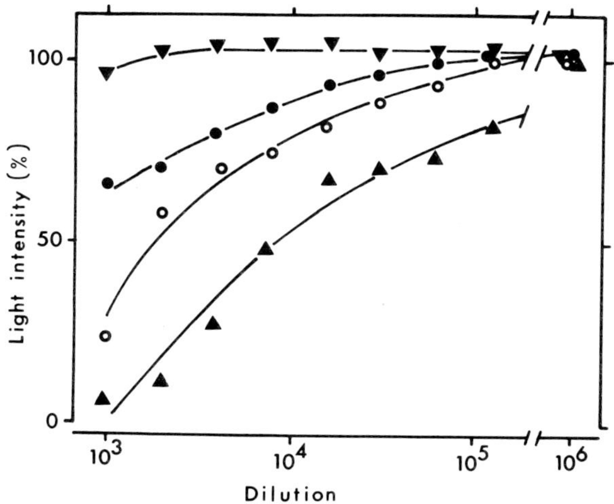


*FIGURE 2. Interference of serum or serum components with the luminol reaction. Presented dilutions also include the final step of $10^2$ in NaOH before addition of reagents. Serum (o), serum albumin (human (●) or bovine (▲), 80 g/l), or lipids ((▼) extracted from serum and resuspended in water to the original volume) were diluted in deionized water and $2 \times 10^{-14}$ mol of hemoglobin was added to each dilution. Inhibition was calculated by comparing the increased light intensity caused by added hemoglobin to the light intensity of hemoglobin in deionized water.*

Inhibition was evident even at serum dilutions of $10^4$. Inhibition could be corrected for using the constant addition technique, but at dilutions of $10^5$ or more the degree of inhibition would be essentially negligible, allowing the use of separate reference samples for calibration. Thus, total dilutions of $10^5$ were chosen for further studies. In practice samples were diluted $10^3$ in deionized water before the final dilution of $10^2$ in NaOH. Even this high degree of dilution resulted in luminescence well within the linear range of the luminol reaction (cf. Fig. 1).

### C. *Comparison of Content of Hemoglobin in Serum Determined by Luminol and TMB Reactions*

Several clinical serum specimens were assayed by the luminol and TMB reactions. Results expressed as hemoglobin (Hb) per litre of serum are presented in Fig. 3. Due to inhibition by serum components the calculations for the TMB reaction were performed by comparison with a standard in serum. The linear relationship between the two methods ($y = 0.66x + c$, $r = 0.88$) deviates from direct proportionality resulting in an apparent higher yield with the luminol

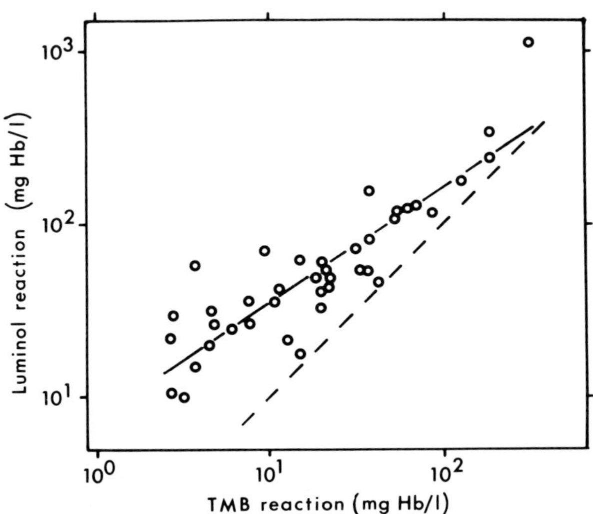


*FIGURE 3. Comparison of content of hemoglobin in serum specimens determined by luminol and TMB reactions. The linear regression line has been indicated (———) and, for comparison also the theoretical directly proportional relationship (----).*

method, particularly of low levels of hemoglobin. There was also considerable scattering of the results around the regression line. This may be partially but not entirely explained by the fact that inhibition of the TMB reaction varied considerably between individual samples.

### *D. Activity of Different Hemoproteins in the Luminol and TMB Reactions*

Hemoproteins, other than hemoglobin, may be present in serum and would be expected to catalyze both the luminol and TMB reactions in relation to their content of protohemin (1,5). The specificity of some hemoproteins possibly present in serum was tested in the two reactions as presented in Table 1.

The relative activity of the hemoproteins was quite constant, based on protohemin content, and would probably not account for the observed differences in results with the two methods (Fig. 3). The results point out that any of these hemoproteins present in serum will be measured as hemoglobin in both reactions. An inhibitory effect due to presence of serum was observed as before with the TMB reaction (3). Activity of free protohemin deviated from the activity of the hemoproteins, perhaps caused by difficulties to dilute this hydrophobic substance.

*TABLE 1. Comparison of Activities, Based on Protohemin Content, of some Hemoproteins in the Luminol and TMB Reactions*[a]

| | *Luminol reaction* | | *TMB reaction* | |
|---|---|---|---|---|
| | *Water* | *Serum* | *Water* | *Serum* |
| *Hemoglobin* | *20* | *22* | *66* | *50* |
| *Myoglobin* | *21* | *23* | *69* | *55* |
| *Catalase* | *22* | *24* | *53* | *53* |
| *Protohemin* | *22* | *31* | *75* | *57* |

[a]*Values refer to solutions which, based on content of protohemin, correspond to 100 mg of Hb/l and are expressed as mV x* $10^{15}$ *or Abs x* $10^{12}$ *per mol of hemin for the luminol and TMB reactions respectively. Samples were made up in serum or deionized water and assayed as described in Fig. 1.*

*TABLE 2. Yield of Protohemin from Serum by Extraction with Acetone/HCl and Determination by Luminol and TMB Reactions*[a]

| Sample | Luminol reaction | | TMB reaction | |
|---|---|---|---|---|
| | Untreated | Extract | Untreated | Extract |
| Serum 1 | 1.9 | 0.8 (42) | 0.8 | 0.7 (88) |
| Serum 2 | 1.0 | 0.5 (50) | 0.5 | 0.3 (60) |
| Serum 3 | 0.8 | 0.3 (37) | 0.5 | 0.3 (60) |
| Serum 4 | 0.8 | 0.6 (75) | 0.5 | 0.3 (60) |
| Hemoglobin | 3.6 | 3.4 (94) | 3.5 | 3.5 (100) |
| Protohemin | 4.0 | 3.7 (93) | 4.0 | 3.7 (93) |

[a]*Values refer to different serum samples, or dilutions of hemoglobin or protohemin in serum corresponding to about 65 mg Hb/l, and are expressed as moles of protohemin x $10^6$. Standards have been corrected for the presence of serum. Results have been calculated by comparison with a standard of protohemin. Values within brackets refer to yield in percent.*

### E. *Extraction of Protohemin from Serum Specimens and Determination by Luminol and TMB Reactions.*

Some serum specimens were extracted for protohemin and assayed by the luminol and TMB reactions in order to determine the presence of other substances reactive in the two reactions. Controls were performed with standards of protohemin and hemoglobin in serum.

The results presented in Table 2 are not altogether conclusive but some activity present in serum is lost by extraction and values are closer between the two methods of determination after extraction than before. Yield of hemin from standard solutions was close to 100%. This indicates the presence of substances with a relatively higher activity in the luminol reaction than the TMB reaction. The substance(s) is not extracted together with protohemin and was not present in the precipitated protein. However, further experiments must be performed in order to verify these results but they offer an explanation for the observed differences of hemoglobin content of serum obtained with the two methods (cf. Fig. 3).

## IV. SUMMARY AND CONCLUSIONS

Hemoglobin may be determined in serum by the sensitive chemiluminescent luminol reaction within the range of 0.3 to 400 mg/l. However, the luminol reaction gave considerably higher values of hemoglobin in serum specimens than a spectrophotometric reference method with tetramethyl benzidine. Interfering material more active in the luminol reaction may have caused this effect. Since also the reference method is non-specific for hemoglobin, the use of another specific method would be suitable in further studies of this application.

## ACKNOWLEDGEMENTS

This work was supported by the Swedish Board for Technical Development.

## REFERENCES


(1) Hanks, G.E., M. Cassel, R.N. Ray, and H. Chaplin, *J. Lab. Clin. Med.* 56, 486 (1960).
(2) Holland, V.R., B.C. Saunders, F.L. Rose, and A.L. Walpole, *Tetrahedron* 30, 3299 (1974).
(3) Standefer, J.C. and D. Vanderjagt, *Clin. Chem.* 23, 749 (1977).
(4) Lijana, R.C. and M.C. Williams, *J. Lab Clin. Med.* 94, 266 (1979).
(5) Ewetz, L., and A. Thore, *Anal. Biochem.* 71, 564 (1976).
(6) Vasileff, T.P., G. Svarnas, H.A. Neufeld, and L. Spero, *Experentia* 30, 20 (1974).
(7) van Kampen, E.J. and W.G. Zejlstra, *Clin. Chim. Acta* 6, 538 (1961).
(8) Smith, K.M. in "Porphyrins and metalloporphyrins" (K.M. Smith, ed.), p. 757. Elsevier, Amsterdam (1975).
(9) Ewetz, L., and A. Thore, *Appl. Environ. Microbiol.* 36, 790 (1978).

# BIOLUMINESCENT IMMUNOASSAY: A NEW ENZYME-LINKED ANALYTICAL METHOD FOR THE QUANTITATION OF ANTIGENS

Douglas W. Reichard
Robert J. Miller, Jr.

U. S. Army Medical Research Institute of Infectious Diseases
Fort Detrick, Maryland

## I. INTRODUCTION

The use of the luciferin-luciferase system offers a rapid and specific method for the determination of adenosine triphosphate (ATP) (1). This specificity should be translatable into a rapid and sensitive enzyme-linked immunoassay system by utilizing ATP-producing enzymes. The choice of enzymes is predicated on the magnitude of $\Delta G$ for the production of ATP and on the availability of both enzyme and substrate. Pyruvate kinase (EC 2.7.1.40) and creatine phosphokinase (EC 2.7.3.2) in the presence of their respective substrates convert adenosine diphosphate (ADP) to ATP at significant rates. We wish to report the preliminary results on the use of pyruvate kinase conjugates in the development of an enzyme-linked bioluminescent immunoassay.

## II. MATERIALS AND METHODS

### *A. Materials*

Pyruvate kinase, phosphoenolpyruvate, luciferin, luciferase, m-maleimidobenzoyl-n-hydroxysuccinimide ester (MBS), and ADP were purchased from Sigma. Rabbit IgG anti-monkey IgG was obtained from Miles and goat IgG anti-rabbit IgG (GAR) from Calbiochem. Microtiter plates were obtained from Cooke.

 ISBN 0-12-208820-4

*B. Instrumentation*

All assays were run on a Chem-Glow microphotomoter equipped with a 50-mV, single-pen recorder (American Instruments).

*C. Reagents*

*1. KPBS.* 0.01 M potassium phosphate buffered saline, pH 7.0.

*2. Luciferin-Luciferase Pre-Mix.* Equal volumes of 0.01 M $MgSO_4$, luciferin (1 mg/10 ml of 0.01 M glycylglycine buffer, pH 7.8), and luciferase (1 mg/3 ml of 0.75 M glycylglycine buffer, pH 7.8) were mixed and allowed to sit 12-24 h before use.

*3. Substrate Solution.* 5 ml ADP (0.2 mg/ml) (purified by DEAE-Sephacel chromatography), 10 ml phosphoenolpyruvate (1 mg/ml in 0.01 M glycylglycine buffer, pH 7.8), and 10 ml of 0.01 M $MgSO_4$.

*D. Conjugates*

Goat anti-rabbit IgG-pyruvate kinase (GAR-PK) conjugates were prepared with MBS following literature procedures (2-4). The conjugate was partially purified on a 1.5 x 85-cm G-200 column equilibrated with 0.01 M KPBS. Fractions which demonstrated high enzyme and high antibody acitivty were used for the assay.

*E. Immunoassay Procedure*

Experimental wells of microtiter plates were coated with 250 µl of rabbit IgG anti-monkey IgG (5 µg/ml in 0.1 M $Na_2CO_3$, pH 9.5). After washing, the plates were treated with 1% BSA to prevent nonspecific binding. Control wells were treated only with BSA to allow background enzyme activity to be determined. Various dilutions of conjugated second antibody were then added to the plate and incubated 1.5 h at 37°C. Plates were then washed and 250 µl of substrate solution were added and allowed to react for 1 h at 37°C. The bioluminescent assay were performed by adding 100 µl of substrate solution to 25 µl of luciferin-luciferase pre-mix, mixing for 5 sec and measuring peak luminescent activity.

## III. RESULTS AND DISCUSSION

Pyruvate kinase and creatine phosphokinase have ΔG for the conversion of ADP → ATP of -7.5 and -3.0 kcal/mol, respectively (5). Therefore, if these enzymes can be covalently coupled to antibody molecules, a bioluminescent immunoassay could be developed utilizing conversion of ADP to ATP and quantitation of ATP by the highly specific luciferin-luciferase reaction. The coupling of pyruvate kinase to antibodies by several means was investigated. Coupling with glutaraldehyde yielded products with no enzyme activity. Since pyruvate kinase contains free thiols not involved in enzymatic activity (6), coupling with MBS was also investigated (2-4). This procedure yielded conjugates that retained enzymatic activity and were acceptable for use in an immunoassay.

Microtiter plates coated with rabbit IgG anti-monkey IgG and BSA were first treated with dilutions of a GAR-PK fraction ($Abs_{280}$ = 0.51). Control wells were only treated with BSA and conjugate dilutions. The luminescence was determined and plotted, after subtracting control values from experimental values. Typical results are shown in Figure 1. Dilutions below 1:160 were too active to be measured in the same time frame as other samples. A 1:640 dilution of the GAR-PK conjugate was then chosen for use in a competitive titration of GAR <u>vs</u> GAR-PK. Microtiter plates coated as previously described were treated with 100 µl of GAR-PK (1"640) and 100 µl of dilutions of GAR. The luminescence is plotted in Figure 2. Again control values have been subtracted from the luminescence data. Control values for each dilution ran 10-15% of experimental values and could probably be lowered by further purification of the conjugate. However, nonspecific binding is easily determined, and, as stated, has been subtracted from all data points. These titration data suggest that a bioluminescent immunoassay procedure is a feasible and reproducible method for antigen detection. Pyruvate kinase and possibly other ATP-producing enzymes can be conveniently conjugated to antibodies and used to develop sensitive bioluminescent immunoassays. These immunoassays would have several advantages: namely, no radioactive compounds are required and the luciferase-ATP reaction is extremely sensitive and highly specific.

Further investigation of these assays is in progress. New conjugates will be prepared and evaluated, especially for the identification of microorganisms.

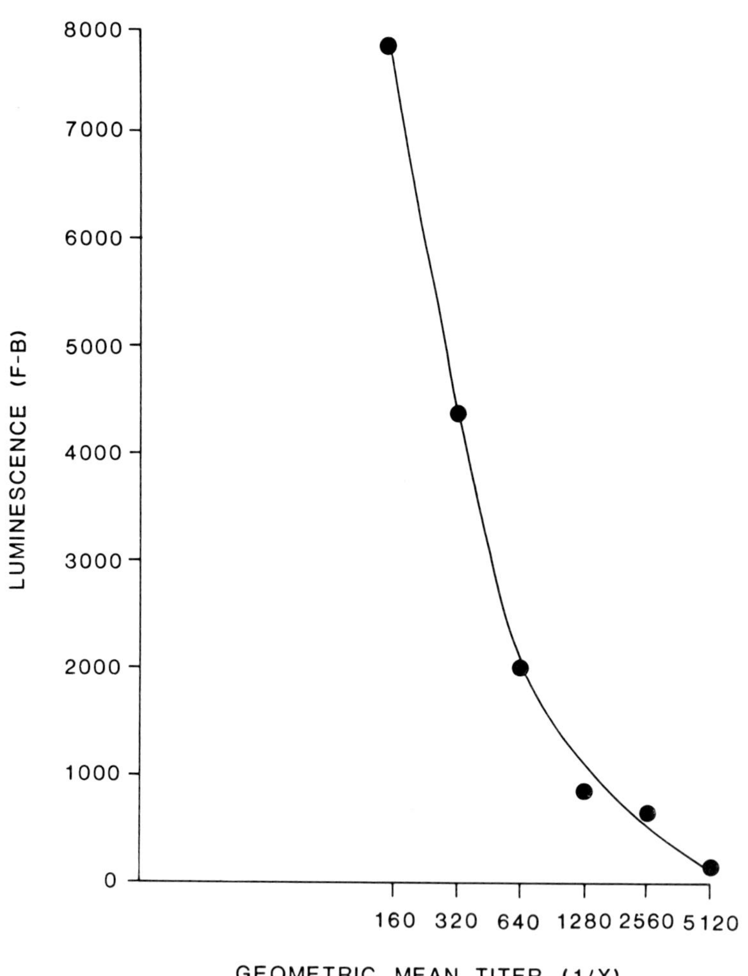


*FIGURE 1. Luminescence for titration of goat IgG anti-rabbit IgG-pyruvate kinase conjugate.*

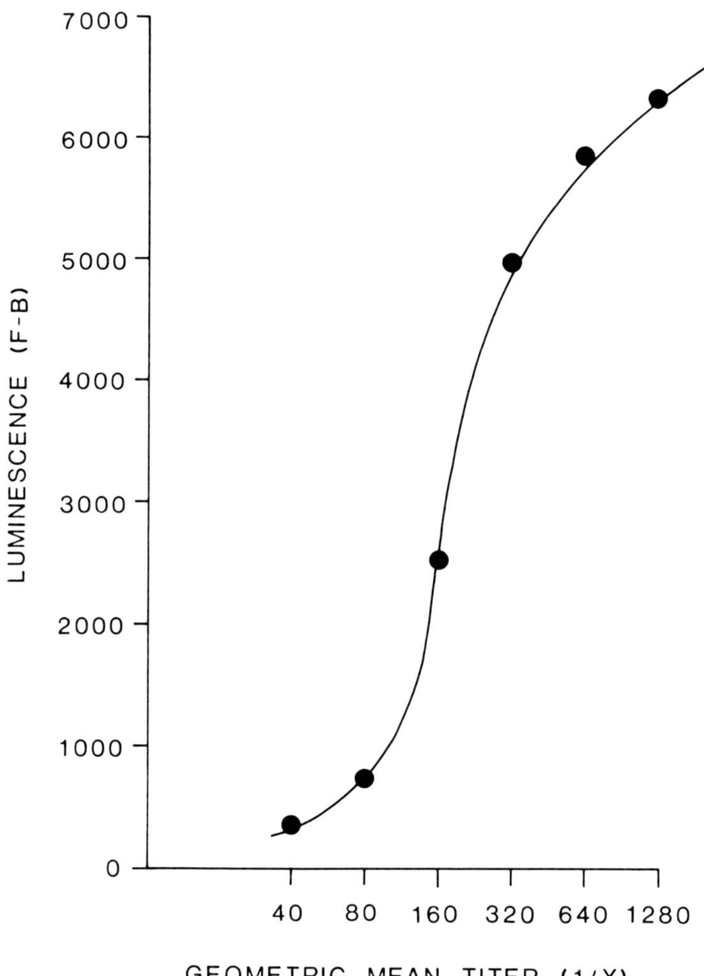


*FIGURE 2. Luminescence for competitive titration of goat IgG anti-rabbit IgG* vs *goat IgG anti-rabbit IgG-pyruvate kinase conjugate.*

REFERENCES

1. Neufeld, H. A., R. D. Towner, and J. Pace, *Experientia* 31, 391 (1975).
2. Kitagawa, T., and T. Aikawa, *J. Biochem.* 79, 233 (1976).
3. O'Sullivan, M. J., E. Gnemmi, D. Morris, G. Chieregatti,

M. Simmons, A. D. Simmonds, J. W. Bridges, and V. Marks, *FEBS Lett.* 95, 311 (1978).
4. O'Sullivan, M. J., E. Gnemmi, D. Morris, G. Chieregatti, A. D. Simmonds, M. Simmons, J. W. Bridges, and V. Marks, *Anal. Biochem.* 100, 100 (1979).
5. Lehninger, A. L., "Biochemistry," 2nd ed., Worth Publishers, Inc., New York, (1975).
6. Cottam, G. L., P. F. Hollenberg, and M. J. Coon, *J. Biol. Chem.* 244, 1481 (1969).

# CHEMILUMINESCENT LABELS IN IMMUNOASSAY

J. Steven A. Simpson
Anthony K. Campbell
J.Stuart Woodhead

Department of Medical Biochemistry
Welsh National School of Medicine
Cardiff, Wales

Adrian Richardson
Russell Hart
Frank McCapra

Department of Molecular Sciences
University of Sussex
Falmer, England

## I. INTRODUCTION

During the past 20 years many of the developments in immunological techniques have resulted from the exploitation of labelled compounds with high activity, which consequently may be detected at low concentrations. The labels have included radioisotopes ($^{125}I$, $^{131}I$, $^{3}H$), enzymes and fluorescent molecules.

The requirements of the label are that it should be capable of sensitive detection, it should not lose activity when coupled covalently to antigen or antibody. The labelled compound must retain immunological activity and be stable for as long as possible when stored under reasonable conditions.

Radioisotopes cannot be highly active and possess a long half life, but, for high sensitivity assays $^{125}I$-labelled compounds have been widely used.

Nevertheless, proteins labelled with $^{125}I$ have a relatively short useful life. Enzyme and fluorescent labels are more




ISBN 0-12-208820-4

stable but in most cases less convenient.

We have examined the properties of some chemiluminescent compounds in order to assess their usefulness as labels in immunoassay systems.

## II. PREPARATION AND PROPERTIES OF LUMINOL LABELLED ANTIBODY

An IgG fraction of a sheep (anti-rabbit IgG) antiserum was labelled with luminol by 3 procedures. Direct coupling was carried out using a diazonium salt of luminol (1). Alternatively, a succinate derivative of luminol was coupled using carbodiimide or mixed anhydride techniques.

Incorporation of luminol was low when carbodiimide or mixed anhydride methods were used. As shown in Table I, incorporation of diazoluminol was high, although the activity of the product indicated considerable loss of quantum yield. This appears to result from polymerization of diazoluminol in such a way that luminescence activity is lost.

*TABLE I. Incorporation of luminol in sheep (anti-rabbit IgG) antibody*

| *Coupling Method* | *Incorporation (mol luminol/mol IgG)* | *Activity (counts[a]/mol IgG)* |
|---|---|---|
| *Carbodiimide* | *0.5* | *$2.5 \times 10^{16}$* |
| *Mixed Anhydride* | *1.2* | *$3.8 \times 10^{16}$* |
| *Diazotization* | *22.0* | *$8.3 \times 10^{16}$* |

[a]*Measured using inhouse photon counter.*

The luminol labelled antibodies were indistinguishable from unlabelled sheep(anti-rabbit IgG) antibodies when tested at several dilutions in a radioimmunoassay using solid phase rabbit IgG and $^{125}$I-labelled sheep(anti-rabbit IgG) antibodies (Fig. 1).

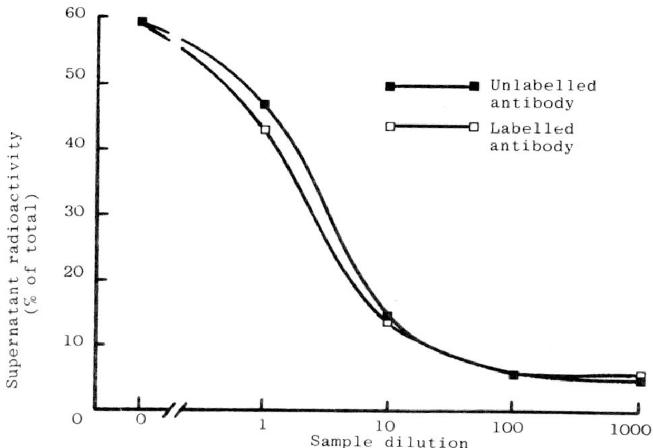


*FIGURE 1. Solid phase radioimmunoassay of sheep-(anti-rabbit IgG) antibody and luminol-labelled sheep-(anti-rabbit IgG) antibody.*

## III. ASSAYS USING LUMINOL-LABELLED ANTIBODIES

A two-site luminometric assay was developed using plastic tubes coated with sheep(anti-rabbit IgG) antibody. Samples containing rabbit IgG were reacted in the tubes overnight, and after washing, luminol-labelled sheep(anti-rabbit IgG) antibody, purified by affinity chromatography was added. Figure 2 shows a standard curve for rabbit IgG obtained with this system. The stability of the labelled antibody is illustrated by the fact that the assay was identical when carried out with luminol-labelled antibodies stored for 16 months at -20°C.

An indirect two-site assay for human α-fetoprotein (AFP) was carried out as follows. Plastic tubes were coated with sheep(anti-AFP) antibody. After reaction with samples containing ARP, rabbit(anti-AFP) antiserum was added and after careful washing, the uptake of rabbit antibody was measured by reacting with luminol-labelled sheep(anti-rabbit IgG) antibody. A standard curve for human α-fetoprotein is shown in Fig. 3. The assay is capable of measuring ARP within the clinically useful range, and results obtained from sera of

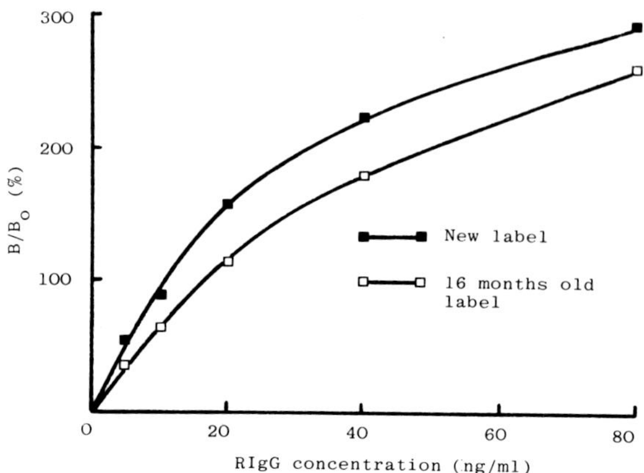


*FIGURE 2. Two-site assay of rabbit IgG using luminol-labelled sheep(anti-rabbit IgG) antibody.*

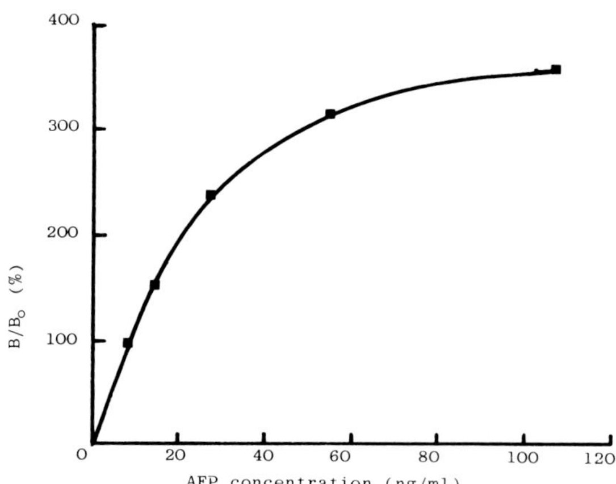


*FIGURE 3. Indirect two-site luminometric assay of human α-fetoprotein.*

pregnant women agree closely with those obtained by conventional radioimmunoassay.

## IV. ACRIDINIUM ESTERS

Acridinium esters are considerably more luminescent than luminol (2). The ester *p*-carboxyphenyl-*N*, 10-methyl-acridinium-9-carboxylate bromide was synthesized by esterification of p-carboxyphenol acetate and acridinium-9-carboxyl chloride, followed by methylation of the nitrogen using "magic methyl" and hydrolysis of the acetate group using hydrobromic acid. Under mild oxidising conditions this compound undergoes the reaction shown in Figure 4.

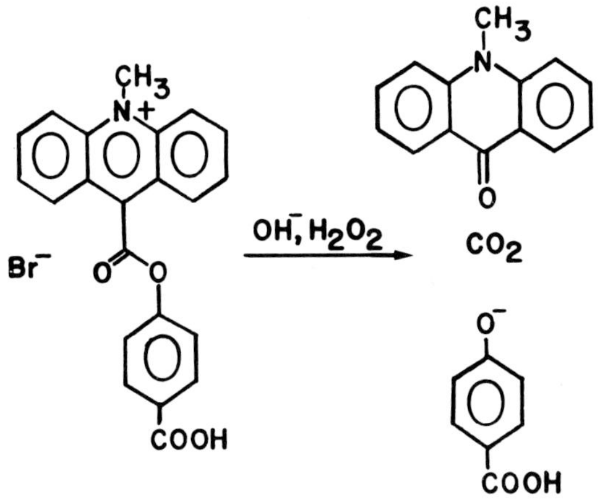

$$\text{Br}^- \;\; \text{acridinium ester (COOH)} \xrightarrow{OH^-,\ H_2O_2} \text{N-methylacridone} + CO_2 + {}^-O\text{-}C_6H_4\text{-COOH}$$

*FIGURE 4. Luminescent reaction of acridinium esters.*

Figure 5 shows the luminescence from a sample of this compound (0.24 pmole) reacted with various concentrations of alkaline peroxide. This illustrates the high signal to noise ratio that can be obtained using appropriate conditions with acridinium esters, in contrast to luminol luminescence which shows a complex reagent concentration dependence.

This acridinium ester was coupled to sheep(anti-ARP) antibody, which had been purified by affinity chromatography using a carbodiimide reaction. A direct two-site assay was carried

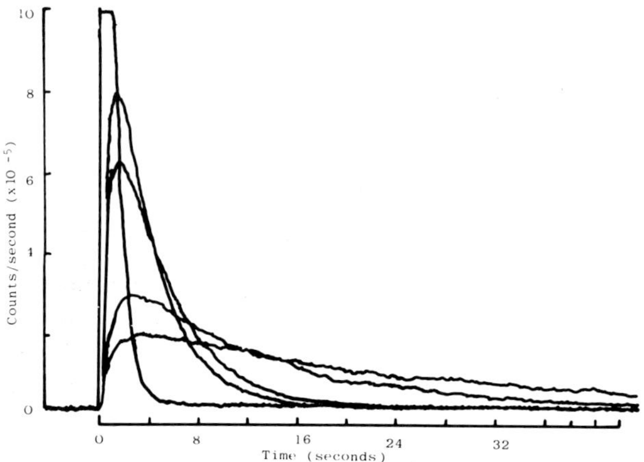


*FIGURE 5. Effect of alkaline peroxide on the luminescence of acridinium esters.*

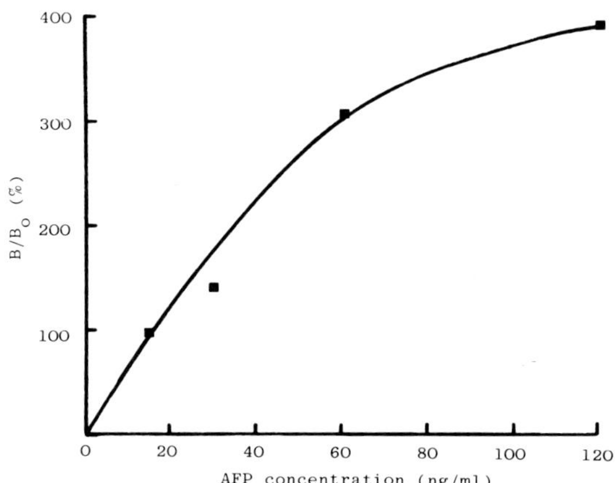


*FIGURE 6. Standard curve for direct luminometric assay of human α-fetoprotein.*

out using this labelled antibody and plastic tubes coated with purified sheep(anti-AFP) antibody. A standard curve for AFP is shown in Fig. 6.

The most highly active conjugates of protein and acridinium ester have yielded activities of 6.8 x $10^{18}$ counts/mol which gives a limit of detection comparable to that of a $^{125}$I-labelled protein. Acridinium labelled proteins remain chemiluminescent for at least 11 month when stored at -20°C.

## V. CONCLUSIONS

Chemiluminescent-labelled antibodies have been prepared using luminol and an acridinium ester. The derivatives are chemiluminescent and immunologically active. Moreover, they are stable for extended periods of time, and can be used to provide sensitive immunoassays for biologically important molecules.

## REFERENCES

1. Simpson, J. S. A., A. K. Campbell, M. E. T. Ryall and J. S. Woodhead, *Nature* 279, 646 (1979).
2. McCapra, F., D. E. Tuft and R. M. Topping, *British Patent No. 1* 461, 877 (1977).

# EVALUATION OF METABOLIC RATES USING BIOLUMINESCENCE ANALYSIS

Peter Wersäll
Sven Brolin
Birger Petersson
Claes-Göran Östenson

Department of Medical Cell Biology
University of Uppsala
Uppsala, Sweden

## I. INTRODUCTION

Since the intracellular energy flux is associated with the conversion of nucleotides, particularly in the respiratory chain, short time hypoxia may be used to estimate the rate of the flux. Bioluminescence is sufficiently sensitive for assay of small samples which are suitable because of their surface-mass ratio, which facilitates an efficient gas exchange.

## II. EXPERIMENTAL

The islets of Langerhans were chosen for the experiments, since it is of interest to obtain information about the insulin and glucagon producing cells in research on diabetes (1, 2, 3, 4). Separate studies of the two cell types were performed by using islets from normal guinea-pigs for assay of insulin producing cells. Islets in which these had been destroyed were used for assay of glucagon producing cells. The

[1]*This study was supported by the Swedish Medical Research Council (Proj. Nrs. 00525 and 02297) and the Swedish Diabetes Association.*




ISBN 0-12-208820-4

destruction was performed with a single injection of streptozotocin into the abdominal cavity (5). The islets which are dispersed as numerous cell bodies in the pancreas were liberated by digestion with collagenase (6, 7) and harvested under a stereomicroscope (8, 9, 10). To allow recovery after the isolation procedure the islets were cultured for one week (11). The cultured islets were transferred to a modified perifusion apparatus which was designed so that an oxygenated perifusion medium used for pretreatment for 5 min could be rapidly replaced with a medium deprived of oxygen (see Fig. 1). The composition of the perifusion medium is shown in Table I. The experiments were terminated by expelling the islets into chilled isopentane (-70°C). After lyophilization, the islets or islet pieces were weighed on a quartz fibre balance. The sample weights varied between 0.1 and 1.0 /μg. Extraction of adenosine nucleotides and subsequent analyses of ATP and the nucleotide sum were carried out as previously described (12, 13). Recording of the analyses was performed by means of microprocessing techniques (14).

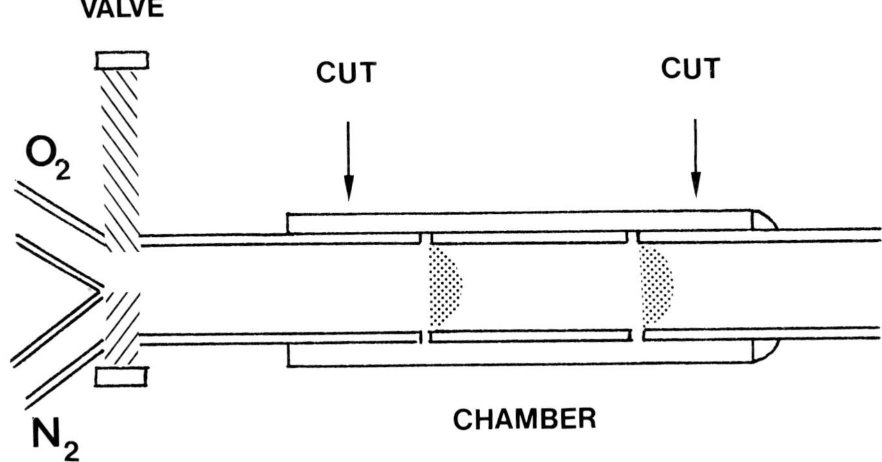


*FIGURE 1. Diagram of apparatus for rapid exchange of perifusion media. The ends of a piece of polyethene tubing are covered by nickel grids (Tebra IGN 400, Analytical Standards AB) and enclosed in another slightly larger tubing. This chamber is connected to a valve for alternating between an oxygenated ($O_2$) and an oxygen deprived ($N_2$) perifusion medium and immersed in a water bath at 38°C[2]. The experiments are terminated by cutting out (arrows) and rapidly freezing the chamber, which is easily renewed.*

*TABLE I. Composition of the perifusion medium*

| Components | | Gas conditions | |
|---|---|---|---|
| | mmol/l | Pretreatment | Hypoxia |
| NaCl | 111 | Oxygen supplied | Oxygen removed |
| KCl | 5 | (95% $O_2$ + 5% $CO_2$) | by nitrogen |
| $NaHCO_3$ | 27 | | bubbling (95% |
| $MgCl_2$ + $6H_2O$) | 0.98 | | $N_2$ + 5% $CO_2$) |
| $Na_2HPO_4$ | 0.7 | | |
| $KH_2PO_4$ | 0.2 | | |
| $MgSO_4$ + $7H_2O$ | 0.3 | | |
| CaCl | 1.35 | | |
| Glucose | 5.5 | | |

## III. RESULTS

In the experiments with samples rich in insulin producing cells a marked decrease in ATP concentration was found after oxygen-free perifusion for 30 sec. Thereafter the concentration dropped slowly (Fig. 2). A tendency to decrease was observed already after 15 sec. The total amounts of adenosine nucleotides remained unchanged. A similar reaction pattern was found for the glucagon producing cells.

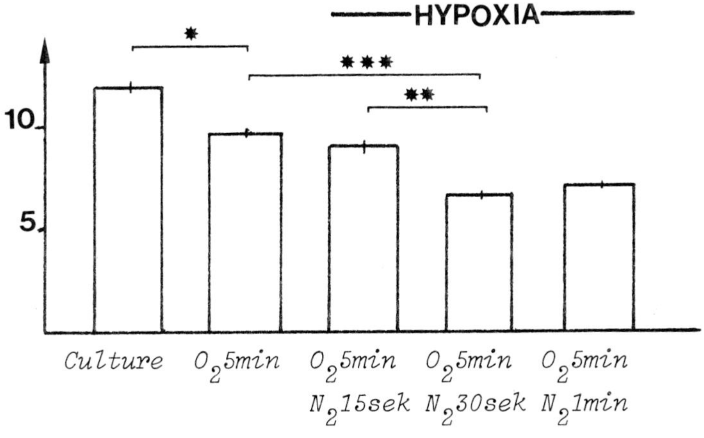


FIGURE 2.

## IV. CONCLUDING REMARKS

The applicability of the experimental is confirmed by the measurements. Preliminary studies indicate that the model can also be used for estimates of other metabolic changes such as the accumulation of NADH and the disappearance of $NAD^+$. The need to use small tissue explants to facilitate satisfactory exchange with the surrounding medium can be achieved because of the high sensitivity of the bioluminescence analyses. The present test of the ATP conversion can be applied not only for bioenergetic purposes but also for checking the condition of cells and pieces of organs. This may be of value to determine a satisfactory tissue preservation prior to transplantations. The test may furthermore be used to reveal metabolic disorders in experiments with tissue explants which are assumed to reflect the *in vivo* state.

## V. SUMMARY

Short time anoxia was used to determine the rate of ATP conversion in cells from the islets of Langerhans. The islets were isolated from the pancreas using collagenase digestion technique and thereafter allowed to recover during culture for one week. They were then transferred to a modified perifusion appartus which was designed so that the gas concentration could be varied rapidly. The experiments were terminated by expelling the samples into chilled isopentane. A decrease in ATP concentration occurred rapidly and a marked reduction was found after 30 sec. The applicability of the experimental model was confirmed by these results. The sensitivity of bioluminescence allowed assay of small samples with a surface-mass ratio which was favourable for gas exchange.

## REFERENCES

1. Östenson, C. -G., A. Ågren, S. E. Brolin, and B. Petersson, *Diabéte & Metabolisme*, 6 , 5(1980).
2. Östenson, C. -G., A. Ågren, A. Andersson, *Biochim. Biophys. Act.*, 628, 152 (1980).
3. Borglund, E., Dissertation. *Acta Universitatis Upsaliensis*, 170, (1973).
4. Brolin, S. E., A. Ågren, and B. Petersson, *Acta endocrinol. (Kbh.)*, in press, (1980).

5. Petersson, B., C. Hellerström, R. Gunnarsson, *Horm. Metab. Res.*, 2 313 (1970).
6. Lacy, P. E. and M. Kostianovsky, *Diabetes*, 16 35 (1967).
7. Howell, S. L. and K. W. Taylor, *Biochem J.*, 108, 17 (1968).
8. Andersson, A. and B. Petersson, J. Westman, J. Edwards, G. Lundqvist, and C. Hellerström, *J. Cell Biol.* 57, 241 (1973).
9. Andersson, A. and C. Hellerström, *Diabetes* 21 *(suppl. 2)* 546 (1972).
10. Olsson, S. -E., A. Andersson, B. Petersson, and C. Hellerström, C., *Diabéte & Metabolisme*, 2, 199(1976).
11. Andersson, A., *Acta Endocrinol (Kbh.)* 83 *(suppl. 205)* 283 (1976).
12. Wettermark, G., L. Tegnér L., S. E. Brolin, E. Borglund, *in* "The Structure and Metabolism of the Pancreatic Islets". Wenner-Gren Center International Symposium Series, Vol. 16 (Falkmer, S., B. Hellman, and I.-B. Täljedal, eds.) p. 275. Pergamon Press, Oxford (1970).
13. Hammar, H., *Acta derm. -venerol. (Stockh)* 52, 251 (1973).
14. Å gren, A., S. E. Brolin, and M. Lindfors, Proc. of Int. Symp. of Analytical Applications of Bio- and Chemi-luminescence, San Diego (1980).

# 1,2-DIOXETANES AS CHEMILUMINESCENT PROBES AND LABELS

H. Wynberg
E.W. Meijer
J.C. Hummelen

Department of Organic Chemistry
University of Groningen
The Netherlands

There are at the present time relatively few techniques for "on-site" investigations of biological systems. Key existing techniques are for example, fluorescent as well as various radioactive labels. A new and promising method would be to use as probes or labels, compounds capable of emitting - under sharply defined conditions - chemiluminescence. The characteristics of the emitted chemiluminescence should give information concerning the site from which it is emitted. When 1,2-dioxetanes are used one can readily control the emission by regulation of the temperature (1).

We are currently investigating two approaches to the use of 1,2-dioxetanes in this context. We have found in the past that 1,2-dioxetanes containing adamantyl groups have good stability characteristics and we have therefore designed our dioxetanes with this in mind.

One approach is to prepare *probes*, the physical nature of which can be varied by means of the substituents attached to the adamantyl nucleus, the substituents being chosen to make the probe compatible with the local environment in which it must function. We envisage the use of such probes in, for instance, membranes (long fatty chains attached to the adamantyl nucleus). On thermal activation (temperatures slightly above physiological are sufficient) chemiluminescence will occur, and this can be analyzed to provide information about the environment around the 1,2-dioxetane.

Another approach is via *labels*, which by means of chemically reactive substituents can be attached by covalent bonds to a target substituent of a specific target molecule. For




ISBN 0-12-208820-4

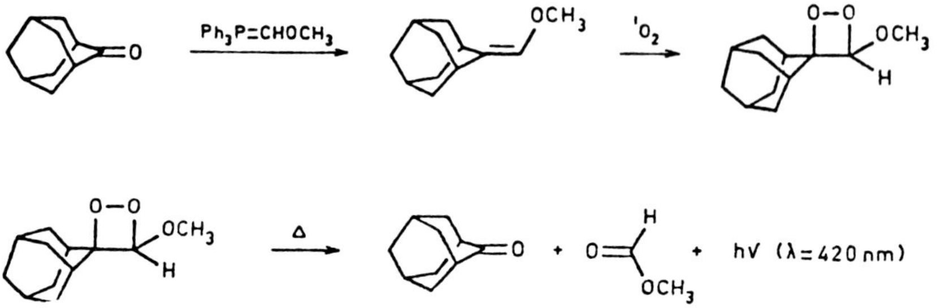

*FIGURE I.*

instance, thiol groups of a peptide can be specifically labeled. Again only thermal activation is required to cause the label to chemiluminesce.

The two approaches to chemiluminescent probes and labels must be worked out with different types of 1,2-dioxetanes. Probes should chemiluminesce at temperatures between 40-80°, and therefore 1,2-dioxetanes of adamantane enol ethers have been developed as outlined in Fig. I (2). Labels should be stable at physiological temperatures and give chemiluminescence at any desired moment. These compounds are prepared from the corresponding functionalized adamantylideneadamantanes by photooxygenation, the general route is outlined in Fig. II (3-5). We will investigate the syntheses and physical characteristics of compounds which can be used as probes and labels, in respectively vesicles and membrane studies (probes) and immonological assays (labels).

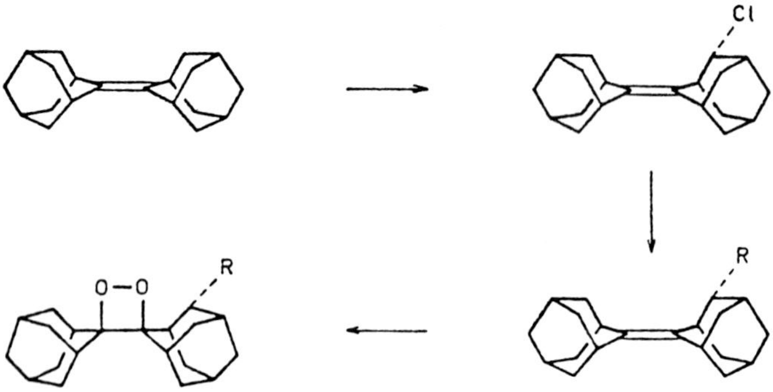

*FIGURE II.*

*labels*

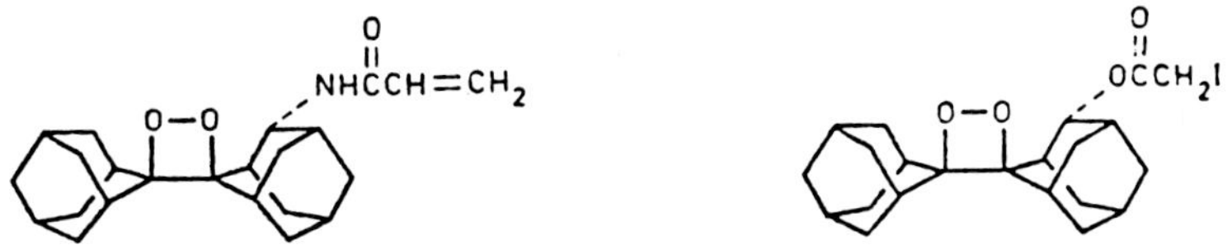

*probes*

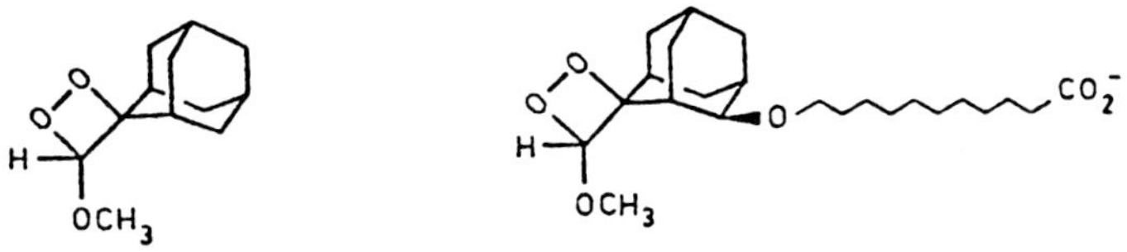

*FIGURE III.*

Examples of approaches are shown in Fig. III. In this figure there are adequate possibilities for regulating, by means of substituents, the thermal stability of the 1,2-dioxetanes and the desired physical properties of the molecules.

## REFERENCES


(1) Wilson, T., *Int. Rev. Sci.: Phys. Chem. Ser. Two* 9, 265 (1976).
(2) Meijer, E.W. and H. Wynberg, *Tetrahedron Lett.* 3997 (1979).
(3) Wieringa, J.H., J. Strating, H. Wynberg, and W. Adam, *Tetrahedron Lett.* 169 (1972).
(4) Bolster, J., R.M. Kellogg, E.W. Meijer, and H. Wynberg, *Tetrahedron Lett.* 285 (1979).
(5) Hummelen, J.C., E.W. Meijer, and H. Wynberg, submitted to *J. Org. Chem.*

# DESIGN OF A CHEMILUMINESCENT AND BIOLUMINESCENT PHOTOMETER

P. Volfin
B. Lécuyer
B. Arrio
A. Dupaix
C. Fresneau

E.R. 118 CNRS, Bât. 432
Université de Paris-Sud
Orsay, 91405, France

The design of a chemiluminescent and bioluminescent photometer depends upon the methods of relating light output to reactants concentration : peak height or peak area. We present a device operating with peak height measurements method.

The mixing time is better than 0.5 sec. The limiting step of the measurement is the potentiometric recorder time response (0.5 sec.). This is supported by the recording of the light emission induced by the injection of ferrous ions in a phosphate buffer. This fast reaction cannot be recorded with commercially available apparatus.

The other main features of our device is its high sensitivity : $5 \cdot 10^{-12}$M ATP, $5 \cdot 10^{-7}$M glucose, $4 \cdot 10^{-13}$M glucose, $4 \cdot 10^{-13}$M horse radish peroxidase. Moreover, its compactness and 12 V battery supply allow measurements outside of laboratories.




ISBN 0-12-208820-4

# BIOLUMINESCENT IMMUNOASSAYS: USE OF LUCIFERASE-ANTIGEN CONJUGATES FOR DETERMINATION OF METHOTREXATE AND DNP[1]

Jon Wannlund and Marlene DeLuca

Department of Chemistry
University of California, San Diego
La Jolla, California

Many clinically important compounds have for years been determined by radioimmunoassays. Recently many new assays have been developed using fluorescence (1-4) or chemiluminescence (5, 6) measurements rather than radioactivity as the means of detection.

We have developed immunoassays for methotrexate and DNP at the picomole level using firefly and bacterial luciferase conjugates of these compounds.

*Methotrexate:*

Methotrexate was covalently linked to purified firefly luciferase using a water soluble carbodiimide. The product contained 2 methotrexates per mole of luciferase and approximately 60% of the enzymatic activity was retained in the conjugate. Antiserum raised againt methotrexate-hemocyanin was fractionated to obtain the IgG fraction. This antibody was then covalently attached to Sepharose. When the luciferase-methotrexate was incubated with the antibody-Sepharose, the enzyme-methotrexate bound to the Sepharose and remianed catalytically active. If free methotrexate was added along with the luciferase-methotrexate there was a competitive binding which was directly proportional to the amount of free methotrexate in the sample. It is possible to detect as little as 2.5 pmoles of free methotrexate using this procedure.

[1]*This work was supported by grants from The United States Army (DAAK 70-79-A-0019) and the National Institutes of Health (GM 24621).*




ISBN 0-12-208820-4

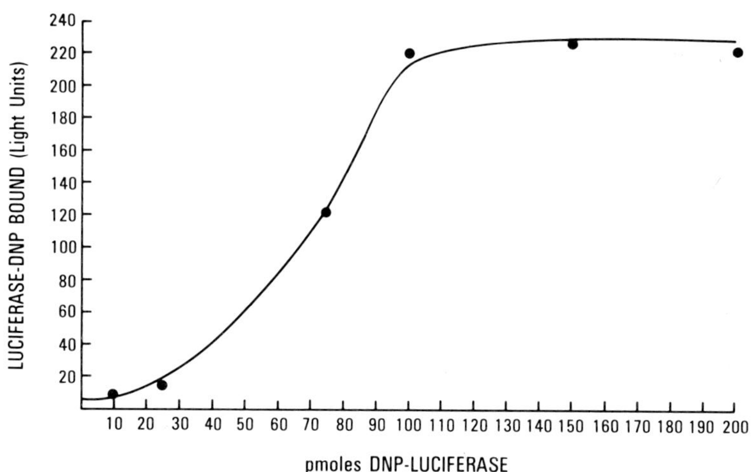


FIGURE 1A.

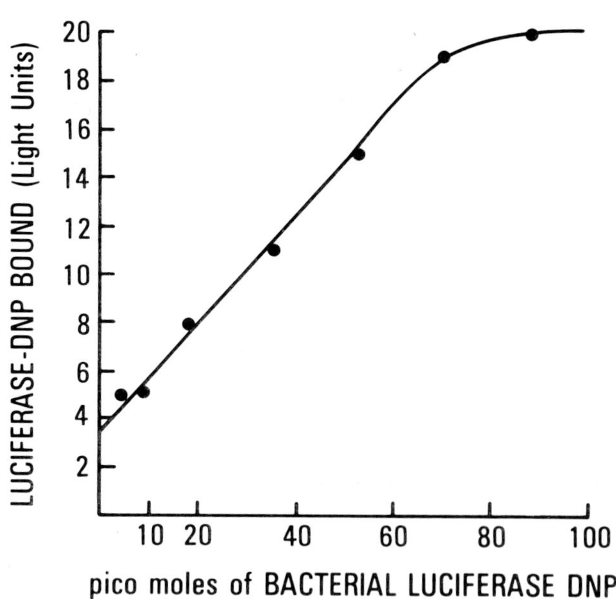


FIGURE 1B.

*FIGURE 1. Binding of luciferase-DNP to antibody-Sepharose: (A) firefly luciferase-DNP, (B) bacterial luciferase-DNP.*

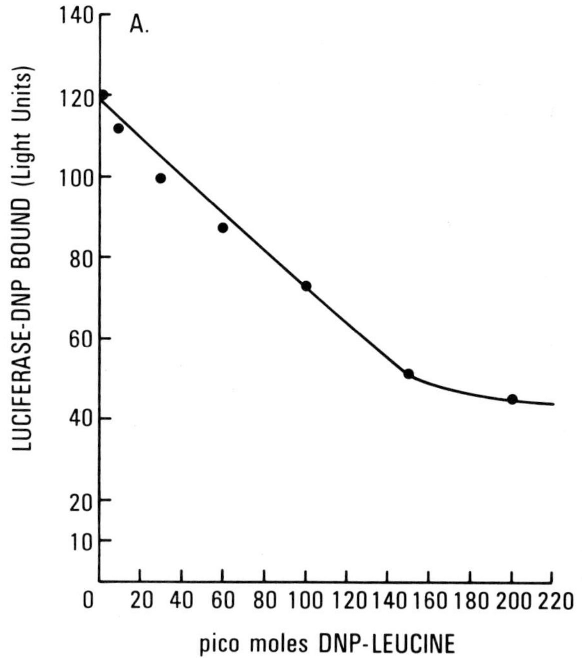


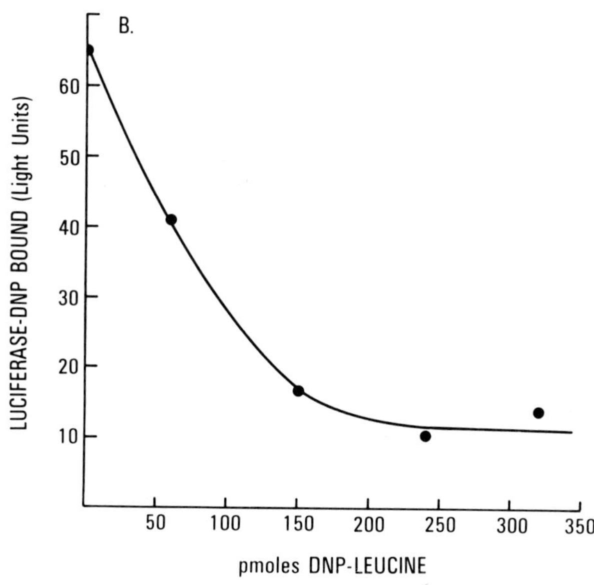


*FIGURE 2. Competitive binding curve of DNP-leucine with luciferase-DNP: (A) firefly luciferase, (B) bacterial luciferase.*

*DNP:*

Bacterial luciferase and firefly luciferase have been reacted with FDNB to produce DNP-luciferase conjugates in a ration of about one DNP per mole of enzyme. Both enzvmes retain 60-70% of their original catalytic activity. Crude anti-serum to DNP was covalently attached to Sepharose. Fig. 1 shows the binding curve of (a) firefly luciferase-DNP to anti-body-Sepharose and (b) bacterial luciferase-DNP.

Fig. 2 shows the competitive binding curve of free DNP-leucine with luciferase-DNP conjugates to the Sepharose antibody; (a) firefly luciferase-DNP (b) bacterial luciferase-DNP. The amount of bound luciferase was measured by bio-luminescence. Using the bacterial luciferase-DNP it is possible to measure 15 pmoles of free DNP. Similarly, with the firefly luciferase-DNP, 10 pmoles of free DNP can be readily measured. Preliminary data obtained with a glucose-6-phosphate dehydrogenase conjugate of DNP and measuring the bound enzyme activity with the bacterial oxidoreductase-luciferase, it is possible to measure femotomoles of free DNP. This system can theoretically be used to measure ligand in the attomolar range. These bioluminescent immunoassays could be used for measuring any ligand of interest if the anti-bodies are available.

## REFERENCES


1. Tsay, Y., L. Wilson, D. Rehfinder, and E. T. Maggio, Am. Soc. Microbio., 1979 Annual Meeting, Los Angeles, CA.
2. Capel, P. J. A., *J. Immunol. Methods,* 5, 165 (1974).
3. Nakamura, R. M., W. R. Dito, and E. S. Tucker III, (eds), *in* "Immunoassays in the Clinical Laboratory, p. 211. Alan R. Liss, Inc., New York (1979).
4. O'Donnell, C. M. and S. C. Saffin, *Anal. Chem.,* 51, 33A, (1979).
5. Halmann, M., B. Velan, and T. Sery, *Appl. Environ. Microbiol.* 34, 473 (1977).
6. Velan, B., H. Schupper, T. Sery, and M. Halmann, in "International Symposium on Analytical Applications of Bioluminescence and Chemiluminescence Proceedings," p. 4321, State Printing and Publishing Co., Westlake Village, CA (1979).
7. Raichard, D. W., and R. J. Miller, Jr., *Fed. Proc.* 38, 1013 (1979).

# ABSTRACTS

## F. Instrumentation and Reagents

# SIX-CHANNEL LUMINESCENCE ANALYZER FOR PHAGOCYTOSIS APPLICATIONS

*Fritz Berthold*
*Helmut Kubisiak*

Laboratorium Prof. Dr. Berthold,
Wildbad, Germany

*Martin Ernst*
*Herbert Fischer*

Max-Planck-Institut für Immunbiologie,
Freiburg, Germany

The observation of chemiluminescence associated with phagocytosis or cell stimulation usually requires the measurement and recording of the sample light output for an extended time period, typically up to one or even two hours. The results are generally presented as histograms showing the relative light output as a function of time. The interpretation of results is usually based upon the comparison of histograms for different samples.

So far, most investigators have used liquid scintillation counters operated in the "off-coincidence" mode, and at ambient temperature (1, 2), while Weidemann et al. (3) used a modified liquid scintillation counter with the sample temperature held constant at 37° C. Cheson et al. (5) describe a procedure where 4 samples are measured in a continuous sequence in a liquid scintillation counter. Andersen and



ISBN 0-12-208820-4

Brendzel (6) designed a special chemiluminescence spectrometer where up to 12 sample vials are placed in a carousel, so that each of the 12 samples can be positioned in front of a photomultiplier tube. The vials are maintained at a present temperature, and reagents are injected into the vial through rubber ports, so that luminescence can be followed immediately after reaction initiation.

We first considered the design of a fully automated version using the carousel approach, but then decided for a concept where several samples can be measured simultaneously rather than sequentially. This has two advantages: higher sensitivity and avoidance of time gaps before any specific sample is measured again.

Our multi-channel chemiluminescence monitor (figure 1) measures up to 6 samples simultaneously in discrete positions using six photomultiplier tubes.

The PM-tubes are operated in the photon-counting mode, since this allows much higher sensitivity and better long-term stability than DC-amplification.

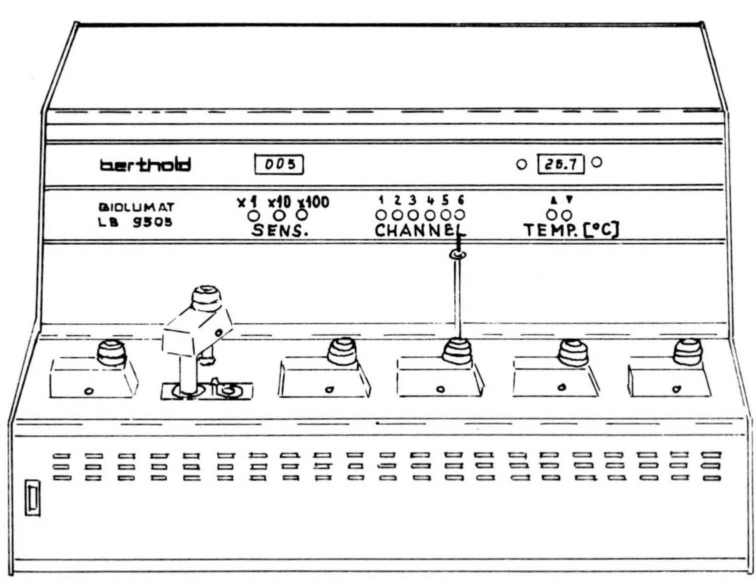


*FIGURE 1. Six-Channel Luminescence Analyzer with 6 measuring positions for simultaneous assay. One position is shown unlocked, one with microliter syringe in injection position.*

The sample vial dimensions are 12 mm dia., 47 mm height. Reflectors ensure optimum light collection on the photocathode. Sample temperature may be set between 25.6 and 51.2° C in steps of 0.1° C, while photocathodes can be cooled down to 6° C, using Peltier-elements, in order to reduce thermal noise.

Each measuring position is equipped with a photodiode, to be used as a light standard at 560 nm, this wavelength being appropriate for phagocytosis associated CL without Luminol "amplification" (1), but also for research using the firefly luminescence for ATP determination.

If Luminol is used, a sealed $^{14}C$ liquid scintillation source emitting at about 420 nm is used as a reference source.

PM-tubes with different cathodes, i.e. Bialkali, S-11 or Rubidium can be interchanged in order to allow an optium match between emission spectra and cathode sensitivity.

Analog and digital outputs are provided. The analog output is connected to a six-channel line recorder, so that count-rate vs. time diagrams are recorded in real-time, each sample being characterized by a different color. An example is shown on figure 2.

Computers may be interfaced on-line to the digital outputs. Only as research progresses will the best programs for data reduction and analysis evolve. Programs presently available with our single-channel instrument BIOLUMAT (7) which may also be useful for the new six-channel instrument include integrated counts between any preset time limits, ratio of different integrals, time at occurrence of maximum slope, curve maximum and half maximum after peak.

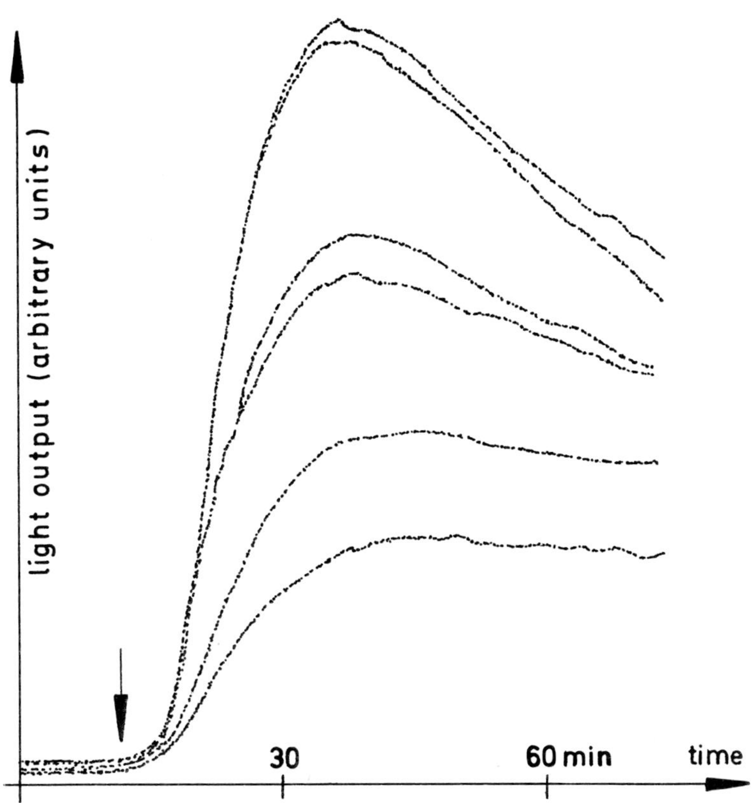


*FIGURE 2. Typical result with Six-Channel Luminescence Analyzer. 6 samples, each containing 100,000 macrophages in Eagle's medium, with 50 mM HEPES and 0.2 mM Lucigenin. At arrow position 5 to 15 µl Zymosan (from 50 mg/ml solution) were added. The histogram was recorded directly on a six-channel recorder (original in 6 colors).*

REFERENCES

1. Wilson, M. E. et al., J. Immunological Methods 23, 315 (1978).

2. Diaz, P. et al., Nature 278, 454 (1979).

3. Weidemann, M. J. et al., FEBS LETTERS 89, 136 (1978).

4. Hatch, G. E. et al., J. Experimental Medicine 147, 182 (1978).

5. Cheson, B. D. et al., J. Clinical Investigation 58, 789 (1976).

6. Andersen, B. R. and Brendzel, A. M. J. Immunological Methods 19, 279 (1978).

7. Laboratorium Prof. Dr. Berthold, Company Brochure.

# UTILIZATION OF MULTI-INJECTION FOR SAMPLE PROCESSING AND AUTOMATIC INTERNAL STANDARDIZATION (AIS) IN LUMINOMETRY

*S. E. Kolehmainen*

Lumac Systems, Inc.
Titusville, Florida

*H. Vanstaen*
*J. Vossen*

Lumac B.V.
Schaesberg, The Netherlands

## I. INTRODUCTION

The expansion of luminometry to routine use has created a need to a high sample output, accuracy and simplicity of procedures. The methods are gradually being used to tedious manipulations, thus it is necessary to increase the automation of sample processing. New commercial reagents and methods have increased the reliability of luminescent system, an internal standardization is recommended. The accuracy of this procedure is dependent on the reagent volume, kinetics of the reaction and the absoluteness of the standard. Unless all samples have a uniform quenching, the internal standardization has to be performed for all samples. This requires tow measurements and doubling of reagent consumption per sample. Too often the quenching is assumed to be uniform in the samples when it is not. This can lead to significant errors in the assays.

A new photon counter, BIOCOUNTER[R]M2010, offers features that simplify and automate different steps in the sample processing, measurement and standardization. This instru-



ISBN 0-12-208820-4

ment has three reagent injectors that provide reproducible dispensing amd mixing of reagents in the dark reaction chamber. The system is controlled by a freely programmable microcomputer. The combination of BIOCOUNTER and the microcomputer provides a multitude of functions that reduce the number of manual steps and other sources of errors in luminescent substrate and enzyme assays, such as:

- automatic extraction of substrates from cells
- automatic addition of coenzymes and enzymes
- timed incubation at controlled temperature
- automatic measuring sequence
- automatic internal standardization (AIS)
- computerized data processing

These features guarantee a high reproducibility of dispensing, measurement of light emission, and they allow internal standardization on every sample without increasing the consumption of the luminescent reagents.

## II. MULTI-INJECTION SYSTEM

Three built-in injectors dispense each 50µl or 100µl of a different reagent at a force that provides reproducible, instantaneous mixing of sample and reagents. The coefficient of variation of the volume is less than 0.5%, thus the accuracy and precision greatly exceed that of manual pipetting. The sequence and time delays of the dispensings can be programmed and controlled with the microcomputer. In coupled assays where a substrate is converted to ATP, NADH, NADPH, or $H_2O_2$ the incubation can be carried out in the temperature controlled reaction chamber over a precisely programmed time. Subsequent luminescence measurement and the internal standardization are all performed in the reaction chamber without having to remove the sample from the chamber until the whole measuring sequence is completed.

The internal standardization procedure can be applied to any type of kinetics because the microcomputer follows the reaction kinetics of the sample and adapts this to the standard too.

In the most sensitive assays fluorescence and phosphorescence of the vials, reagents and the sample can cause errors. This is eliminated in the procedures used by the BIOCOUNTER as the samples do not get exposed to light at any stage during the measuring sequence.

## III. PRACTICAL APPLICATIONS

### A. Measurement of viable cell count

The viable cell count based on the bioluminescent assay of ATP requires an extraction step prior to the measurement of luminescence and the standardization. All these steps can be performed in an automatic manner with BIOCOUNTER. The sequence of reagent dispensings and measurements are shown in Fig. 1.

A 100μl sample of bacterial supension (process water) is added to a cuvette. The cuvette is placed into the reaction chamber, and the chamber is closed by pushing down the dispensing arm that accommodates a nozzle of the three injectors. The measuring procedure is started by activating the microcomputer, which then takes over the operations of the photon counter. The first phase begins by the injection of the extraction reagent (NRB$^{R}$) into the sample, followed by a 10-second delay for the completion of ATP extraction. The second phase starts with the injection of the lucifer-luciferase reagent (LUMIT$^{R}$) into the extracted sample, proceeded by a 2-second delay and subsequent 10-second integration of the light intensity of the sample ATP. After the measurement of the sample itself, the third injector dispenses a known quantity of ATP standard, and after a two-second dealy the instrument intergrates the light emission for an other 10-second period. Upon completion of the second integration the computer calculates the concentration of ATP in the sample and converts that to the number of viable cells per milliliter.

### B. Other assays

The multi-injection system simplifies the operations and increases the accuracy of all luminescent assays where several reagents are needed for the measurement, such as substrate assays with a coupled system, enzyme assys, and the measurement of the adenylate energy charge (ATP,ADP,AMP).

## IV. CONCLUSIONS

The multi-injection system offers automatic sample prodessing in the stat assays, and it improves the accuracy of luminescent assays in several ways. Automatic dispensing of reagents, while the sample is in the reaction chamber, eliminates pipetting errors, provides exact delay-times for the incubation and extraction, allows an internal standardization for every sample, and eliminates problems caused by fluorescence and phosphorescence of vials, reagents and sample.

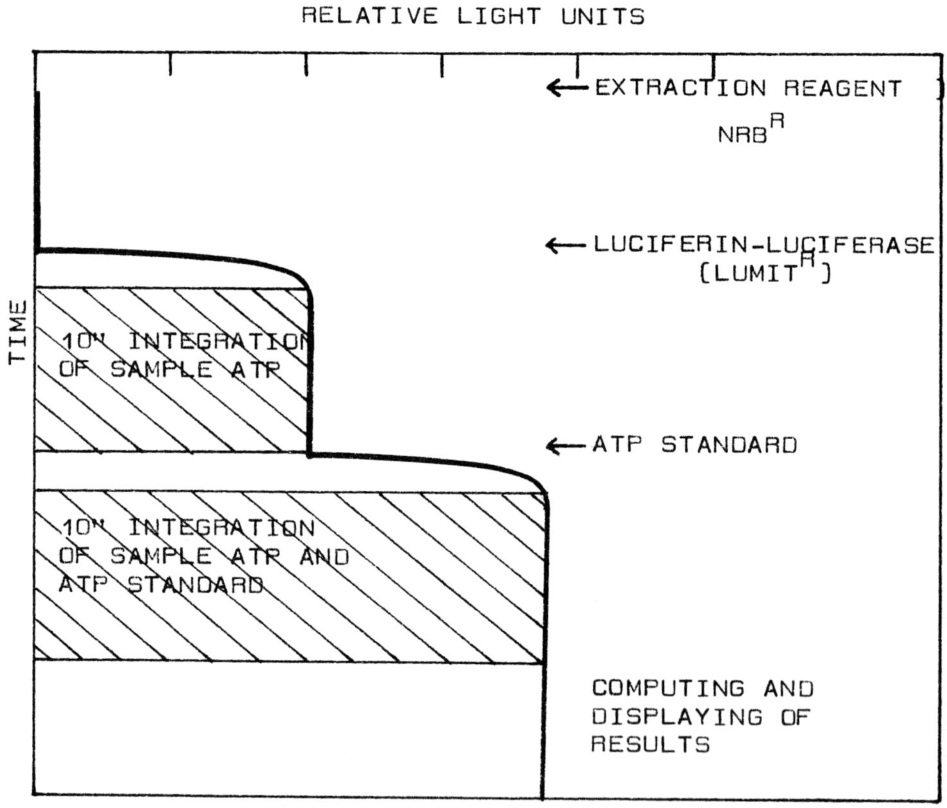

*FIGURE 1. Automatic extraction of ATP and internal standardization by injection of three reagents.*

# PRELIMINARY REPORT ON THE MECHANISM OF BIOLUMINESCENCE IN THE OCEANIC SQUID *SYMPLECTOTEUTHIS*

G. Leisman[1]

[1]Marine Biology Research Division A002
Scripps Institution of Oceanography
University of California, San Diego
La Jolla, California

F. I. Tsuji[1,2]

[2]Department of Biological Sciences
University of Southern California
and VA Medical Center Brentwood
Los Angeles, California

## I. INTRODUCTION

Although a sulfated form of an imidazo-pyrazine compound, which structurally resembles *Renilla* luciferin and the light-emitting chromophore of aequorin, has been proposed as the "luciferin" in the *Watasenia scintillans* bioluminescence reaction, the biochemistry of the luminescence reaction in self-luminous squids is still not understood (1-5). Attempts to obtain a fully active luminescence system *in vitro* have been thus far unsuccessful. In order to obtain some understanding of the molecular mechanism of the reaction, cell-free extracts of the large dorsal luminous organ of the oceanic squid, *Symplectoteuthis oualaniensis*, have been studied.




ISBN 0-12-208820-4

## II. MATERIALS AND METHODS

Specimens of the squid were collected by hand jigging from a ship off the coast of Waianae, Oahu, Hawaii. They were placed in individual tanks supplied with fresh running sea water and sacrificed as needed. The large dorsal luminous organ was removed by cutting with a pair of scissors. Later experiments showed that the excised luminous organ could be stored in a deep-freeze without significant loss of activity, so the large majority of experiments were carried out with frozen organs. The results did not vary with freshly obtained and frozen organs. The yellow photogenic tissue comprising the surface layer of the organ was dissected away and ground in 5 ml of cold 0.05 M Tris-HCl buffer, pH 7.2, using an all-glass homogenizer immersed in an ice bath. The homogenate was centrifuged at 10,000 x g using an Eppendorf Model 3200 centrifuge and the pellet was washed once in 0.05 M Tris-HCl buffer, pH 7.2, before being resuspended in the buffer. Washing once with buffer was found to slow down the rate of loss of activity of the suspension. The suspension was kept in an ice bath during the duration of the experiment. Light intensity was measured with an SAI Model 2000 ATP photometer and recorded on an Esterline Angus Miniservo strip chart recorder. The ATP photometer was calibrated using the light standard of Hastings and Weber (6). Reagent grade chemicals and glass-distilled water were used throughout the experiment. Various potentially stimulatory compounds were dissolved in 200 µl of 0.05 M Tris-HCl buffer, pH 7.2, directly into 100 µl of the supernatant or pellet suspension.

## III. RESULTS AND DISCUSSION

Light emission from the supernatant was observed to be almost negligible, whereas the pellet suspension emitted light of constant low level intensity which gradually decayed over a period of 5 hours. The light emission was presumed to be coming from subcellular particles produced by the grinding inasmuch as electron microscopic examination of the pellet showed no intact cells or organelles to be present. Mixing and substitution experiments revealed that the potassium ion was a powerful stimulant of light emission from the suspension, producing a flash which was followed by a long-lasting glow; on the other hand, the supernatant was inactive. Injecting the same volume of Tris-HCl buffer

*TABLE 1. Effects of Different Cations on Light Intensity*

| Cation[a] | Initial Maximal Light Intensity[b] |
|---|---|
| $K^+$ [c] | 7.45 |
| | 7.78 |
| $Na^+$ | 3.20 |
| | 3.10 |
| $Li^+$ | 0.19 |
| | 0.19 |
| $Mg^{++}$ | 0.42 |
| | 0.21 |
| $Ca^{++}$ | 0.15 |
| | 0.08 |
| $K^+$ [c] | 2.79 |

[a] *All cation solutions were prepared by dissolving their chloride salts in 0.05 M Tris-HCl buffer, pH 7.2; the final concentration after injection was 333 mM.*

[b] *In light units; one light unit was equivalent to 2.13 x $10^{10}$ quanta/second.*

[c] *Last potassium assay was conducted two hours after the initial potassium and sodium determinations; during this interval the activity of the suspension decayed to approximately 37% of the initial activity. The other cations were tested during this two hour period.*

without potassium or other ions into a control suspension of the same volume caused only a slight decrease in light intensity due to volume dilution. At equal concentrations, potassium ion stimulated an initial maximal light intensity about 2-3 times greater than sodium ion (Table 1). Lithium, calcium, and magnesium ions had only a slight stimulatory effect, whereas the ammonium ion had a moderate stimulatory effect, on light emission. The results obtained by injecting potassium ions at two different intervals into aliquots of the same suspension provided an estimate of the spontaneous loss of activity due to standing in the ice bath. This amounted to a loss of about 67% inactivity over a period of two hours (Table 1). Injections of EGTA, $NADH_2$, and $NADPH_2$ had no effect on light emission. Sudden decrease in pH from 7.2 to

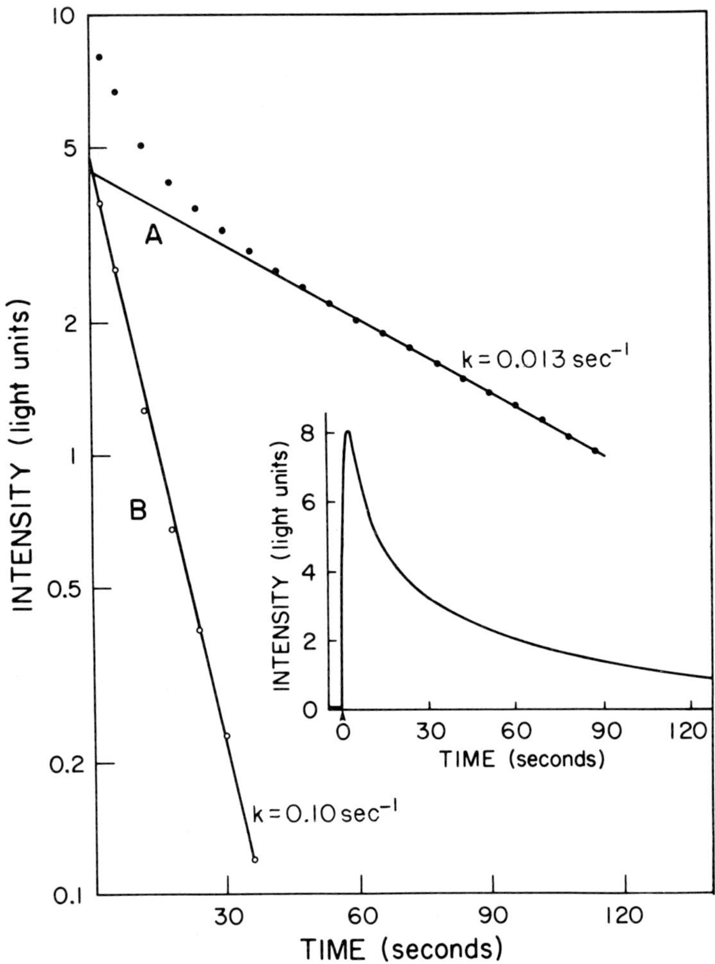


*FIGURE 1. Decay kinetics of potassium stimulated bioluminescence. The inset shows injection at time zero (arrowhead) of potassium chloride dissolved in 0.05 M Tris-HCl buffer, pH 7.2, into pellet suspension yielding a final concentration of 600 mM potassium ion. The subsequent decay curve is shown plotted as the logarithm of the light-intensity versus time. The linear portion of this decay curve has been extrapolated to time zero (curve A) and the difference values used to plot curve B. One light unit equals 2.13 x $10^{10}$ quanta per second. k's are first order decay constants.*

4.0, induced by injecting an HCl solution, resulted in a slight drop in light intensity, while a sudden increase in pH from 7.2 to 9.5, caused by injecting an $NH_4OH$ solution, also produced a decrease in light emission.

The light emission resulting from injecting potassium ions into the suspension appeared to be resolvable into two components, one decaying faster exponentially than the other (Fig. 1). Similarly, the light stimulated by sodium ions also showed a two component decay. The optimum potassium ion concentration for light emission was between 0.5-0.6 M. It is of interest to note that Girsch, *et al.* (5) observed a simple first order decay of light emission, which was spontaneous and long-lasting, from intact light organ granules of the related squid, *Ommastrephes pteropus*.

Attempts to demonstrate the luciferin-luciferase reaction proved unsuccessful, as has also been reported by others (3,5). Neither the injection of a fresh extract of the organ into a decayed suspension nor the injection into a homogenate of a methanol extract evaporated to dryness under argon atmosphere and resuspended in Tris-HCl buffer resulted in increased light emission. The inability to separate a luciferin and luciferase from photogenic tissues of the animal suggests that the essential light-emitting components are membranebound.

Flushing experiments with oxygen-free argon, prepared by passing 99.99% pure argon over hot copper filings, demonstrated that the reaction has an absolute requirement for molecular oxygen. Similar results have been reported by Herring (5) in the case of *O. pteropus* bioluminescence.

## IV. CONCLUSIONS

The present results suggest a strong dependent relationship between potassium and/or sodium ions and light intensity in the *Symplectoteuthis* bioluminescence reaction. The reaction possibly involves two light-emitting species, one decaying more rapidly than the other. The essential components of the reaction appear to be membrane-bound. The reaction requires molecular oxygen. The dependence on potassium and/or sodium ions suggests a possible mechanism for control of the bioluminescence reaction by the animal.

ACKNOWLEDGMENTS

The authors express their gratitude to Dr. Richard E. Young, Department of Oceanography, University of Hawaii, for arranging for laboratory facilities on board the R/V *Kana Keoki* and for advice and encouragement during the course of the work. This work was supported by N.S.F. research grant No. PCM 79-21658 (F.I.T.) and by O.N.R. contract No. N00014-80-C-0066 (Kenneth H. Nealson).

REFERENCES

1. Goto, T., H. Iio, S. Inoue, and H. Kakoi, *Tetrahedron Lett. 26,* 2321-2324 (1974).
2. Inoue, S., S. Sugiura, H. Kakoi, K. Hasizume, T. Goto, and H. Iio, *Chem. Lett.,* pp. 141-144 (1975).
3. Inoue, S., H. Kakoi, and T. Goto, *Tetrahedron Lett. 34,* 2971-2974 (1976).
4. Inoue, S., H. Taguchi, M. Murata, H. Kakoi, and T. Goto, *Chem. Lett.,* pp. 259-262 (1977).
5. Girsch, S. H., P. J. Herring, and F. McCapra, *J. Mar. Biol. Ass. U. K. 56,* 707-722 (1976).
6. Hastings, J. W., and G. Weber, *J. Opt. Soc. Am. 53,* 1410-1415 (1963).

# A NOVEL RAPID-MIXING APPARATUS FOR STUDIES OF CHEMILUMINESCENT REACTIONS

*Jerzy D. Meduski*
*Jerzy W. Meduski*
*James De La Rosa*

Nutritional Research Laboratory
University of Southern California
School of Medicine
Los Angeles, California

## I. INTRODUCTION

Since the 1920's when Hartridge and Roughton (1) introduced flow techniques as a substitute for "pipette and shake" numberous instruments were designed to mix and measure reactants in solution quickly. The trend in these past years has been for faster and smaller volume instruments. These have worked well for absorbance and fluorescence measurements; however, in the case of chemiluminescent reactions the need for large volumes and light-tight housings made most of the designs unsuitable. In these pages we shortly describe our instrument whihc was designed specifically for chemiluminescent studies and possesses some novel features.

## II. EXPERIMENTAL: DESIGN AND CONSTRUCTION

The apparatus presented here consists of a two-syringe drive assembly, (Fig. 1, A) and a channel assembly made using two teflon plates, (Fig. 1, B). These two parts are made light-tight by an aluminum cover (not shown).



ISBN 0-12-208820-4

### *A. The Syringe Drive Assembly*

Two air-tight syringes (Hamilton Co. 1000 series; Fig. 1, 3) are mounted to the main aluminum support frame. These syringes are filled through the teflon three way valve (Hamilton Co. 3mff3, Fig. 1, 4) using large reagent filled syringes attached directly. The drive sytringe plungers held by adjustable brackets to the brass drive block (Fig. 1, 2). The drive block runs in a groove on the main support frame, driven by the air cylinder (Bimba Mfg. model D2693A, Fig. 1, 1). Air valves (Miller Fluid Power Corp.) were used with compressed nitrogen from a tank to power the air cylinder for slow upward filling and rapid downward drive. The downard drive could expel total fractional syringe volumes.

### *B. The Fluid Channel Assembly*

Two teflon plates form the fuluid channels in this assembly. The top plate has three Kel-F female Luer threaded adapters attached. Two of these are for connecting the drive syringe valves (Fig. 1, 5) to the fluid channels. The channels are formed by the opposition of the teflon plates (Figs. 2, B & C). The observation channel (Fig. 2, C) is sealed by the quartz window in the bottom aluminum plate (Fig. 2, D). Fluid, after passing the observation channel into a resevoir attached to the fluid drain (Fig. 1, 6).

These components are joined by several screws and bolts. For high pressure use the teflon plates are sealed with teflon gaskets. The entire rapid mixing apparatus is attached to a cooled photomultiplier housing (Products for Research, TE104TS-RF, Fig. 2, E). After the drive syringes are filled an aluminum cover is sealed onto the bottom plate (Fig. 2, D) and then the air cylinder is operated externally.

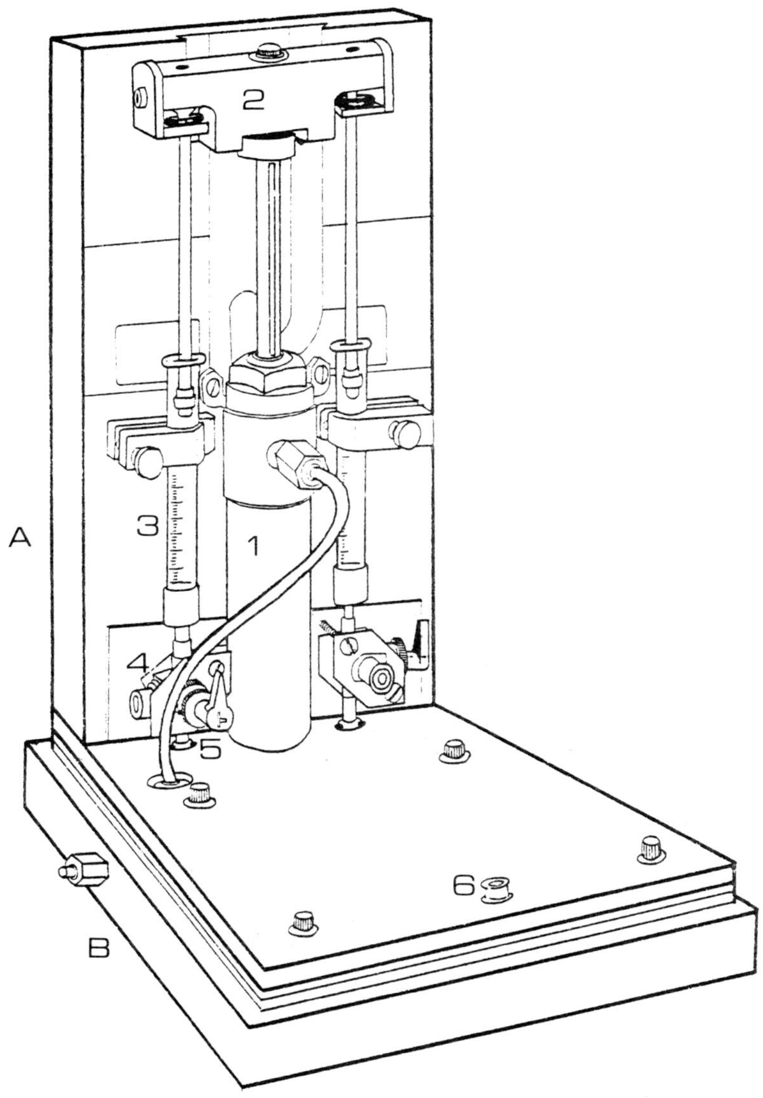


*FIGURE 1. Rapid Mixing Apparatus, Detailed view. Syringe drive assembly, A, consists of: air cylinder (1), drive block, (2), air-tight syringes, (3), a Luer three-way teflon valve, (4), connected to a Luer-to-thread adapter, (5), going into the channel assembly, B, with fluid drain, (6).*

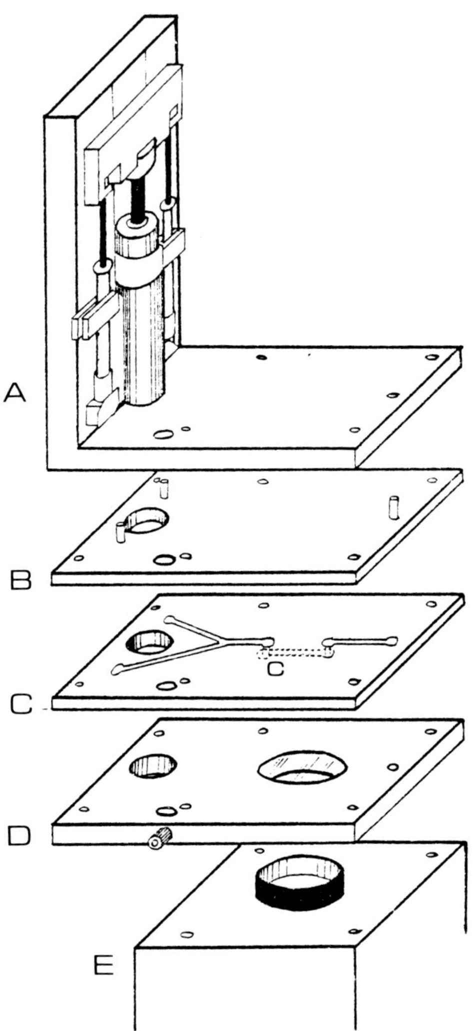


*FIGURE 2. Rapid Mixing Apparatus, Exploded View. Syringe drive assembly, A, teflon plate with three female Luer mounts, B; teflon fluid channel plate, C, with observation channel on underside, C; bottom plate with quartz window, D; photomultiplier tube housing, E.*

## III. DISCUSSIONS

The apparatus described here has been constructed using only inert materials in parts which come in contact with reactants. It may be disassembled and cleaned very easily.

Our apparatus contains only pneumatic and not electrical regulatory devices: no electrical noise is generated.

The versatility of the apparatus is increased by introducing exchangeable syringes and teflon plates. The capacity of the syringes available for use can vary from 0.1 ml to 5.0 ml. The teflon plates have many configurations. A plate (Figure 2, C) may have a minimum volume of 0.2 ml and a straight observation channel or a large volume and zig-zag observation channel. Several mixing devices may be built in.

Operation with 0.2 ml observation volume, 5 ml drive syringes, and 80 cm. $sec^{-1}$ drive velocity gives 6 msec dead time.

## IV. REFERENCE

1. Hartridge, H. and Roughton, F. J. W., Proct. Roy. Soc. London, A104, 376 (1923).

# SIGNALS OF CHEMILUMINESCENCE EMITTED BY SPLEEN CELLS AND BONE MARROW MACROPHAGES AFTER STIMULATION WITH MITOGENS AND PARTICULATE SUBSTANCES

Sybille Müller
Sabine Falkenberg
Robert A. Fromtling
Anne M. Fromtling

Robert Koch-Institut
Bundesgesundheitsamt
Berlin (West), Germany

Volker Klimetzek

Department of Microbiology
Freie Universität Berlin
Berlin (West), Germany

## I. INTRODUCTION

It is now generally accepted that an optimal antigen-induced immune response or its suppression requires the collaboration of T- and B-lymphocytes and macrophages (1). The mechanisms by which such complex functions are carried out remain poorly understood. The generation of activated specimens of $O_2$ ($OH^\cdot$, $O_2^-$, $H_2O_2$) has been suggested to be involved in the early events of lymphocyte-macrophage interactions (2,3).

Production of activated specimens of $O_2$, as measured in the luminol-dependent CL[1], has been found in stimulated thymocytes, lymphocytes, macrophages and granulocytes (3,4). Since these signals of CL are characteristic for each cell

[1]Chemiluminescence



ISBN 0-12-208820-4

type, it should be possible to study the kinetics of activation in a cell mixture and discriminate between the different cell populations involved. We, therefore, studied the induction of the luminol-dependent CL in spleen cells of normal and pathologically altered mice by mitogenic and particulate substances and compared these signals of CL with those of pure bone marrow derived macrophages.

## II. MATERIALS AND METHODS

Spleen cells were prepared from 6 to 8 week old normal and uremic BALB/c mice. Uremia was induced by injection of 0.15 ml glycerol (90%) intramuscularly 24 hr prior to sacrifice. Bone marrow derived murine macrophages were grown on teflon-coated dishes for 8 to 15 days and removed at various times as described by Klimetzek and Remold (5).

The different cell populations were suspended at a concentration of 1 x $10^7$ cells/ml either in carbonate-free Dulbecco's MEM buffered with 50 mM HEPES (pH 7.2) or in Dulbecco's PBS. Both media were supplemented with 10% FCS. the luminol dependent CL was measured in the Biolumat LB 9500. For each assay, 0.5 ml of the cell suspension were preincubated with 10 ul luminol (5-amino-2,3-dihydro-1,4-phthalazinedione, Sigma, München; 1 mg/ml in PBS) for 10 minutes at 37$^{\circ}$C. The CL was evoked by adding Con A (Serva, Heidelberg), PHA (Wellcome, Burwedel), latex beads (Serva, no. 41951), opsonized live *Staphylococcus aureus* (SG 511, donated by Dr. Wecke, Berlin) and zymosan (donated by Dr. Ernst, Freiburg) at various concentrations. Inhibition studies of induced CL were done with SOD[2] (3000 units/mg protein, Sigma), catalase (39,000 units/mg, Serva) and α-methylmannoside (Sigma). The magnitude of CL was expressed as counts per minute using a calculator (Hewlett-Packard 97 S I/O, Frankfurt) connected to the Biolumat.

## III. RESULTS

Spleen cells of normal mice stimulated with the mitogens Con A and PHA showed two slightly different signals of CL (Fig. 1a). In both cases, there was an initial rise in CL within one minute, which decreased after 4 to 5 minutes. This activity was always more prominent after adding PHA than Con A The first peak of activity was then followed by a second peak

[2]Superoxide-dismutase

of CL, whose maximum was reached after 8 to 10 minutes in the case of Con A and was diminished after 20 to 30 minutes, whereas in the case of PHA, the maximum was found only after 30 to 40 minutes. The particulate substances latex, zymosan and *S. aureus* induced always identical signals of CL, varying only in their magnitudes. A typical example is shown in Fig. 1b. It consisted of a short Cl signal of about 4 to 5 minutes (identical to that found after stimulation with mitogens) and a second broad peak of about 20 minute duration. The amount of CL emitted by the cells was dependent on the amount of mitogens or particulate substances added, as shown for the concentration dependence of CL response to Con A (Fig. 2). The CL of normal spleen cells induced by all tested concentrations of mitogens and particulate substances could be suppressed by the prior addition of 0.2 ml/ml SOD and 100 mM α-methyl mannoside but not by 0.8 mg/ml catalase.

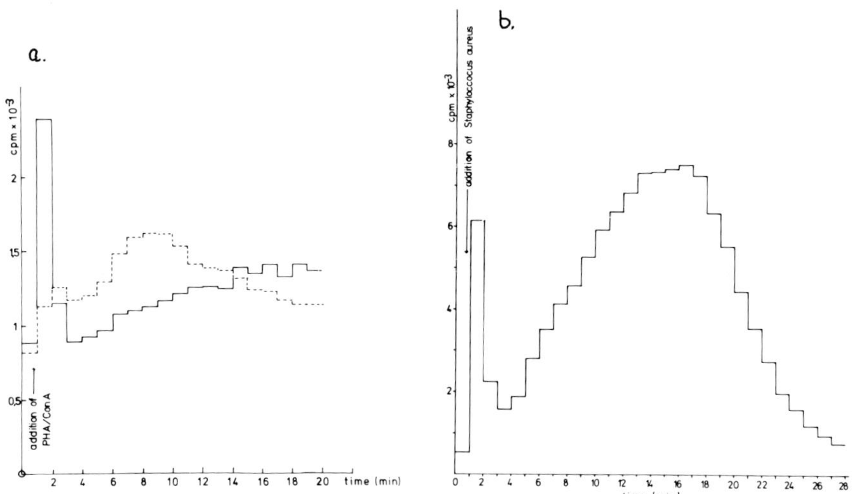


*FIGURE 1a. CL responses of spleen cells ($10^7$ cells/ml) after stimulation with PHA (——) or Con A (---). Concentration of both 0.1 mg/ml.*
*FIGURE 1b. CL responses of spleen cells to 5 x $10^8$ S. aureus/ml.*

When spleen cells of uremic instead of normal mice were used a 5 to 10 fold increase in CL levels was found. The characteristic of the shape of CL signals, however, did not change. The CL of uremic spleen cells showed the same behavior to inhibition by SOD, mannoside and catalase.

The addition of the same concentrations of mitogens and particulate substances to 8 to 15 day old bone marrow derived macrophages also induced luminol-dependent CL signals in these cells. These differed, however, characteristically in shape, duration and, in part, in the sensitivity to inhibition by SOD, α-methyl mannoside and catalase. In macrophages the mitogens PHA and Con A induced identical signals of CL. They consisted of only one peak of 4 to 5 minute duration which was found within one minute after addition of the mitogen (Fig. 3a). No second peak of CL could be found in 8 to 15 day old macrophages.

The mitogen-induced signal (μ-signal) was sensitive to inhibition by 100 mM α-methyl mannoside and 0.2 mg/ml SOD, but not by 0.8 mg/ml catalase. The particulate substances latex and zymosan induced identical signals which consisted of two peaks (Fig. 3b), a short one of 4 to 5 minutes duration, which was evoked within one minute after the addition of the substances, and a second broad peak, which declined after 15 to 20 minutes. The first peak was inhibited by 100 mM α-methyl mannoside and not by catalase (μ-signal).

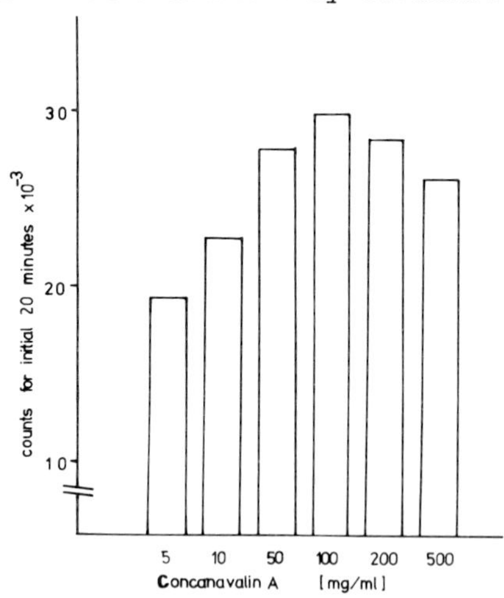


*FIGURE 2. Amount of CL emitted by spleen cells ($10^7$ cells/ml) after stimulation with 5 - 500 mg Con A/ml.*

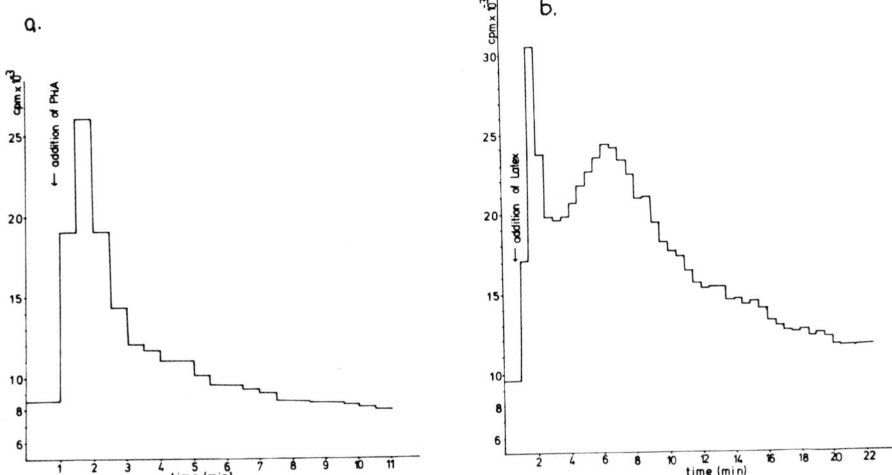


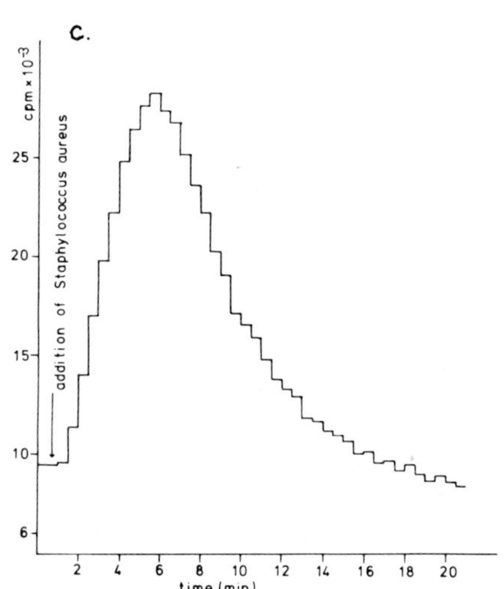


*FIGURE 3. CL signals of 11 day old bone marrow derived macrophages ($10^7$ cells/ml) in response to stimulation by (a) 0.1 mg PHA/ml; (b) 200 latex particles/macrophage; and (c) 5 x $10^8$ S. aureus per ml.*

The second peak, on the other hand, was inhibited by 0.8 mg/ml catalase and not by α-methyl mannoside (κ-signal). Both peaks were, however, inhibited by 0.2 mg/ml SOD. Addition of *S. aureus* induced in macrophages only one braod signal of CL whose maximum was reached after 6 minutes (Fig. 3c). It declined after 16 to 20 minutes. Inhibition studies with catalase and α-methyl mannoside showed that the signal of CL consisted mainly of the catalase-sensitive κ-signal.

When spleen cells and bone marrow macrophages were incubated with PHA or Con A prior to the addition of the particulate substances latex or zymosan, the induction of CL by these particulate substances could be inhibited. There were, however, striking differences in the amount of mitogens needed to inhibit the induction of CL by particulate substances. The addition of 0.2 mg/ml Con A to macrophages suppressed completely the CL signals in response to latex 200 beads/cell) and zymosan (1 mg/ml). In spleen cells the CL signal induced by the latex (200 beads/cell) could be inhibited by the prior addition of 0.2 mg/ml Con A, whereas the CL signal induced by zymosan (1 mg/ml) required a Con A concentration of at least 1 mg/ml for its inhibition.

## IV. DISCUSSION

Characteristic signals of CL were obtained by incubating spleen cells or macrophages either with the T-cell mitogens PHA or Con A or the particulate substances latex, zymosan and *S. aureus*.

In pure bone marrow macrophages three different signals of CL were found. The different pattern of these signals can be explained by the differential composition of two elementary signals (μ- and κ-signal) which are characterized by shape and sensitivity to inhibition by mannoside and catalase.

In spleen cells two main types of CL signals were found. The first peak of the two types of CL signals is sensitive to mannoside and therefore seems to be exclusively emitted by splenic macrophages. If the μ-signal was inhibited by mannoside, the following peak of the CL signals was also lost, showing that macrophages are essentially involved in the activation of lymphocytes. Mitogens also induced in spleen cells a second peak not found in macrophages. Since pure mitogen-stimulated lymphocytes emit only weak signals of CL in comparison to granulocytes or macrophages (3), we conclude that the second peak seen in spleen cells resulted from macrophage-lymphocyte interactions. The differences in CL response to Con A and PHA may be attributed to the activation of different T-cell subpopulations. Also, the second peak seen in spleen cell preparations after stimulation with particulate substances is not solely a macrophage emitted signal, since it cannot be inhibited by catalase, as found for the corresponding peak of pure stimulated macrophages. It can therefore be concluded that particulate substances, which preferentially act on macrophages, lead also to cell-cell interactions, which can be monitored in the luminol-dependent CL assay.

## ACKNOWLEDGEMENTS

This research project was supported by the Georg and Agnes Blumenthal Stiftung and the Deutsche Forschungsgemeinschaft.

## REFERENCES

(1) Rosenthal, A.S., J.T. Blake, J.J. Ellner, D.K. Greineder, and P.E. Lipski in "Immunobiology of Macrophages" (D.S. Nelson, ed.) p. 131. Academic Press, New York (1976).

(2) Metzger, Z., J.T. Hoffeld, and J.J. Oppenheim, *J. Immonol.* 124, 983 (1980).

(3) Wrogemann, K., M.J. Weidemann, U.P. Ketelson, H.Weckerle, and H. Fischer, *Eur. J. Immunol.* 10, 36 (1980).

(4) Allen, R.C., *Biochem. Biphys. Res. Commun.* 47, 679 (1980).

(5) Klimetzek, V., and H. Remold, *Cellul. Immunol.* 52 (1980) in press.

CHEMILUMINESCENCE OF LUCIGENIN IN MICELLAR SYSTEMS

Constantine M. Paleos
George Vassilopoulos
John Nikokavouras*

Chemistry Department
Nuclear Res. Center "Demokritos"
Aghia Paraskevi Attikis
Athens, Greece

Recently there has been considerable interest in the structure of micelles (1,2) as well as on their effect in catalyzing certain reactions (2,3,4). Micelles are easily and reproducibly prepared and have been used for the organization or rather the preferential solubilization of various substrates. In the light of these orientational and/or solubilization effects, catalytic (2,3,4), photophysical (5,6), photochemical (7,8) and primarily biological processes (2) may be rationalized. Specifically, for the last processes, micelles have served as models. Conversely, it is also possible to use physicochemical evidence to deduce the structural and dynamic properties of micelles.

It seems therefore reasonable to investigate chemiluminescent reactions in micellar media. In this first report, the classical lucigenin light reaction was selected for study for the following reasons: (a) There are no gaseous reactants involved that might cause foaming of the surfactant solution; (b) Water-insoluble reaction intermediates might be solubilized by the micelles. (c) Although N-methylacridone (NMA) (9,10) is well-established as the primary excited product, probably arising from decomposition of a 1,2-dioxetane (10-12), energy transfer to other species results to chemiluminescence spectra which do not match with the NMA fluorescence spectrum. It was hoped, therefore, that if the decomposition step, producing the excited NMA, were taking place inside a micelle, energy transfer might be hindered due to seclusion of the primary excited species, consequently,



ISBN 0-12-208820-4

the chemiluminescence spectrum would more closely resemble that of NMA fluorescence; (d) If the reaction would occur within the micelle, information could be obtained, regarding the effect of the polarity of the medium (core or Stern region) on the chemiluminescence efficiency; (e) Although, the present system is not particularly well-suited, it could be possible at high lucigenin per micelle ratios (more than one lucigenin molecule per micelle) to observe excimer emission and (f) Chemiluminescence could offer just another method for determining the critical micelle concentration (cmc) of surfactants.

The chemiluminescence intensity-time (I-t) diagrams as well as the chemiluminescence, fluorescence and absorption spectra were obtained as described earlier (13). The I-t diagrams were recorded on addition of sodium hydroxide solution (1 ml, 5N) and hydrogen peroxide (1 ml, 10%) to lucigenin in the appropriate aqueous cetyltrimethylammonium bromide (CTAB) solution (20 ml, $2 \times 10^{-4}$M). Each of these experiments was followed by a blank in the absence of CTAB (homogeneous experiments) and the results were compared.

The chemiluminescence intensity was not higher in the micellar CTAB solution but the duration of emission was increased, resulting in a dramatic enhancement in chemiluminescence efficiency as compared to the one of the homogeneous solution. This is shown on Table I where the ratio of quantum yields under micellar ($Q_{mic}$) and homogeneous ($Q_{hom}$) conditions is shown as a function of CTAB concentration and also as a function of lucigenin to micelle ratio. The same results are shown diagramatically in Fig. 1.

These results indicate that the presence of micelles in surfactant solutions is necessary for increased quantum efficiency and that in CTAB solutions below cmc, the efficiency is lower than that of the homogeneous aqueous solution. In addition, this increased efficiency is associated with a significant number of micelles in solution or otherwise, to a ratio of lucigenin per micelle equal to two or less.

For the rationalization of these results the dioxetane mechanism is considered in conjunction with the "porous cluster micellar model" proposed lately by Menger (1). This model is spherical with a rather disorganized structure in which the so-called Stern region (previously called Stern layer) apart from the polar heads of the surfactant, also includes a significant fraction of the alkyl chains which are free to move about. In this region, there is considerable alkyl chain-water contact rather than an ion double layer shielding the non-polar nucleus. Thus, deep grooves are formed where the water and/or various substrates, could reach even beyond the sixth carbon atom of the chain. This area,

*TABLE I. Ratio of $Q_{mic}/Q_{hom}$ as a function of CTAB concentration or of lucigenin/micelle.*

| CTAB[a] (M) | $Q_{mic}/Q_{hom}$ | Lucigenin[b] / Micelle [c] |
|---|---|---|
| $5 \times 10^{-4}$ | 0.73 | - |
| $2.5 \times 10^{-3}$ | 1.5 | 8.00 |
| $5 \times 10^{-3}$ | 2.0 | 4.00 |
| $7.5 \times 10^{-3}$ | 2.7 | 2.60 |
| $1 \times 10^{-2}$ | 3.1 | 2.00 |
| $1.75 \times 10^{-2}$ | 3.0 | 1.14 |
| $2.5 \times 10^{-2}$ | 3.7 | 0.80 |
| $5 \times 10^{-2}$ | 3.3 | 0.40 |
| $1 \times 10^{-1}$ | 4.0 | 0.20 |

[a] *The cmc of CTAB is $9 \times 10^{-4}$ (14).*
[b] *Lucigenin concentration was always $2 \times 10^{-4}$M.*
[c] *This is the average number of lucigenin molecules per micelle, calculated on assuming an aggregation number of CTAB equal to 100 (15).*

by analogy with 4-dodecyl-1-methylpyridinium iodide may possess comparable polarity to ethanol (2,4), whereas the core, occupying only a 15-20% fraction of the micelle has a dielectric constant similar to that of pure hydrocarbons. It may now be assumed that NMA is located in the deep grooves of the Stern region, formed by the decomposition of an intermediate dioxetane; the excitation efficiency of the dioxetane decomposition is probably favoured by this medium which is less polar than water. In any case, it was not possible to attribute the increased chemiluminescence efficiency in the micellar media to increased fluorescence efficiency. Such an effect was not apparent from our fluorescence spectra and previous results (16).

The location of NMA in this area is verified by the absorption spectra. Thus, the NMA absorption spectrum, as the light reaction proceeds in 0.01 M CTAB solution, resembles

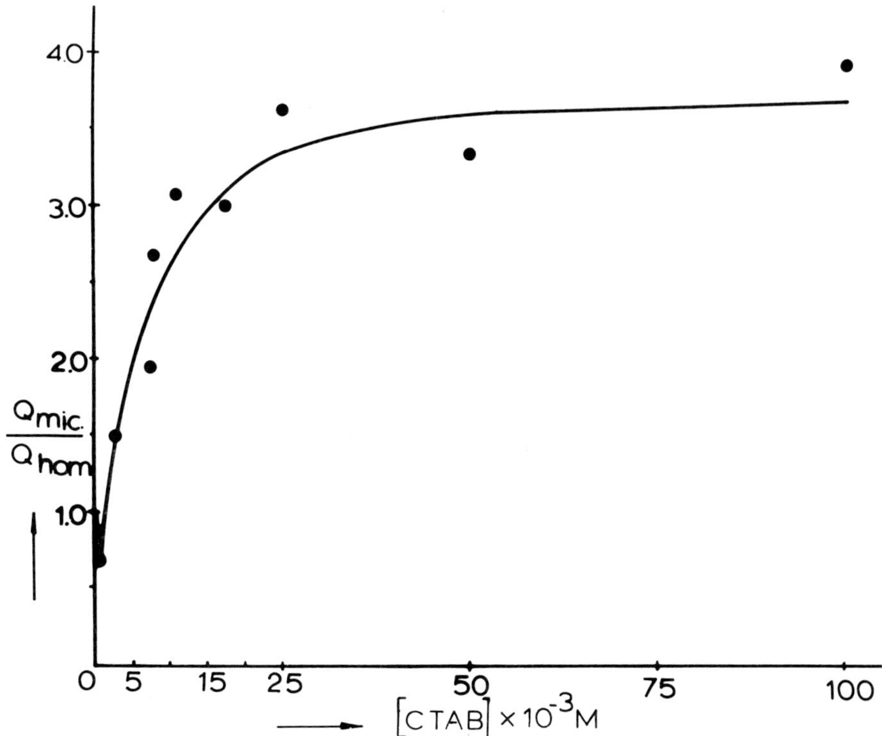


*FIGURE 1. Variation of $Q_{mic}/Q_{hom}$ as a function of CTAB concentration.*

that in ethanol and not those in hexane or water; as the polarity of the Stern region is comparable to that of ethanol, NMA has to emit from this area.

As predicted, the primary emission of NMA in the micellar medium was detected, from the very beginning of the light reaction, in its chemiluminescence spectrum. This is shown in Fig. 2. The effect was more pronounced at higher CTAB concentration or small lucigenin to CTAB ratios; even more affected in this manner were the fluorescence spectra, of the reaction mixture, in the course of the light reaction.

The addition of increasing amounts of sodium chloride in micellar experiments, further increased the quantum efficiency by about 30%, a fact which might be attributed to less quenching, due to the substitution of bromide by chloride ions in the Stern region.

Furthermore, the slope changes in Fig. 1 clearly indicate that chemiluminescence, or at least this reaction, could form the basis of another method for determining critical micelle

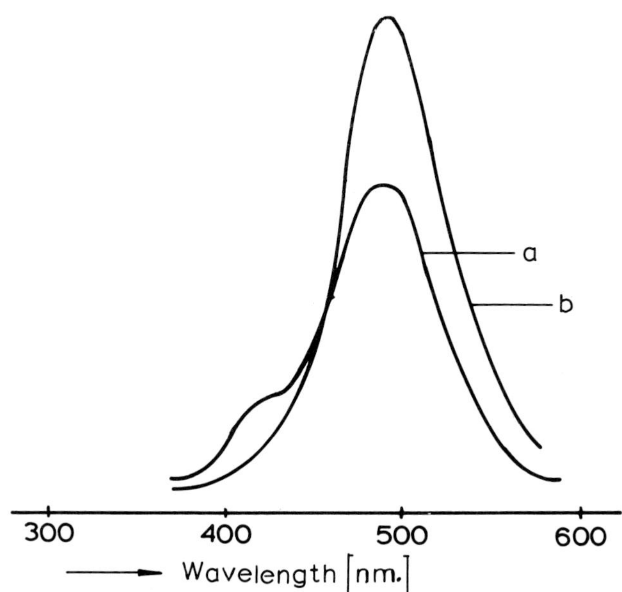


*FIGURE 2. Chemiluminescence spectra of the lucigenin light reaction in aqueous 0.1 M. CTAB (a) and aqueous (b) solutions.*

concentrations of surfactants.

Finally, the failure to detect excimer chemiluminescence at high lucigenin to micelle ratios should be attributed to the unsuitability of the primary emitter.

REFERENCES


(1) Menger, F.M., *Acc. Chem. Res.* 12, 111 (1979).

(2) Südhölter, E.J.R., G.B. Van de Langkruis, and J.B.F.N. Engberts, *Recl. Trav. Chim. Pays-Bas,* 99, 73 (1980) and references cited therein.

(3) Bunton, C.A. in "Techniques in Chemistry", Vol. X. Part II (J.B. Jones, C.J. Sih and D. Perlman, eds.) p. 731, J. Wiley and Sons (1976) and references cited therein.

(4) Cordes, E.H., *Pure and Applied Chem.* 50, 617 (1978).

(5) Thomas, K.J., *Acc. Chem. Res.* 10, 133 (1977).

(6) Kalyanasundaram, K., *Chem. Soc. Rev.* 7, 453 (1978).

(7) Turro, N.J.,K.C. Liu, M.F. Chow, *Photochem. and Photobiol.* 26, 413 (1977).
(8) Turro, N.J., *J. Amer. Chem. Soc.* 100, 7431 (1978).
(9) Totter, J.R., *Photochem. and Photobiol.* 3, 331 (1964).
(10) McCapra, F., and D.G. Richardson, *Tetrahedron Lett.* 3167 (1964).
(11) McCapra, F. and R.A. Hann, *Chem. Comm.*, 442 (1969).
(12) McCapra, F. in "Essays in Chemistry", Vol. 3 (J.N. Bradley, R.D. Billard, and R.F. Hudson, eds.) p. 101, Academic Press (1972).
(13) Nikokavouras, J. and G. Vassilopoulos, *Monats* 109, 831 (1978).
(14) Mukerjee, P. and K.J. Mysels in "Critical Micelle Concentration of Aqueous Surfactant Systems", Nat. Stand. Ref. Data Ser., Nat. Bur. Stand., USA (1971).
(15) Sepulveda, L., and R. Soto, *Makromol. Chem.* 179, 165 (1978).
(16) Legg, K.D. and D.M. Hercules, *J. Amer. Chem. Soc.* 91, 1902 (1969).

# PICO-ZYME F AND PICO-EX FOR MEASURING CELLULAR ATP ON A PICO-LITE LUMINOMETER

Cindy Sanville

Clinical Investigations
Packard Instrument Co., Inc.
Downers Grove, Illinois

## I. INTRODUCTION

All living cells contain adenosine triphosphate or ATP. This is the primary energy donor for most cellular processes. The measurement of ATP therefore gives us a handle on testing the presence and viability of cells in a medium. ATP is easily detected using firefly luciferase:

$$\text{ATP} + \text{luciferin} + O_2 \xrightarrow[\text{luciferase}]{Mg^{++}} \text{AMP} + \text{light} + \text{products}$$

Instrumentation is presently available that will detect and measure the light produced by this reaction.

ATP is usually located within a cell. Luciferase is not normally permeable to cell walls or cell membranes. In order to measure intracellular ATP, cells must be lyzed or made permeable to ATP so that luciferase can react with the ATP. This may be done in a variety of ways such as boiling buffers, acid extractions, chemical extractions, sonication, and others. Most of these methods involve time and difficulty.

In order to simplify cellular ATP measurements, Packard Instrument Company has developed reagents and instrumentation for doing such luminescence assays. PICO-ZYME F is a highly purified firefly luciferase containing luciferin and other cofactors. PICO-EX is a reagent that allows ATP to be released from a cell within seconds so that it may come in contact with PICO-ZYME F to produce light. The PICO-LITE is an instrument specially designed to do luminescence measurements including cellular ATP.




ISBN 0-12-208820-4

## II. MATERIALS AND METHODS

PICO-ZYME F, PICO-EX S (for ATP extraction from somatic cells), PICO-EX B (for ATP extraction from bacterial cells), and PICO-CHEC (an ATP preparation) were produced by Packard Instrument Company. All measurements were done on a Packard PICO-LITE Model 6100.

Cellular ATP measurements were done essentially as described in Packard's "ATP Detection System" Bulletin(1). ATP extractions were done by adding an equal volume of PICO-EX B or PICO-EX S (depending on cell type) to a cell suspension. Usually 50 µl PICO-EX B or PICO-EX S was added to 50 µl of cell suspension. Twenty microliters of this ATP extracted preparation was assayed in a 6 x 50 mm cuvette by adding 50 µl of PICO-ZYME F to the 20 µl of extracted cells. Samples were counted 10 seconds, then 10 µl of PICO-CHEC (1.0 x $10^{-6}$ molar ATP) was injected into the cuvette and counted another 10 seconds. The injection of PICO-CHEC served as an internal standard to correct for light quenching and enzyme inhibiting factors.

## III. PICO-ZYME F

PICO-ZYME F was shown to have a low decay, high response curve enabling measurements to be made without critical timing (Fig. 1). This also allows the use of an internal standard in the same cuvette. Firefly luciferase in a more crude form has a much more rapid decay (Fig. 2). PICO-ZYME F also produces a linear response between 1.0 x $10^{-11}$ moles and 5.0 x $10^{-16}$ moles ATP when measured on a PICO-LITE (Fig. 3).

## IV. PICO-EX

PICO-EX is an efficient reagent. It was shown to extract greater than 80% of the ATP from various cell types compared to boiling tris (Table I) with linear results over several orders of magnitude (Figs. 4 and 5).

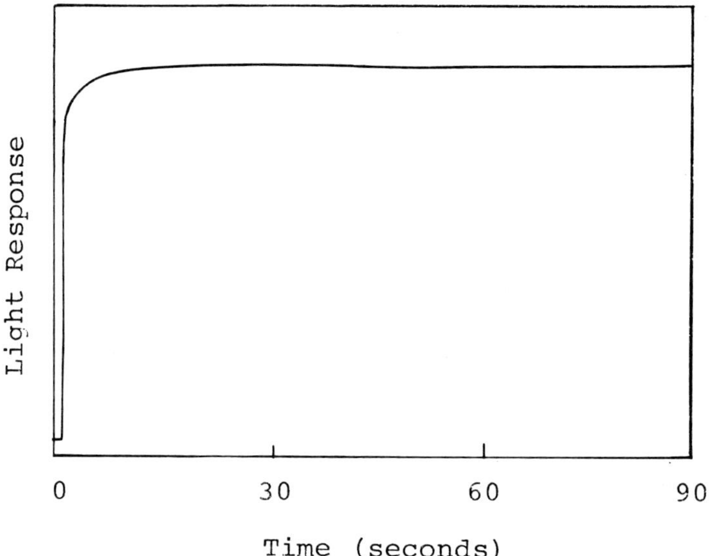


*FIGURE 1. PICO-ZYME F light response with 1 x $10^{-11}$ moles ATP.*

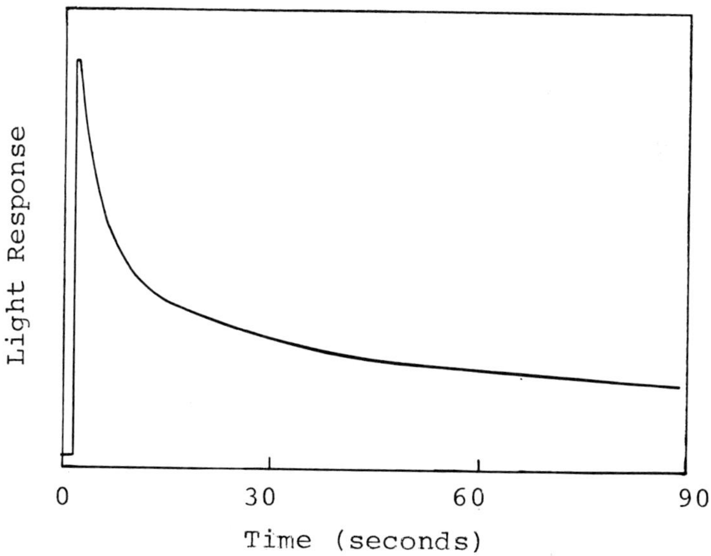


*FIGURE 2. Crude luciferase light response with 1 x $10^{-11}$ moles ATP.*

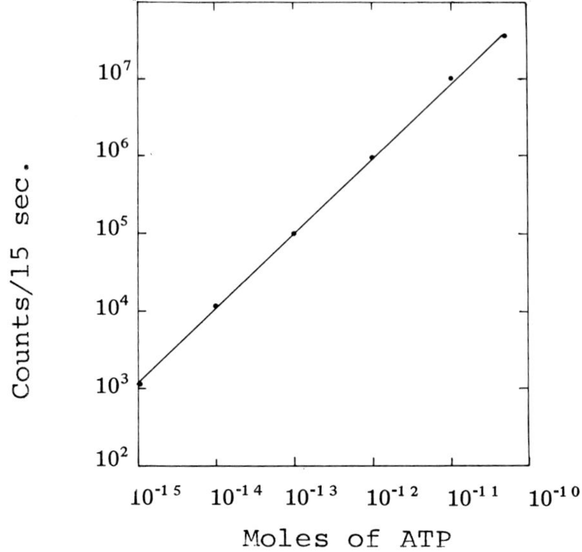


*FIGURE 3. ATP standard curve using PICO-ZYME F*

*TABLE I. Extracting Efficiency of PICO-EX B or S*

| *Cell Type* | *% Extracting Efficiency* | *PICO-EX Used* |
|---|---|---|
| *Red Blood cells* | *93%* | *S* |
| *E. Coli* | *83* | *B* |
| *K. pneumoniae* | *85* | *B* |
| *E. cloace* | *96.6* | *B* |
| *P. vulgaris* | *110* | *B* |
| *P. acnes* | *102* | *B* |
| *Group B Streptococcus* | *218* | *B* |
| *S. cerevisiae* | *138* | *B* |

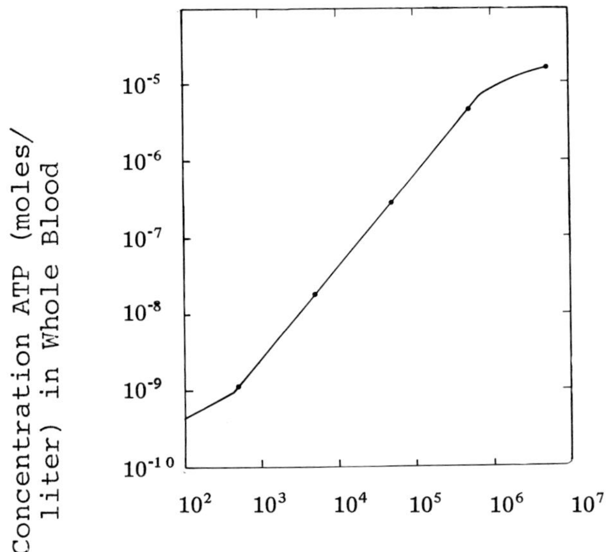


*FIGURE 4. Linearity of PICO-EX S for extraction of ATP from RBC.*

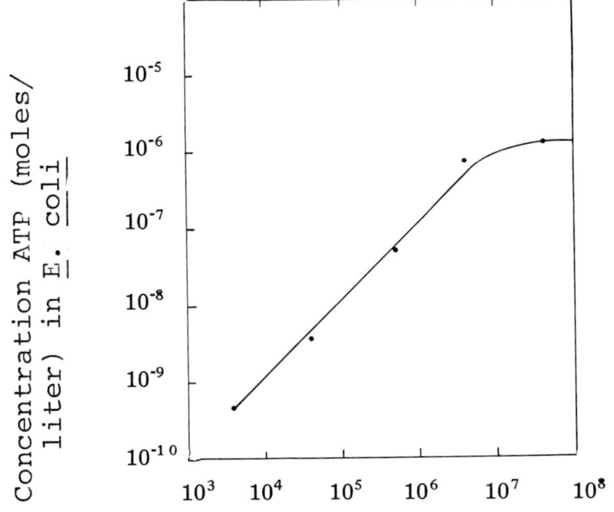


*FIGURE 5. Linearity of PICO-EX B for extraction of ATP from E. coli.*

## V. PICO-LITE

The PICO-LITE is a microprocessor controlled instrument containing 7 preset programs. Program 7 is an internal standard program that proved particularly useful for cellular ATP measurements. The instrument makes the calculation $C_1/(C_2 - C_1)$, where C is the counts due to ATP in the unknown sample being measured, and $C_2$ is the counts due to the addition of a known amount of ATP or an ATP standard ($ATP_s$) into the unknown sample (Fig. 6). The value printed by the PICO-LITE is ATP in unknown sample X $ATP_s$. ATP concentration can be entered as a constant if the same ATP standard concentration is used for each assay.

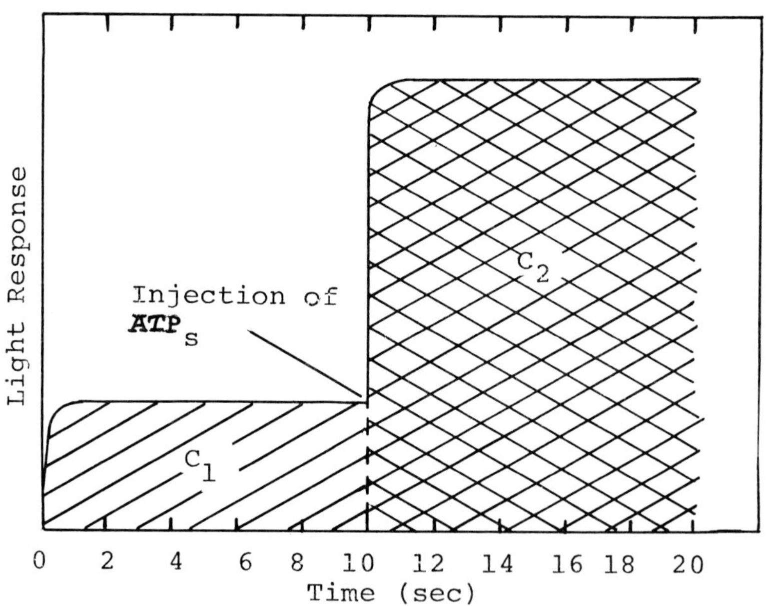


*FIGURE 6. PICO-LITE internal standard program.*

## VI. GROWTH CURVES

To study the theory that ATP is proportional to viable cell number as determined by plate counts, two types of bacteria were separately grown in Eugon broth and monitored over a period of time. *E. coli*, a gram (-) organism and Group B *Streptococcus*, a gram (+) organism were used in this experiment (Figs. 7 and 8). These bacterial cells were incubated at 37°C on a shaking water bath for six hours. Bacteria were refrigerated at this point for 72 hours. This was done to see if changing cell growth would show a similar effect on ATP levels. Although ATP/cell levels are not constant through the entire growth cycle, ATP levels do follow plate counts very well.

*FIGURE 7. Growth curve of E. coli.*

*FIGURE 8. Growth curve of Group B Streptococcus.*

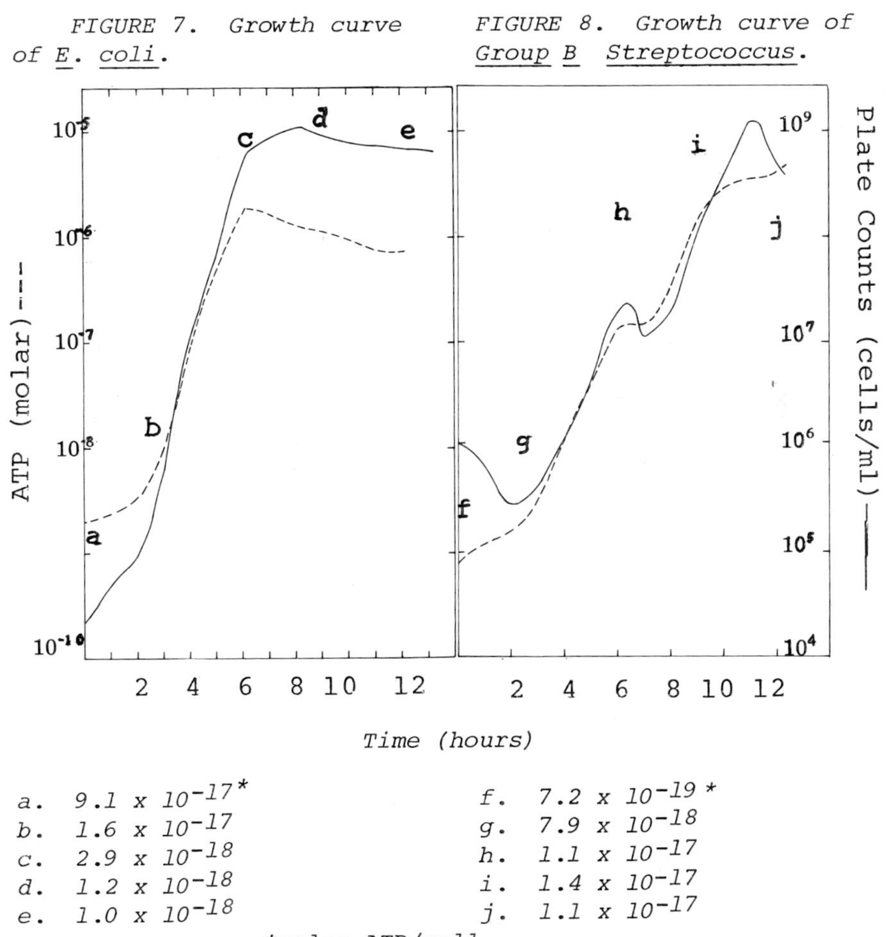


a. $9.1 \times 10^{-17}$*
b. $1.6 \times 10^{-17}$
c. $2.9 \times 10^{-18}$
d. $1.2 \times 10^{-18}$
e. $1.0 \times 10^{-18}$

f. $7.2 \times 10^{-19}$*
g. $7.9 \times 10^{-18}$
h. $1.1 \times 10^{-17}$
i. $1.4 \times 10^{-17}$
j. $1.1 \times 10^{-17}$

*moles ATP/cell

## VII. DISCUSSION AND SUMMARY

A luminescence adenosine triphosphate assay provides ease, speed, and economy in monitoring or detecting viable cells.. PICO-ZYME F, a purified firefly luciferase-luciferin reagent, provides a low decay light response that can be used for convenience and economy. It eliminates the need for critical timing and the need for separate internal standard assays. PICO-EX B and S allow rapid chemical extraction of ATP with greater than 80% extracting efficiency for most cells. The PICO-LITE with its microprocessor control and preset programming allows quick and easy calculation of ATP levels from samples. With thes reagents and instrumentation, an adenosine triphosphate assay usually doesn't require more than a few minutes. This compares quite favorably with traditional plating techniques which take 24-48 hours, and with microscope counting methods which are tedious and more difficult to do.

Adenosine triphosphate levels follow bacterial cell number very well through a growth cycle even when cell growth is slowed or stopped by refrigeration. If this is true of most other organisms under various conditions, cell measurement by adenosine triphosphate luminescence assay becomes a valuable tool for many areas of cell measurement.

The list of uses for cell measurement by adenosine triphosphate is lengthy. Because of the amount of research being done in this area, some of these uses have been presented in two recent review articles (2,3) as well as other sources (4, 5, 6). As a result of this research and the improved instrumentation and reagents now commercially available, work in cellular adenosine triphosphate measurement can be expanded and improved.

## REFERENCES

1. Sanville, C. Y., Packard "ATP Detection System," Packard Instrument Company, Downers Grove, IL (1980).
2. Gorus, F., and E. Schram, *Clin. Chem.* 25:4, 512 (1979).
3. Whitehead, T. P., L. J. Dricka, T. J. N. Carter, and G. H. G. Thorpe, *Clin. Chem.* 25:4, 1531 (1979).
4. Borun, G. A. (Ed.), "Proceedings of the Second Biannual ATP Methodology Symposium," SAI Technology Company, San Diego, CA (1977).

5. Chappelle, E. W., and G. L. Picciolo, (eds.), "Analytical Applications of Bioluminescence and Chemiluminescence," (NASA Publ. No. SP-388), National Aeronautics and Space Administration, Washington, D.C. (1975).
6. DeLuca, M. A. (ed.), "Methods in Enzymology," vol. LVII "Bioluminescence and Chemiluminescence," Academic Press, New York, NY (1978).

# HIGH QUALITY BACTERIAL LUMINESCENCE REAGENTS FOR FMN MEASUREMENTS

Tenlin S. Tsai

Clinical Investigations
Packard Instrument Co., Inc.
Downers Grove, Illinois

## I. SUMMARY

Clinical and analytical applications using bioluminescent measurements have gained increasing attention recently because the bioluminescent methods offer greater simplicity and sensitivity. High quality bioluminescent reagents, therefore, are in demand for exploitations of basic research and applications in bioluminescence.

In this paper we describe the preparation of a highly purified bacterial luciferase. The lyophilized bacterial luciferase reagent, PICO-ZYME B, is now commercially available. The specific activity (relative light units/second/mg of enzyme protein) and the specificity (devoid of FMN reductases activity) of PICO-ZYME B have been studied and compared to be superior to those of bacterial luciferases from other sources. We also characterized the kinetcs, linearity and stability of PICO-ZYME B used in a bioluminescent assay for FMN in which all the assay conditions were optimized.

## II. INTRODUCTION

Bacterial luciferase catalyzes the oxidation of reduced flavin mononucleotide in the presence of molecular oxygen and a linear, saturated long-chain aldehyde (1,2). The bioluminescent reaction is briefly shown in the following reaction:




ISBN 0-12-208820-4

$$FMNH_2 + O_2 + RCHO \xrightarrow{\text{bacterial luciferase}} FMN + RCOOH + H_2O + light$$

FMN reductase activity is closely associated with the crude bacterial luciferase (3) and hence NADH and NADPH can also be used as substrates in the following bioluminescent reaction to produce light with the presence of FMN (4):

$$NAD(P)H + FMN + H^+ \xrightarrow{\text{FMN reductase}} NAD(P)^+ + FMNH_2$$

In this way, FMN or NAD(P)H or, by use of coupled reactions, enzymes and substrates which are involved in FMN($H_2$) and NAD(P)H-producing or consuming reactions can be theoretically determined by bioluminescence. Various clinically important enzymes and substrates have been worked out in a bioluminescent assay employing bacterial luciferase (5,6,7,8). To name a few: creatine kinase; lactate dehydrogenase; alcohol dehydrogenase - ethanol; malate dehydrogenase - malate and oxalo-acetate; glucose-6-phosphate dehydrogenase - glucose-6-phosphate and glucose; urease - urea and ammonia.

## III. PURIFICATION OF BACTERIAL LUCIFERASE

PICO-ZYME B, a highly purified bacterial luciferase, was purified from *Beneckea harveyi*-M17 strain. We followed the purification procedure of Gunsalus-Miguel (9) with minor modifications.

The bacteria cells were harvested at peak bioluminescence induction period. The crude extract from cell lysis and sonication was adsorbed to DEAE-cellulose and batchwise eluted with increasing concentration of phosphate buffer. The active fraction was fractioned with ammonium sulfate before chromatographing on DEAE-cellulose and eluting with gradient phosphate buffer to separate the bacterial luciferase from reductases. The resolved bacterial luciferase fraction was then put through DEAE-Sephadex and Sephadex G-100 chromatography to obtain higher purification and specific activity. The final purified product was then processed and lyophilized to give an optimized stable bacterial luciferase reagent, PICO-ZYME B. Preliminary studies indicate that this freeze-dried reagent is stable for a minimum of three months. The activity of the bacterial luciferase also maintains at a constant level for that period of time when the hydrated PICO-ZYME B is kept at a refrigerated temperature (4°C-8°C).

PICO-ZYME B is a highly purified bacterial luciferase reagent which contains less than 0.1% NADH-specific and 7% NADPH-specific FMN reductase activity. The comparison of the specific activity of PICO-ZYME B with that of the bacterial luciferases from other sources is given in Fig. 1. When using the same amount of enzyme protein, as determined by method of Lowry (10), PICO-ZYME B gives 40%-90% more light production in the bioluminescent FMN assay and, therefore, offers more sensitivity.

## IV. OPTIMIZATION OF FMN ASSAY USING PICO-ZYME B AND PICO-LITE

The assay was performed in a 6 x 50 mm cuvette with a total volume of 120 μl. Add first 40 μl of assay premix (1 part of 1% v/v decanal with equal volume of Triton X-100 and 40 parts of 0.35M phosphate buffer, pH 6.7 with $10^{-3}$M DTT) to the cuvette. Then add 30 μl of the PICO-ZYME B (enzyme diluted to a final protein concentration of 4 μg per ml of assay mix). The cuvettes were then placed inside

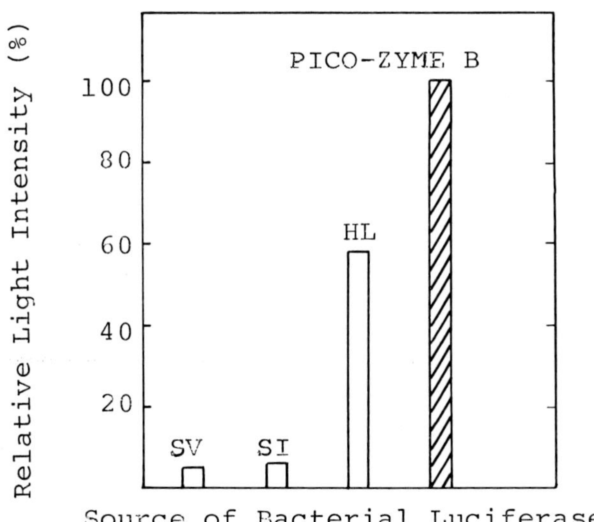


*FIGURE 1. Comparison on specific activity of bacterial luciferases from different sources.*

the detector chamber of the PICO-LITE (a luminometer by Packard Instrument Co., Inc., Downers Grove, Illinois). The program and functions were set on the PICO-LITE analyzer depending on the design of each individual experiment. With a microsyringe fifty µl of 5 x $10^{-5}$M $FMNH_2$ in $10^{-3}$ EDTA (keep reduced photochemically) was injected through a rubber septum to each cuvette to initiate the reaction and simultaneously the AUTO-START readings. The activity of bacterial luciferase is measured as the initial maximum light intensity or the total light produced during a fixed period of time with background and blank activity being accounted for.

### *A. Kinetics*

The kinetics of the bioluminescent reaction in assay of FMN is shown in Fig. 2. The light production reaches its peak within 2 seconds, then quickly tails off. Without decanal in the reaction, the light production is only 5% of that with the presence of the long-chain aldehyde.

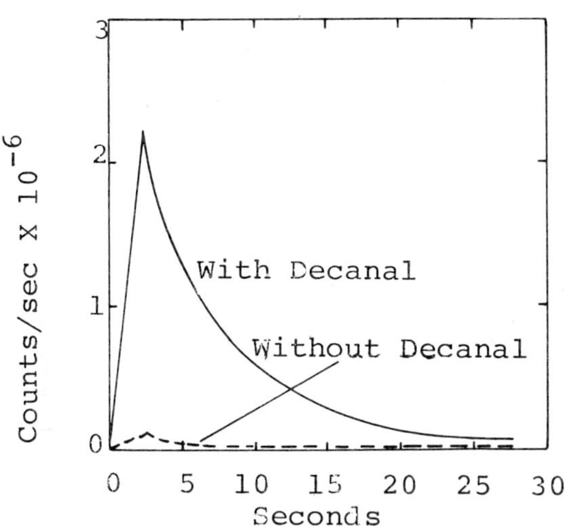


*FIGURE 2. Effect of decanal on light output of bacterial luciferase used in the assay of FMN.*

*B. Temperature and DTT*

The assay is optimal when it is performed at 15°C and adding $10^{-3}$ dithiothreitol (DTT) to the phosphate buffer helps to preserve the activity of the bacterial luciferase (Fig. 3).

*C. Concentration of Bacterial Luciferase*

Amount of bacterial luciferase used in the assay of FMN will also affect the kinetics (Fig. 4a) and the light production (Fig. 4b) of the reaction. The presence of more than 5 μg/ml bacterial luciferase in the assay mixture will delay the appearance of the peak light production (Fig. 4a). However, enzyme protein less than 4 μg/ml will result in decrease in light production and the light produced is a linear function of the protein concentration of the bacterial luciferase (Fig. 4b). Based on these observations, the amount of bacterial luciferase used in the assay should be carefully controlled in order to obtain reproducible results.

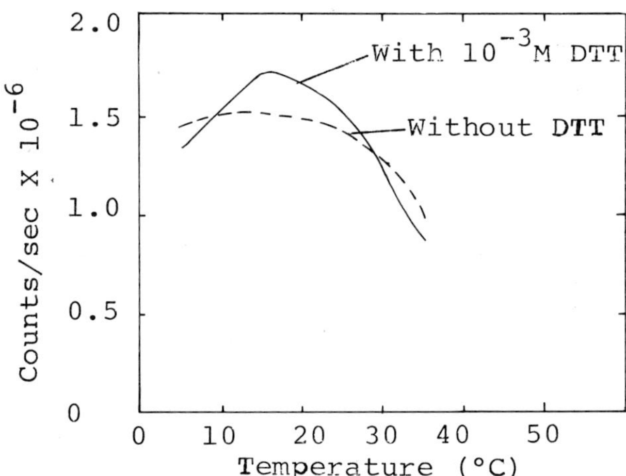


*FIGURE 3. Effect of temperature and dithiothreitol (DTT) on light output of bacterial luciferase used in the assay of FMN.*

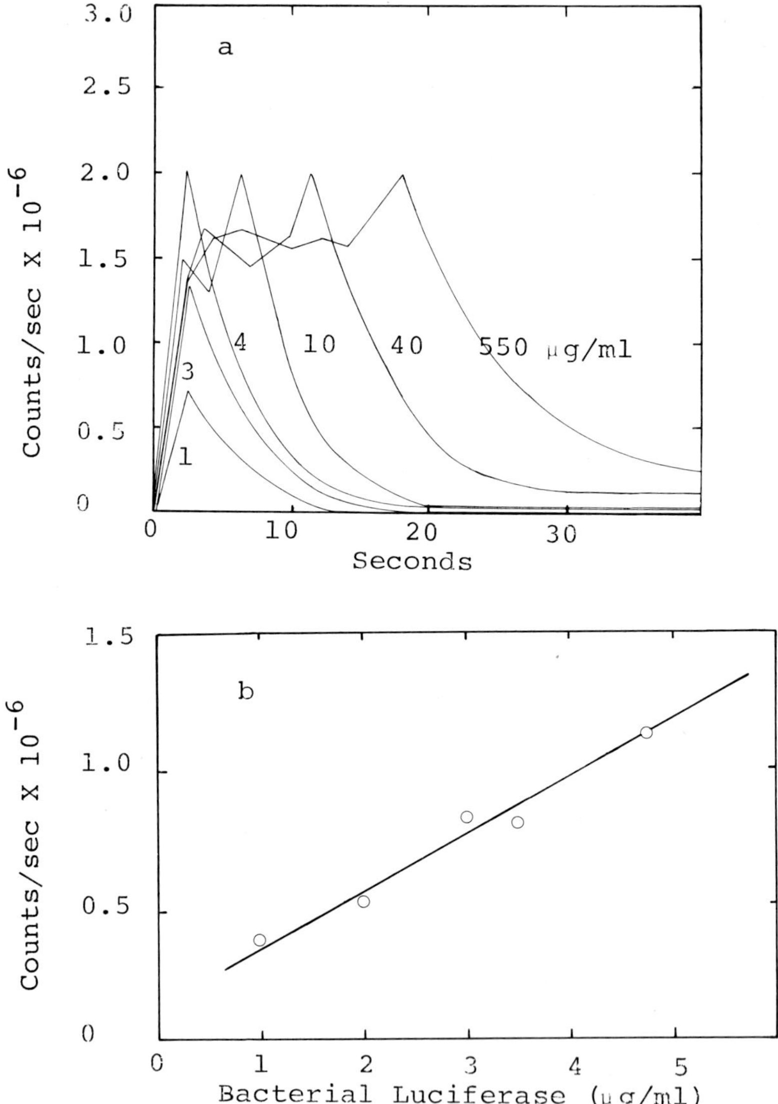


*FIGURE 4. Effect of bacterial luciferase concentration on light output in the assay of FMN.*

## V. CHARACTERISTICS OF PICO-ZYME B IN A BIOLUMINESCENT ASSAY OF FMN

### *A. Linearity*

PICO-ZYME B can be used to quantitate FMN in the range of 0.1 picomole to 1,000 picomole. The linearity of this assay is shown in Fig. 5. This bioluminescent assay offers the sensitivity to measure FMN down to femtomole range and the capacity of five orders of magnitude.

### *B. Stability*

The stability of PICO-ZYME B in the assay for FMN was tested and the results were plotted in Fig. 6. The bacterial luciferase remains at its original activity for up to 50 hrs in room temperature.

The stability of PICO-ZYME B adds another advantage to this bioluminescent assay when large numbers of samples have to be assayed.

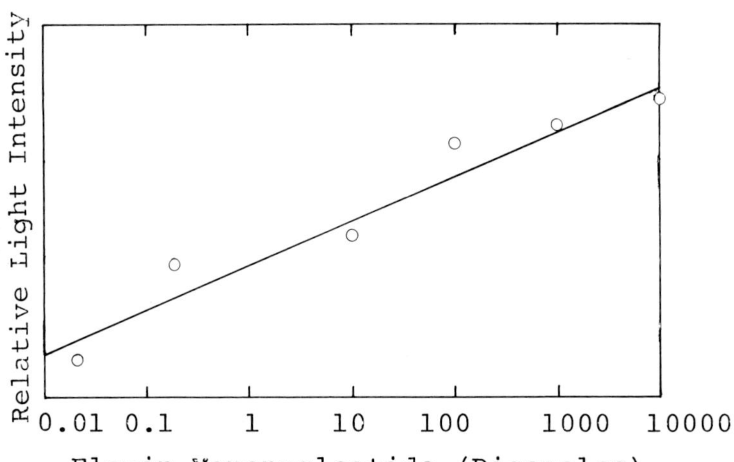


*FIGURE 5. Linearity of FMN assay using bacterial luciferase bioluminescence system.*

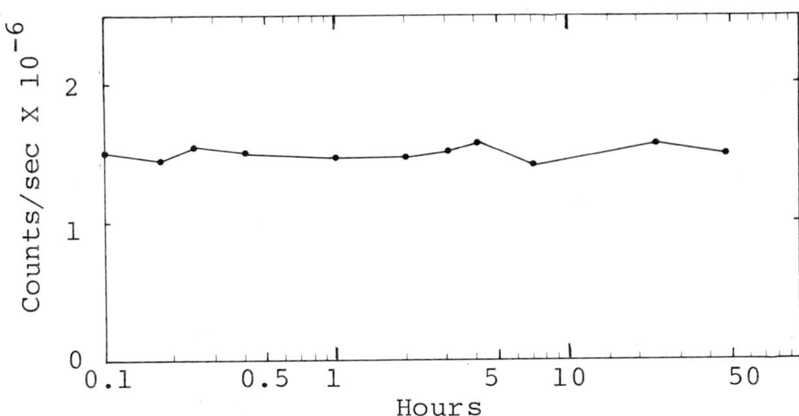


*FIGURE 6. Stability of bacterial luciferase in room temperature.*

## VI. CONCLUSION

The applications of bioluminescence as an analytical tool have been a subject of great interest, yet have not been widely used because of the lack of proper instrumentation, good quality luminescent reagents and some inherent technical difficulties associated with the method.

The availability of a high performance bacterial luciferase reagent and an optimized bioluminescent assay for FMN contribute partly to the implementing of the bioluminescent techniques in clinical applications.

## REFERENCES

(1) Hastings, H.W., *Ann. Rev. Biochem.* 37, 597 (1968).
(2) Dunn, D.K., G.A. Michaliszyn, I.G. Bogacki, and E.A. Meighen, *Biochemistry 12*, 4911 (1973).
(3) Duane, W., and J.W. Hastings, *Mol. Cell Biochem.* 6, 53 (1975).
(4) Shimomura, O., F.H. Johnson, and Y. Kohama, *Proc. Natl. Acad. Sci. USA* 69, 2086 (1972).
(5) Stanley, P.E., *Liq. Scintill. Counting* 3, 253 (1974).

(6) Schram, E. in "Liquid Scintillation Counting, Recent Developments" (P.E. Stanley and B.A. Scoggins, eds.), p. 383. Academic Press, New York (1974).
(7) DeLuca, M.A., ed. in "Methods in Enzymology" Vol LVII, p. 181. Academic Press, New York (1978).
(8) Schram, E., and P.E. Stanley, eds in "Proceedings of the International Symposium on the Analytical Applications of Bio- and Chemiluminescence," Brussels (1978).
(9) Gunsalus-Miguel, A., E.A. Meighen, M.Z. Nicoli, K.H. Nealson, and J.W. Hastings, *Biol. Chem.* 247, 398 (1972).
(10) Lowry, O.H., N.J. Rosebrough, A.L. Farr, and R.J. Randall, *J. Biol. Chem.* 193, 265 (1951).

# A COMPARISON OF COMMERCIAL FIREFLY LUCIFERASES[1]

JoAnn J. Webster
Jenq C. Chang
Jeffrey L. Howard[2]
Franklin R. Leach

Department of Biochemistry
Oklahoma State University
Stillwater, Oklahoma

## I. INTRODUCTION

Several commercial firms are now supplying firefly luciferase reagents. For the purpose of developing a sensitive ATP quantitation procedure for studying ground water pollution we compared a number of these commercial reagents. Some of the results from those comparative experiments are presented herein to allow other investigators to choose the most appropriate luciferase preparation for their needs.

## II. MATERIALS AND METHODS

Table I lists the sources, lot numbers, and some other properties of the various commercial luciferase preparations used. Samples for evaluation were provided by Mr. Mike Ringrose of Analytical Luminescence Laboratory, by

[1]*Supported in part by EPA Grant R 804613 and Oklahoma Agricultural Experiment Station Project 1640. This is manuscript J-3815 from the Oklahoma Agricultural Experiment Station.*

[2]*Present address: Johns Hopkins University, Baltimore,MD*




ISBN 0-12-208820-4

Boehringer-Mannheim, by Mr. Arne Mhyrman of LKB, by Ms. Elle Williams of Packard, and by Mr. Dave Murphy of SAI. Boehringer-Mannheim also supplied luciferin. Different Lumac lots were later supplied by Dr. S.E. Kolehmainen. The other luciferase preparations were purchased.

The dry weights were determined by weighing the contents of a bottle. A luciferase (LU) produces 1000 counts (1.4 x $10^7$ photons) per minute with 1 ng of ATP. The luciferin present in a preparation was determined by measurement of absorbance at 325 nm and using the molar extinction of 1.82 x 10 $M^{-1}$ $cm^{-1}$ (1).

Protein was determined by absorbance at 280 nm (2), by the method of Lowry et al. (3) by fluorescence measurements using o-phthalaldehyde (4), and by dye binding using either bromosulfalein (5) or Coomassie blue (6) all with bovine serum albumin as the standard. Because each protein estimation is based on determining some particular property which is not constant in all proteins and since the commercial preparations contain various additives, some of the protein determinations gave obviously elevated values. These were not used in obtaining the average that is presented in Table I.

Total organic solids were determined by Johnson's procedure (7) using sucrose as the standard. Sulfhydryl groups were determined using p-chloromercuribenzoate according to the procedure of Boyer (8) with cysteine as the standard. Arsenic was determined by atomic absorption using a Perkin-Elmer Model 403 instrument. Pyrophosphatase (EC 3.6.1.1) was determined by the method of Lowry (9). Adenylate kinase (EC 2.7.4.3) activity was determined by measuring ATP production from 4.9 µM ADP in the luciferase reaction system. Nucleoside diphosphate (XDP) kinase (EC 2.7.4.6) activity was determined by measuring the net production (corrected for adenylate kinase activity) from 1.72 µM GTP and 4.4 µM ADP. One unit of each of these enzymes produces 1 nmol of ATP per minute.

Light production was measured in a SAI Technology Model 3000 ATP photometer or in an Aminco SPF-125 spectro-fluorometer equipped with a Chem-Glow reaction chamber. The rapid kinetic measurements were obtained using a two opposing jet mixing device. The results were displayed on a Tektronix Type 564 or 5116 storage oscilloscope for Polaroid photography.

The standard reaction mixture contained (final concentrations) 0.025 M Tricine buffer pH 7.8, 5 mM $MgSO_4$, 0.5mM EDTA, 0.5 mM dithiothreitol, and the amount of luciferase shown in Table I.

| Product Information | | | Amount Per Assay | | Chemical Properties | | | | | Enzyme Activities | | | Light Production | | |
|---|---|---|---|---|---|---|---|---|---|---|---|---|---|---|---|
| Manufacturer | Designation | Lot # | Mg Dry Wt | LU | $LH_2$ µM | Protein µg/mg | TOS µg/0.1 mg | -SH µg/0.1 mg | As µg/mg | PPase | Adenylate kinase Units/LU | XDP kinase Units/LU | Inherent Light No Add. | Inherent Light 50 µg $LH_2$ | x by $LH_2$ |
| A. Lantern Extracts | | | | | | | | | | | | | | | |
| Calbiochem-Behring | Calbiochem | 630145 | 1.44 | 254.0 | 8.9 | 80 | 18 | 0.23 | 170 | 0.21 | $8.5 \times 10^{-4}$ | $1.4 \times 10^{-3}$ | 0.1 | 33 | 163 |
| Sigma Chemical | FLE-50 | 1176-8075 | 0.44 | 292.0 | 12 | 480 | 11 | 0.5 | 123 | 0.283 | $1.1 \times 10^{-4}$ | $2.6 \times 10^{-6}$ | 0 | 4.4 | 33.8 |
| B. Partially Purified (not sold as crystalline) | | | | | | | | | | | | | | | |
| Analytical Luminescence Lab: | Firelight | 6581 | 0.13 | 86.8 | 31 | | | | | 0.00012 | $<2.7 \times 10^{-9}$ | $<2.7 \times 10^{-9}$ | 0 | 0.2 | 4.2 |
| E. I. DuPont | DuPont | 80378 | 0.4 | 64.0 | 7.1 | 38 | 97 | 0.1 | 0.04 | 0.0088 | $1 \times 10^{-7}$ | $6.0 \times 10^{-7}$ | 0 | 2 | 13 |
| LKB-Wallac | LKB | 103 | 2.3 | 49.4 | 65 | 410 | 45 | 0.1 | 0.06 | 0.00038 | $<4.9 \times 10^{-11}$ | $>4.8 \times 10^{-3}$ | 0.6 | 0.8 | 1.2 |
| Lumac | Lumit HS | 3104 | 0.3 | 8.8 | 114 | 350 | 75 | 23 | - | 0.018 | $1.9 \times 10^{-6}$ | $1.1 \times 10^{-4}$ | 0 | 0.3 | 25 |
| | Lumit PM | 3101 | 0.1 | 5.9 | 80 | 630 | 78 | 22 | 0.6 | 0.0012 | $<4.5 \times 10^{-7}$ | $>4.5 \times 10^{-7}$ | 0 | 0.1 | 8 |
| Packard Instrument | Pico-zyme | 609071 | 1.1 | 23.8 | 11 | 197 | 81 | 0.8 | 0.13 | 0.0004 | $<8.2 \times 10^{-7}$ | $>1.7 \times 10^{-6}$ | 0 | 1.1 | 10 |
| SAI Technology | SAI | 6581 | 0.2 | 77.9 | 31 | | | | | 0.00097 | $2.8 \times 10^{-9}$ | $8.2 \times 10^{-10}$ | 0 | 0.2 | 5.6 |
| C. Crystalline | | | | | | | | | | | | | | | |
| Boehringer-Mannheim | Boehringer | 01 | 0.005 | 21.9 | 0 | | | | | | | | 0 | 0 | |
| Sigma Chemical | Type IV | 776-3928 | 0.002 | 43.9 | 0 | 640 | 70 | 2 | 6 | 0 | $5.5 \times 10^{-9}$ | $2.3 \times 10^{-8}$ | 0 | 0.5 | |

Table I. Properties of Commercial Firefly Luciferase Preparation

## III. RESULTS

### *A. Comparative Properties*

Table I summarized several measurements comparing various luciferase preparations. The luciferase preparations are divided into three classifications: 1) Lantern extracts, 2) partially purified luciferases but not sold as crystalline, and 3) crystalline luciferases. In the preparation of luciferase-luciferin reagents many of the manufacturers add various components. These are not always listed on the data sheet supplied with the preparation. Polyacrylamide gel electrophoresis of most preparations showed the presence of multiple protein bands with most containing a band that migrated concurrently with bovine serum albumin. The dry weight and protein measurements, therefore, do not necessarily indicate luciferase. The amount of luciferase used in the assay is shown as LU (luciferase units) which is the response to 1 ng of ATP.

The luciferin concentration measurement is based on absorbance at 325 nm - other compounds could interfere. Both crystalline luciferase preparations show a complete dependence on added luciferin which is the basis for the indicated absence of luciferin. The contents of total organic solids, sulfhydryl groups, and arsenic in the various preparations are presented. It appears that the Lumac preparations contain a high concentration of a sulfhydryl containing compounds. The crude preparations contain arsenate.

The purification procedures used for the partial purification eliminate most of the contaminating pyrophosphatase, adenylate kinase, and nucleosidediphosphate kinase. With aging the inherent light production is low in most preparations; supplementation with an addition 50 μg of luciferin increases the inherent light. All of the crude and partially purified preparations except LKB were stimulated by additional luciferin.

### *B. Comparative Responses*

Figure 1 shows an ATP dose-response curve for each of the commercial luciferase preparations (this experiment was without added luciferin since this is the protocol of their intended use -- the two crystalline preparations were, of course, supplemented with 50 μg luciferin). A family of parallel lines results. The relative positions of the lines can be shifted by changing the concentration of either luciferase or luciferin or both. A similar family curve was generated by changing the concentrations of either luciferase or luciferin or both. Thus, the order of the luciferase preparation does not reflect an absolute position.

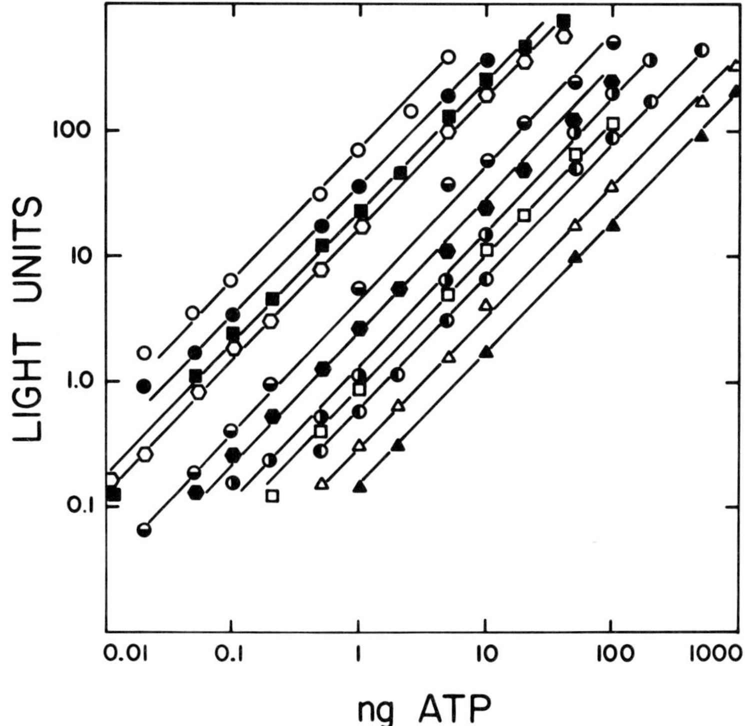


*FIGURE 1. Response of different luciferase preparations measured at different ATP concentrations. The light production for each of the luciferase preparations was determined in the standard assay system. The amounts of luciferase-luciferin preparation used are shown in Table I. The luciferase preparations are: ○ , Sigma Type IV, ● , LKB ⊖ , DuPont; ⬢ , Boehringer-Mannheim; ◑ , Sigma FLE-50; □ , Pico-zyme; ◐ , Calbiochem; △ , Lumac HS; ▲ , Lumac PM; ■ Firelight; and ⬡ , SAI.*

### C. *Kinetics*

DeLuca et al. (10) have compared the kinetics of light emission from purified and crude luciferase. They find that by using appropriately chosen experimental conditions crude and purified enzymes have identical kinetics. Figure 2 shows the kinetics of light production by the various preparations at three ATP concentrations. At low ATP concentrations the light production rises to a maximum and remains fairly constant for about 2 minutes. With high ATP concentrations there is a peak light emission followed by a significant decrease in the light production. Results are shown for a short time interval after mixing (2 seconds). Comparable results for light production were obtained over a two-minute time course (11).

## IV. DISCUSSION

The commercial availability of partially purified firefly luciferase preparations has increased the specificity of routine ATP determinations. The crude extracts produce light with other nucleotides because of the contaminating enzymes, adenylate kinase and nucleosidediphosphate kinase. These are markedly reduced by the purifications procedure yielding preparations of suitable specificity.

Karl and Holm-Hansen (12) have reported that additional luciferin increases the response given by Sigma FLE-50 lantern extracts and that aging of the supplemented extracts prior to use reduces the inherent (without ATP) light production. Additional luciferase and/or luciferin increases the light output from a given ATP concentration. The inherent light when supplemented with luciferin, that is the background emission, is an important determining factor in the use of the preparation.

## V. SUMMARY

The availability of several types of firefly luciferase reagents allows the investigator to choose from among several manufacturers the preparation which will best suit the experimental circumstance. The degree of nucleotide specificity, the range of ATP concentrations to be determined, and whether

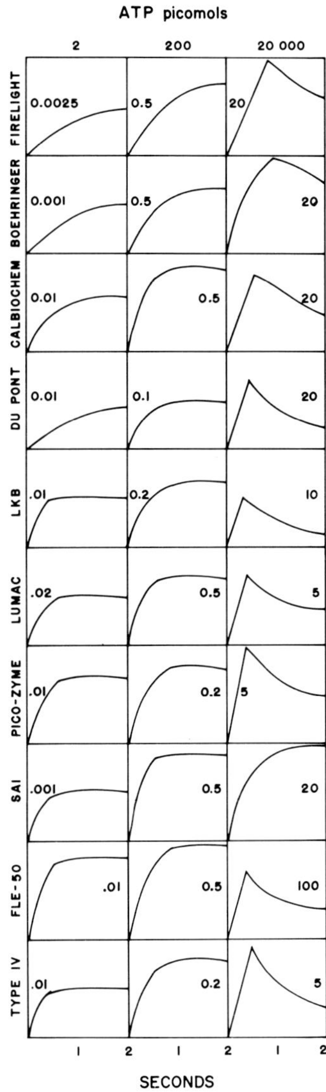


*FIGURE 2. Rapid reaction pattern of various luciferase preparations. Light production was initiated by a rapid mixing method using opposing jets of (A) luciferase, luciferin (30 µg), $Mg^{2+}$, EDTA, and DTT and (B) ATP. The traces were obtained from a Polaroid photograph of the oscilloscope recording.*

to supplement with additional luciferin are all factors to be considered. The response to a given quantity of ATP can be changed by altering the luciferase and/or luciferin concentrations.

REFERENCES


(1) Morton, R.A., T.A. Hopkins, and H.H. Seliger, *Biochemistry 8,* 1598 (1969).
(2) Warburg, O. and W. Christian, *Biochem. Z. 810,* 384 (1941).
(3) Lowry, O.H., N.J. Rosenbrough, A.L. Farr, and R.J. Randall, *J. Biol. Chem. 193,* 265 (1951).
(4) Kutchai, K. and L.M. Geddis, *Anal. Biochem. 77,* 315 (1977).
(5) McGuire, J., P. Taylor, and L.A. Greene, *Anal. Biochem. 83,* 75 (1977).
(6) McKnight, G.S., *Anal. Biochem. 78,* 86 (1977).
(7) Johnson, M.J., *J. Biol. Chem. 181,* 707 (1949).
(8) Boyer, P.D., *J. Am. Chem. Soc. 76,* 4331 (1954).
(9) Lowry, O.H. in "Methods in Enzymology 4" (S.P. Colowick and N.O. Kaplan, eds.), p. 366, Academic Press, New York (1957).
(10) DeLuca, M., J. Wannlund, and W.D. McElroy, *Anal. Biochem. 95,* 194 (1979).
(11) Webster, J.J., J.C. Chang, J.L. Howard, and F.R. Leach, *J. Appl. Biochem. 1* 471 (1980).
(12) Karl, D.M. and O. Holm-Hansen, *Anal. Biochem. 75,* 100 (1976).

# LUMINESCENT OSCILLATIONS IN MICROAEROBIC BATCH CULTURES OF *PHOTOBACTERIUM PHOSPHOREUM*

Richard A. Wecher
Ralph J. Bushnell

Department of Biology
Sonoma State University
Rohnert Park, California

## I. INTRODUCTION

Bacterial luminescence has been postulated to be an adaptation to oxygen-poor environments allowing limited respiration under conditions that otherwise necessitate anaerobic energy production (1,2). Mechanisms controlling luciferase synthesis (e,4,5) and some aspects of both the activity and biochemistry of luciferase (6,7,8) indicate a metabolic role in *Photobacterium* species. Strains of *Photobacterium phosphoreum* have been shown to synthesize luciferase constitutively (2) and to increase its synthesis in response to low oxygen (9). Habitats from which *P. phosphoreum* are commonly isolated (10,11) are probably oxygen-poor.




ISBN 0-12-208820-4

## II. METHODS

We examined the growth and *in vivo* luminescence of asynchronous batch cultures of *Photobacterium phosphoreum* grown under microaerobic conditions in sterile 20 cc scintillation vials. Photoemission of up to 20 such cultures was easily monitored with a Beckman LS 150 Liquid Scintillation Counter. Luminescence integrated over .1 to 1 minute periods was recorded as cpm via teletype. Operation of the counter was regularly checked with calibrated carbon-14 sources and with background sources.

Parallel cultures were prepared to estimate bacterial growth and the luciferase content of cultures.

The bacterial strain used was isolated from refuse of a local fish market. Axenic cultures obtained by successive re-streaking were identified as *P. phosphoreum* based on sugar fermentations and growth temperatures (10,11). Both the isolation and the maintenance of stock cultures were carried out on Difco Photobacterium broth with 1.5% agar added.

Experiments were performed on cultures grown in an unbuffered complex media modified from the "Luminous Media" of Reichelt and Baumann (11). This was prepared by mixing 5 g yeast extract, 5 g tryptone, and 3 ml glycerol in 1 l of Instant Ocean brand artificial sea water of one-half normal concentration. This media with 2.0% agar added was used for viable plate counts.

Bacterial numbers were estimated turbidometrically in a Spectronic 20 at 460 nm. Absorbance was measured in cultures grown in sterile borosilicate cuvettes designed for this machine. Absorbance from .004 to .487 units was nearly linear with colony forming units of plate counts. An absorbance of approximately .237 was equivalent to $1 \times 10^8$ colony forming units.

*In vitro* luciferase activity was estimated in crude cell extracts by an assay modified for small volumes (19,20). Extracts prepared by osmolysis of cell pellets were obtained by centrifuging parallel cultures grown in sterile centrifugation tubes. *In vitro* photoemission was measured in a Chem-Glo photometer (American Instruments).

Cultures were prepared by inoculating up to 500 ml of media with late log-phase cultures. The newly-inoculated media were divided among sterile scintillation vials and other sterile glassware. Microaerobic cultures were layered with 1.0 to 1.5 cm sterile mineral oil after these transfers.

Closely comparable results were obtained for light output by replications from batches prepared from such a single flask. Incompletely analyzed variations, arising from age of

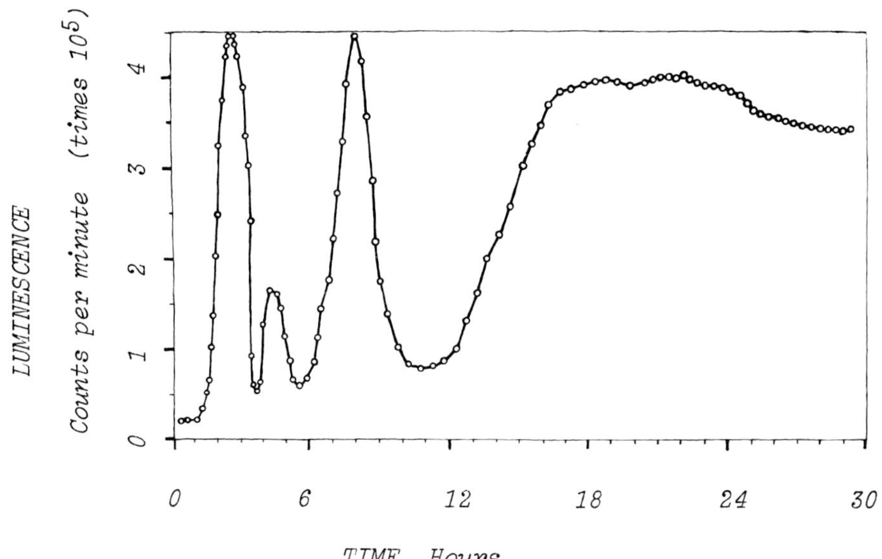


*FIGURE 1. Oscillations in peak light output as frequently seen in the development of luminescence of Photobacterium phosphoreum in the microaerobic environment. The occurrence of these early peaks can easily be missed if readouts of light production are taken at hourly or longer intervals. The results shown here are just a few of the points from readouts every minute (recorded as CPM) over the first 12 hour period. After the first 12 hours longer periods of counting were recorded.*

inoculum, exposure to light, and from variations in media were encountered. We therefore make our comparisons within groups of scintillation vial cultures derived from the same inoculated batch.

## III. RESULTS

For a period of two to three hours after starting the cultures in scintillation vials the luminescence remained at a low level of approximately 25,000 counts/min (Fig. 1). Then photoemission increases to a value some 17 times this (450,000 CPM) only to soon decrease to the lower level.

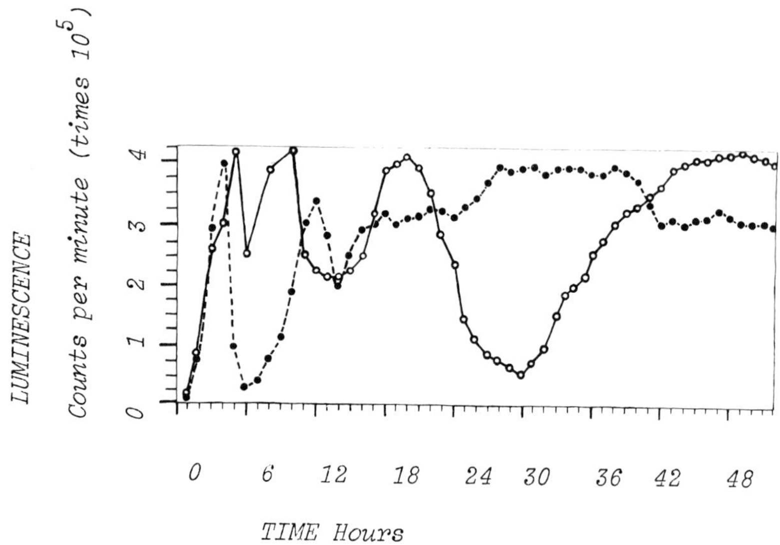


*FIGURE 2. A comparison of the luminescent oscillations seen in scintillation vial cultures grown under microaerobic or ambient-oxygen conditions. Open circles joined by the solid line represent the average light production of six microaerobic scintillation vial cultures, layered with mineral oil. Note the prominent oscillations for the first 12 hours and the broader oscillations as the cultures age. The solid markers joined by the broken line record the averaged light output of four cultures where oxygen was <u>not limited</u> by mineral oil layering. Here the oscillations tend to become less prominent after the first twelve hours.*

At or about nine hours from the start of the culture a second major peak is attained, to be followed by a gradual decline over three hours toward the lower level. By 18 hours elapsed time, luminescence climbs again to the higher level.

These patterns of early rapid rise and steep decline of light output (oscillations) were similar for the first 12 hours of growth under either ambient-oxygen or microaerobic conditions (see Fig. 2). Oscillations were less pronounced in older cultures where oxygen was not limited than those seen in microaerobic cultures.

Bacterial numbers and photoemission were recorded for up to five days (see Fig. 3). The luminescent oscillations of older microaerobic cultures were characteristically of lower amplitude and frequency than early oscillations. Early

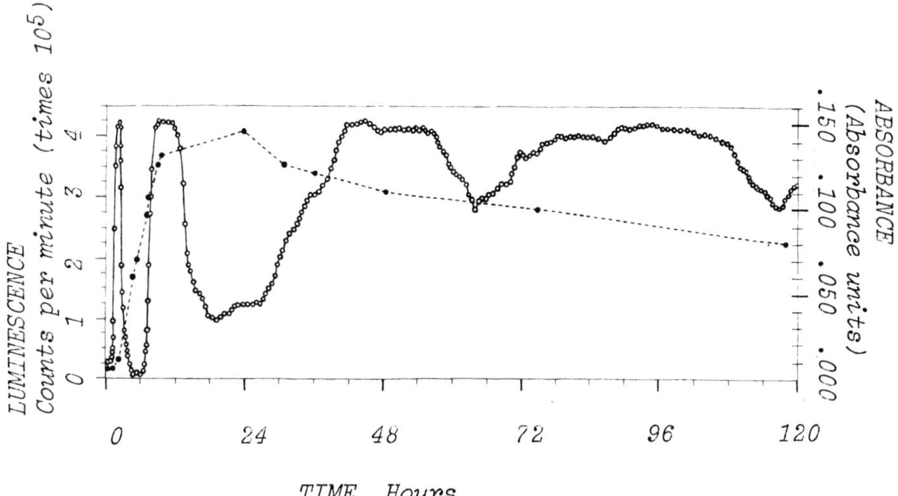


*FIGURE 3. The open circles joined by a continuous line show the course of light output (CPM) over a period of five days under microaerobic conditions as described in the text. Note the rapid steep early oscillations of counts per minute as compared to the later broader oscillations. The solid markers show population growth in parallel cultures measured turbidometrically. Note that the curve of bacterial growth does not closely parallel that of the light emission.*

luminescent peaks were associated with the start and late portions of log growth. Broader luminescent peaks characterized the decline phase.

Luciferase content of cultures was monitcred for up to 30 hours. The content of luciferase extracted from cultures closely followed the increase in the bacteria they contained (not shown).

## IV. DISCUSSION

Under the above described growth conditions, luminescence oscillated; periods of constant low light output alternated with bursts of luminescence. Two or more luminescent peaks were observed in all cultures. Although this pattern of photoemission was easily replicated, light output at any time was apparently not directly related to either bacterial numbers or to their luciferase content.

It is possible that luminous oscillations represent changes in microbial metabolism during growth, wherein respiratory substrates and light producing pathways are rapidly developing. In the early stages of population growth, induction of one or more enzymes may be limiting so that a period of time is required to attain a balance for sustained high level light output. Other possible mechanisms include one or more rate-limiting substrates, and an allostearic control on the activity of luciferase.

It is important to point out that should the light production be monitored at two to three hour intervals the early rapid oscillations can easily be overlooked.

Literature reports with two exceptions (12,13) of oscillations in bacterial metabolism indicate short periods (15,16,17) resembling the short, early oscillations of light output that we observed. Metabolism during synthesis of the bacterial luminescent system was recently reported (18).

We have used the system described here to investigate the effects of metabolic inhibitors and the ionic environment on the development of the luminescent system in *Photobacterium phosphoreum*. The results will be described in reports being prepared.

## V. SUMMARY

*Photobacterium phosphoreum* continues to grow and luminesce when oxygen tension is reduced.

Both the amplitude and frequency of oscillations in light production decreased as cultures aged.

In cultures where oxygen was not limited by mineral oil (ambient oxygen cultures), oscillations occurred during logarithmic growth but later ones did not appear.

The changes in microbial numbers do not parallel those occurring in light output.

## REFERENCES

1. Nealson, K. H. and J. W. Hastings, *Microbiol. Rev. 43,* 498-518 (1979).
2. Watanabe, H., N. Mimura, A. Takimoto, and T. Nakamura, *J. Biochem. 77,* 1147-50 (1975).
3. Nealson, K. H., A. Eberhard, and J. W. Hastings, *Proc. Nat. Acad. Sci. (USA) 69,* 1073-76 (1972).
4. Watanabe, H., A. A. Takimoto, and T. Nakamura, *J. Bio-*

*chem. 82,* 1707-14 (1977).
5. Nealson, K. H., T. Platt, and J. W. Hastings, *J. Bacteriol. 104,* 313-322 (1970).
6. Balarishnan, C. V., and N. Langerman, *Arch. Biochem. Biophys. 181,* 680-682 (1977).
7. Watanabe, T., and T. Nakamura, *J. Biochem. 72,* 647-53 (1972).
8. Mangold, A., and N. Langerman, *Arch. Biochem. Biophys. 169,* 123-33 (1975).
9. Nealson, K. H., and J. W. Hastings, *Arch. Microbiol. 112,* 9-16 (1977).
10. Reichelt, J. L., P. Baumann, and L. Baumann, *Arch. Microbiol. 110,* 101-120 (1976).
11. Reichelt, J. L., and P. Baumann, *Arch. Microbiol. 94,* 223-330 (1973).
12. Sturtevant, R. P., *Int. J. Chronobio. 1,* 141-146 (1973).
13. Rogers, L. A., and G. R. Greenbank, *J. Bacteriol. 19,* 181-190 (1930).
14. Degn, H., and D.E.F. Harrison, *Biochem. Biophys. Res. Com. 45,* 1554-1559 (1971).
15. Berzhanskaya, L. Y., I. I. Gitel'zon, A. M. Fish, R. I. Chumakova, and V. Shapiro, *J. Biophysics 18,* 293-301 (1973).
16. Harrison, D.E.F., and S. J. Pirt, *Gen. Microbiol. 46,* 193-211 (1967).
17. Harrison, D.E.F., *in* "Biological and Biochemical Oscillators" (Chance, B., E. K. Pye, A. K. Ghosh, and B. Hess, eds.), pp. 399-410. Academic Press, New York, 1973.
18. Karl, D. M., and K. H. Nealson, *J. Gen. Microbiol. 117,* 357-368 (1980).
19. Coffey, J. J., *J. Bacteriol. 94,* 1638-47 (1967).
20. Hastings, J. W., T. O. Baldwin, and M. Z. Nicole, *in* "Methods in Enzymology" (M. DeLucca, ed.), pp. 135-152 (1978).

# Index of Authors


A

Abakerll, Rosangela B., 81
Addink, R., 507
Agren, Ambjorn, 575, 591
Ahmad, Mustag, 435, 491
Allen, Robert C., 63, 531
Arrio, B., 515, 525, 619
Ayer, Kevin, 397

B

Baeyens, W., 335
Baldwin, Thomas, O., 121, 155
Baquet, Fernand, 516
Barnard, G., 311, 351
Becvar, James E., 147, 397, 583
Bergstrom, K., 659
Berthold, Fritz, 699
Biggley, W. H., 21
Blinks, John R., 243
Boehringer Mannheim GmbH., 260
Bognar, Andrew, 129
Boguslaski, R. C., 283
Branchini, Bruce R., 467
Brolin, Sven E., 287, 575, 591, 681
Brown, Craig, 2
Bruice, Thomas C., 391
Burlingame, A. L., 113
Burr, Bean, 397
Bushnell, Ralph J., 319, 763

C

Caldini, Anna L., 477
Campbell, A. Bryce, 397
Campbell, Anthony K., 601, 673
Carlson, A. D., 569
Carreira, L. A., 103
Castranova, Vincent, 327, 385
Chang, Jeng, C., 755
Clough, Roger L., 81
Collins, W. P., 311
Cormier, Milton J., 225
Curtis, G. D. W., 485

D

Daw, Richard A., 601
Dekkers, H.P.J.M., 569
De la Rosa, James, 639, 715
Dellinger, Barry, 2
De Luca, Marlene, 179, 693
De Moerloose, P., 335
Dominguez, Oscar, 397
Dougherty, John J. Jr., 121
Dupaix, A., 515, 526, 691
DuPont, de Nemours, 260

E

Eberhard, A., 113
Eberhard, C., 113
Eggert, Hartwig, 617
Ernest, Martin, 609, 617, 699

F

Falkenberg, Sabine, 721
Fattah, F. Abdel, 335
Fischer, Herbert, 609, 617, 699
Foote, Christopher S., 81
Freeman, Thompson, M., 347
Fresneau, C., 515, 526, 691
Fromtling, Anne M., 721
Fromtling, Robert A., 721

G

Gast, Robert, 103
Ghisla, Sandro, 97, 403
Goetz, Alexander, 639
Goto, Toshio, 203
Grant, Gary G., 409
Grayeski, Mary Lynn, 623

Gruber, Wolfgang, 209
Gundermann, Karl-Dietrich, 17

H

Hallett, Maurice B., 601
Hamman, J.P., 21
Haneda, Yata, 257
Hart, Russell C., 601, 673
Hastings, J.W., 97, 403, 417
Hemmerich, Peter, 97, 403
Henry, D. Linkley, 443
Herring, Peter, 527, 601
Hines, C. M., 55
Hostak, John, 397
Howard, Jeffrey L., 755
Hummelein, J. C., 569, 689
Hunter, Deborah J., 531

I

Idahl, Lars-ake, 627
Irwin, R. M., 103

J

Jahrling, P. B., 383
Johnston, H. H., 485
Jones, George, 327, 385

K

Kasha, Michael, 2
Kato, Taizo, 617
Kenyon, G. L., 113
Kim, J. B., 351, 357, 651
Klimetzek, Volker, 721
Kohen, F., 351, 357, 651
Kosower, E. M., 365
Koka, Prasad, 103
Kolehmainen, S. E., 705
Kubisiak, Helmut, 699
Kurfurst, Manfred, 97, 403

L

Lang, Helmut, 609
Lazoroni, J. A., 443
Leach, Franklin R., 497, 755
Lecuyer, B., 515, 526, 691
Lee, John, 103
Lehrer, Robert I., 81
Leisman, Gary, 709
Lemasters, John J., 197
Linfors, Gunilla, 453
Lindfors, Marianne, 575
Lindner, H. R., 311, 351, 357, 651
Lundin, Arne, 187, 453

M

Mahomedy, Amina, 397
Marquez, Daniel, 397
Marschner, Thomas M., 467
Matamoros, Maris, 45
McCapra, Frank, 673
McCarthy, J. P., 383
McElroy, W. D., 179
Meduski, Jerzy D., 639, 715
Meduski, Jerzy W., 639, 715
Meighen, Edward, 129, 409
Meijer, E. W., 569, 687
Merritt, Margaret V., 121
Messeri, Gianni, 477
Miller, Robert J., 667
Miller, Robert Jr., 645
Montemurro, Angelina M., 467
Moreels, Eric, 491
Müller, Sybille, 721
Mullin, Jerry, 623
Myhrman, Arne, 453

N

Nikokavouras, John, 729
Nichols, Wright W., 485
Nealson, K. H., 113

O

Olsson, Thomas, 659
Oppenheimer, N. J., 113
Östenson, Claes-Goran, 681

P

Paleos, Constantine, M., 729
Pazzagli, Mario, 477
Peden, David, 45, 327, 385

Petersson, Birger, 681
Phillips, Ruth M., 327
Piccard, Jacques, 516
Presswood, Robert, 403

R

Rausch, Steven K., 121
Reichard, D. W., 383, 645, 667
Richardson, Adrian, 673
Riendeau, Denis, 129
Rietschel, Ernest Th., 617
Ringrose, Michael, 385
Ryall, Malcolm E. T., 601

S

Sanville, Cindy, 735
Schippers, P. H., 569
Schmidt, Steven B., 23
Schram, Eric, 435, 491
Schroeder, H. R., 55, 283
Schuster, Gary B., 23
Seitz, W. Rudolf, 347, 623
Seliger, H. H., 21
Shepherd, Peter T., 391
Simpson, J. Steven A., 673
Skan, A. G., 281, 282
Slessor, Keith, N., 409
Small, E. Daune, 103
Sobocinski, P. A., 383
Somiya, Hiroaki, 561
Stahler, Fritz, 209
Stanley, Philip E., 275
Staudinger, Hansjurgen, 609
Stevens, Paul, 75

T

Taylor, Phyllis B., 497
Thore, Anders, 659
Tasi, Tenlin S., 745
Tsuji, F. I., 709
Tu, Shiao-Chun, 161, 425
Tuzzi, Paola, 477

U

Ulitzer, Shimon, 139

V

Van Dyke, Chris, 385
Van Dyke, Cynthia, 45, 385
Van Dyke, Knox, 45, 327, 385
Vanstaen, H., 705
Vassilopoulos, George, 729
Visser, A.J.W.G., 103
Vogelhut, P. O., 55
Volfin, P., 515, 526, 691
Vossen, J., 705

W

Wampler, John E., 249
Wannlund, J., 693
Ward, William W., 235
Watanabe, H., 417
Webster, JoAnn J., 497, 755
Wecher, Richard A., 763
Weiser, Irith, 139
Welch, William D., 75
Wersäll, Peter, 681
Wettermark, Gunnar, 287
Whitehead, T. P., 303
Wokalek, Heinrich, 617
Woodhead, Stuart J., 673
Wulff, Karl, 209
Wu, Li-Huey, 147
Wynberg, H., 569, 687

Y

Yannai, Shmuel, 139

Z

Ziegler, Miriam M., 155
Zygowicz, Elizabeth, 623

# Subject Index

## A


Acetone, dioxetane formation of excited, 5
Acridinium esters, luminescent reaction of, 677
Acriflavin, 140
Adenine nucleotides, determination by firefly luciferase, 190
Adenlyate kinase, 194, 478
  activity of pancreatic islets, 594
Aequorea bioluminescence, 225, 230
Aequorin, 230
  application of, 243
  as a Ca++ indicator, 246
Aequorin luminescence
  Ca++ relations, 244
  effects of various cations, 245
Affinity chromatography gels, 470
Alcohol dehydrogenase, 169, 418
Aldehyde dehydrogenase, 131, 169
  specificity of, 133
Aldehydes
  biosynthesis of, 134
  oxidation and reduction, 131
Aldehyde pheromones, 410
  detection by bacterial luciferaes, 411
  in insects, 413
N-alkylamino luminol, 19
4a-(alkylperoxy) flavins, chemiluminescence, 391
American Instrument Company, 260
Ames Test, 140
5-Amino-2,3-dihydro-1,4-phthalazinedione, 609
Aminoluciferin, 41
$LH_2$-AMP, synthetic, 180
AMP-luciferin-luciferase, 478
c-AMP phosphodiesterase, 194
Analytical Luminescence Laboratory, 260
1,5-anilinonaphthalenesulfonate, 199
Antibody, luminol labelled, 674
Anti-human IgG, immunoassay for, 58
Antonik Laboratories, 260
Apyrase, 194
Arachidonate, 48
Arachidonate-based chemiluminescence, 385
Arachidonate metabolism, secondary chemiluminescence, 50
*Argyropelecus hemigymnus,* counterlighting hypothesis, 517, 525
Aristostomias, 527
ATP
  assay of, 276
  automatic extraction, 708
  monitoring, 189
  phosphoribosyl transferase, 194
  sulphurylase, 194
ATPase, 194
Autoinducer
  from *P. fischeri,* 144
  properties of, 119
  purification scheme, 116
  structure of, 118
  synthesis of, 117
Autolumography, 583
Avidin, 284
7-azaindole
  H-bonded dimer of, 8
  tunneling, 32


## B


Bacterial bioluminescence, 96
  accessory enzymes, 129
  applications to entomology, 409
  postulated molecular mechanism, 369
Bacterial bioluminescent reaction, different chain length aldehydes, 147
Bacterial biomass, estimation by the firefly assay, 189
Bacterial luciferase
  active center, 155
  active center studies, 121
  oxygenated intermediate, 161
  red absorbing flavin, 403
  response to aldehyde pheromones, 412
  stability, 752
Bacterial luciferase/oxidoreductase, immobilized, 278
Bacterial luminescence
  determination of dehydrogenase, 417
  FMN measurements, 745

Bacteriuria, screening, 443
*Beneckea harveyi,* 130
  autolumographs, 585
  electrogram, 585
  emission, energy diagram for, 152
  luciferase, 148
*Benthalbella,* 561
Bentonite clays, harvest of luminous bacteria, 397
Benzpyrene, 35
Berthold Laboratorium, 260
Bioluminescence
  aequorea, 225, 230
  analysis, metabolic rates, 681
  applications of, 275
  assay, in cell biology, 287
  assays, insulin and glucagon producing cells, 591
    microphysiology in, 287
    microprocessing techniques, 575
  bacterial, 96
    accessory enzymes, 129
  circular polarization, 569
  Dinoflagellates, 505
  *Diplocardia longa,* of, 251
  earthworm, 249
      distribution of, 250
  electron transfer mechanisms, 365
  emission spectra for FMN, 98
  emissions with different flavins, 99
  firefly, 177, 178
  fish, 505
  general, 505
  potassium stimulated, 709
  *Renilla,* 225
  *Symplectoteuthis,* 709
  urine screening, 417
  worms, 505
Bioluminescence test
  different mutagens, 141
  for mutagenic agents, 139
Bioluminescent immunoassay, 667
  creatine phosphokinase, 669
  determination of methotrexate, 693
  pyruvate kinase, 669
Bioluminescent photometer, design of, 691
Bioluminescent, screening, 443
Biotin, 55
Biotin-isoluminol conjugate, 284
Bi-photonic absorption, 4
Biprotonic phototautomerism, 4, 10
Blood, chemiluminescence, 619

## C

Calbiochem-Behring Corporation, 260
Calcium-activated photoproteins, 243
Calmodulin, 49, 89
6-[carboxymethoxyacetyl-n-(6-aminohexyl)-n-ethylamino]-2,3-dihydrophthalazine-1, 4-dione, synthesis of, 57
Carcinogenic metabolite, 21
Carnidazole, 336
Catalase, 92, 539
Cell nucleus, internal probing with chemiluminescence 379
Cephalopod photophores, red fluorescence, 527
Cetyltrimethylammonium bromide, 19
Charge-transfer excitation, 4
Chemically initiated electron-exchange luminescence 28
Chemilumigenic labels, 285
  immunoassays, 283
Chemiluminescence, 17
  4a-(alkylperoxy) flavins, 391
  blood, 629
  cell nucleus, 379
  effect of r-formyl-methionyl-leucyl-phenylalanine, 330
  electron transfer mechanisms, 365
  enzyme-catalyzed, 17
  flow injection analysis, 623
  general, 505
  granulocyte, 75, 617
  immunoassay, 55
    plasma cortisol, 651
  lucigenin, 63, 729
  lucigenin-dependent, 69
  luminol, 40
  lysosomes, 378
  macrophages, 609
  metabolic events, 49
  mitochondria, 375
  mitogens, 721
  neutrophils, 45
  non-enzymatically cataylzed, 17
  phagocyte, 63
  platelet, 617
  silver, 639
  *Streptococcus faecalis,* 531
  trypticase soy broth, 536
Chemiluminescent immunoassay
  detection of virus/antibody aggregates, 645
  estriol, 360
  labels, 673

Chemiluminescent photometer, design of, 691
Chemiluminescent probes
  1,2-dioxethanes, 687
  site-specific, 373
Chemiluminescent reactions
  alcohols and aldehydes, 319
  rapid mixing apparatus, 715
Chloragocytes, 555
Chloragosomes, 555
*Chlorophthalmus albatrossis,* 562
  bacterial light organ, 561
Chlorhydrins, 84
Cholesterol, 84
  oxidation of, 35
Cholinesterase, sepharose-isoluminol, 633
Ciell mechanism, 38
Circular polarization, bioluminescence, 569
*Coccorella,* 561
Coelenterate green-fluorescent proteins
  properties of, 235
  purification and characterization of, 235
Coellenterate luciferin, 604
Coelomocytes, *Pontodrilus bermudensis,* 543
Coenzyme, A., 194
Concanavalin, 328, 609
Coronene, 393
Cortisol, chemiluminescence immunoassay, 653
Cortisol-aminopentyl ethyl isoluminol, proposed structure, 652
Creatine kinase, 182, 194, 276
  bioluminescence microassay, 477
Creatine kinase assay
  comparison of bioluminescence and photometric, 213
  firefly luciferase, 209
  standard condition for, 211
Creatine-phosphate, 478
Cyclothone, 525
Cyclo-oxygenase, 385
*Cypridina,* 97
  luciferase, 203
    pink-colored, 204
    Uv spectrum of the pink-colored, 205
  luciferin, 203
Cytochrome P 35, 450


## D


Dexamisole hydrochloride, 336
Diadenosine pentaphosphate 210
Dicoumarol, 199
Dielectric relaxation
  7-azaindole, 10
  indole, 10
  solvent cages, 9
6,7-dihydroflavin, chemiluminescence, 507
Dimethyl-6-aminochrosene, 24
10, 10'-dimethyl-9, 9'-biacridinium, 64
7,8-dimethyl-10-(w-carboxypentyl) isoalloxazine, 166
Dimethyldioxetanone, chemiluminescence of, 23, 25
6,7-dimethyl-8-ribityllumazine, 104
Dinitrophenol, 199
The p'-{(2-[N-(2,4-dinitrophenyl) amino] ethyl} -$p^4$-5(5'-adenosine) tetraphosphate 193
1,2-dioxetane, 22
  chemiluminescent probes, 687
  decomposition of, 4
  mechanism, 40
Dioxetanone, decomposition of, 23
9, 10-diphenylanthracene, 393
*Diplocardia longa,* bioluminescence of, 251, 557
Diphenyl anthracene, fluorescence of, 31
1, 3-diphosphoglycerate, 194
Dipole cage, ordered, 11
Dipole cage effects,random vs. ordered, 12
Diple moments, induced, 10
DL-carnitinyl maleate-isoluminol, 375
DNA-intercalating agents, 140
DMP-luciferase, 694
*Dyakia striate,* 257


## E


Earthworm bioluminescence, 249
  analytical applications, 255
  physiology of, 249
Earthworms, luminescent, physiology and biochemistry of, 252
Econazole nitrate 336
Ectoplasmic granulocyte, 556
Electret state, 13
Electroluminescence, 289
1,4-endoperoxide, 22
Enzyme immunoassay techniques, 305
Erythrocyte ATP, 192
Estriol, chemiluminescence immunoassay, 360
Estriol-16α-glucuronide
  aminobutyl ethyl isoluminol, 353
  chemiluminescence immunoassay, 351

Etomidate, 336
Exciplex luminescence, 22
Excitation
  atoms, 2
  dissociative, 8
  isomerizational, 8
  resonance, 6
Excited state
  generation of, 2
  proton-transfer, 9
  solute, 8
Excitons in molecular pairs, 4

## F

Fatty acid, reductase
  function of, 135
  stimulation of luminescence by, 133
α-fetoprotein, luminometric assay of, 676
Filolobopodial granulocyte, 554
Firefly, 97
Firefly assay, overview of analytical applications, 191
Firefly bioluminescence, 177, 178
Firefly bioluminescence, and spectrophotometry, correlation between, 482
Firefly light emission
  product inhibition, 182
  time course of, 182
Firefly luciferase
  affinity chromatography-based purification, 467
  assay of NAD/NADH, 627
  comparison of commercial samples, 755
  DNP, 694
  effect of sodium arsenate, 182
  electrogram autolumographs, 587
  inhibition by carbonylcyanide *m*-chlorophenyl hydrazone, 201
  inhibitors in urine specimens, 485
  linked to glass beads, 184
  linked to sepharose, 184
  product inhibition of, 200
  purification of, 453
  purification scheme, 469
  reactions catalyzed by, 179
  reaction kinetics, 435
  sensitivity of ATP determination, 497
  time-course of light reaction, 491
  turn-over, 437
  use in bioassays, 182
Firefly luciferase bioluminescence, dynamic measurement of ATP, 197
Firefly luciferin
  CIEEL mechanism stability, 516
Firefly luminescence, application of, 187
Firefly reagents, comparison of, 188
Fish, luminescence, 522
Fishphotophores, red fluorescence, 527
Flavin hydroperoxide, 100
Flavodoxin, 110
Flubendazole, 336
Fluorescence, emissions with different flavins, 99
FMN measurements, bacterial luminescence, 745
$FMNH_2$, oxidation of, 97
FMN reductases
  NADH specific, 131
  NADPH specific, 131
n-Formyl-methionyl-leucyl-phenylalanine, 328

## G

6-p-Gluconate, 478
Glutamate dehydrogenase, 418
Glycerol kinase, 194
cGMP phosphodiesterase, 194
Gonadal steroids, antibody-enhanced chemiluminescence, 357
Granulocyte, binding of perturbant, 46
Granulocyte chemiluminescence, 45, 50, 75, 383, 617
Granulocyte membrane, binding of chemotactic peptide, 48
Granulomatous disease granulocytes, 51
Granulomatous disease
  hexose monophosphate, 327
  transmembrane potential, 327
Green fluorescent protein, 227, 231
Granylate kinase, 194

## H

*Hemiplecta weinafliana,* 260
Hexokinase, 194, 478
*Histioteuthis,* 527
  photophores, 529
  5α-hydroperoxide, 82
  5α-hydroperoxycholesterol, formation of, 165
  4a-hydroperoxyflavins, effect of dodecanol, 165
  7-hydroxy-6,7-dihydroluminflavin, 507
  3-hydroxy-flavone, 9
  hydroxy flavone, fluorescence, 32
  3-hydroxyflavone, photautomerism, 32
  5-hydroxyflavone, 9, 32
  7-hydroxyflavone, 9

## I

*Ichthyococcus,* 527
IgG conjugates, chemilumigenic, 9
Imazalil, 336
Imidazoles, phosphorescence of, 335
Imidazolopyrazinone, 243
Immobilized luminescence immunoassay, 184
Immunoassay
anti-human IgG, monitored by chemiluminescence, 61
bioluminescent, 667
chemilumigenic labels, 283, 673
immobilized luminescence, 184
luminescent, 192
using luminol, 55
Infrared transition, 6
Inhibition of calmodulin, by trifluoroperpazine, 90
Insect pheromones, 409
Iodoacetamide, 122
ISO-ATP, 180
ISO-FMN, 98
Isoluminol-carbamyl chloride, 380
N-ISO-valeryl-3-aminopropanal, 253, 343

## K

Kaempferol, 32
Kinetics of firefly light emission, 180

## L

Lactate dehydrogenase, 418
*Lestidium,* 561
Leukocyte activation, 89
Leuckocytes, chemiluminescence of, 72
Levamisole hydrochloride, 336
$LH_2$- -AMP, 180
Lipoxygenase, 385
Linocyte, 546
Liposomes, 605
LKB-Wallac, 260
Lobopodial granulocyte, 555
Luciferase
absorption spectra, 406
alkylation, 156
binding of $FMNH_2$, 122
bound $FMNH_2$, circular dichroism spectra, 158
bound peroxyflavin, 97
catalyzed reactions, sequence of events occurring for firefly, 181
effect of Bentonite, 401
effect of chymoteypsin on nitroxide-labeled, 125
N[1-$^{14}$C] ethylmaleimide labeled, 124
flavin intermediate, 100
function of growth, 135
inactivation by iodoacetamide, 157
methotrexate, 184
nitroxide-labeled, 123
peroxyflavins, fluorescence properties of, 97
Renilla, 225
synthesis of, 130
titration with methyl methanethiolsulfate, 159
-w-carboxypentylvlavin, 162
Luciferin binding protein, Renilla, 228
Luciferyl adenylate, 179
Luciferyl sulfate, 267
Lucigenin, 33, 64
chemilumigenic substrates, 71
chemiluminescence, 63, 729
-dependent chemiluminescence, 69
light reaction, chemiluminescence spectra, 733
polymorphonuclear leukocytes 65
Lumazine fluorescence, 106
Lumazine protein, 169
polarization of fluorescence, 109
properties of, 103
purified, 105
spectral properties of, 109
Lumazine, spectrum of, 109
Lumichrome, 13
Luminescence, summary and future applications, 301
Luminescent cofactor immunoassay, 307
Luminescent earthworms, physiology and biochemistry of, 252
Luminescent enzyme immunoassay, 307
Luminescent enzyme multiplied immunoassay, 307
Luminescent immunoassay, 307
Luminescent fish, vertical distribution, 521
Luminescent immunoassays, 192, 307
Luminescent oscillations, *Photobacterium phosphoreum,* 763
Luminescence analyzer for phagocytosis, 699
Luminescence, immobilized immunoassay, 184
Luminescence of Renilla, $Ca^{2+}$-triggered, 229
Luminol, 33, 64
chemiluminescence, 17, 40
determination of hemoglobin, 659
substrates, 71
reaction, interference of serum, 662
Luminol-dependent chemiluminescence, 45
serum opsonic activity, 76

Luminol reaction
catalase, 664
hemoglobin, 664
myoglobin, 664
photohemin, 664
Luminometry, 705
Lumisomes, 230
Lymphocytoid, 547
Lysophosphatidylcholine, 48
Lysosomes, internal probing with chemilluminescence, 378
Lytic events, 192

## M

Macrophages, chemiluminescence, 609
Macrophagous lymphocytoid, 548
Magnetic-dipole mechanism, 38
*Malacosteus,* 527
Meleic hydrazides, 18
Marwell International AB, 281, 282
Mebendazole, 336
phosphorescence, 340
Megascolecidae, 543
Mercaptoacetyl-isoluminol, 380
Metalloporphyrins, chemiluminescence of, 27
Methyotrexate, 184
-hemocyanin, 184
Methotrexate immunoassay, 693
Methylcholanthrene, 35
Methyl methanethiolsulfonate, 122, 156
Metomidate hydrochloride, 336
Miconazole nitrate, 336
Microsomal oxidation, 35
Microperoxidase, 58
Microsomal oxidation, 35
Microperoxidase, 58
Millipore Corporation, 281, 282
Mitochondria, internal probing with chemiluminescence, 375
Molecular exciton effects, 4
Monomethylaminoluciferin, 41
2-morpholino, 98
Multiple excitation, 4
Mutagenic agents, 139
Mutagens
detected by bioluminescence test, 141
mode of action, 144
Myctophids, 525
Myctophiformes, 561
*Myctophum,* 561
Myeloperoxidase, 82, 85, 538
-deficient cells, 92
Myocardial infarction, 214

## N

NADH:FMN oxidoreductase, 131
NAD(P)H, assay of, 277
N-Di-nitrophenyl-β-alanine, 193
New Brunswick Scientific Company, Inc., 281, 282
Neuscopelus, 561
Nematogenia panamaensis, 557
N, N-dimethylaminoluciferin, 41
N-N-octylmaleimide, 122
N-(3-oxohexanoyl)-3-aminodihydro-2(3H) furanone, 114

## O

Obelia, 267
Obelin, 602
measurement of free $Ca^{2+}$, 601
Oligochaetes, 249
Opsonophagocytosis of *E. coli,* 79
Oxidative phosphorylation, 191
Oxidation of aldehydes, 131
Oxidative phosphorylation, 182
Oxidoreductase, 277
Oxidoreductase, NADH:FMN, 131
Oxygen, singlet molecular, 5
Oxygen probe, tetrakis alkyl aminocthylene chemiluminescence, 347
Oxygenase reactions, chemiluminescence of, 21
Oxyluciferin, 180
Oxylylene peroxide, 18

## P

*Pachystomias,* 527
Packard Instrument Company, Inc., 281, 282
Peroxyflavin, luciferase-bound, 97
Peroxyflavins-luciferase, fluorescence properties of, 97
Peroxy-luciferyl AMP, 41
Perylene, 393
Phagocyte chemiluminescence, 63
Phagocytosis, 303
chemiluminescence of, 77
Phenothiazine, 89

Pheromone, bioluminescent analysis of airborne, 414
Pheromones
aldehyde, 410
insect, 409
Phorbol myristate acetate, 328
Phorbol-12-myristate-13-acetate 64
Phosphatidyl choline, 48
Phosphatidylethanolamine, 48
Phosphatidyl-N-monomethylethanolamine, 48
Phosphodiesterase, 193
Phosphoenolpyruvate, 194
Phosphoglycerokinase, 194
Phospholipid methyl-transferase I and II, 48
Phosphorescence, 31
imidazoles, 335
spectrometry, 335
Phosphoribosyl ATP, 194
*Photinus pyralis,* 33
peak emission, 180
*Photobacterium fischeri,* 99
autoinducer
identification of, 113
purification of, 113
synthesis of, 113
*Photobacterium leiognathi,* 139
*Photobacterium phosphoreum,* 99, 103, 564
luminescent oscillations, 763
Photoproteins
calcium-activated, 243
entrapment of, 603
measurement of free $Ca^{2+}$, 601
temperature stability of, 269
Photophosphorylation, 191
Photosynthesis, 182
*Photuris lucicrescens,* 569
*Photuris versicolor,* 569
*Phrixothrix,* 34
Pico-Lite luminometer, 735
Platlet
chemiluminescence, 617
metabolism, 388
*P. phosphoreum* autoinducer, assay procedure, 115
Polarization cage, 11
effects, 7, 10
*Polymetme,* 527
Polymorphonuclear leukocyte, 63
lucigenin-dependent chemiluminescence, 65
singlet oxygen production, 81
Polynoid worms, bioluminescence of, 526
*Pontodrilus bermudensis,* coelomocytes, 543
Pregnanediol-3α-glucuronide aminohexyl ethyl isoluminol (AHEI) conjugate, proposed structure of, 312
Pregnanediol-3α-glucuronide
chemiluminescence immunoassay, 311
dose-response curve, 314
Proflavine, induction of luminescence, 143
Progesterone, 55, 284
assay, 358
chemiluminescent, structures for, 358
11α-hemisuccinate, 284
Prostagladin, 48
Proton-transfer spectroscopy, 13
*Pyrocistis lunula,* spectroscopy of bioluminescence, 515
Pyrene, 393
Pyruvate kinase, 194

## Q

*Quantula striata,* luminous organ, 258
luminescent land snail, 257
Quercetin, 32

## R

Radiationless electronic transitions, 2
Radioimmunoassay for steroids, 295
Radioiodination, 646
Radioimmunoassays, 184
solid phase, 675
Random-dipole cage, 11
Reduction of aldehydes, 131
Renilla
bioluminescence, 225
green fluorescent protein, physical properties of, 238
luciferase, 225
reaction catalyzed by, 226
luciferin binding protein, 228
luciferase-luciferin reaction, effect of green fluorescent protein, 227
luminescence of $Ca^{2+}$–triggered, 229
polyps, 230
Rhodamine, 393
*Rhodospirillum rubrum,* 192
Riboflavin synthetase, 105
Rubrene, 24, 393
fluorescence of, 31

## S

SAI Technology Inc., 281, 282
Salicylate hydroxylase, 425
*Salmonella typhimurium,* 139
Sepharose-isoluminol, assay of serum cholinesterase, 633
Serum cholinesterase, assay of, 634
Sigma Chemical Company, 281, 282
Singlet oxygen, 17
Solute excited state, 11
Solute ground state, 11
Solvent cages, dielectric relaxation, 7
Solvent cage, spectroscopy of, 2
Spectroscopy in solvent cages, 6
Steroid immunoassay, 369
Steroid-isoluminol conjugate, 312
Stomiatoid fishes, photophores, 528
*Streptococcus faecalis,* chemiluminescence, 531
Superoxide dismutase, 539
Super-radiance, 4
Symbiotic luminous bacterium, electromicrograph of, 566
*Symplectoteuthis,* bioluminescence, 709
*Symplectoteuthis oualaniensis,* 709

## T

Tetraanisylporphyrin, chemiluminescence of, 27
Tetradecanoic acid, 133
Tetrakis alkyl aminoethylene chemiluminescence, oxygen probe, 347
Tetramisole hydrochloride, 336
Thalassicolin, 602
2-Thio, FMN, 98
Thyroxine, 55
  isoluminol, 284
2,6-toluidinonaphalene sulfonate, 199
Trifluoroperpazine, 89
  inhibition, calcium-reversal of chemiluminescence, 50
    6,7,8-trimethyllumazine, 393, 507
Triplet-triplet annihilation, 4
Trypticase soy broth, chemiluminescence 536
Turner designs, 281, 282

## U

Uracil-isoluminol, 380

## V

*Valenciennellus,* 527
Vitatect Corporation, 281, 282

## W

*Watasenia scintillans,* 709
Woodward-Hoffman rules, 17

## Z

Zymosan, 92
  induced chemiluminescence, 620